UFO PROPULSION DYNAMICS

UFOS AND GRAVITATIONAL MANIPULATION

Paul Potter

Adventures Unlimited Press

UFO Propulsion Dynamics

Copyright 2016
by Paul Potter

All Rights Reserved

No portion of this study may be reproduced in any form without written permission of the author.
No portion of this study is permitted to be used – under any circumstances – for the purpose of warmongery.

First published in 2008 as
Gravitational Modification of Domed Craft

ISBN: 978-1-939149-58-9

All Rights Reserved

Published by:
Adventures Unlimited Press
One Adventure Place
Kempton, Illinois 60946 USA
auphq@frontiernet.net

www.adventuresunlimitedpress.com

ANTI-GRAVITY PROPULSION DYNAMICS

UFOs and Gravitational Manipulation

New Edition

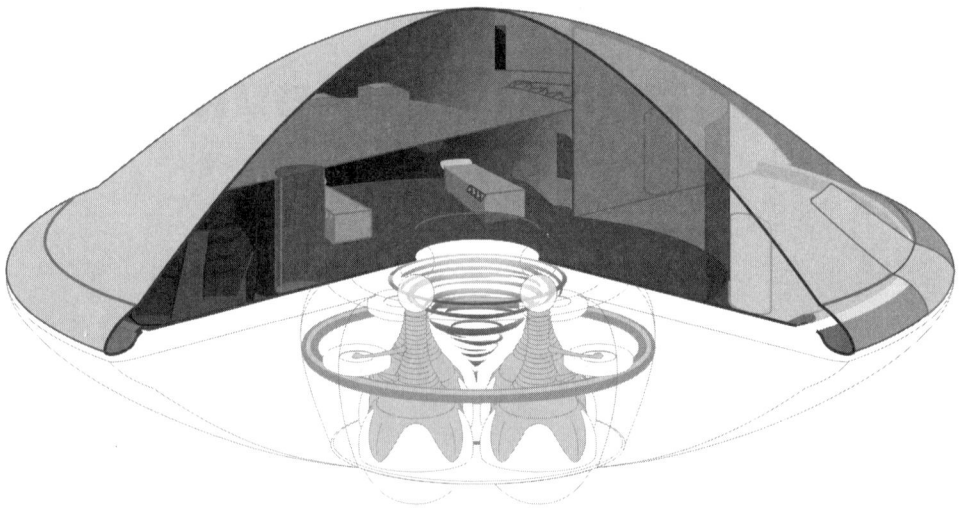

All rights reserved. No portion of this study may be
reproduced in any form without written permission of the author.
No portion of this study is permitted to be used—under any
circumstances—for the purpose of warmongery

Table of Contents

Foreword: (Written by Betty Ann Luca)   vii

Acknowledgments: xi

Illustrations: xiii

Introduction: xix

Disclaimer: xxv

Pre-emptive Statement: xxvi

Chapter 1: The Andreasson-Extraterrestrial Paradigms
1.1 Polymechanical Engineering Studies from Distant Galaxies   1
1.2 Valuable Memories of Alien Technologies from Abductees   8
1.3 The Curious Observations Recalled   10
1.4 Debriefings Attended by Aerospace Scientists   22
1.5 We Are Alone... in our Crown of The Creation Syndrome   25
1.6 Damage Limitation Strategy of Defense Departments   27

Chapter 2: Helical Field Tube Technology of Propulsion
2.1 Domed Craft Inside a Surrounding Energy Envelope   32
2.2 Propulsion of a Beam-drive UFO   35
2.3 Film Footage (from Mexico) of Hyper Jumps   36
2.4 Extraterrestrial Musings Between the Stars   39
2.5 Jets Produced from Rotating (Kerr) Black Holes   43
2.6 Mechanisms for Production of Electron-Positron Avalanches   48
2.7 Rotation Force and Momentum Transfer from a Black Hole   52
2.8 Magnetic Flux Break-Reconnection Ionization (Reconnection Pt I)   57
2.9 The Quest for Stability in the Development of a Filament-tube   65
2.10 Collimation and Hoop Stress   68
2.11 Electrodynamic Anchoring of the Magnetic Field   76
2.12 Williams and De Felice on Negative Radial Momentum   78

Chapter 3: Holy Grail of Astrophysics: The Black Vortex
3.1 Introduction to the Crystal-glass Spheres   83
3.2 Schoolroom Earth or the Universal School of Thought?   86
3.3 Oscillations of the aaUFO's Spheres of Crystal-glass   87
3.4 Plasma: The Electronic State of Matter   95
3.5 Ultra-violet Ionization Mechanisms   95

3.6 Exciton Excitations Around Dielectric Spheres — 96
3.7 Electron Fusion and Gyration — 97
3.8 Generation and Transfer of Charges from the Toroid Fluid — 98
3.9 Formation of the Toroid's Magnetic Field Lines — 101
3.10 Toroid's Magnetic Flux Lines Pulled into Shape — 105
3.11 Breaking and Reconnection of Flux Lines (Reconnection Pt II) — 108
3.12 Top Sphere Emissions Giving off Bars of Light — 109
3.13 Gyration of a Cyclotronic Storing Field — 112
3.14 Three Intensities of the Magnetic Field to Aid Emissions — 115
3.15 Magnetic Funnel Creates a Black Vortex at the Heart of the aaUFO — 115
3.16 Vacuous Force—Light Converts to Gravity — 118
3.17 Clues Forthcoming in the UFO's Quasi-Black Hole Drive — 119
3.18 On the Coherent Amplification from its Black Vortex — 124

Chapter 4: Ferroelectric Emission Sphere Fabrication
4.1 Photon Generator of Gravitational Forces — 127
4.2 Electrical Mechanisms Working in Quartz-Glass — 129
4.3 Magneto-Electric Oscillations from Magnetostriction Materials — 130
4.4 Magnetostrictive and Ferroelectric Ceramics in Spheres — 132
4.5 Ferroelectric Emissions and Polarization Reversal Emissions — 135
4.6 Colored Luminescence from Within the Spheres — 137
4.7 Quartz or Glass, Crystalline or Amorphous? — 139
4.8 A Crystal's Acoustoelectric Phonon Wind — 141
4.9 Sphere Fabrication with Sol-Gels of Silicons and Lanthanoids — 143
4.10 Surface Facets to Enhance Electron and Phonon Exchange — 146
4.11 Whirling Zones Inside Focused Acoustic Orbs — 148
4.12 Permeability Switching Caps Around Top Spheres — 149

Chapter 5: Gravitationally-Polarized Power Spheres
5.1 Polarization Changes in Lower Spheres Orbiting Black Vortex — 153
5.2 Rotating Forces Enclosed Within UFO's Ergosphere A — 161
5.3 Displaced Alignment of Crystalline Lattice Cells — 162
5.4 Fitting the Black Vortex into the Craft's Electronic Circuit — 167
5.5 Negative Phase Velocity of Metamaterials — 170
5.6 Plasmonic Pump from a Thin Wire Lattice Network — 174
5.7 Thick Wire and Diamond-lattice Plasmonic Frequencies — 176
5.8 Faceted Spheres Inside Microwave Spinning Fields — 177
5.9 Summary One: Glass Spheres that Think they are Superconductors — 181
5.10 Summary Two: Wire Network in Megaconductor Dielectric Spheres — 184

Chapter 6: The Craft's Electric Circuit
6.1 Radial Momentum Induced into the Telemeter Wheels — 199
6.2 Pathways of Conductivity for an Electrical Circuit — 203
6.3 Electrical Energy from the Black Vortex — 204
6.4 Array of Tube-guides with Crystal-like Firing Lenses — 205
6.5 Orthogonal Magnetic Field Turns Beams Helical — 209
6.6 Pumping up the Microwave Beams — 210
6.7 Firing Through the Toroid Wall — 212
6.8 The UFO Takes on Water to Breakdown into Electronic Fluid — 214

6.9 Pulse Firing into the Toroid to Ionize its Fluid	215
6.10 Fluid Polarized by Black Vortex at Inner Rim	218
6.11 Inner Ring-tube Controlling Mechanism	226
6.12 Active Governing Control Through the Inner Ring-tube	227
6.13 The Soft Polymer-plastic of the Inner Ring-tube	229
6.14 Adjustability of the Toroid Fluid's Permittivity	233

Chapter 7: Black Vortex of Immense (Silent) Power

7.1 Intake and Outflow of an Astrophysical Hole in Space	235
7.2 How the aaUFO's Black Vortex Inhales and Exhales	240
7.3 Artificial Engineering of a Functional Black Vortex	242
7.4 Real-time Engineering of a Functional Black Vortex	244
7.5 Lining up the Sphere-sets to the Central Vortex	249
7.6 Seeing a Complete Range of Frequencies Inside the UFO Craft	252
7.7 Gyrating Storage Field Feeds Rim Field (Current Ring)	253
7.8 The Gully Paradox of a Small Gap Leading to Bigger Things	258

Chapter 8: Gravitational Pull and Push Forces

8.1 Acoustic Black Vortex Gives UFOs the Stealth of Silence	259
8.2 Black Vortex Establishes Horizontal Gravitational Force	260
8.3 Difference in Aeronautics With Gravitational Buoyancy	261
8.4 Base Disc Divides Gravitational Field Dynamics	265
8.5 Negative (Repulsive) Gravity Below the Base Disc	266
8.6 Beyond the Buoyancy of Radial and Negative Gravitational Forces	270
8.7 On My Way Home After a Hard Night's Abduction...	272

Chapter 9: Superradiance Amplification inside UFOs

9.1 Rim's Gyrating Current Ring Inductive-coupling to Vortex	275
9.2 White Wall Repulsion Force Inside Rim	280
9.3 Fourth Amplification: Superradiance	283
9.4 Modelling an Ergosphere for a Slower Light Speed	286
9.5 Moving Fluid Slows Light Speed Even More	289
9.6 UFOs Establish Electromagnetic Anchor Within Their Ergosphere	292
9.7 Superradiance Plus: Another Plasma Pump (Reconnection Pt III)	293
9.8 Stabilising the Back EMF on the Black Vortex	297
9.9 Did the Greys Mention Superradiance During 1967 Abduction?	298

Chapter 10: Gravitational Momentum Field Projections

10.1 Toroidal Frustum of Hoop Stress	301
10.2 Gravitational Momentum Sent into a Curved Geometry	304
10.3 Gated Rim Ejects Electromagnetic-Gravitational Wave-packets	306
10.4 Dome's Curved Geometry Compresses the Wave-packets	309
10.5 Sixth Amplification: The Dome's Apex Control of the Tube's Bore	312
10.6 Before Einstein's Cosmological Constant Caper	318
10.7 Repulsive Gravitational Force and Negative Spatial Curvature	321
10.8 Using the Repulsive-Attractive Forces	327
10.9 Chirping the Semi-full Swing-backs into Wave-packets	331

10.10 Perturbations to Vary the Optical Transparency of Spacetime — 335
10.11 A Successful Mechanism to Harvest ZPF Energies By — 336

Chapter 11: Mass Diminution and Time Dilation in UFOs
11.1 A UFO is a Sealed Environment Around a Quasi-black Hole — 339
11.2 Dimension Anomalies Experienced by Abductees — 346
11.3 Deopulating Within a Mass-density Gradient — 347
11.4 Matter Transportation Field Projections — 360
11.5 de Broglie Wave Mechanics and the Planck Scale — 363

Supplementary Chapter:

Supplement 12: Area 51's S-4 UFO Reinterpreted
12.1 Lazar's Superheavy Gravitational Field — 365
12.2 The Bob Lazar S-4 UFO Through the Looking Glass — 369
12.3 The Central Reactor's White Hole — 372
12.4 Fishing for the Gravity A Force — 376
12.5 Bob Lazar's Technology Reconfigured to Make More Sense — 377
12.6 Quasi-black Hole's Delicate Stomach — 378
12.7 Bob Lazar's Element 115 is Here to Stay... its Official! — 385
12.8 Constant Energy Delivered from Polarized Ununpentium 115 — 388
12.9 Proton Cyclotron Beneath the Dome — 394
12.10 Pitching in From Three Angles to Load the Dome — 398
12.11 Holy Greyl Found Hidden Inside Lazar's Frequency Parameters — 405

Section 13: Afterword... on The Transformation Phenomenon — 411

Section 14: Notes to Chapters — 417

Section 15: Bibliography of References and Web-links — 503

Foreword (by Betty Ann Luca)

Having worked with the author off and on, for five years, I am pleased and happy to present this introduction to Paul Potter's inspirational material and excellent book, Gravitational Manipulation of Domed Craft. It is a first time publication, and a must have... for people interested in what's really happening in the field of UFO phenomenon.

Paul's total commitment and extraordinary ability to develop awesome answers to the phenomenal UFO propulsion system is sheer genius. His knowledge and interest of physics is remarkable. This study of the AA craft's structures has led him into magnetohydrodynamics, fluid dynamics, electrochemistry, electro-kinetics, and plasma propulsion physics. Paul has managed to unlock secrets of universal powers hidden and utilized in the interior of the craft, and the exterior of the craft's outer shell.

Through this book he shares the bountiful blessings of discovery, acquired knowledge, and brilliant research that sets his work high above the usual information on propulsion systems in today's market. Such blessings did not come easy and were motivated by struggle in body, soul, and spirit. Like many inventors, Paul has not been a stranger to personal hardship, sacrifice, and financial stress. Life often acquires some type of suffering in order to bring one to the gate of balance, expectation and success. In my opinion Paul has achieved that true measure of success through his work on the craft.

To the Reader... I believe it is not by chance this book is in your possession. You are definitely a serious thinker heading toward the fantastic reality of futuristic flight. Your thirst for answers to the sorely ignored subject of powered craft such as unidentified flying craft or objects will soon be satisfied. And Paul's success will have led the way with all his ground work. This new information is not a book about multiple sightings, nor alien astronauts who easily operate such unbelievable machines. But it is about the machine and its incredible power source, and how it evolves. Its volume of spectacular ideas, and use of scientific channels will challenge your mind to observe every detail that pulls together a workable finished product. The power for the AA craft.

Governments, military, and corporate industries have secretly tried to back-engineer retrieved UFO craft, and are still doing so to this day. Because of Paul's tireless research and designs we can began to comprehend the strange and original complexities of an extremely advanced propulsion system that powers these alien ships. Until now, there was little news or gathered understandable knowledge available to the general public, that might reveal how the UFO craft and its source of energy was capable of such out-of-this-world travel. Well... Get ready to be blown away! Answers we've all been seeking concerning this sought-after information will be found in the pages of this book! Paul's relentless scientific research into electromagnetism, superconductivity, dispersed charge particles, atomic and fusion reaction, magnetic fields, helical plasma, multiple poles, toroid magnetic fields, flux lines, crystal technology with quasi-piezoelectric glass kept him focused on the Andreasson Affair craft's crystal orbs, and internal structures until he achieved a tremendous breakthrough!

His idea is that a created high-powered energy release in the lower section of the craft between its stems and orbs would automatically circulate internally and then outward around the exterior skin of the craft, which would produce form and control, and easily distribute an even flow of movement (from the system's multi-unit propulsion center) to power, lift, and move the ship in any direction.

But what about the mysterious UFOs that silently haunt our skies, and tend to leave us all dumb-founded, wondering... how is it possible! How does it fly! Many people desperately need and deserve an answer! Its strange ability to produce fear or overwhelming curiosity after a personal sighting, may some day become a thing of the past, if the blueprints that fill the volume of this book could receive the important scientific support it truly deserves. It would be great if funds became available to build a working prototype of the AA machine! As you study author Paul Potter's carefully assembled, thoroughly researched ideas, conclusions, and beautiful illustrations, you're left with amazement! The total work quickly sweeps away any skeptical thought such as... it isn't possible... to WOW!

Hopefully the written body of technical evidence, and pictorial renditions will astound and challenge the layman, the scientist, and professionals as well to accept the incredible results Paul has accomplished.

Working together with him has been an exciting, productive, and rewarding journey for me. Being an artist myself, I often caught my breath with exuberant joy over the wonderful pictorial illustrations, and functional descriptions of Paul's work. Each snail-mail and e-mail that arrived allowed the flow of enthusiasm, insight, and more pin-point questions to materialize. Indirectly his determination served as a catalyst, that actually forced my mind to remember new, previously forgotten details such as the stairs of the craft disappearing into the door, or curved floor, allowing the perimeter trough to have complete clearance around the ships inner entrance.

Paul has studied many types of reported craft for their merit and possible use. And after his previews of those important cases which contributed to finding many similar, or similarly combined, propulsion characteristics, each of those studies has lead him to further enhance the probability of reaching successful conclusions.

However, after reading *The Andreasson Affair* book, and reflecting on the drawings, his AA craft project began. The alien saucer-like machine I witnessed during a 1967 encounter, caught his attention, and created a desire to find and prove its power source. Although large barriers such as an ocean, and continent, were in the path of face-to-face inquiry, we managed to establish report through written correspondence which in the beginning moved very slow. As time progressed letters kept coming with multiple questions about the craft's crystal orbs, colors, lights, size, patterns, shapes, sounds, metals, structures, and other possibilities. Paul was so into details that would help him figure out the energy system I could barely keep up. As soon as he received my letter, his work exploded with more ideas, conclusions and possibilities! His scientific knowledge, education, and intelligence was way out of my league. At times he would mention electrical names, and terms, nuclear physics, types of waves, technical abbreviations and observations, and engineering jargon, that was so unfamiliar to me, I was stymied. Paul must have realized my apprehension over the situation, and immediately, and thoughtfully backed-up, and took the time to explain things in a simpler way toward what he was meaning. Believe me it was through prayer, and by Gods grace I began to understand, and joyfully recognize the absolute genius of his work! Absolutely! As you view his designs, and artistic illustrations you will appreciate such clear visual aids. Truly... A picture is worth a thousand words. Many of his drawings map the invisible energy lines in operation, moving in different directions, yet encompassing the entire craft. To say the least, the incredible computer images are very impressive.

As you shall see, this vehicle's reality is on the cutting-edge of our near future!

Get ready to be thrilled with the possibilities such discovery may bring into the world, and within your life-time! As God has already promised, "All things are possible, for those who believe. And His Son has also stated, greater things shall you do, then I have done!" Are you ready to see those words of truth materialize? I believe there will soon come a time, we will no longer wonder how the phenomenal UFO saucer type ships perform. But that we will see and fly man-made ships, after the magnificent concept and design of Paul Potter's AA craft, for ourselves!!

Betty Ann Luca, 2006

Acknowledgements

Special thanks to Betty Ann Luca and Bob Luca, both of whom have helped tremendously with the finer points of extraterrestrial thinking.

Thanks to Mahmoud Yousif, Dana Redfield, Xan Phillips, Patricia Thomas for your contributions, experience, encouragement, humor and excellent coffee.

Special thanks to Jeff Savage and Rich Robson for their technological nous (and thanks also to Warren York) for explaining the extensive technology behind the Onion Drive craft, without which we would never have discovered what sort of physics so many of the known UFOs, including the Bob Lazar UFO, use and how they work!

Thanks to John Searl for those precious words of encouragement, and also to Gary Voss from The Ranch (indeed, thank you Gary for getting me to start this book in the first place).

Special thanks to Big John, whose ear I have bent on many occasions in order to further my knowledge of the finer points of electronics.

Special thanks to Dr. Valeri Dvuzhilny whose dedication to the academic study of UFO artifacts has been truly inspirational.

Last of all, thanks to the unseen ones who care in their own special ways, and who have achieved an amazing task here, and who have higher hopes for our future.

Illustrations:

Figure 1. Outside image of UFO featured in Andreasson Affair books

Figure 2. These glass shoes appear to work through sound wave technology (from "Andreasson Affair—Phase Two" fig.34)

Figure 3. Virtual energies making up recycling life forms (from "Andreasson Affair—Phase Two" fig.33)

Figure 4. Reflected and refracted rays making up an engineered environment that even this Grey seems to be alien to (from "Watchers II" fig.8)

Figure 5. Betty has her shoes replaced to insulate her from this environment (from "Watchers II" fig.10)

Figure 6. In the mist a Grey asks Betty to put on special shoes (from "Watchers II" fig.6)

Figure 7. Energy in the envelope surrounding a domed craft will follow a helical path of reducing radius of curvature

Figure 8. Electrical field lines perpendicular to magnetic field lines

Figure 9. UFO's propagating beams of energy and moving into them

Figure 10. In this video the UFO moves in jumps up into the sky. Here I show the frames where three of the jump sequences begin

Figure 11. Inter-stellar magnetic fields of a black hole and accretion disk

Figure 12. Magnetic fields and magnetosphere of Andreasson UFO

Figure 13. The well collimated M87 galaxy jet

Figure 14. Comparison of the NGC6251 jet and a frame from one of the beam propulsion UFOs filmed over Mexico

Figure 15. A black hole, ergosphere, and accretion disk twisting the inter-stellar field lines

Figure 16. Ionization, annihilation and photon production outside a quasi-black hole

Figure 17. Sphere-sets of top spheres, lower spheres and telemeter wheels

Figure 18. Emissions gyrating around magnetic field lines

Figure 19. Each of the four sphere-sets grabs a quadrant of the toroid's magnetic flux lines

Figure 20. In the Searl rotors a frustum field is first generated

Figure 21. Current sheets as a spiral corrugation in an MHD pulsar wind

Figure 22. Helical field created by a Searl rotor from charged particles produced through magnetic field breaking and reconnection. Also the Z-pinch effect on a rotating field

Figure 23. Depending on how its magnetic helical field is configured a domed craft can propel a field of charged particles parallel to its longitudinal axis or in an azimuthal curling trajectory

Figure 24. The flow velocity and density fields of a protostar. Is the energy for the jets being pushed into those areas or pulled?

Figure 25. The UFO creates its magnetic anchor inside the craft at the rim

Figure 26. The pitch angle of a helical field

Figure 27. Ejected energy curls around craft's dome in a Fibonacci curve

Figure 28. UFO harvests energy from the ZPF field through the acoustic waves around the lower spheres

Figure 29. Re-routing of the upper and lower magnetic fields of the toroid forms two intensities around the lower spheres

Figure 30. The leading Grey switches the outer shell to transparent mode to show Betty the sphere-sets working inside that UFO (from "Andreasson Affair—Phase Two" front cover)

Figure 31. Some of the mechanisms which direct energy toward the top spheres, note the acoustic waves bouncing off the inner wall of the toroid

Figure 32. Charged particles will move in certain directions under the influence of fluctuating magnetic field lines. Standard throughout this study will be the circular motions depicted in these charts

Figure 33. The toroid fluid and the vortex will flow in a counter-clockwise direction, but the positive charges of the diffuse layer will flow in a clockwise direction

Figure 34. Negative charged particles will flow out from center while positive charges will flow into center

Figure 35. The intensity of magnetic flux lines pulled into stem coils will cause high excitation of charged particles around top spheres

Figure 36. As the sphere-set assembly rotates the magnetic flux lines will break and then reconnect thus propelling any charged particles

Figure 37. In the middle of the sphere-sets is empowered a whirling vortex (from "Andreasson Affair" fig.8)

Figure 38. The sphere-set is comprised of a top sphere, a stem coil, a wheel held by conducting arms, a revolving clamp and a lower sphere. In this drawing the lower sphere clearly shows a wire network inside it (from "ASBT-II" drawing 42)

Figure 39. On a microscopic scale magnetic flux lines have several stages of existence, magnetic strips can break down into bubble-like domains

Figure 40. Crystal structures of crystallized silica and amorphous silica

Figure 41. On several occasions the extraterrestrials took time to show Betty a particular facet of their energy storage system

Figure 42. The two counter-rotating flows meet at the photon orbit ring—exactly where the top spheres orbit around the vortex

Figure 43. The stem coils direct the craft's magnetic fields through the lower spheres as well as through the top spheres

Figure 44. The central ergosphere forms a powerful engine with its central black vortex

Figure 45. Gravitational forces from the black vortex will displace the more massive nuclei away from the center of the electron orbit

Figure 46. Displacement of positive cations inside the negative oxygen anions

Figure 47. Difference in curvature of influence at top and bottom of black vortex

Figure 48. One of the drawings of the extraterrestrial's crystalline power system resembles cation displacement within a cellular structure

Figure 49. The aaUFO's electrodynamic circuit follows the Blandford-Znajek electric circuit for a black hole. Where the rotating flow negotiates around the four sphere-sets an oscillation will occur in the vortex's mid-section

Figure 50. Charged particles that will be forced to rotate around the outside of these spheres will move in an undulating trajectory. At different diameters their undulations will be varied and so will set up a harmonic resonance

Figure 51. Spheres revolving into intense magnetic field at bottom of stem coils will increase the effective mass of electrons in cells

Figure 52. Low-resistance gold wire encased inside quartz filaments braided together in bunches of 20 to 30 filaments were found to be formed into a mesh network encapsulated inside a carbon-glass material

Figure 53. From the detailed analysis of the artifacts left behind by the Dalnegorsk UFO, and from the observations made about the Andreasson UFO, it has been possible to reconstruct the basic design of the power spheres used by these UFOs

Figure 54. The lower section of the aaUFO is made up of three different compartments wherein three (or four counting the vortex) fields of energy rotate at different velocities

Figure 55. Ergosphere 'A' is a very compact space where energy will be constantly drawn into so that its rotational velocity will ever-increase

Figure 56. As the conducting tubes align with the toroid wall they will also align parallel to the toroid's magnetic field lines, this will cause the oscillating energies to turn in a helical path

Figure 57. The telemeter wheels fire ionizing beams into the fluid moving around the toroid

Figure 58. The sphere-set assembly will rotate by the force of the black vortex to fire pulses into fluid

Figure 59. Gravitational forces will be induced upon the heavier atomic nuclei of the fluid's molecular structure, through the heavy atoms that make up the base disc and the toroid casing

Figure 60. Microwave energies pass through the conducting tubes and become amplified in the maser-like crystals before being fired through the toroid wall into the toroid's fluid

Figure 61. At the photon orbit ring the rotational forces will be so energetic that the incoming positive particles will collide through Penrose-Williams collisions and will be ejected forcibly away toward the outer rim of the aaUFO

Figure 62. Interior decks of the aaUFO craft (from "Andreasson Affair" fig.11)

Figure 63. The stricken aaUFO is revitalized by a sister ship after the lower UFO's electrodynamic power system and central vortex short-circuited (from "Watchers" fig.15)

Figure 64. Showing the commutator-type arrangement which enables the clamp and lower sphere to revolve independently of the stem coil, but to be in constant electrical contact

Figure 65. Betty's drawings of the quarter-section of the aaUFO which facilitated the entrance and the waiting chamber, note the gully that curves around the outside of that chamber (from "Andreasson Affair" fig.10 & fig.39)

Figure 66. Using a flexible material (such as a magnetically shaped polymer) the deck can be raised to allow energy to rotate out of the planar waveguide into the gully to create a current ring around outside of craft

Figure 67. The UFO observed and worked on by Bob Lazar at Area 51 would propagate a rotating field of repulsive (or negative) gravitational force

Figure 68. Some of the negative gravitational force can be induced into the lower spheres, and a shearing force will be established beneath the craft to produce electron-positron pairs

Figure 69. This shows the negative gravitational force propagated below the aaUFO craft and how the helical filament-tube can be established at any part of the upper dome's curve

Figure 70. The electric field will extend straight up if the magnetic field lines are spiraled around the upper dome

Figure 71. The central vortex, the inner ring-tube, and the outer rim's current ring will all be inductively-coupled together—this setup will dampen the reverse-electromotive forces generated when the wave-packets leave the craft's rim

Figure 72. Many mechanisms will generate spontaneous emissions especially in the planar waveguide

Figure 73. Electromagnetic and gravitational waves radiating away from the central vortex will pass through the toroid fluid and be slowed down, they will also be refracted along different paths by the moving fluid

Figure 74. The breaking and reconnection of magnetic field lines will generate pulsing shock waves through each of the four quadrant fields

Figure 75. Four quadrant fields will undergo break-reconnection shock-explosions to accelerate the seed fields and generate further emissions simultaneously

Figure 76. Wave-packets of gravitational energies and angular momentum will curl around the upper dome and into a helical filament-tube

Figure 77. The position of the rim gate and its degree of arc will be determined by what direction the craft needs to go in and how much energy is needed to go into each wave-packet

Figure 78. The physical contours of the rim can determine the bandwidth of wavelengths which will make up the wave-packet. At the apex a channel can be established in a precise direction by a simple electromagnetic beam

Figure 79. These light beams (of red or infra-red photons) were noticed in one of Betty's abduction experiences (from "Andreasson Affair—Phase Two" fig.38)

Figure 80. This model of the universe is both a solution to the flat space problem and it can be seen as a diagram to show how UFO's manufacture both positive and negative gravitational forces above and below them

Figure 81. The wave-packets ejected from the UFO will have both negative gravitational force and positive gravitational force, the former will be an expansive force and the latter will be the trailing attractive force

Figure 82. The electrodynamic arrangement of energy inside the aaUFO is analogous to the mechanism of chirped pulse amplification

Figure 83. How did the Greys, and Betty Andreasson, AND the tall Elder all fit comfortably into this UFO? This drawing shows Betty being brought before the Great Door beyond which she saw Beings of Light (from "Watchers II" fig.37)

Figure 84. A UFO establishes a skirt between itself and the ground and then engineers a mass-density differential between itself and the lower craft. Inside the lower craft when it shrank in size were two humans and a crew of Greys! (from "Watchers" fig.16 and "ASBT-II" drawing 52)

Figure 85. The secret to a 'TARDIS Effect' inside a UFO has to do with its sealed environment (see also figure 54 of this study)

Figure 86. The energy orb surrounding the Dalnegorsk UFO would have provided a barrier, or shield, between the UFO's higher ($k>1$) density and the earth's ($k<1$) mass density

Figure 87. Betty arrives into a room where Greys are watching molecules! Does this mean that electromagnetic energies were slowed down because that environment's permittivity had been changed? From "Watchers II" fig.53

Figure 88. Either the UFO's vent hole opened up by a huge amount to allow this to happen, or Betty's and the Grey's physical forms underwent diminution from an alteration to the mass-density environment surrounding them. Note how this beam is on the INSIDE of the spheres. This image is from "Watchers" fig.9

Figure 89. The pass-not shield around the reactor dome was not actually generated by the reactor—it was only where those gravitational forces met

Figure 90. Anti-protons when they collide with matter are the ideal fuel for the quasi-black holes. Quasi-black holes are the ideal power system to fully contain their annihilation reactions

Figure 91. Very cleverly the extraterrestrials have designed their UFO structures to transmit both electrical force and gravitational perturbations

Figure 92. The alien's ununpentium 115 fuel discs had to be stacked in a special way before they were fused together and milled into a wedge

Figure 93. With the aid of the magnetic fields when the protons are fired at, and collided with, the antiprotons created will shoot downwards while the protons will shoot upwards

Figure 94. The box below the reactor will house a simple cyclotron accelerator

Figure 95. To keep the antiprotons away from their channel walls the gravitational polarization tracks should follow the same path as those channels, so as to give each channel wall a micro-magnetic blanket

Figure 96. Stressing the magnetic field lines around this UFO will produce electron-positron pairs which will then be induced into the gyrating beams below the craft (the beams do not pass directly through the hull)

Introduction

One of the most fundamental reasons as to why UFOs were initially deemed unreal by scientists was the fact that their ability to generate variable amounts of gravitational forces could not be accepted as being possible. After all, to an advocate of medieval Newtonian theory, gravitational forces cannot oscillate and nor can those forces be artificially generated because to those thinkers nothing other than what is already inherent in mass can determine that mass's gravitational force. Hence notions such as anti-gravity or gravitational manipulation proved to those unenthusiastic scientists as heretic a notion as that of the earth's surface being curved rather than it being flat.

It was for this very same reason that when black holes were first discovered they were metronomically assumed to be stationary affairs for, to those adherents of Newton's Law of Universal Gravitation it would have made no difference whether those black holes rotated or not, because, by Newtonian thinking the black hole's influence would solely be determined by how massive it was, period. History, of course, has proved those scientists very wrong for when Kerr in 1963 published his theory on rotating black holes and that theory was developed by Roger Penrose in 1969 and Zeldovich in 1972 the idea that energy could be extracted from the shearing rotational forces of black holes and that gravitational forces can additionally be produced by those rotating energy fields was born and here to stay.

But did this rotational revelation bring about new ways of scientific thinking across the board? If one considers how space-craft designers still worked along the lines of that most tried-and-tested of principles, that of forward force being proportional to stupendous amounts of backward force, then the answer has to be a disappointing...NO! The new ways of thinking about how gravitational forces can be generated through rotating energy fields and subsequently turned into propulsion forces didn't filter down to the scientists working for the big aerospace corporations, who are comfortable only when working with *linear* forces, and so they continued to develop their combustion engine rockets, which, even now in the guise of the USAF's latest top secret Aurora spyplane with its pulse-detonation engines that can travel up to a very frictional 3,600 mph, these aircraft are still variations upon the basic models of explosive combustion. And, these aircraft (and spacecraft) are substantially hampered by the fact that they can only travel in a very straight line under such technologies (whereby if they attempted to turn through a tight curvature they would disintegrate).

Unfortunately, the penalty afforded to radically new departures from the old established beliefs is to be called a 'kook' scientist, or worse, to be registered as someone who propounds 'exotic science.' Indeed, to see how the system of scientific progression actually works is a real eye-opener, for to scan through physics journals over many years and follow how a particular scientific concept progresses through time, it will be seen that that progress is made in quite a gingerly manner, by ample reference to earlier established studies and then by introduction of the present scientist's research findings in the tentative hope that someone else will duplicate that scientist's latest discovery and acknowledge its worth. Six steps forward, followed by five back.

So, with the above points in mind, a UFO propulsion system following the same principles as a quasi-black hole obviously faces stiff opposition from those who raise their eyebrows to departures from existing scientific precepts.

Therefore, it is hardly surprising that the term UFO in the minds of physicists, who worry over their credibility and more especially over their future funding, carries with it a certain amount of scorn and derision which might weigh down any research proposals with enough supra-gravity to prevent UFO propulsion from ever getting off the ground… For instance, I can just imagine Stephen Hawking's reply to this sort of comment, with his voice-box in its squeaky monotones deriding UFO propulsion technologies as being "…the works of cranks and weirdoes." Perhaps, to make this scientific research paradox more noticeable I should make a one guinea wager with Hawking that if he were ever invited into the underground hangars of America's Area 51, and offered their alleged selection of off-world craft from which he might choose just one to dissect and inspect for himself, that after his request was honored he wouldn't then regard those physicists and engineers… as cranks and weirdoes!

But then theoretical Physics, does not conclusively inspect a theory and register it as true or untrue, it merely registers whether a theory should be deemed acceptable or unacceptable by majority agreement toward its probable workability. It was because of this quirk of procedure that Einstein's theories of relativity initially sat forlorn in the wilderness for many a year before they became accepted by the majority (except for when Hubble's telescope proved Einstein's assumption of the universe being static and unchanging was wrong, resulting in Einstein's famous admission to his 'greatest blunder'); and why Harry Kroto's vision of the Buckminster Fullerine structure of carbon-60 took so many years to be accepted; and why the Russian Talyarkhan's sonofusion was so openly attacked before his findings eventually became accepted as agreeable theory through sufficient nods of the heads in the late twentieth-century. And, of course, Stephen Hawking himself faced being called a crank and a weirdo by his peers when he was forced to calculate and then admit in the 1970s, that rotating black holes do indeed radiate an out-flowing energy. But that is what science and physics are about. At no one moment is there a definitive paper which proves all. A study builds upon existing knowledge so that future studies of that same subject can evolve.

Therefore, from the very earliest moments of my scientific study of UFOs, some twenty years ago, I decided to be very careful about how I would use recognized expressions of energy mechanics because such recognized expressions tend sometimes to be associated with detrimental hard-and-fast notions already preconceived in the readers mind, this is especially true when a new science or a novel concept is being introduced which smacks in the face of existing industry. For instance, the expressions such as anti-gravity, or electro-gravitic, or ion-propulsion, are all likely to register as more appropriate to Hollywood-fictional screenplay writers than as technologies to pursue by physicists for their possible commercial viability at the likes of America's NASA, Boeing, and SpaceX aerospace corporations, the Russian Federal Space Agency (RKA), or at the ESA (European Space Agency), or at JAXA (Japan's Aerospace Exploration Agency).

Better then will it be, for this study to have its technological expressions accompanied by numerous graphic illustrations, rather than them being left naked and vulnerable to misinterpretation by pre-assuming minds. This will be why my study will shy away from long and pedigreed words and instead be lavishly punctuated with as many cut-away diagrams and plain-worded technical explanations as possible. I also know first-hand what it is like to study hundreds upon hundreds of physics papers where the authors don't support their ideas with suitable diagrams—please note, in my opinion all physicists should learn to draw and present their work in the fashion similar to that of a Roger Penrose or a Richard Feynman.

With this in mind a very important feature of this study will be to offer clear representations of many new and radical notions regarding propulsion dynamics without reams of mathematical equations. Some may say that physics is encapsulated in mathematics, but I feel physics is encapsulated beneath the inadequacies of mathematics—I cite as a for-instance space-science's predisposition toward linear propulsion forces, perhaps because the mathematics which govern rotational-confinement forces is deemed too difficult for scientists to work with...?

What I have also avoided as much as I can in this book are referrals to previously printed books on UFOs and ETs. This was a lesson I learned from my website management days, especially regarding the Andreasson Affair books and Linda Moulton Howe's book *Glimpses of Other Realities—High Strangeness* (vol 2), because good books on UFOs can become very unobtainable for numerous reasons. Making a point that relies on the reader referring to a book they will struggle to find is a pointless exercise, so instead most of the technical points in this study will be backed up by referrals to scientific articles and physics papers that can be readily obtained from universities, public libraries or from the internet. Research papers referred to throughout this study can be found at the very comprehensive (and free) Cornell University on-line library at http://eprintweb.org/S/search, or they can be Google-searched for by using either their "arXiv:" type reference (i.e. arXiv:gr-qc/9712010) which I have included in their bibliography entries or by searching the author's name and the numbers associated with it.

More background knowledge leads ultimately to more forward knowledge. But you don't go very far searching through this field before you realize that there is a very active movement to 'dull down' through misinformation the whole phenomena of UFOs, ETs, and alien abduction, or to pigeon-hole these with such exacting proportions as if all things were fully known about them, proving them one way or the other to exist or to not exist. I personally find it quite laughable that the odd person here and there is so dogmatic in their quest to illustrate ET's non-existence, but I have to say I'm saddened for some skeptics whom will fervently devote their whole lives to exploits which refute ALL extraterrestrial involvement in this earth realm's development, or on the other hand, that they believe (with such piety) that if ET's existence is to be acknowledged then these unearthly beings are un-Godly and must be malevolent toward mankind across the whole board of human dynamics. May these skeptics, critics, and scaremongers ever subsist though, for I believe everyone has their task to do in this theatre called earth, and they have great value unto themselves to dampen down the general public's inquisitiveness and in keeping the general public misinformed at this evolutionary time. Indeed, along this line of thought, the reader may wonder why it is that out of all the subjects in the world that might be classified perhaps as 'sensational' or 'mysterious' that the subject of UFOs and extraterrestrials is the one stand-alone subject which has been controlled by such a disproportionate amount of denial by government departments, and why newspaper and media program editors are so insistent about refusing to cover newsworthy stories on this subject (and I give as an example the eye-witness accounts of UFOs that were seen hovering over Chernobyl after its nuclear power plant exploded and were observed taking away its harmful radioactive emissions). Some may say that if ever there was a case for truth generating the raw fear of reprisal amongst media corporations shackled by governments—then this is it.

So what this book will set out to do is present a scientific look at the viability of observations about UFOs. It will take facts from witness reports and from abductee reports, and attempt to understand extraterrestrial physics principles, and compare those to physics principles known to the scientists of this world. And in the process this book will construct a workable aircraft through its pages, and offer quite a few surprises to those who believe that the world is not ready yet for gravitational manipulation.

This study has without doubt been very arduous, but what I personally find so rewarding about the work presented in this book is that the ET's multiple-functioning engineering, that has been recognized from recurrent design-features within so many different UFOs and witnessed by so many different people, has been fully displayed throughout the chapters of this book as corresponding to bona fide workable theory. In other words, even if the reader is skeptical about whether or not UFO's and extraterrestrials can truly exist without their acceptance, because the overall designs presented throughout this book are so wholly workable and can result in actual aircraft being engineered and flown inside and outside of earth's atmosphere, then the extraterrestrial element numerously referred to throughout this book can actually be relegated to a supplementary source of this technology—not, as it used to be, the only source.

So, I hope more than anything the reader will be impressed with the way these extraterrestrial mechanisms are presented throughout these pages, not so much from my basic form of artistry but from the perspective of the extraterrestrial's design prowess. Interesting design is contagious, and certainly it could well have the effect of changing the way its experiencer participates in his or her own design projects thereafter. Be warned! Because I will attempt to emphasize design function as best I can as a sub-plot throughout this study.

Disclaimer

The electrodynamic engine described in this study is known to produce large amounts of electrically charged particles at extremely high voltages, it will produce high energy extreme-ultraviolet radiation (and in certain areas of the craft X-ray and gamma-ray radiation), and it will also produce changes to the mass-density of its localized environment. All these energies can be extremely dangerous to human beings if little or no respect is given to adequate safety measures.

The author hereby disclaims any liability for injury to persons or property that may result due to the construction and use of the domed craft alluded to throughout this study.

In this wise the author would recommend that any project to build these UFO craft should be undertaken only in a university environment, such as in a physics department, or in a well facilitated laboratory run by professional engineers, with adequate protective-wear for personnel and safety encapsulation between all personnel and the craft under critical stages of development and construction. Most especially, with respect to the craft's central ergosphere, the author would also strongly recommend that even before the 'nuts and bolts' of its construction were planned it should FIRST be envisaged how that laboratory environment could facilitate a governing mechanism that could if necessary orchestrate a fast-acting, organized, and safe, shut-down procedure for after the central vortex has established itself and is running.

The author has issued this disclaimer and warning because there have been suggestions by enthusiastic experimenters that they might build one of these craft in their garage. My response to that idea will always be an unequivocal—DONT!

Some of the excellent safety documents for experimenters to read and take note of can be found at:
http://www.pupman.com/safety.htm and at:
http://www.tesla-2.org/safety.html

Pre-emptive Statement

I have endeavored to honor and validate the principles presented in this study very vigorously, as the reader will note from the voluminous research notes presented at the end of this book, however, as with any new and radical departure from old scientific beliefs, I fully expect that many of its conclusions will 'step on the toes' of some scientific egos (I know it has already). I will therefore say here, in this study's defense, that I present this study of energy dynamics with the utmost honesty and declare that whosoever deigns to derogate any part of it in a non-constructive way will do so only in proportion to their own self-aggrandizing egotism and dilettante ignorance of the broadest possible acknowledgement of the great multiplicity of physical, astrophysical, electro-chemical, aero-dynamical, and electronic-engineering principles, cosmological dynamics, and of density-gradient mechanics presented in this study.

Paul E Potter (July 2008)

1: The Andreasson-Extraterrestrial Paradigms

Contents:
1.1 Polymechanical Engineering Studies from Distant Galaxies	1
1.2 Valuable Memories of Alien Technologies from Abductees	8
1.3 The Curious Observations Recalled	10
1.4 Debriefings Attended by Aerospace Scientists	22
1.5 We Are Alone… in our Crown of The Creation Syndrome	25
1.6 Damage Limitation Strategy of Defense Departments	27

1.1 Polymechanical Engineering Studies from Distant Galaxies

To me, the feature that makes the extraterrestrials so scientifically advanced ahead of us is the profound approach they have to energy transmutation where, as will be alluded to throughout this study, they look for and dissect energy-changing mechanisms on so many more levels of operation than we do; and then they build a machine which inter-couples a whole multiplicity of those mechanisms into one poly-interactive unit.

For an example of what I mean regarding this multiplicity here is an analogous comparison: consider if you will how we on earth design diverse bits of machinery and equipment mostly with one purpose in mind, for example we have a carburetor which converts a liquid into a vapor which is then mixed with air, or we have a starter motor that turns the crankshaft of a car, or a dynamo that propels electric charge along a wire, and so on and so forth. And then we bolt many dozens of these separate units together to make a car, which in turn mostly does just three things—conveys us from A to B, constantly depreciates to waste our money, and constantly pollutes the atmosphere to waste our race. And so we on earth collect many separate one-function-machines and then we flit between one and another of them in order to occupy our lives.

But the ETs build their scientific machines as a co-ordination of many different functions stripped down to their most basic dynamics, and then they install the basic requirements of those numerous dynamics into the very structure of their craft so that one mechanism simultaneously aids another mechanism and then another, and so on and so forth. Take for example the Andreasson craft's toroid shell, which is the donut-shaped container that not only forms the lower hull of the craft but it also forms part of the radial planar waveguide through which microwave pulses are directed from the craft's center outward to the craft's outer rim: That same toroid at its inner wall (as can be seen in many of the diagrams throughout this study) is given a specific curve which on one side of that wall serves to direct electric charges towards the toroid's sharp-edged circular rim, and on the other side of that wall it serves as a parabolic reflector: And the fluid contained inside that toroid will be used as a power source, but also, as a medium through which electromagnetic waves will be slowed down when they refract through it.

One structure: multiple functions: greater efficiency. This hopefully will be what this book will be all about, design efficiency, and the intention will be to show this multi-functioning aspect of their electro-mechanical systems within the design of the UFO featured throughout this study. So in comparison, sure, in every sense of the word we on earth have *Vorsprung durch Technik* (leading through technology), but the extraterrestrial's technology is leveled to an entirely higher plane than that which we presently aspire to on this planet.

This polymechanical factor so omnipresent in the way the ETs make their UFO craft, was a point that didn't go unnoticed by Col. Philip J. Corso in his book *The Day After Roswell* which explains his role in back-engineering (or rather, the farming out to US electronics labs) of alien artifacts and strange exotic technology for the US Army. Corso mentions that the military analysts and engineers whom were given the task of secretly retrieving and back-engineering the Roswell remains of that UFO craft were initially perplexed as to why they couldn't find that UFO's engines anywhere.

Its easy to see why they were perplexed, that UFO craft was flying through a heavy storm and got struck by a lightning bolt that caused its internal working fluid to explode which split open the craft's lower hull section, the craft lost its main propulsion drive and careered ground-ward only to rise up again and coast back into the air by using the gravitational buoyancy it still had imbued into its structure: But that craft with a gaping hole in the side of its hull and with the contents of its hull on fire was, for all intents and purposes, a lame duck.

The members of the four-being-crew that survived the initial mid-air explosion were probably trying to make for White Sands military base in order to seek medical help, but the alien crew or the alien buoyancy only kept the UFO in the air for another three miles or so and the craft came down at a location on a heading toward White Sands, a short distance from Roswell [note: 1].

White Sands was where America's first rockets were developed under the German Von Braun and launched for the first time on April 16 1946, and the testing at White Sands of these V2-type rockets continued all through 1946/7. It was also around this time that the 509th Bomb Group was stationed at the Walker Army Air Force base at Roswell which, coincidentally, was the only airbase in the whole world to possess an arsenal of nuclear bombs with B29's primed and ready to go [note: 2]. Hence an immediate News blackout was enforced by the highest of orders all the way down from Washington, and the weather balloon placebo that was given to the world was to dull down the whole UFO incident and get White Sands out of the world's newspapers and TV news broadcasts. White Sands military base was as top secret and as sensitive as you could get in the US, it was the Area 51 of its day, and the top US brass didn't want any civilian sensationalists sniffing round it!

But from the fact that rocket engineering was then the newest and most advanced technology available in the world, those military engineers were only searching through the Roswell UFO debris looking for signs of that craft being rocket powered. Just as Colonel Corso explains, those analysts in the 1940s had no real idea of what did actually power those strange UFO craft and so, because they couldn't find anything they could recognize as being part of a rocket engine they automatically assumed that the UFO's power drive must have either fallen out somewhere, or that it had been self-destructed by order of the last surviving alien crew-member.

It was because the Roswell debris included no apparent engine that most of that UFO's structural artifacts collected by the army personnel then went into store, and as Corso explained, they were all but forgotten about for at least a decade. It was only later, in the 1970s, that some bright spark worked out that the engines didn't fall out after all, and that the craft's central power generators only looked like engines while that UFO is powered-up… For when they are decommissioned and powered-down there is nothing there that looks remotely like a conventional engine.

This is not saying that the ETs had developed some deft 'magic arts' that rendered their engines invisible when not in use, far from it. It just means you have to know what you are looking for before you can find it, and, you need to venture that for a drive-unit to provide rotating forces that drive-unit doesn't necessarily have to be an ensemble of moving parts encased inside a big chunk of metal. But that, I might add, is what this study is going to be all about, for in a similar fashion neither does the central power generator of the Andreasson UFO, which is featured throughout these chapters, look anything like an engine when it is inanimate and powered down [note: 3].

Obviously, the necessity for compactness is the governing reason behind the ET's preference for multi-functioning, for their craft have to convey their pilots, engineers, doctors, and nurses (etc), safely through earth's atmosphere and back into space again, or at least back out to the far reaches of earth's atmosphere to rendezvous with larger motherships or into a grouping-type arrangement high above us.

What the above account emphasizes more than anything else is that scientists who work on power systems of the norm on this planet, are not necessarily cognizant of how the power drive systems engineered on other worlds appear to work. As has been the case with astrophysicists, who when they first discovered black holes in space they struggled initially to understand exactly what their purpose was and how they worked, especially with regard to how black holes can generate jets that can span thousands of miles through space, and only now after some forty years of speculative discussion are these theorists coming to a working hypothesis for these most powerful engines of the cosmos. The reader will observe as this study progresses that black hole physics and UFO physics are closely allied, which in itself is a very new discovery and is one that has never before been openly discussed among physicists.

What complicates the UFO back-engineering's recognize-and-discover process mostly though is the fact that there isn't just one type of UFO craft; and there is the fact that a UFO doesn't just fly from one piece of sky to another using one form of propulsion. And so an awareness of the diversity of propulsion configurations, leading to what types of flight maneuvering can be performed by the various types of UFO, is a very necessary prerequisite to discovering what energy manipulation principles the different groups of ETs are using in their different craft. But this is not overly difficult, knowledge of these is quite forthcoming through earnest studies and can of course come from a long study of the well documented evidence now available... from (A) what has so far been observed and reported by those who have encountered UFOs; from (B) those who have filmed them flying in our atmosphere; from (C) those who have sifted through the NASA footage which show UFOs; from (D) those who have taken real photos of real UFOs; from (E) those who have been abducted and conveyed to strange-looking destinations inside UFOs; and even more so (F) by those who have worked in government 'back-engineering' programs at places like Area 51 in the US; and (G) by those who have passed UFO artifacts from the military to electronics labs in the civilian sector for technological appraisals (as was the case with the Col. Corso mentioned above). The purpose of back-engineering studies is to form opinions about the strange signatures of energy transmutation and gravitational manipulation that are the hallmarks of UFOs, so that these signatures can be recognized alongside known scientific principles. And even though those associated principles may be rarely known and in some cases under-developed on earth they do provide enough clues to show that the mysteries of UFO technology are mostly resolvable—by those technically-minded and versed in a broad, and I do emphasize broad, range of disciplines scientific and otherwise.

If you haven't the time and patience to look through the literature and film footage that is out there on UFOs I will summarize the main attributes of a range of UFOs; as having the ability to fly and hover in silence, to accelerate to tens of thousands of miles per hour in a matter of seconds without sonic booms, to travel from this atmosphere into both local space and far space (with no need for heat shields), to travel through instantaneous 90-degree bends impervious to G-force reactions, to suddenly appear or disappear from visual sight and be on and off radar surveillance at will, to create light-spheres of synchrotron emissions around themselves (ie as a globe emitting bright light), to enter into water without speed reduction or perturbation to the surface water, and to detect hostile attacks at their moment of intention of perpetration (ie just before anything is physically fired at them). UFOs have the ability to propagate space-time geometry that will trap electromagnetic waves such as light, and gravitational waves, so that they can activate localized gravitational fields which can transport material objects over short distances (ie ground to ship and vice versa), and UFOs can even manufacture a topological envelope which can cause material objects to undergo dilation or diminution (to become larger or smaller in size). The extraterrestrials can also propagate energy fields upon humans that will arrest the motoneuron mechanisms in human beings (to paralyze human muscles at will). Metals inside UFO craft can be switched from opaque to translucent or to even become illuminated (although this is something our scientists on earth are just now starting to develop too). And associated with the UFO and ET sciences is also a range of gate-hole phenomena where intense energy fields are manifested in thin air and then materialized into bridging-holes through which beings enter into and leave from this material realm [note: 4]. In similar fashion (but mostly without the spectacle of intense energy fields) the same bridging-holes can occur at existing spatial barriers, at our normal everyday material walls such as walls inside buildings, where bridging-holes are temporarily formed through which alien beings seem to be able to pass into and out of our environments (the latter suggests, if nothing else, that some of our preconceptions about 'solid' materials needs a little more education).

As will be explained further in the following chapters of this study much of the above phenomena involves the manipulation of energy envelopes which are transposed from the UFO craft to a distant location exterior to that craft. As this is a very novel form of energy manipulation for our stage of scientific development this technology, which can be discovered in astrophysical principles I would hasten to add, will be featured throughout several of the chapters in this study.

As far as material fabrication and structural design are concerned there is a wealth of technological advances waiting to be discovered here, although, while this field is mostly alien (please excuse the pun) to anything we might presently engineer on this planet, there have been parallels found in earth's sciences which might offer great advantages in those areas if pursued vigorously enough to anywhere near the stage the ETs have taken them (these are the areas of acoustic and vibrational structuring of matter and the way materials are oscillated with energy) [note: 5]. Indeed, these fields of material fabrication and structural design are certainly taken much more seriously by the ETs than by the scientists of commercial industry on earth, and these fields have obviously been investigated much more rigorously by the ET designer than the scientific designers of earth, as the reader will see.

My discoveries though offer just a small cornerstone toward scientific advancement, and it would be naive of me not to suggest that greater scientific minds, once they latch on to the more detailed analysis of extraterrestrial technology (as it becomes more available through industrial enterprise), will no doubt want to progress more openly their commercialization of extraterrestrial technology through leaps and bounds. In the meantime, and while much misinformation is still perpetuated through the, quite understandable, ignorance of news and science media producers, and is prevalent and waiting to misdirect the unsuspecting researcher down the longest of garden paths, may I offer a few pointers as to what UFO technology is not. I will allude to this because there have been some really wild assumptions over the years made about how a UFO powers its propulsion (by TV producers mostly who have seen too many episodes of Flash Gordon perhaps, and even by some so-called UFOlogists); and on this matter I will hazzard a guess that a UFO does not carry an electrical battery, nor a set of Tesla coil accumulators, nor a nuclear fusion-reactor, nor is rocket-powered (so that it should carry low-intelligence high-octane fuels in any way shape or form), nor would it carry any super-conducting magnets, and nor for that matter would a UFO need to carry an array of separate dynamo power generators, and nor will it have a separate engine through which it energizes its gravitational forces—for indeed—the UFO's structure is its engine, its power generator, and its gravitational force amplifier!

Moreover, designed into the UFO's structuring is, as I say, this multi-functioning that enables the composite whole to work to such a high degree of efficiency that, with respect to its electro-dynamic power circuit, the entire craft can function as a mega-conductor or a mega-accumulator or a mega-amplifier without its pilot even having to actively control these functions. So clever are the extraterrestrials that their electrical current is transmitted through the actual structure of their UFOs (rather than through looms of wires), and energy-threshold levels are designed into the very materials which form that electrical circuit (ie as we also incorporate into our semiconductor circuits), and so these materials essentially determine the moment when electronic triggering takes place from one stage to the next in their craft's energy-development processes. The energy dynamics I have discovered would always be oscillatory or cyclic and so whole chains of reactions, thresholds-triggerings, and power amplification effects take place within a UFO with a metronomic regularity that might be likened an up-scaled '555' electronic timer circuit! Back-electromotive forces are aplenty in UFOs, and although these 'rogue' forces are the chagrin of electronic engineers here, they are integrated into these craft's electronic circuits as an integral, and wholly necessary, part of the UFO's energy accumulation mechanisms.

Hopefully though, more than anything else, this study will show the reader that UFO craft are engineered and structured as multi-purpose and multiple-function polymechanical machines, and as I discovered in my jaunts through the various UFO designs (such as the Andreasson UFO, the Bob Lazar UFO, the Adamski UFO, the Roswell UFO, the Onion Drive UFO, the Grangemouth UFO, the upright cylinder UFOs, and even the non-UFO designs of Searl's IGV craft), there is no structure in a UFO which has just one function relating to it doing one task, and no structural assembly inside a UFO is merely perfunctory (to borrow a phrase from Bob Lazar).

As above, one might conclude that this novel-to-us polymechanical technology was born out of necessity, from the desire to put the whole of the UFO's power generation and propulsion drive mechanisms into a very small volume, so as to develop large amounts of energy quickly and without wasteful dissipation, and advantageously using the medium which surrounds that craft to draw on as a viable source of energy to replenish its own energy supplies as and when necessary. Or... it may just be that the designers for these little Greys are just incredibly CLEVER at building these UFOs!

Essentially, I hope through this study there will be inspired a new breed of multiple-function designers who will design from their own ingenuity power-drive systems that will show a much higher degree of efficiency than at present, and that they might after reading this study ask the question of any component of any machine, "Yeah? What else does it do...?"

1.2 Valuable Memories of Alien Technologies from Abductees

How did all this information about UFOs and the extraterrestrial nous for science and engineering come to the fore?

Not through official channels that's for sure! And certainly not through the words of a government spokesman or spokeswoman. They came mostly from ordinary people who've had personal experiences with the extraterrestrials through what is known as alien abduction, where those people were taken by the ETs and shown various attributes of alien life and alien technology as befitting to the level of intellect of the person they took. If the reader is interested in this subject then I advise them to go and inquire about this vast field of study like everyone else who has been involved in it, for it surely is outside the remit of this book to repeat such information, which is abundantly presented much more eloquently elsewhere [note: 6].

Suffice it to say that, for this study, the main source of the technical diagrams and artful representations that I have been working with in order to present a clearer explanation of extraterrestrial science and technology has mostly been those of Betty Luca, or Betty Andreasson, as she was known when she first introduced herself in 1975 to Dr. J. Allen Hynek the foremost investigator of UFO phenomena in response to Hynek's newspaper request for personal UFO experience information.

Betty Andreasson Luca was abducted at various stages through most of her life, from her childhood years in the 1940s up through to the 1980's, and ever since her first experience she has been gifted with the ability to draw in great detail the sights she has both experienced herself and has been given to portray at various later times by other beings. Betty's type of experience with the extraterrestrials have been featured in the works of Whitley Strieber, Dr. J. Allen Hynek, and Dr. David Jacobs. Indeed, the rich content of Betty's experiences suggests that the term abducted is not one that should also be associated with the words, against her own will, for whoever has read at least one of the five books detailing her experiences may readily acknowledge that her experiences, often accompanied by one or more of the small Greys, are more perceptible as instructive adventures than grueling traumas. If the reader is aware of the usual procedures of physical inspection carried out by extraterrestrials on all abductees, with regard to the medical examinations and painful procedures administered with little apparent care toward the abductee's pain threshold, all of which have led to no apparent reward, then compared to every other mortal taken by these aliens Betty was for the most part given the Royal Tour [note: 7].

Betty also came across those humanoid beings known as Elders whom seem to be the guardians and instructors of the Greys. And mostly Betty was given the privilege of the Royal Tour of things extraterrestrial because of the way she dealt with her human emotions throughout all the types of experience that most other abductees have been scared witless by. But I think another reason also was their trust in Betty to accurately recall, and to portray into drawings the strange scenes she experienced, for the benefit of others as is evidenced in this study which features some of those drawings. Her first published experience (although not Betty's first chronological experience) of the little Greys entering into her house and into her life tells the reader of a group of Greys shuffling through the door of her house, in fact quite literally shuffling THROUGH Betty's closed door—as if they entered from a parallel dimension! Then in this account published in "The Andreasson Affair", there follows her being given the loan of a blue book by the Greys, which she accepted in exchange for her bible (which the leading Grey immediately replicated into several copies and handed over to his colleagues) [note: 8]. The blue book was unusual in abduction terms for it being an actual book with pages, and showing technical schematics and diagrams of parts of the UFOs she would often be taken away in, it also contained numerous mathematical formulae, much of which she has since forgotten, but much which she published through her own ASBT books. This particular blue book was lent to Betty for a matter of days only and was then taken back by the ETs... Fortunately, many of the experiences that Betty (Andreasson) Luca recalled with the aid of hypnotherapy sessions carried out during the 1970s, 1980s, and 1990s, have been published in five books [note: 9], and Betty herself has published a further two booklets on a specific series of messages given to her by the extraterrestrials on the future events awaiting the human race. Through her many sessions of hypnotherapy Betty gradually discovered that she, her husband, and many of her immediate family had been taken by the ETs on experience-abductions, and that she personally had experience-abductions ever since she was seven years old. Which coincidentally was the age of the author's first experience-abduction.

1.3 The Curious Observations Recalled

To give the reader some idea of what strange technologies can be found in the recollections of UFO and ET experiencers, I'm going to feature this small episode of Betty Luca's experiences with the ETs in this section because it shows how from small acorns big oak trees can grow. This particular scene was recalled by Betty in 1992 and its details were published in Ray Fowler's book "Watchers II" pp67-82.

What I think is being depicted here has to do with how sound waves can be turned into light and ultra-violet energies. Sonoluminescence as in the findings of Seth Putterman and others is a way in which a bubble of gas in a watery fluid can focus and amplify acoustic wave energy a trillion-fold to produce short flashes of light in the visible light band (and the near-UV band) of the electromagnetic spectrum [note: 10]. This phenomena has huge potential for further development but lately has smacked into some rather unsavory scientific politics which suggests that is not currently being used to anywhere near its full potential (and so as it can't be exploited its still only regarded as a novelty). But if it were used in such a way so as to create a train of oscillations, of UV ionizing radiation, then the system has great potential in ionizing the air, and of course as a generator by using the UV radiation it produces via the photoelectric effect to release electrostatic charges from material objects.

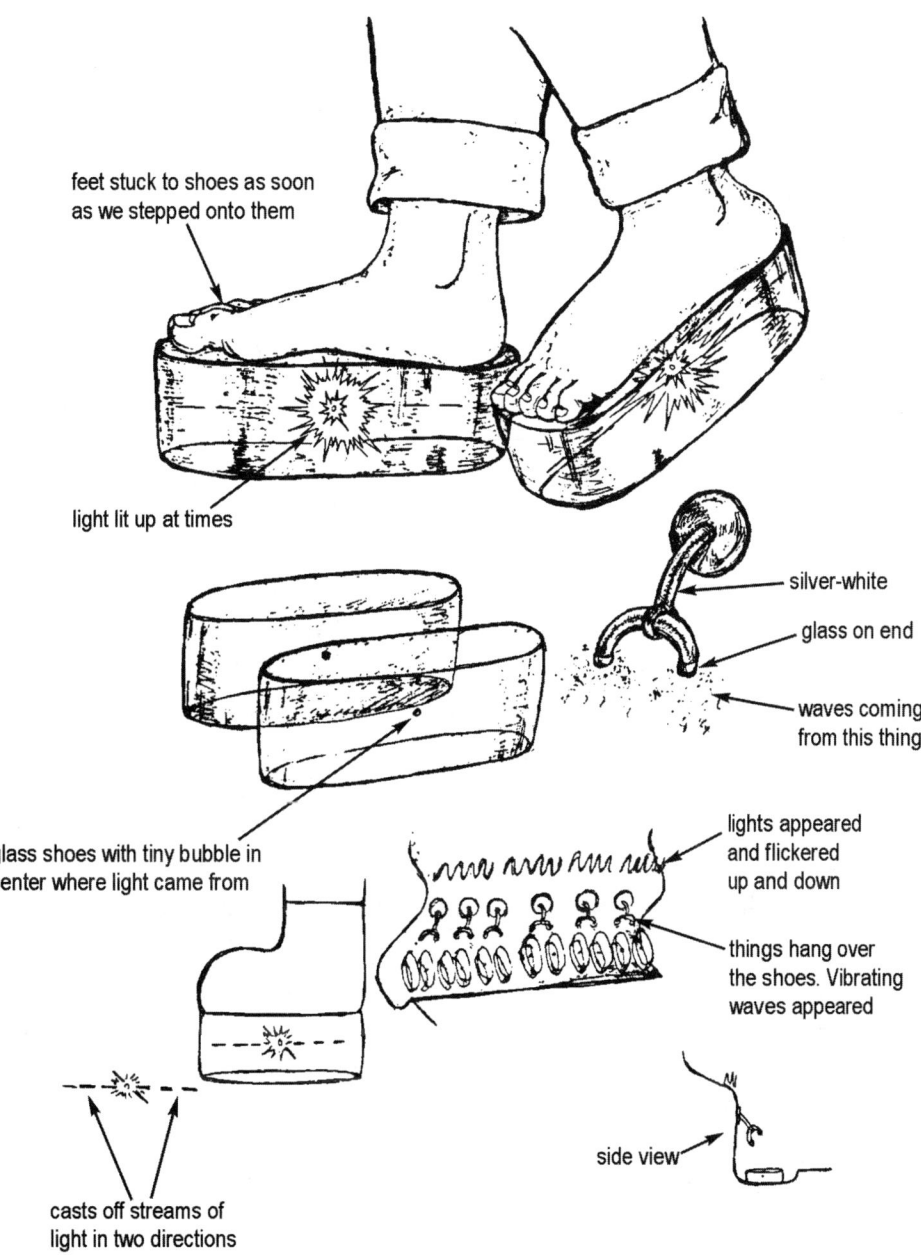

Figure 2. These glass shoes appear to work through sound wave technology
(from "Andreasson Affair—Phase Two" fig.34)

This I believe is one possible explanation as to what seems to be happening in these transparent shoes that Betty was given to wear by the extraterrestrials. Because as this drawing shows (see figure 6 below) the little Grey is pointing to the charging-up process of these shoes, showing Betty that they are being charged up on a stand with some sort of energy that Betty has suggestively been made to draw as an array of 'visible' oscillating waves.

But in consideration of the fact that a pulsing 'spark' of bright light was seen to flash from a small central sphere in the middle of those transparent shoes, and to then stream outwards in both directions in a thin line outward to both ends those shoes, then this is exactly what sonoluminescence, and its sibling process sonofusion, does. The later development, sonofusion, has even more potential and works with sound waves (at about 20 kHz) being pumped into a fluid (such as water) which causes small bubbles to form and to implode at a pre-determined rate. As that implosion, in both sonoluminescence and sonofusion, carries with it the electromagnetic waves of sound the frequency of those waves compresses as the bubbles implode and the sound waves transform into light waves. That the shoes stayed on Betty's feet, because of this energy process there is obviously at lot more that this sound-fusion process can deliver if it is encompassed into a purposely-designed structure—as these shoes appear to be. And these images suggest that the ETs are utilizing some sort of converging-non-uniform electric field being established along the axis of the shoe.

Although, by taking this concept further, and using a bit more ingenuity, the initial sonofusion implosion-explosion activity could be directed through a spiral motion inside these very thin hollow tubes running through the center of these shoes, to form rotating energy fields that spin so fast, as they expand along the tubes, that they ionize the surrounding air. Similarly, it might be that the sonofusion implosion is being used to forcibly draw in air from outside the shoe through these tiny hollow tubes as a negative-pressure field which spins through those tubes. If so, then there may be a possible link to that rotating energy field generating its own localized attraction force (similar to the gravitational force effect noted in chapters 9 and 10 of this study).

As an aside, Betty Luca in the booklets she published of scientific notes she received from the extraterrestrials there is a whole electric-power-generator schematic for a sonofusion-style turbine—so obviously this is an area where the earth's scientists have been dragging their feet somewhat but where the extraterrestrials have seen its worth.

What is also interesting in this set of figures, is that the environment associated with those strange shoes seemed to have been experienced in the near ultra violet range of the electromagnetic spectrum... The hints alluding to this are shown early into Betty's encounter when she notices that the craft's lights are giving off a magenta-pink light ("Andreasson Affair—Phase Two" p75), and that in some instances she could feel this magenta light hurting her eyes (just as if there was sand in them). That she does not mention any pain from this hue later into her encounter may be indicative that her eyes had become used to the UV-lit environment (or that the short wavelength UV was replaced by a longer wavelength and not so harmful).

Another contributory factor to this assumption of a frequency shift might be that Betty says she had to pass through both a red and a green environment on her way to this UV frequency world, or crystal forest, as Betty called it. This suggests that her physical senses were hyper-tuned into a higher frequency through graduated steps, first into the red frequency and then into the green frequency. This pre-organizing was obviously because the ETs wanted Betty to experience quite a different life-sustaining environment, different from what she was used to on earth, with presumably the intention on the part of the ETs that she would later pass this unique information on through her drawings, and through her hypnotic-regression recollections, to others (as is happening here). The reason why the ETs should go to all this trouble in these episodes is not wholly clear to the author, and even though I can offer here a few suggestions as to why the ETs gave Betty such a strange environment and such a cornucopia of strange experiences, for the true implications to present themselves about this environment there will need to be a little more input added to them (and this input may come from the ETs at a later date, perhaps).

In that quite different environment Betty was shown what can presently only be described as a holographic vista, seemingly kept in place by the aid of projected energy (light) rays. That these rays took the form of refracted rays may be some sort of clue that needs to be associated with this episode, to suggest that the immediate environment she was in had a different permittivity and permeability (meaning, a different refractive index) to the environment we normally live in upon earth. Or in other words, that it was a specially manufactured environment quite artificial to that of the 'real' world we presently experience on this planet [note: 11]. Because, notice the feature-less beings in these scenes, Betty has seen them before and has drawn them before, these were the same feature-less beings she watched playing games on board a mothership when she herself was out-of-body watching them ("Watchers" pp158-171). They are not so much bodies as more like spiritual essences. Betty describes these feature-less beings as light-forms and I would recognize these light-forms as being like the spirit or the divine essence of a being. These light-forms could be perceived when a person was undergoing an out-of-body experience, or perhaps if their physical body was ramped up into a higher vibrational frequency, which is the sort of thing Eastern mystics and Adepts of Buddhism can do to reach higher spiritual levels of perception [note: 12]. The latter point may be seen to be all the more important when one considers that on previous occasions, when Betty has been the guest of the Greys and the Elders, her spiritual essence has been made to leave her body so that she herself may accompany those Greys or those Elders, but on this occasion both Betty and the little Grey which accompanied her obviously remained in their physical bodies.

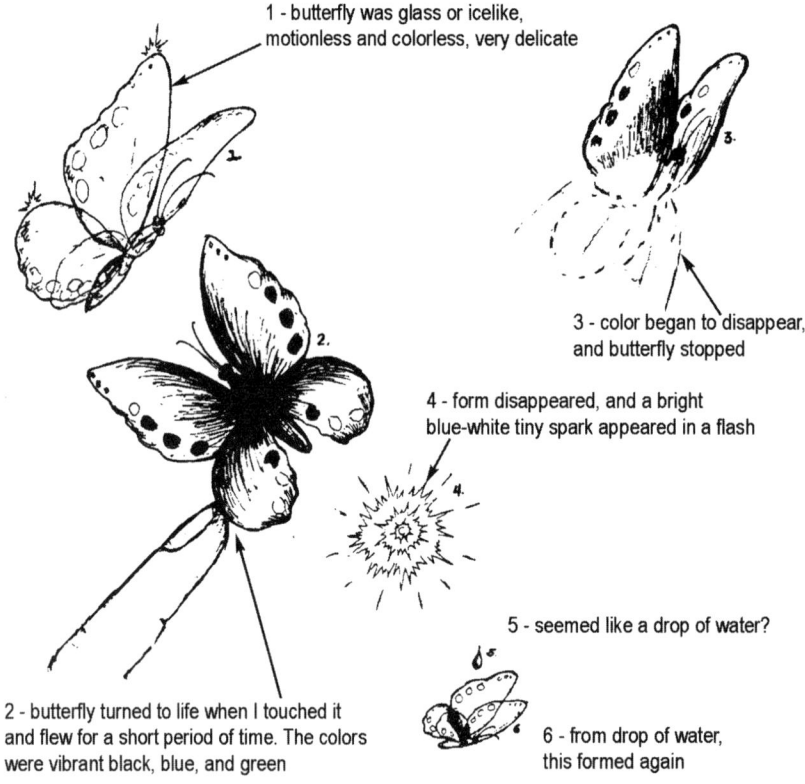

Figure 3. Virtual energies making up recycling life forms (from "Andreasson Affair—Phase Two" fig.33)

Further clues can be found inside this holographic vista of a higher frequency which supports the ultra-violet hypothesis [note: 13]. These are given in Betty's descriptions of the strange crystal environment where everything to her looks like crystal and is somewhat opaque in appearance—rather than being naturally colored—as was the case with the glasslike butterfly. Betty seems to contaminate this butterfly by touching it with her earth-frequency finger, because before she touched it it was inanimate and colorless, but after she touched it she could see it take on its more earthly colors, before it transformed yet again into a flash of energy and then back into another butterfly ("Andreasson Affair—Phase Two" p123).

This whole scenario smacks of a particular realm I'd heard of from other abductees, whom had described communicating with certain aetherial and alien intelligences about a strange new environment... as has been spoken about by Dana Redfield, another author whom has also researched the abduction phenomena, and has progressed to a level which can understand that there might be a benign link between extraterrestrials and humans. In Dana's quest to find out more about why the ETs are now here she began recording hundreds of communications from evolved beings, which seemed to be aetherial or spiritual beings just as much as they might be described as extraterrestrials. Some of these communications received by Dana, like I say, mentioned a 'NEXT dimension' of space-time where a more aetherial vibration (than that of our present existence) will become available for some of this human race to continue into (as is explained in greater detail in "The ET-Human Link" by Dana Redfield).

Whether or not these two scenarios are two different approaches to the same future environment, or some sort of taster for human beings who might want to discover more about extraterrestrial-created environments, no one is yet sure. Certainly though, this scenario of Betty's is most like an artificially manufactured environment (for which the only available comparison I can think of is a holographic one like the holodeck environments presented in *Star Trek—The Next Generation*). But from these recollections it doesn't seem fully permanent, it appears only to be a temporary creation, like a model of a real environment—wherever that is. As it appears in these drawings with all the refraction rays beaming down through it, and because it appears that the extraterrestrial Elders (or whomever they are) are in spiritual form when they are sculpturing that manufactured environment back into shape to repair the damage done to it, then it all seems specifically centered around something which can only be engineered, and repaired, by highly evolved spiritual entities.

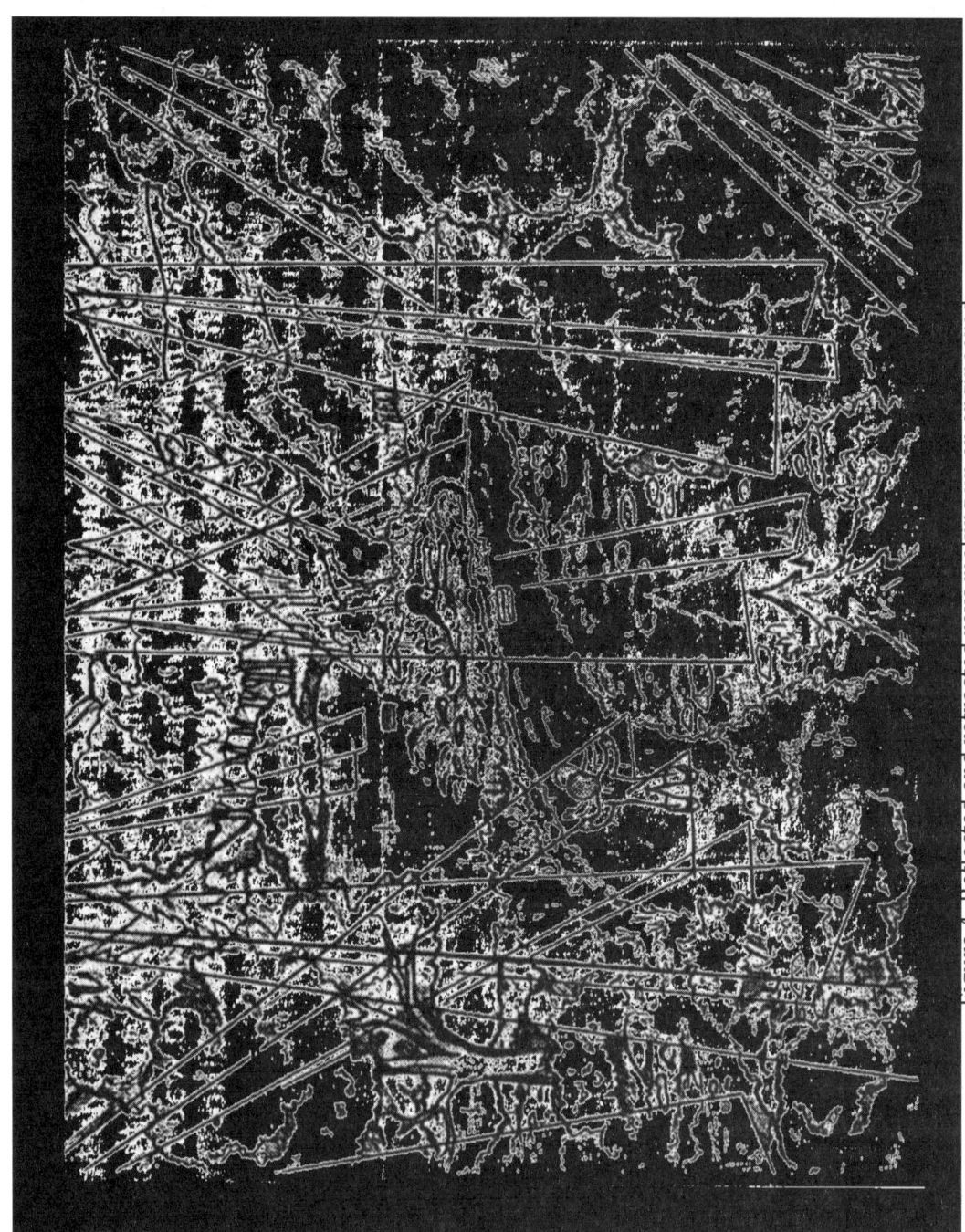

Figure 4. Reflected and refracted rays making up an engineered environment that even this Grey seems to be alien to (from "Watchers II" fig.8)

The scenes drawn from Betty's recollection before this (figure 4) one show a pleasant natural scene of trees and a footpath that Betty and the little Grey were walking along quite happily, but when Betty decided she'd run along that path and when she subsequently fell off her special sonofusion shoes, that's when all the refraction rays switched on. After loosing her special shoes and stumbling over Betty then fell into the lake next to the path, but the water she had fallen into was solid and glass-like, and as she felt around herself she thought she was in a gaping hole in that lake. Then strangely, she could feel the glass-like appearance to the water change to it being more like liquid. And then she recalled that she began to feel wet and cold in the lake. After this, while recalling how she crawled out of that hole, she remembered that the grass on the bank as soon as she touched it turned from being crystal-glass-like to it feeling like proper green-colored grass! Yes, this is altered awareness, but it also indicates the transformation phases matter will go though (through the liquid and solid phases) while remaining, essentially, the same form of matter.

But further, when the light-forms came to her aid, it was as if they were standing close-by watching all this unfold, observing Betty's reactions to this mind-perplexing situation, and then when they'd seen enough they came to Betty's aid. But that they could then stand on the surface of the lake to repair it throws a person's pre-conceived ideas about such a situation, should it be compared in the context of our world, right out the window! Just as perplexing is the next scene (in figure 5) which shows more light-forms that have come along with a glass barrel to place Betty in while they continue to repair the crystal lake, out of this barrel was projected laser-light beams and the light-forms seem to be using that barrel for the purpose of insulating Betty's vibrational frequency, which must have been different to theirs and different to this manufactured environment, until they could transfer her back onto her insulating shoes.

Otherwise, the other thing that puzzled me about this incident was that while this environment's refractive index may well have been different to our real world, why was that environment real-looking before Betty damaged it, but was then permeated by an array of refraction rays as soon as Betty did damage it? I did wonder if this whole incident was strangely reminiscent of the plot from the movie The Truman Show! And at the end of the day, were the ETs educating Betty or were they learning something from Betty? Was this whole episode arranged for Betty to test the integrity of the ET's specially manufactured environment, to test how a human's vibrational frequency might interact with some new dimension being prepared? Or was this to test how a human can cope when surrounded by non-sensible stimulation?

But this, it should be noted, highlights the quintessence of alien abduction (and the problem for the abductee-experiencers) because when an abductee is taken to 'off-world' environments they will seldom recognize any of it, and indeed their mind will remember very little of these episodes because their mind just won't know how to file its details. Such an incident will have little similarity to anything previously recorded in the person's memory bank, with which the mind can associate these strange details—so an incident like this doesn't get recorded in the mind in the usual way.

Figure 5. Betty has her shoes replaced to insulate her from this environment (from "Watchers II" fig.10)

This addresses the principle in psychotherapy that the mind cannot file something if it can't associate with it in some way some sort of descriptive label. And if the mind can't recognize something it sees and it can't label it how will the mind find it later and remember it? For, if it perceives something it has not the remotest chance of recognizing that mind will have two choices, it either throws itself into a spin and overloads, or it dumps that strange episode's memory-details into a blank file and shoves that blank file into a remote corner of its memory-banks with no labels and no retrieving tags attached to it. The memory will be there in the mind, but the special triggers to get that file to open, to be reviewed so that those memories can be recalled into the person's conscious mind, will never occur (particularly while that person is back on earth living a life of 'normality').

Here is the problem for abductees, the retrieving tags for their experiences amongst the aliens will have blanks on them, so when the mind scans its memory bank these memories will be undetectable. The incidents in these memories will never get restimulated because there will be very few circumstances on earth that will be anything like the extraterrestrial environments. And so, for all intents and purposes the episode that person's mind had experienced will effectively be obliterated from that person's grasp. In fact, this mechanism guarantees that the episode will be a 'missing time' blank in that person's memory, and the alien's know this, they know full well how the human mind works. But, cleverly, this mechanism will guarantee that those memories can be retrieved at a later date if that person were ever re-united with whomever or whatever environment was originally involved in the generation of their experience-memories (ie that person's mind will rarely remember an abduction episode on its own in an earth environment, but it could be restimulated if that person ever returned to that particular alien environment again, or if that person was later re-introduced to those strange beings again).

And so here, in a nutshell, is one of the major memory-controlling mechanisms used by the extraterrestrials on humans. And to a certain degree what hypnotherapy does is re-introduce the environment of the ETs back into the abductee's mind (but with the added comfort of the safe surroundings of here on earth), so that very carefully, whatever experience-memories have been filed into the mind's out-of-the-way corners, can be searched for and eventually found.

Indeed, the crystal lake incident remembered by Betty through hypnotherapy is also unique for the fact that not only Betty was given the special insulating shoes to wear but so too was the Grey accompanying her, which suggests that the little Grey in his physical body was also alien to this particular type of environment. So did the shoes offer a bridging function between our bandwidth of vibration and the frequency of this strange environment? If so then does this suggest that there exists a mass-frequency spectrum... or a gradient to the space-time metric?

One final point to make here about Betty's recollections, or indeed any abductee's recollections, of these extraterrestrial-related incidents, and this refers back to the head of this section to where Betty was pre-prepared for the crystal forest, and that is that there is an important point to make about the whole memory recall process through hypnotherapy and how it works. If the memory in the process of being extracted has scant labeling, with few descriptive tags, its going to come out most probably in dribs and drabs. How can the mind know it has found a complete episode when there are so few remembering-tags attached to it? A perfect example of this fragmentation of a memory can be found in what Betty initially recalled about the crystal lake incident, which came up first in a hypnotherapy session in 1980, when her recollection about the shoes she was given to wear was recorded then as only,

"And we're walking along, and as you walk it seems like that little one, that person there, his foot lights, that glass thing's underneath lights. 'How come we have to have these things on our feet?' I asked him. And he says they are necessary." ("Andreasson Affair—Phase Two" p124).

But later in 1992, after Betty had remembered a substantial amount of other incidences to do with the extraterrestrials, her commentary about this crystal forest scene in her hypnotic sessions ran like this;

"There's...mist all around, but there's...these things that come out and there's the light in waves, like light rays. There's these glass—I don't know if they're glass. They're clear and they're just [like] clasp-on shoes.

"That looks so familiar down here. It's those glass shoes and those things sticking out. I don't know...exactly, but there's light, and there's a crystal forest around here somewhere. And we're putting on our [glasslike] shoes. They just stick to our feet..."
("Watchers II" pp67-68)

As these two recollections stand, their raw comments might remind the reader of Walt Disney's animated version of Alice in Wonderland... but it was because many of Betty's later recollections referring back to earlier ones of her experiences with the extraterrestrials that more details of those scenes could then be brought forth to provide a better clarification of the earlier episodes, and then when those episodes were compared with Betty's scene-sketches, drawn under hypnosis, they provided an excellent base to work from and to ascribe scientific principles to.

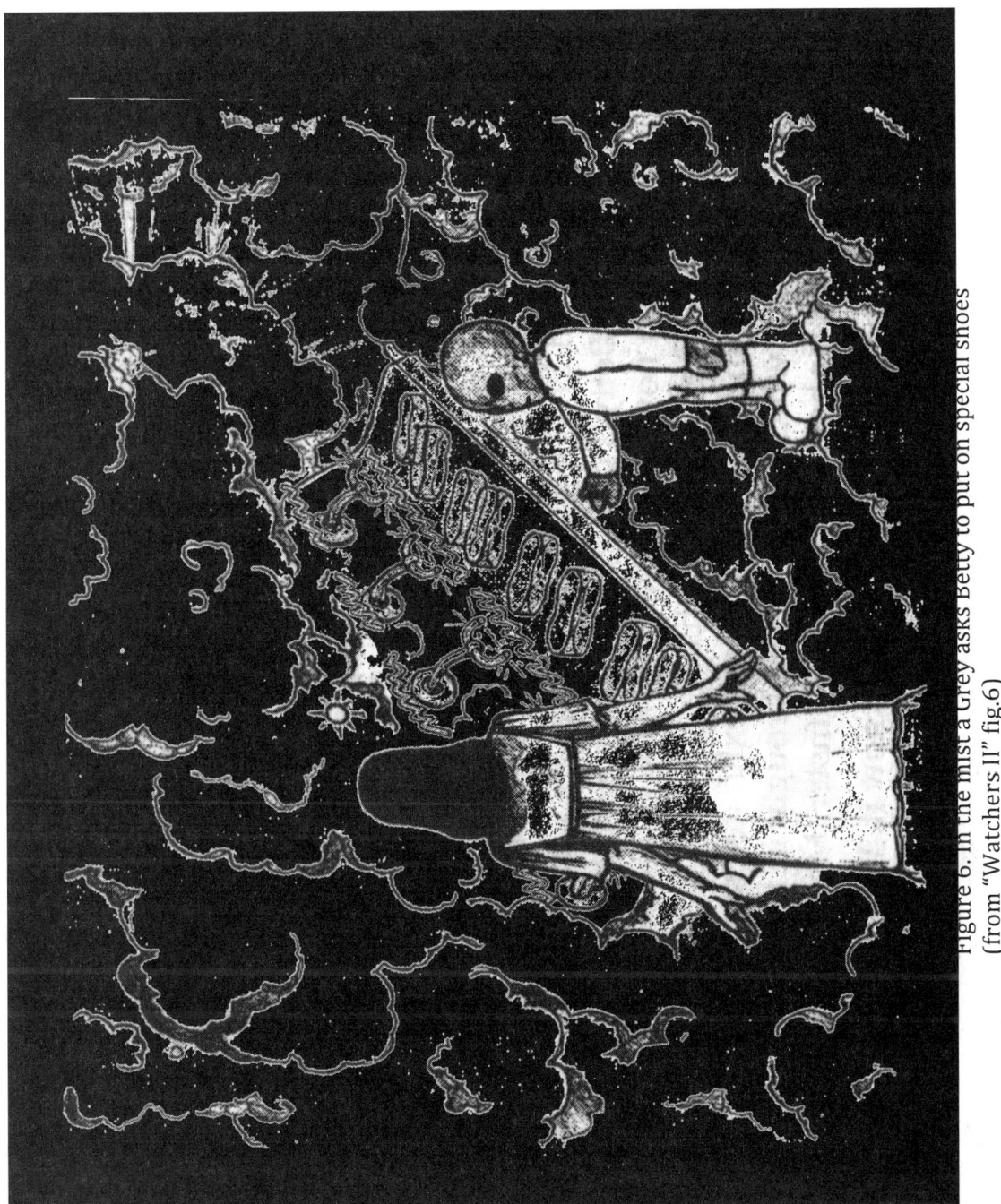

Figure 6. In the mist a Grey asks Betty to put on special shoes (from "Watchers II" fig.6)

Subsequent research to check out these characteristics has then led to workable conjectures being made (as has been the case with the association of the technological field of sonofusion with the glass shoes mentioned above), about the energy dynamics involved, or, as was the case with the crystal forest scene, about the possibility of manufacturing a topological environment (which may well turn out to have the characteristics of a different metric of mass-density to the one we exist in on earth—there will be more about this in later chapters). And, of course, the exact same process has been afforded to the electrodynamic characteristics of the UFOs that had been observed by Betty and where their technical details have been brought forward by this experienced abductee. So anyone who says abduction experiences are not worthwhile and extremely fruitful can only be saying this to thwart scientific advancement [note: 14].

1.4 Debriefings Attended by Aerospace Scientists

At one point Betty's hypnotic sessions were attended by some very learned scientific investigators whom, through the course of those sessions would suddenly become very animated about what mechanisms Betty was describing [note: 15]. Obviously they were scientists at the top of their field and at the cutting-edge of the most up-to-date research projects, and they were fishing for clues about technologies they'd heard about in UFOs but couldn't understand. Because on one occasion one particular mechanism which was being described by Betty caused great excitement among these scientists. From the flurry of remarks passing between those scientists at hearing these descriptions it was obvious that they had recognized something very special in the words Betty was passing on to them, and was being shown to them in the diagrams Betty was drawing under hypnotic trance. The description in question entailed the very bright emissions of energy that Betty described seeing coming out of the top spheres, of the rotating sphere-set assembly at the center of the UFO which the Greys took the trouble to show her during her 1967 abduction. These details of how some of that UFO craft worked were obviously enough to suggest new scientific advances could be made, and that these scientists knew how some of the power generation mechanisms inside that UFO worked.

Officially, none of these scientists returned to announce what they had discovered in Betty's descriptions of extraterrestrial technology. At that time though, during the late 1980s, the scientific community was abuzz with a recent discovery known as ferroelectric emissions, but research had stalled and scientists although they knew how the basic process of it worked, were not getting the desired results from the materials they were testing. And here was Betty Luca describing the most amazing account of super-abundant emissions being ejected out of this UFO's glass-like spheres, and describing the very goal that many contemporary scientists were striving for [note: 16]. I don't have the whole list of these men, but apparently one of the scientific investigators being kept informed about whatever Betty and her husband Bob were describing during the later hypnotherapy sessions (as mentioned in "Watchers" p84) was none other than John F. Schuessler, project manager for McDonnell-Douglas Aerospace, a company which later merged with Boeing and which had in past years produced the F-15 Eagle and F/A-18 Hornet military aircraft.

Throughout many of her hypnotic regressions Betty was known to bring forth scientific formulae, sometimes from her recollections of being with the ET scientists, and sometimes these formulae came direct from the ETs themselves, or by them using Betty's voice-box while her body was in hypnotic trance (a known phenomena in spiritualist-channeling and psychic-communication procedures). These formulae coming through Betty would mostly be in riddles, understandable only to those who already had peripheral knowledge of that particular science—the ET's have very strong reservations it seems against handing out concepts about their sciences to those scientists who live by GREED with little thought for the rest of mankind. And one such series of chemical formulae was given to Betty in a number of messages that were originally received toward the end of the 1970s and through the early 1980s (these were later published in the "Andreasson Affair" books, and later still in Betty Luca's own publications of the ASBT booklets). These strange formulae delineated chemical procedures completely un-known at the time, but as one in particular promised to yield a new form of carbon it was carefully studied by very enthusiastic scientific experts.

One group of messages was taken up during the 1980s and was deciphered independently by two different researchers, one from the Caribbean Marine Research Center, National Undersea Research Program (Florida), another was from a university of Colorado. These messages were a complete riddle to Betty herself, because of their strange prose, but to the chemist they could be understood; they spoke of a chemical combinational hierarchy and of elements in the periodic table, they spoke of structures such as Buckanosa, which was later recognized to be a solid reference to Buckminster fullerene 3-dimensional atomic-structuring. When the latter association was made and the messages finally revealed their chemical procedures these two researchers waxed lyrical about the great discoveries they had made.

In fact, the story of how the buckminsterfullerene structure could be associated to atomic structure is a very interesting one: Harry Kroto, from the University of Sussex, in England, appears to have been the first to have noticed this odd new carbon in 1985, along with his associate Wolfgang Kratschmer at the Max Planck Institute in Germany. They nicknamed this new carbon as 'junk' because originally they didn't quite know what it was (all they knew was that it was a carbon with 60 clusters which at the time was considered very strange). When Kroto went over to America and met up with Rick Smalley (at the Rice University, Houston) they conducted tests to create more of this 'junk'. But it was here when they made the momentous discovery that this new carbon had a soccer-ball structure to it, and then when they refined its theoretical structure by ascribing to it the geodesic structure of the buckminsterfullerene they suddenly realized why it had that precise number of clusters—it was a 3-dimensional molecule! This was a brand new addition to the family of carbon. However, this discovery was far too radical for other scientists to believe, and as is quite usual amongst the scientific fraternity, they ridiculed these claims as being nonsense, and so that's as far as that research project then went. So the C60 trail ended, or at least it laid dormant for five years.

The reader, of course, will note from this study's bibliography that Ray Fowler's books of the *Andreasson Affair* were being published just before and during the 1980's... And it wasn't until some theoretical physicists (from America, I do believe) suddenly began publishing papers in the late 1980's, on how buckminsterfullerenes were now considered the perfect form of structure, that C60 research resumed once again. When that research on C60 did resume and a way was found (in 1990 through Kratschmer and Don Huffman at university of Arizona) to produce crystallized C60 then the new carbon was here to stay, and thankfully Harry Kroto was one of those awarded with the Nobel prize for chemistry (in 1996). After that the Bell laboratories in America found that by linking C60 with potassium then a superconductor molecule could be produced.

It is not known how influential Betty's messages of scientific prose and formulae was in that convoluted discovery, although indubitably, it was the extraterrestrials who used the earliest attributable reference to "Buckanosa" in their allegorical prose to denote what the new carbon's chemical structure should look like.

Col. Corso alluded to the fact that at one time it was also quite openly admitted that many MUFON groups in America were frequently headed by US aerospace engineers, for the sole purpose of transposing UFO-related technologies into new ideas for the aerospace industry. These new ideas were allowed to infiltrate into both the civilian sector and the military sector around the 1970s and 1980s. Corso gave an amusing account in his book *The Day After Roswell* (pp98-99) of when he and some of the guys at the 509th at Roswell, tried on for themselves one of the alien's special headbands which inter-linked in with the craft's electronic operational-control circuit, and which transmitted as control signals into the craft's circuit the electromagnetic pulse-wave-activity coming from the alien's brain. This technology later found its way into the HUD tele-mental control helmets used by military helicopter pilots, and became a technological advancement that spread over to both Britain and Russia during the 1990's. So here was a case for ET technology permeating into the world which came directly from Corso's UFO back-engineering connections, through the US electronics corporations of Bell laboratories, and Sperry-Rand, and through the aerospace giants Hughes Aircraft Corp. That it took all of fifty years to do so tells its own story [note: 17].

1.5 We Are Alone... in our Crown of The Creation Syndrome

It has always amazed me that whoever went to all the trouble to orchestrate and film the infamous 'alien autopsy' (the source of the comedy of the same name which Ant and Dec produced in 2006) that they showed the viewer some of the duller points of a human midget's internal organs, and that they had depicted on the postmortem table a body with a human stomach!

For if you think about it a body with a stomach will then need to have a digestive track, and also a throat and a mouth. I'm certainly not an expert on all the biological details of *every one* of the different extraterrestrial races that have arrived here for our event, but from all the abductee accounts I've studied I have always assumed that a Grey (which is the type of alien this film should have been portraying) has no mouth and nor does it have any teeth! And nor do the Greys which have been predominantly featured in hundreds of abductee reports have to deal with the excretion of waste. So a stomach inside an alien would be as useful as an underwater cup of coffee!

Greys only have a small cavity behind a very small mouth, and take in nutrition through the pores of their skin. As we know from observations made by abduction-experiencers when on board UFO craft, and in particular from one report where a person who was left unsupervised and was allowed to wander into the alien's living-quarters and who managed to grab a quick glimpse of an unsuspecting group of Greys taking on nutrition, in an apparatus which appeared to spray a fine mist at their bodies which their bodies immediately soaked up without spillage. This particular account was collected by the very credible Dr. D.M. Jacobs of Harvard and so, if we can assume that the extraterrestrial Grey absorbs nutrition through the pores of his or her skin, then this would explain why there is no stomach needed, and why the mouth of a Grey is non-functional for, as we also should know, the Grey prefers to talk directly into a human's brain through telepathic communication, and it doesn't use a mouth through which to talk [note: 18].

My main thought about the faked alien autopsy film was that it seemed to 'uglitize' the extraterrestrials, to make them look like the bad guy perhaps, like Hollywood did when they brought out calvalry-and-indians movies back in the 1950s, maybe there was an assumption that all aliens should be acknowledged as being more hostile than us, or less beautiful and less spiritual than us. As strange inferiors perhaps they could then be regarded as less threatening to us?

Funnily enough, this factor is echoed by Beatriz Gato-Rivera in her eloquent paper on her "Subanthropic Principle" (p9) with its insight into a multidimensional empire within the cosmos that few people of this world have ventured to think about.

What I found so illuminating in this Beatriz Gato-Rivera conjecture about life in the cosmos was that it concluded with the rather provocative realization that although we humans have never bothered to investigate the possibility of alien civilizations we automatically BELIEVE that there should be none, and also automatically assume that whosoever claims that there might be civilizations more advanced and intelligent than us should be ridiculed or shouted down. Such assumptions, according to its author, are part of a deep psychological prejudice that has been programmed into us humans and which is termed the Crown of the Creation Syndrome... for obvious reasons [note: 19].

1.6 Damage Limitation Strategy of Defense Departments

Perhaps prejudice is the marker for intelligence. It has been well documented that black unmarked helicopters would buzz the homes of the Luca's back in the 1970s, solely for the purpose of intimidation (as recorded in "Andreasson Affair—Phase Two" pp210).

Betty Luca also told me that a specific series of technical schematics she had 'received' in the 1970's, through an automatic writing technique she and the aliens used, which featured extraterrestrial energy development mechanisms, were stolen from her house when she and her husband were away on a day-trip (although after this break-in, when the Luca's returned home, they found that no money and not one of their other valuables had been taken by whomever did break in).

These diagrams were totally unique and very specific in their content, in that they formulated the energy dynamics of gravitational force manipulation through the use of rotating energy fields (explaining why UFOs are fundamentally circular). They would have been prized with the highest value by any government-run laboratory, for instance, involved in the back-engineering program of UFOs. That they were stolen in such 'targeted' circumstances (and just after the Luca's had driven away from their house) would suggest that the Luca's home had been bugged or had been the subject of intensive surveillance, of course, and that the thieves had stolen these schematics to order.

Thankfully though, it has lately been discovered from the one surviving drawing of this coveted series (because this drawing was stored separately at the time of the above-mentioned break-in, and so it has remained in Betty's possession), that the aliens had been using these schematic drawings to explain their hitherto unknown propulsion processes that, incidentally, are wholly at odds with the sort of physics principles our scientists use (particularly the chemical-thrust type of linear forces the scientists at NASA have relied so heavily upon for the last fifty years). And so, against the odds, this shared form of communication between the aliens and humans did succeed in doing its job after-all, and this alien form of gravitational force manipulation will be explained in due course throughout chapters 8, 9 and 10 of this study.

So some have thought it important to try to stop the scientific information that came with Betty Luca's abduction story getting out to the general public. But then, judging by what claims Col. Philip J. Corso made in his book about seeing the aliens and alien aircraft as being able to out-think and to out-maneuver anything or anyone in the US military, it doesn't need a Sherlock Holmes to work out that the US Department of Defense would have gotten extremely agitated about what was then being 'disclosed' in the late 1970s through all the books being published on Betty (Andreasson) Luca's abduction-experiences, in which books Betty was describing in detail a UFO's power-drive that had, for all intents and purposes, the same power-drive system of electrodynamics and gravitational propulsion as the Roswell craft had in 1947. But all abductees would do well to realize that government departments have a vested interest in the suppression of all things alien, and they have no doubt told scientists, doctors of psychiatry, military spokespersons, academic journal editors, and newspaper editors at key times to 'play dumb' by whatever means they can.

But I can understand the ethos to hush things up. Because, as there is no hazardous limit of G-forces upon the crews of UFOs, essentially because the UFO takes with it its own source of gravity, and it would be possible for these craft to use a propulsion mechanism of force-entrainment which would mimic the same sort of ejection dynamics our scientists have observed in astrophysical jet expansion, and these entrainment field velocities for mass-particles (where those particles are pulled rather than being pushed) are far in excess of linear pushing forces, then a UFO could be towed along by those force-entrainment fields at velocities as high as 100 km/sec (or 360,000 kmh)!

So with UFOs traveling at such high velocities, and having the ability to turn on a dime (again, because they carry their own gravity field with them), they would represent a type of aircraft that would be impossible to catch by any of earth's military aircraft, considering that the fastest run-of-the-mil patrol jets of Britain-America and Russia have top speeds of only around two-Mach or 1600 kmh [note: 20].

So UFO's are kinda too fast for military planes to chase after and to combat against, right?

Unfortunately, because the civil servants in the various defense departments around this world haven't worked out for themselves WHY the ETs are here they have assumed that there is every chance the ETs could be hostile, and so they assume that the public would panic if they ever found out that the ETs can out-run, out-maneuver, and out-think any of this world's defense strategists.

And if I based my data regarding extraterrestrials on fear, paranoia, and wild supposition, then if I were running a nation's DoD I too would be running a "mustn't admit there are UFOs program." And so in consideration of the above equation of 360,000 kmh divided by 1600 kmh then I too would instruct all-and-sundry that worked in that DoD, from the top professors down to the tea makers, to never ever admit, acknowledge, or even look at the UFO phenomenon, because I'd know that a government person who has nothing to argue with is a person who has already lost that argument. So at all costs UFOs would have to remain un-acknowledged and non-existent, and that's for all time.

Unfortunately, this will ALSO be why UFOs can never be openly developed by any commercial aerospace company that is obliged to respect their government's undermining authority—you just won't get government funding for any research-and-development program concerning UFO propulsion dynamics! This is the scientific paradox—plenty of new and exciting science to study but exhaustive suppression stacked against anyone who attempts to perform those studies...

2: Helical Field Tube Technology of Propulsion

Contents:
2.1 Domed Craft Inside a Surrounding Energy Envelope   32
2.2 Propulsion of a Beam-drive UFO   35
2.3 Film Footage (from Mexico) of Hyper Jumps   36
2.4 Extraterrestrial Musings Between the Stars   39
2.5 Jets Produced from Rotating (Kerr) Black Holes   43
2.6 Mechanisms for Production of Electron-Positron Avalanches 48
2.7 Rotation Force and Momentum Transfer from a Black Hole   52
2.8 Magnetic Flux Break-Reconnection Ionization (Reconnection Pt I) 57
2.9 The Quest for Stability in the Development of a Filament-tube 65
2.10 Collimation and Hoop Stress   68
2.11 Electrodynamic Anchoring of the Magnetic Field   76
2.12 Williams and De Felice on Negative Radial Momentum   78

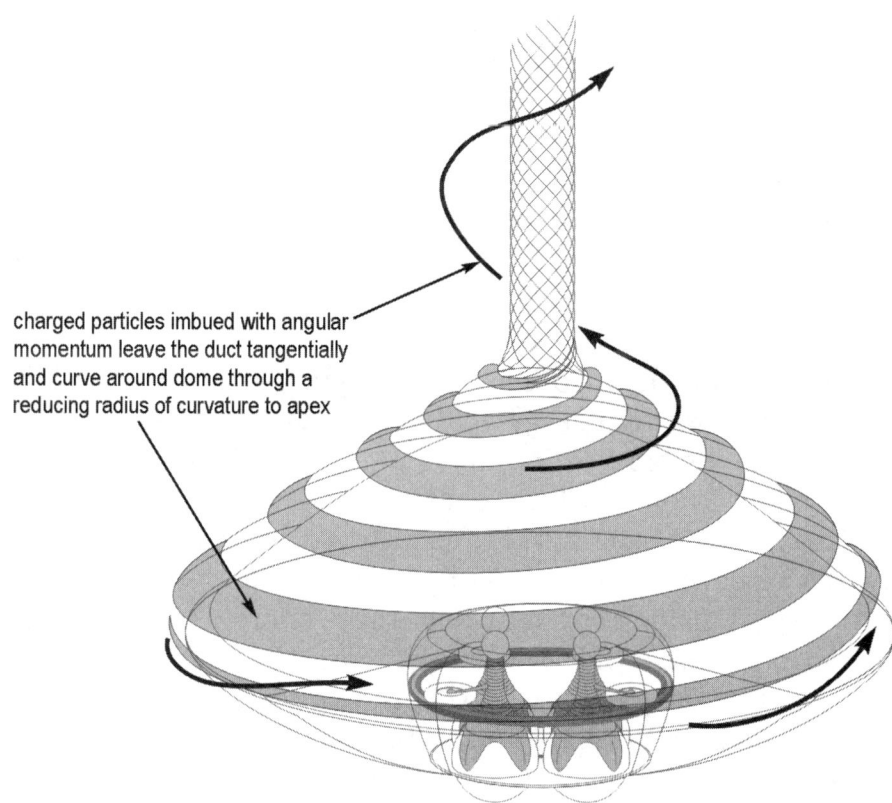

charged particles imbued with angular momentum leave the duct tangentially and curve around dome through a reducing radius of curvature to apex

Figure 7. Energy in the envelope surrounding a domed craft will follow a helical path of reducing radius of curvature

2.1 Domed Craft Inside a Surrounding Energy Envelope

What I intend to do in this chapter is give the reader a brief look at the energy field created around a UFO. Not particularly as to how that craft might use energy to move through the air (that will be covered through subsequent chapters), but just to see how a craft can create an energy field that might parallel the sorts of energy fields which exist in space. In order to do that therefore, this chapter will also look at what astrophysical energy propulsion models there are and how they work, and look to see if there are any clues within those models which might provide an insight into UFO propulsion technology.

As this study will be formulated more as a design proposal for the engineering of a UFO craft, to be developed and constructed in a laboratory environment, rather than as a brief explanation of how one particular UFO craft might work, I will not be sticking to just one type of UFO and offering the reader a list of pretty comparisons to comfortably-known space technologies. I will be featuring extensively the "Andreasson Affair" UFO (henceforth referred to as the aaUFO) as alluded to in the previous chapter, but I will also and very purposefully be drawing attention to characteristics exhibited in as many other types of UFO as is fitting to include with the approval of the overall goal of this study, that being the development of a craft which can be engineered with complete workability [note: 1]. This cross-referencing to other types of UFO will be important also because I would like to involve as many UFO witness reports as I can, to emphasize that those numerous reports have provided invaluable contributions to the understanding of ET technology. Long may those reports persist to seek scientific attention.

One of the interesting facts I've learned about UFOs, and UFO reports, is that a real UFO will always have some sort of energy field surrounding it when it is in the air. It will either have a fairly visible flow of air or a gyrating flow of electrically charged particles surrounding it, or it will be encased inside an energy-envelope that pulses with colored (light) emissions. These light wave emissions though will not necessarily follow the exact route of the UFO's propulsion energies, merely will they be ejected from those propulsion energies where those energies turn through a sharp curve (as will happen around a UFO's perimeter rim). The propulsion energy goes one way and the photons it generates, through a process called curvature radiation, will go another way. In turn those colored light emissions would be affected by refraction as they pass through the rotating medium surrounding that UFO, and while that wall of vibrant energy will refract different light colors into different directions, then it would seem to me that whatever image was projected of that craft through all that deflection would only be apparent as a fuzzy and distorted one (so a photograph that purports to be of an extraterrestrial UFO which shows that craft's contours as crystal clear are most probably fakes).

It then follows that the powered-up UFO while it exists inside an energy envelope would become either more energized, or less energized, through some sort of electromagnetic-field-energizing system that the craft's crew will be in control of. This could conform, for instance, to the characteristics of a simple electromagnetic field, such as a toroidal field or a poloidal field, in which charged particles will be imbued with Lorentz force and will be induced to spin around the craft in a direction perpendicular to the lines of such an electromagnetic field [note: 2]. And depending upon what level of power that energy envelope is energized to then the charged particles free to circulate around the UFO will move at a low velocity or at a high velocity.

This also means that when they are in the air hovering or maneuvering UFOs will have a vibrant electromagnetic field surrounding them which in effect will manufacture its own electric charge density environment, or envelope, which will be projected outward (to an outside observer) as a moving field of visibly excited energy. In later chapters there will be further mention of how this excited energy field can be made to distort the visual image of whatever is inside such an energy envelope, to make that craft's image appear to progressively dissipate until, for all intents and purposes, it seems to have become invisible inside that energy field.

Good examples of vibrant energy fields surrounding a UFO have come from the videos of UFOs filmed over Mexico during the 1990s, which show UFOs hovering in the air with very fuzzy outlines and faint bursts of colored light coming from the rim areas of those craft. This fuzzy imagery is of course a damn shame for that tabloid newspaper photographer who wishes to obtain the ultimate proof of UFO existence, but it is wonderful for researchers interested in how these UFOs work. For the energy signatures that are perceivable can tell a great deal about what is going on both outside and inside a UFO, and quite frankly I for one would rather have the latter person advantaged than the former [note: 3].

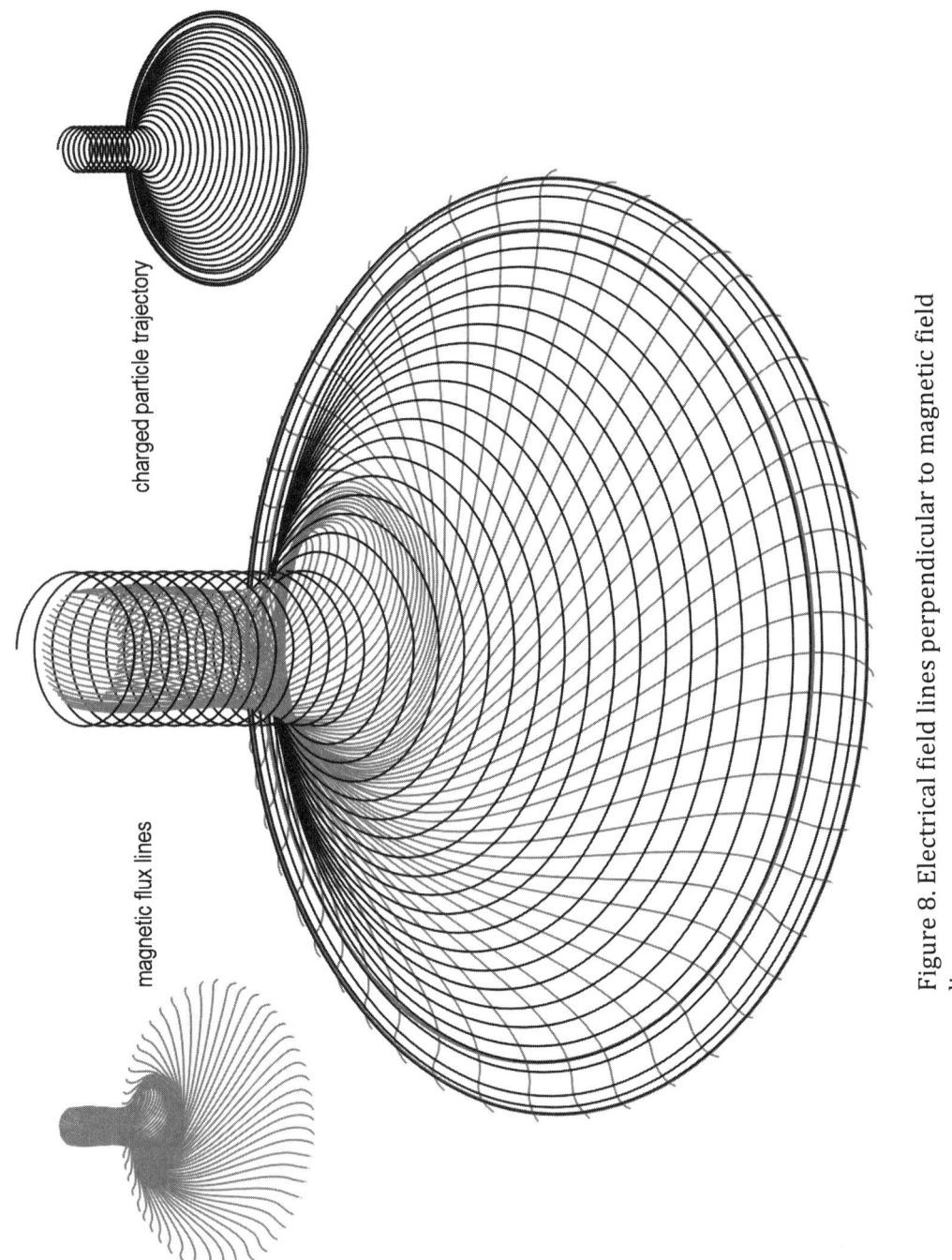

Figure 8. Electrical field lines perpendicular to magnetic field lines

2.2 Propulsion of a Beam-drive UFO

But the real starting point for this study will be one of the most spectacular phenomena of UFO propulsion, and that is the beam-drive or hyper-jump where the UFO extends a forward beam ahead of it and then traverses along that beam across the sky, speeding into the beam like a torch attracted into its own rays of light.

Therefore, along with the basic intention of investigating how the UFO's energy envelope can be comprised this chapter will also inquire about, in order to further understand, how this energy envelope might then be extended ahead of the craft, and how that extension to the craft's energy field might be configured for propulsion. So that the 'beam drive' system used by some UFOs, which appears to be modeled upon a propulsion technology that pulls the craft into it rather than to push it, will be looked at in great detail. This lore of motion will of course be one that the reader will know full well is the complete opposite of how rocket science works, which is understood to work by generating huge amounts of combustion and by blasting it one way in order to force the rocket to move the other way, much like an explosive device.

Exactly how the beam-drive works through what parameters, and how such a pull-drive system might be configured will obviously benefit greatly from being compared to known electromechanical phenomena, but also to known mechanisms investigated in astrophysical studies, and to known phenomena discovered through particle acceleration physics. Therefore, very specifically, this chapter will also endeavor to highlight some of the helical-vortical-beam drive mechanisms from the above-mentioned astrophysics studies, and couple them to the most useful energy generation, particle acceleration, and gravity manipulation mechanisms known from physics studies. This will be done in concert alongside the known phenomena of UFOs, by closely considering the factual observances reported in eye-witness accounts relating to UFO's physical structures, UFO's energy signatures, and a UFO's unique methods of maneuvering (ie zig-zagging through 90-degree turns). It will then focus on the energy requirements of UFOs and the configuration parameters gleaned through this detailed investigation, of energy envelopes that a UFO may establish around itself and may extend from it, and how it would be possible for such beam craft to energize a propulsive force through the whole length of that extended beam which would pull the UFO craft at near-relativistic speeds impervious to ambient gravitational forces.

Figure 9. UFO's propagating beams of energy and moving into them

2.3 Film Footage (from Mexico) of Hyper Jumps

The above graphic is an artist's impression of an extraterrestrial UFO propulsion system which came to light after real-time UFOs were filmed in daylight over Mexico and were subsequently broadcast on Mexican and British TV in the early-1990s [note: 4]. British viewers will remember that the editors for Jaime Maussan, the newsman who collected all these recordings for Mexican TV, noticed that when the video footage was slowed to a frame-by-frame speed this footage clearly showed some sort of 'hyper-jump' action being performed by the UFOs where a long tube would be energized in the atmosphere ahead of the UFO craft, into which the craft then jumps at incredible speed, and as the craft comes to the end of that energy tube it sends forward another such tube for it to ride into again. This it will do repeatedly, thus allowing the UFO to cross a vast expanse of the sky and all within a matter of a few seconds.

As it turned out this hyper-jump was not an isolated occurrence and such jumps seemed to have been performed several times in the skies over Mexico and were filmed several times by numerous people—all from different angles. Another of these jumps from Jaime Maussan's compilations can be seen to extend a long filament-tube above itself, to traverse it, and then to send another filament-tube into the air above it, to traverse that, and then to send another and another and another.

This sequence comprised of no less than nine consecutive jumps, which took the craft so high into the sky that it disappeared into the upper stratosphere [note: 5].

But what was just as interesting about this particular UFO was that the video footage showed that a few seconds before this UFO had sent out its successful beam (into which it traversed), it sent out an earlier energy beam that seemed not to connect with the air ahead of it, and which appeared to be much fainter than the successful beams which did connect between the craft and the forward air.

This suggests that there are determining factors which make this beam-drive system work, or to not work, and that these determining factors must be the variable condition of, or characteristics of, the particular ambient medium which the UFO is travelling through (in other words, it suggests that the atmospheric air ahead of that craft can be 'worked' by the UFO as if it were a fluid that can be layered—and that those layers of air can be trapped one upon the other somehow). And so this suggests the first big clue which needs to be investigated, because if there are variables to this propulsion drive system then it means that the aerodynamics of the UFO's propulsion has to conform somehow to the specific characteristics of our atmospheric air. So, one, what are the parameters possessed by our atmospheric air which the UFO's energy field has to adjust to? And then, the second clue will be that the propulsion occurs through jumps of finite duration, and this suggests that the UFO is using "packets" of propulsion energy rather than a continuously generated field of propulsion energy.

Certainly then, there in those Mexico city UFO videos, as well as in these three pictures taken from one of those videos featured on the next page, is the beginning to a study which will eventually solve the question of how UFO propulsion can work—these first two clues to pursue—are raring to go and are fully loaded with questions, and these questions can form a framework with which the reader can search into the fundamentals of extraterrestrial physics.

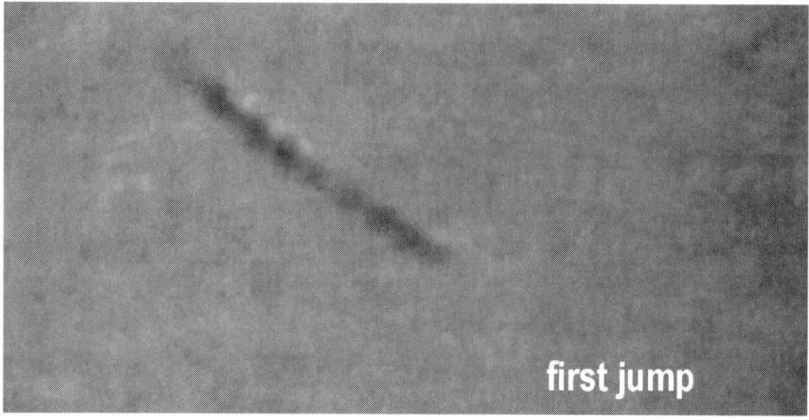

Figure 10. In this video the UFO moves in jumps up into the sky. Here I show the frames where three of the jump sequences begin

And so, as I say, it is advantageous to gather information from as many sources as possible, especially when they all tie-in together so conclusively, because what one UFO report will lack in clarity can probably be explained in another report from an entirely different part of the world. The above two clues in UFO propulsion will provide the main driving force for this chapter. It also bodes well though to work with these particular hyper-jump images in mind, because these video-clips are wholly available for all to see on video and DVD compilations that have been available for years and have been seen by thousands of people world-wide. That these video-clippings are of such high quality giving excellent frame-by-frame detail of the beam-drive phenomenon is an added bonus.

2.4 Extraterrestrial Musings Between the Stars

As this study is probably the first to attempt to look through the eyes of the ET physicists it will set new precedents by giving them home-advantage and look at energy systems that are in existence and in operation right now outside of earth's atmosphere. And this new precedent is well justified because if we should surmise that the ET will not have learned his or her physics from the scientists of Earth then it would make perfect sense for us to look toward the dynamics of *galactic energy fields*, for the very models from which UFO drive systems would have been designed.

More specifically, what I am proposing for this comparison of a UFO's beam-drive is that the mechanisms of astrophysical jets should be studied with the view to understand the electrodynamic and hydrodynamic principles involved in their helical energy fields, to discover whether or not those principle could be utilized by a UFO craft, through which that craft could then engineer ahead of itself a field of energy which it could use as propulsion. So what the key factors are to these astrophysical jets, of how they are structured to keep their strength and stability, for instance, will be assembled here and those working principles when established and further developed, will then be used to progressively design a method by which a craft can engineer for itself miniaturized helical tubes. These helical tubes, that will span for the UFO for much shorter distances than in the astrophysical models, might then allow that craft when it sends electrodynamic energy packets into them, to traverse through long distances of space or through short distances within earth's atmosphere (and earth's other mediums).

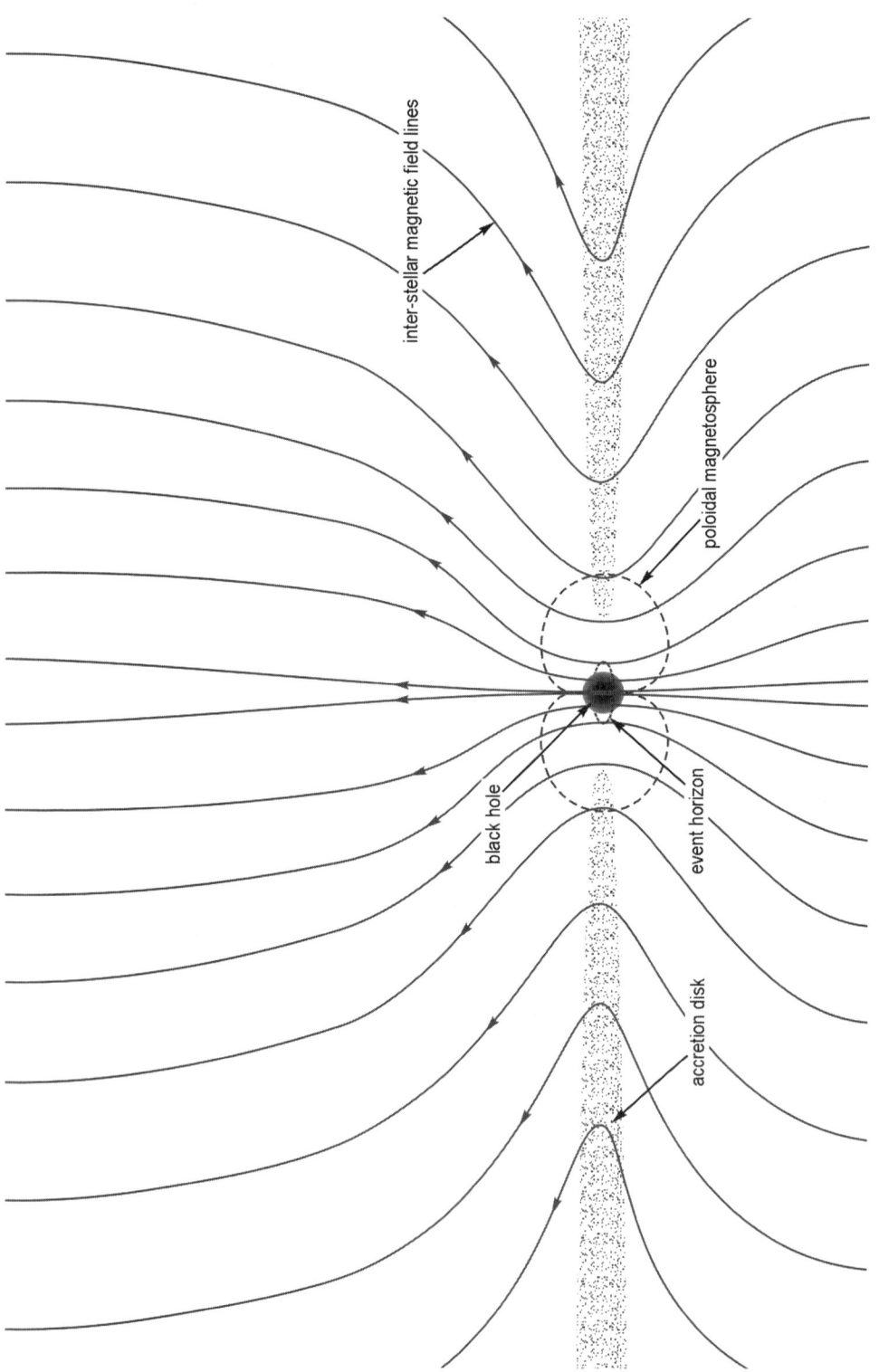

Figure 11. Inter-stellar magnetic fields of a black hole and accretion disk

With the benefit of radio telescopes we have been gathering information about the astrophysical jets in space since the 1920's and so we know a great deal about not only the jets but also the galactic drive systems that actually power them. These power systems can be briefly summarized as being the result of a gas cloud out in space condensing into a rotating disk, or series of coiling energy-sheets flowing around what appears to be a dense mass (such as a black hole). And upon drawing into itself the magnetic field lines of space and then wrapping those magnetic field lines around itself in a dynamo-like motion, such a system then becomes an engine which generates huge amounts of electrically polarized particles. And further, while coiling those magnetic field lines around itself that engine's charged particles will be accelerated (by the Lorentz forces of that coiled field) as a plasma flowing inside and around those coiling magnetic field lines, so that this rotating plasma will subsequently leave that black hole system as a relativistic-velocity jet. This will happen mainly because as the gas cloud of the rotating disk continues to generate energy, through its dynamo-like motion, that energy will always move inward toward the axis of that disk's rotation, because that's where the magnetic field lines will be most intense (and of course because of the gravitational pull from the dense mass at the center of that rotating disk). In fact, the dense mass or black hole *should* be able to deal with this influx of energy but it can't because there is so much of it, and matter-energy can only enter into a black hole through the horizontal rotating path established by its accretion disk (and because the charged particles in that gas cloud can only move perpendicular to the magnetic field lines), and so the incoming energy stacks up just around the outside of the black hole and instead of being swallowed up by that black hole it develops into a state of extreme excitation. This excitation and excess of energy can't go anywhere other than up into the coiled-around magnetic field lines, hence the jets [note: 6].

The fact that some of the energy-generating and jet-forming mechanisms of an astrophysical jet can be found in UFO drive systems will be of great interest to the science-minded UFO researcher. But, looking at this another way, that some UFO drive systems when inspected and compared with astrophysical dynamics will show previously unrecognized energy dynamics, particularly in reference to gravitational force generation, then there might also be afforded a greater understanding of the galactic drive systems that exist in space too, and this should be of great interest to the astrophysicist

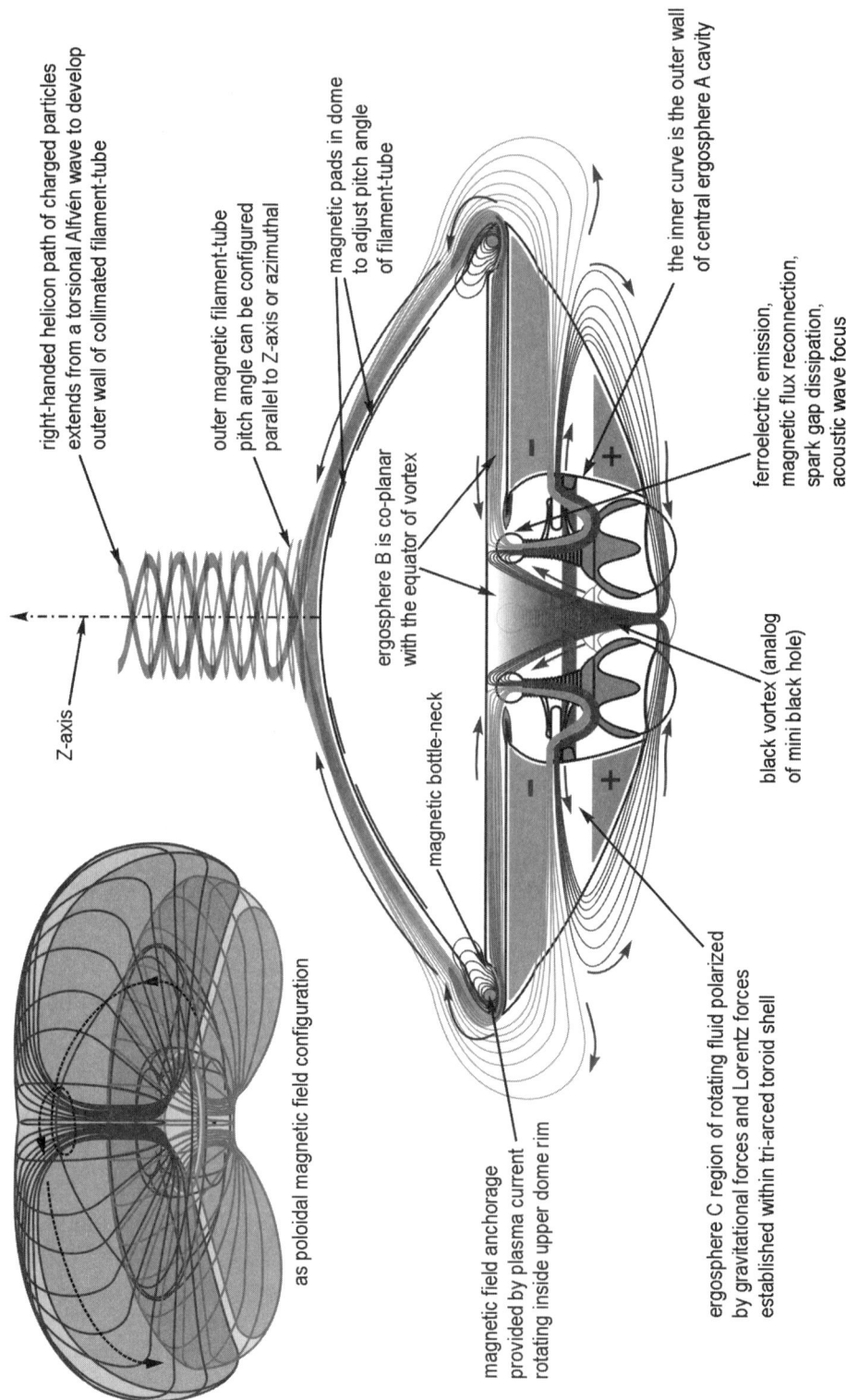

Figure 12. Magnetic fields and magnetosphere of Andreasson UFO

Astrophysical comparisons that are readily apparent within the dynamics of the "Andreasson Affair" UFO (aaUFO) are; the strong toroidal magnetic field, the ergosphere, the rotating vortex, the accretion disk's shearing field, the poloidal magnetosphere, the magnetic flux rotating-jet filament-tube, the inertial frame dragging, the avalanche-ionization processes by magnetic flux-line reconnection, the emission of synchrotron radiation, and the magnetic field anchor point... to name but a few: As well as a mechanism similar to a black hole that will strongly amplify and constantly supply for the craft's propulsion both angular momentum and gravitational force.

That even one of these mechanisms can be found in a UFO's power drive system suggests that the ET physicists have worked along scientific principles known to this world, but I think there will also be found evidence that these physicists may have a few further tricks up their skin-tight sleeves to add to the above list, which we on earth haven't dared to look at yet, but could benefit greatly from. But, to reiterate the points made above, in order for us to extract a technology from these engines of space, like the ETs, the big question will be, how much do we know—about how much the ETs know—about astrophysical jets? And for that matter, what do WE know about the jets in space?

2.5 Jets Produced from Rotating (Kerr) Black Holes

The latter question is easily answered because, luckily, the debate over astronomical jet mechanisms has been hotly pursued on this planet for over 40 years, so much so that this science now forms a very substantial body of research both in theory and simulations [note: 7]. David L. Meier of the Jet Propulsion Lab (California Institute of Technology), for instance, in experiments to simulate MHD accelerated jets modeled from astrophysical jets has found out how highly-magnetized, rapidly-rotating ultra-relativistic jets can be established and how they are self-stabilizing due to the inertia of the rotating magnetic field itself—and as long as the jet-tube, says Meier, is strongly magnetically-dominated then its long filament-tube structure will remain stable and intact (indeed, subsequent research by many others has established a very strong case for the stability of a filament-tube surrounded by an infinitely-long toroidal magnetic field). Meier's model is along the lines of an accretion disk magnetosphere with field lines twisting inward toward its black hole and outward along its axis of rotation to create the jet as a torsional Alfvén wave which curls around and extends, through what is known as a Poynting-flux dominated mechanism, along the longitudinal axial line (the z-axis) away from the black hole [note: 8].

What is most interesting about Meier's hypothesis is where he points out that there are essentially two parts to the helical jet, the inner core (of an ultra-relativistic spine) and its outer sheath (of a mildly-relativistic tube), giving the M87 jet (Virgo A) as one good example of an outside sheath field surrounding a core jet [Meier 2003]. And then, through the process of differential rotation which twists the disk's field the resultant magnetic-pressure-density-gradient will accelerate the field's plasma from the disk into the helical field or jet which will extend outward along the longitudinal axis of the disk and black hole (see figure 13 of the M87 jet). Stability of the helical field is established from an interplay between the radial-to-axial (centripetal) magnetic tension balancing against its outward-forcing centrifugal force. Great importance is placed upon this inward-flowing force called the "hoop" stress, as it pinches inward and collimates the curling jet.

Figure 13. The well collimated M87 galaxy jet

Indeed, the stability of a rotating helical energy field will be the much sought after pre-requisite for any UFO to establish in its extended energy field, and much of that stability will come from the helical nature of the magnetic flux lines of the filament-tube, AND from the helical nature of the electrically-charged particles which will be accelerated up along the outer wall of that magnetic tube. And as it will be seen further into this chapter, the ability to maintain a balance between the inward hoop stress and the outward centrifugal force in a helical-tube, will be that which determines the stability of that helical-tube, and will be one of the most important factors to be considered when it comes to the duplication of such an energy system for these new domed craft [note: 9].

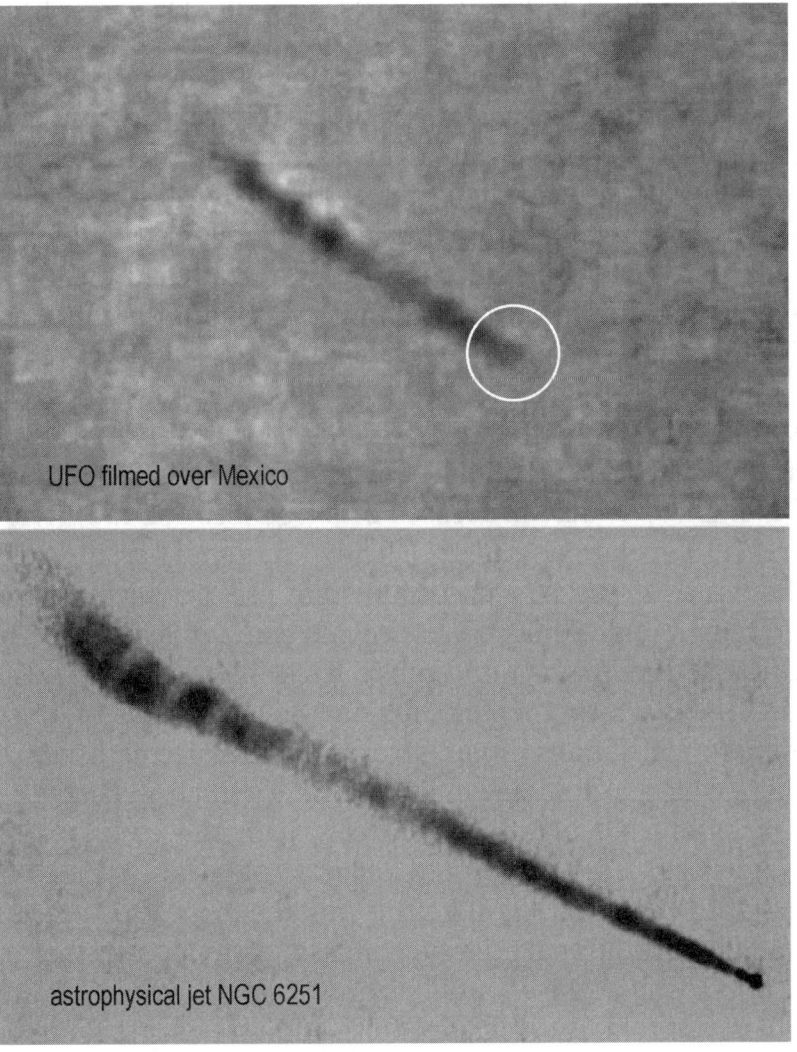

Figure 14. Comparison of the NGC6251 jet and a frame from one of the beam propulsion UFOs filmed over Mexico

Just to emphasize the enormity of these energy acceleration systems out in space the reader might like to look at the jets of Cygnus A, HH30 or NGC6251, particularly where, as in some cases, these astrophysical jets extend to megaparsec lengths (mpc ~3 x 10^{24} cm) yet quite incredibly have opening angles as narrow as just one degree [note: 10]. In particular though, the best example of such a jet in space is NGC 6251 which clearly has good collimation (of only 4 degrees) and an impressive length of more than 10^{13} Km! For an interesting comparison in both their forms see the image of one of the frames from the Mexico UFO's energized filament-tubes, and the image of astrophysical jet NGC 6251 which is millions of miles in length (as compared in figure 14).

Can we see here a comparison with a UFO drive mechanism? In the sequence of frames in the Mexico UFO videos whatever was at the bottom of the tube (circled) traverses this tube at very high velocity and then after momentarily pausing seems to rise again into another more distant tube, and then again to shoot upward through a higher curve, and when the UFO traverses that tube it pause momentarily again before jumping up yet another tube until finally the craft disappears out of sight into the upper stratosphere of clouds. Interestingly the video shows that the initial tube remains intact in the air until the craft reaches the top of it, at which moment (while the next forward tube is being energized) the initial tube behind the UFO slowly disperses into the air.

While the above example of a UFO's filament-tube will shed its mystery later in this study (through chapters 5, 8 and 10), further examples (and more mathematical equations) of the astrophysical processes of jet generation are given in studies by [Tsinganos] and [Li-Lovelace] [note: 11]. The [Tsinganos] study explains in simple mathematics how the azimuthal accretion field at first grows outward from the axis of rotation, and then it shows how the more extreme forces exerted by the shearing dynamics of the disk then return and move back toward the axis, both above and below the black hole, which then results in the formation of the cylindrical jet outflows (complete with their hoop stresses). This latter study works on the principle of there being a poloidal magnetic field which generates rotational and magnetic pressures which lead to tensile forces inside the jets.

Of interest to UFO research will be the way the azimuthal accretion field when it has gained sufficient rotational energy, will also push upon the magnetic field lines to bunch them into a wall at the event horizon of the black hole (as mentioned in the brief summary above in the previous section), adding Lorentz force to the radial-to-axial flow of charged particles toward the axis which coils the charged particles around those magnetic field lines [Khanna]. This follows the oft mentioned observation in black hole physics that the flux lines of a magnetic field which links the black hole to the accretion disk can only link to the black hole on the outside of its event horizon: Because, as is shown in many studies, no magnetic flux lines can actually exist inside of the black hole and so those flux lines will bunch up outside of that black hole's event horizon—which is obviously a power amplification mechanism which can be taken advantage of [Punsly 2001]. Also explained in [Tsinganos] is the important observation that very weak collimation of that cylindrical jet field will result if its rotational velocity is too low (giving as a reference the rotational velocity of our sun), whereas a rapidly rotating body (rotating 10 times faster than our sun) would result in a much more tightly collimated and more stable jet. This principle, that the rotational velocity of the filament-tube can determine the stability of that filament-tube, will obviously be one of great importance, especially if one considers that the more an energy field is rotated the more its electromagnetic forces will be amplified (and too its gravitational forces according to [Zeldovich 1972]). So here already are three factors that can be used in UFO dynamics; a) as the magnetic flux will be more intense around the event horizon, so there will be an electro-mechanical potential-difference between the inside and the outside of that field; b) that there will be a threshold rotational speed where the filament-tube's collimation goes from unstable to stable; and c) that the amplification of the UFO's propulsion force inside its filament-tube will be proportional to the bore-diameter of that tube which it extends ahead of it (this very important factor will be further expanded upon in later chapters).

This 'tuning' of the filament-tube to determine propulsion has already been proposed in Jeff Savage's model of the Onion UFO power-drive system, whereby the hoop stress of its filament-tube would be wound-in to form a tightly-collimated filament-tube so that its UFO can trigger a shock-force type of acceleration, or alternatively if that UFO's rotating beam's hoop stress is slackened off so that its filament-tube is spread out into a larger bore diameter then that craft will undergo a slower acceleration.

So very early into this study there is substantial progress in understanding how a UFO works, and in the later chapters of this book there will be some interesting conclusions made about these dynamics.

2.6 Mechanisms for Production of Electron-Positron Avalanches

Actually, just after I embarked on this study of UFOs and astrophysical dynamics I was told by one or two researchers that space was empty and that because of it being a vacuum it was completely void of any energy. Apparently, to a certain sub-set of seemingly well-educated people this is common knowledge. But my argument to those people is this, well if space is a vacuum and is empty what the hell are they standing on?

I find the mechanisms of energy-production in space immensely fascinating. Such energy is as raw as it can be. Ejections of powerful bursts of energy in space can literally be the harbingers of both life and death. So what I shall do next in this chapter is give a lot of attention to the various energy production mechanisms found in space, for the purpose of bringing in similarly configured mechanisms observable in many UFO structures and in many UFO energy signatures. In this way I hope to keep the reader glued to the marvels of ET technological design, and of course awake!

Contrary to the *I-fink-its-empty* brigade's view of space, in reality space energies tend to come into existence through violent transformations, such as happens when breakdowns occur in the fabric of space-time when an electromagnetic stress is placed upon that seemingly nebulous fabric. These breakdowns then produce avalanches of electrons and positrons out of that so called vacuous nothing. Physicists ascribe this electron-positron pair production to the notion that space is in actual fact a thriving mill of virtual energies with unlimited potential, and obviously the jets coming out of black holes are living proof of that abundant energy production. Certainly it is well known through astrophysical studies that there are a whole range of ways the space-time fabric can be stressed, as there are a whole variety of mechanisms which lead to the production of avalanches of electron-positron pairs. Knowledge about these methods of energy production is very extensive mainly because they had to be well researched in order to explain the huge generations of charged particles which go into the formation of plasma clouds found throughout space, and to explain the large electric currents which are required to power the numerous ejection events, which range from minor outbursts, such as our sun's solar flares, to the gargantuan galactic jets that blast through space for millions of miles much like the NGC 6251 jet cited above.

But it is important to emphasize here that this inspection of the engines of space for sources of energy particles primarily involves the galactic energy systems that develop from cold plasmas. Or that they will involve the cold bipolar gases that subsequently pass through various thermodynamic changes during their accelerations, as in the [Blandford-Payne] and [Uchida-Shibata] theories. This is to differentiate these energy production sources from the 'hot' systems such as is found in AE Aquarii which while this white dwarf does have a magnetic 'propeller mechanism' with which it eject its huge amounts of matter, its plasmas are in the region of 2000K degrees and are readily ionized by intense heat [note: 12]. So rather than follow earth-science's preponderance for seeing mega energy production only through thermodynamics I shall make the bold boast that the extraterrestrial physicist appears to prefer using non-thermodynamic mechanisms of particle production, which generate charged particles through dissipation, Compton and inverse-Compton scattering, magnetic field reconnection, and curvature radiation ionization. Thankfully many of these particle production mechanisms are favored throughout the numerous studies of black holes, accretion disks, and astrophysical jets.

One in particular of these cold mechanisms is mentioned in the [Blandford-Znajek] paper and is one which will be of interest in UFO dynamics. The mechanism is called spark-gap ionization. It suggests that a spark gap will develop across all concentric electric field lines in a radially expanding electric field—which, if you can imagine an electric field emanating from a central sphere, for instance a rotating black hole, and having higher and higher electrical charge as that field expands away from the central sphere, then that electrical field will be seen to be composed of a whole number of concentric spheres of electrical energy. It will then be noticed also that an electrical potential will exist between each concentric sphere. The latter image becomes interesting when electrons and positrons are placed into those curved concentric field lines between those spheres. Because as in the case of a single electron (or positron), it will be accelerated along those curved lines by these concentric potential-differences and it will reach a high enough level of energy to then radiate gamma-ray photons away from it (by the curvature radiation process). These emitted gamma-rays, shooting off at a tangent, will then collide with the other electrons and positrons, for instance, traversing the next outer concentric sphere along their curving magnetic field, to then produce more electron-positron pairs from those collisions and more gamma-rays, and these in turn will do the same to collide with other particles, and so the process will repeat and the result will be masses and masses of electrons and positrons shooting off in opposite directions all around that magnetic field. This process, happening to all electrons and all positrons in each concentric layer all at the same time, will cause a continual production of charged particles, which is a process known as electron-positron avalanching. The same process will happen to concentric rings of electric fields, of course. For this reason alone circular energy generators will always be more efficient than angular ones.

Obviously, the reason why I mention this astrophysical procedure here is because the same procedure for particle production can be found in the aaUFO's internal fields of rotating energy which will also have a series of concentric electric fields, which will vary in potential between its central vortex and the craft's outer rim. So the aaUFO's rotating energy fields will have the potential to also produce electron-positron pairs when its field's rotational momentum reaches a critical velocity (which will be detailed in later chapters). It should be noted also that the Searl Energy Generator will have this concentric field arrangement in its toroidal energy field (a factor which will be gone into more thoroughly elsewhere in chapter 4 of this study) [note: 13].

Study by study astrophysical researchers have shown different preferences for different ionization mechanisms. For instance, dissipation ionization can come about by co-rotating viscous-shearing fields or by the breaking and reconnection of magnetic field lines, and these form just two of the methods of energy production from a very large range of methods favored throughout the [Blandford-Payne] study of electron-positron production; while the studies of [Punsly-Coroniti 1990b] and [Punsly 2001] emphasize a particular [Blandford-Znajek] gamma-ray and UV photon Compton-collision method of producing electron-positron pairs. This latter model of a black hole dynamics has its plasma aligned to the interstellar magnetic field lines that thread through the equatorial plane of the accretion disk just outside the event horizon, so that the plasma evolves into a disk, which could be described as a condensate of electron-positron pairs (and which forms into two streams of counter-rotating plasma centered at the equatorial plane of the black hole). Just to remind the reader of the process, it will be from this plasma being highly conductive that it will be expected that the interstellar magnetic field lines will become frozen into that plasma which rotates within the accretion disk, and as the disk rotates it will drag and twist those magnetic field lines, pulling them together. But, as will be observed a little later into this study it will be the first two ionization mechanisms mentioned above (as favored in the Blandford-Payne study) that will feature the most prominently in the aaUFO's electrodynamic system.

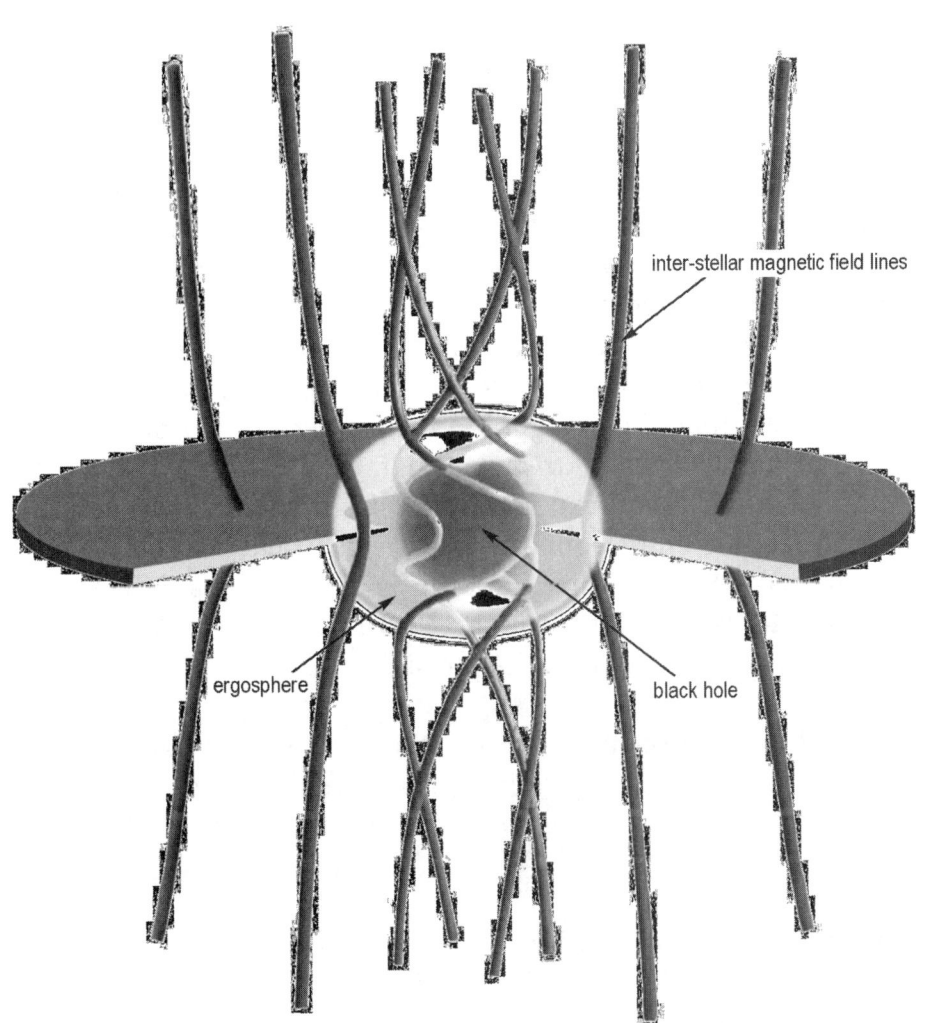

Figure 15. A black hole, ergosphere, and accretion disk twisting the inter-stellar field lines

Intriguingly, while I have pointed to the fact that the reported evidence of UFOs suggests 'cold' ionization processes, there is some evidence of thermodynamic processes which do take place in the craft's energy conversion mechanisms at later stages (as will be explained more fully in chapter 9). As for instance in the UFOs filmed over Mexico city in the 1990s which clearly show a heat-haze trailing behind some of the slowly moving UFOs. This heat may be associated with the visible effects of synchrotron radiation which appears to be emitting from around the UFO's rim.

In the Grangemouth UFO photograph which depicts a UFO at night passing very close above the photographer there are distinct heat trails coming from its three main areas of energy emissions—which also appear to be synchrotron-like emissions [note: 14].

2.7 Rotation Force and Momentum Transfer from a Black Hole

At this stage I should point out, especially to prospective developers, that with all the recent discoveries about energy generation dynamics in astrophysical systems, it matters not one iota whether the reader should believe in UFOs or not believe in UFOs. Likewise, it doesn't really matter whether or not an ET physicist has designed the flying disc associated with this extensive study, or whether it was designed by a few scientists with active imaginations in a laboratory, for the details of this study have been brought forward to such a high degree of workability now that belief or disbelief in the ET-hypothesis will not lessen the construct-ability or workability of these craft (to this end I hope the reader will have noticed that I will be backing up every step of this working hypothesis with substantial scientific reference, so that, for all intents and purposes the UFO has now been taken out of the clouds and has been placed squarely into the physics lab).

However, I would like to think that extraterrestrial engineers did construct these UFOs and model them on the astrophysical principles that exist out in space, and so I will continue this study according to their non-terrestrial brilliance and polymechanical ingenuity.

And one example of that ingenuity toward transporting astrophysical mechanisms into their craft can be seen in the intriguing method of energy-transference, that was no doubt seen to occur *first-hand* out in space by our very-keen-eyed ET physicist, being used by those extraterrestrial engineers to fuel their craft's rotating engine. Whereby, on the inside of this UFO's central ergosphere immediately surrounding the craft's rotating quasi-black hole, the freshly created UV and gamma-ray hard photons (of synchrotron radiation, inverse-Compton-scattering and electron-positron-annihilation emissions) occurring just outside the event horizon, are drawn from the rotating energies that comprise that ergosphere and are spiraled down into the black hole, and this spinning down results in them being 'lost' forever [note: 15].

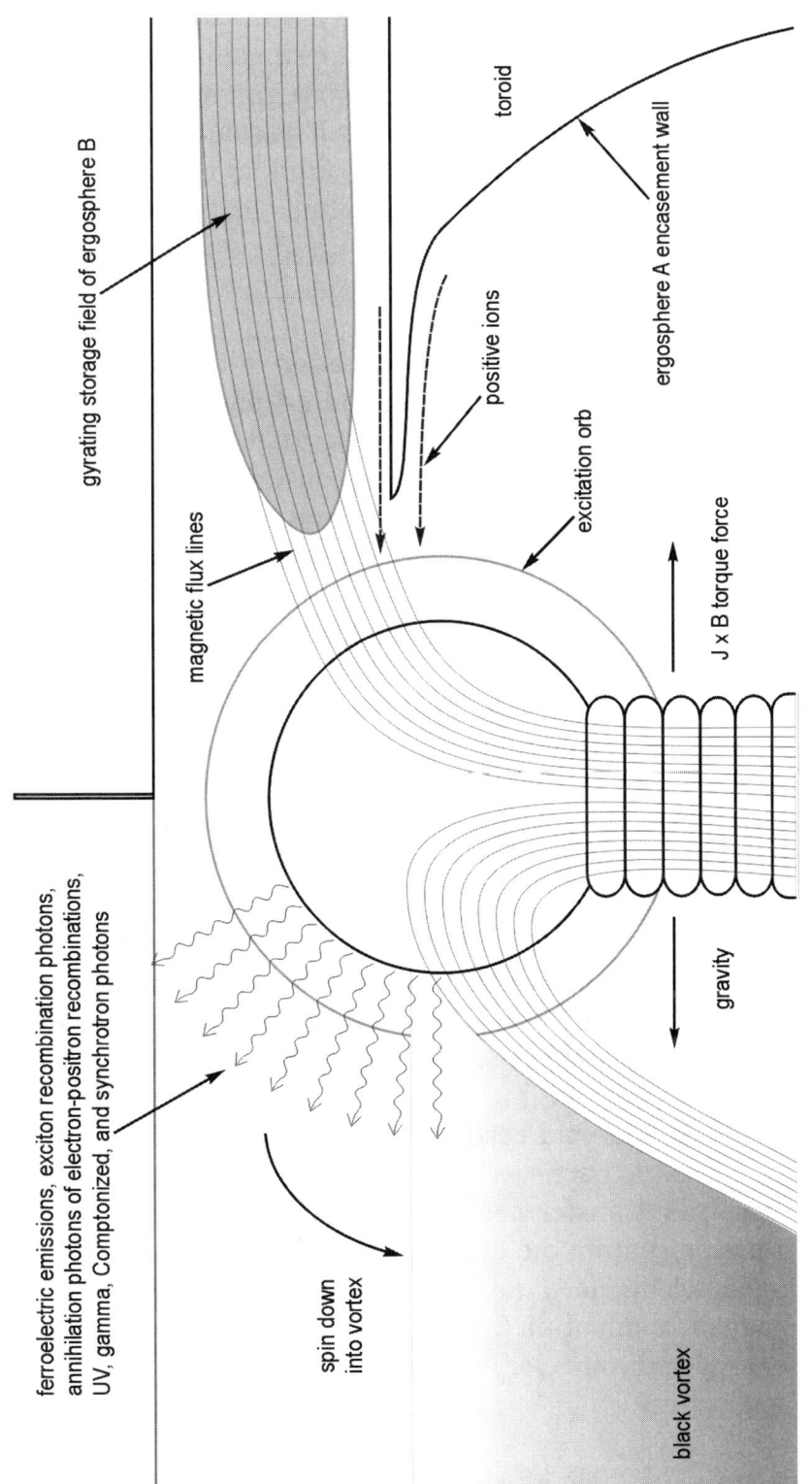

Figure 16. Ionization, annihilation and photon production outside a quasi-black hole

But then energy is never really lost, all it does is get transferred from one form into another form (or from one place to another place). And so in this case, even though these sub-atomic particles are spinning down into this black hole there will result a corresponding generation of spin momentum and gravitational pull which will radiate away from that black hole and which will permeate through to its adjoining power systems. And because of that induction of spin and gravitational momentum being transposed into various fields of rotating charged particles, there will then occur ionization processes through which there will subsequently be generated even more electron-positron pairs.

This transfer of energy from the black hole, the reader will remember, is what happens in descriptions of the accretion disk (in section 2.4 above), and this stems from the fact that the flux lines of the co-mutual magnetic field will thread through both the ergosphere surrounding that black hole and through the accretion disk. Meaning that the extra Lorentz force generated around the inboard flux lines (around the event horizon) will feed Lorentz force into the outboard flux lines (in the accretion disk) [note: 16].

So the consequence of this will be that the original photon mass and inertia given up and supposedly 'lost' into the black hole will not be lost completely, for it will subsequently induce more rotational force and freshly generated plasma into the energy fields rotating around the outside of that black hole. This emphasizes that the electrodynamics of a black hole equate to that of the perfect self-sustaining engine.

Furthermore, in this symbiotic relationship there is established an energy-exchange process which quite clearly shows that force and energy is being extracted from the black hole. Indeed, this exposes the secret of the black hole's efficiency, and it also will show how that efficiency can be improved upon. Certainly, the assumption that if something is dragged into a black hole it should equally be seen as being lost unconditionally, is a tabloid-reader assumption, and in the real world the black hole intake of sub-atomic particles should actually be seen as merely one type of energy which becomes transformed into several other types of energy, which, for the present situation (within the dynamics of UFOs being aircraft that are predominantly circular and which need to defy gravity)—then the extraction of gravitational force and angular momentum from its rotating engine will be infinitely more useable forms of energy, and this would be a very welcome exchange process indeed [note: 17].

On this very subject, of black hole efficiency, it is most intriguing to find that estimates of a black hole's efficiency as a mass-conversion-to-energy system that turns mass into useable energy are very good, ranging from 6% for a non-rotating hole to a 32% efficiency for a rotating hole (while nuclear fusion slumbers woefully behind them at less than 1% efficient) [note: 18]. In this respect you could almost as equally say that gravitational force converts light photons (as the fuel of a black hole) into a J x B force (the torquing force of accretion disk rotation) [note: 19].

Intriguingly, this factor of energy-extraction from a black hole may also be reversible, in that gravitational force can be equated as the force which is generated through a differential in angular-momentum existing in the same spatial frame. This is something already borne out by the findings of [Zeldovich 1972], Einstein, and Rosen, whom have postulated that rotational forces on mass-energy fields will generate and amplify gravitational force in those energy fields (and this discussion will take on further weight in section 10.7's note 17) [note: 20].

Indeed, some scientists who have been working on the miniaturization of black hole dynamics have associated the energy swallowing process of a black hole to that of cannibalism, and have seen the center of our galaxy, for instance, as the best example of this process, whereby the constant stripping away of energy from mass-objects that orbit too closely to that black hole end up contributing to the dynamism of that black hole [note: 21].

It is significant then that there is a remarkable similarity to that process of cannibalism around a black hole, to that featured inside the aaUFO around its central vortex, because it seems that this same photon spin-down arrangement to feed energy into its rotating engine, is right there in the center of this craft, and which seems to have been structured so as to copy the most essential characteristics of what a black hole does.

And so it is very apparent that inside this UFO there is possibly a miniaturization of that cosmic cannibalism process, whereby the four top spheres of the aaUFO with their light bars of intense photon emissions (designed to orbit very closely around the outside of the event horizon of the craft's central vortex) are very purposely and orderly supplying that quasi-black hole with a constant 'feed' of sub-atomic mass and photonic energy.

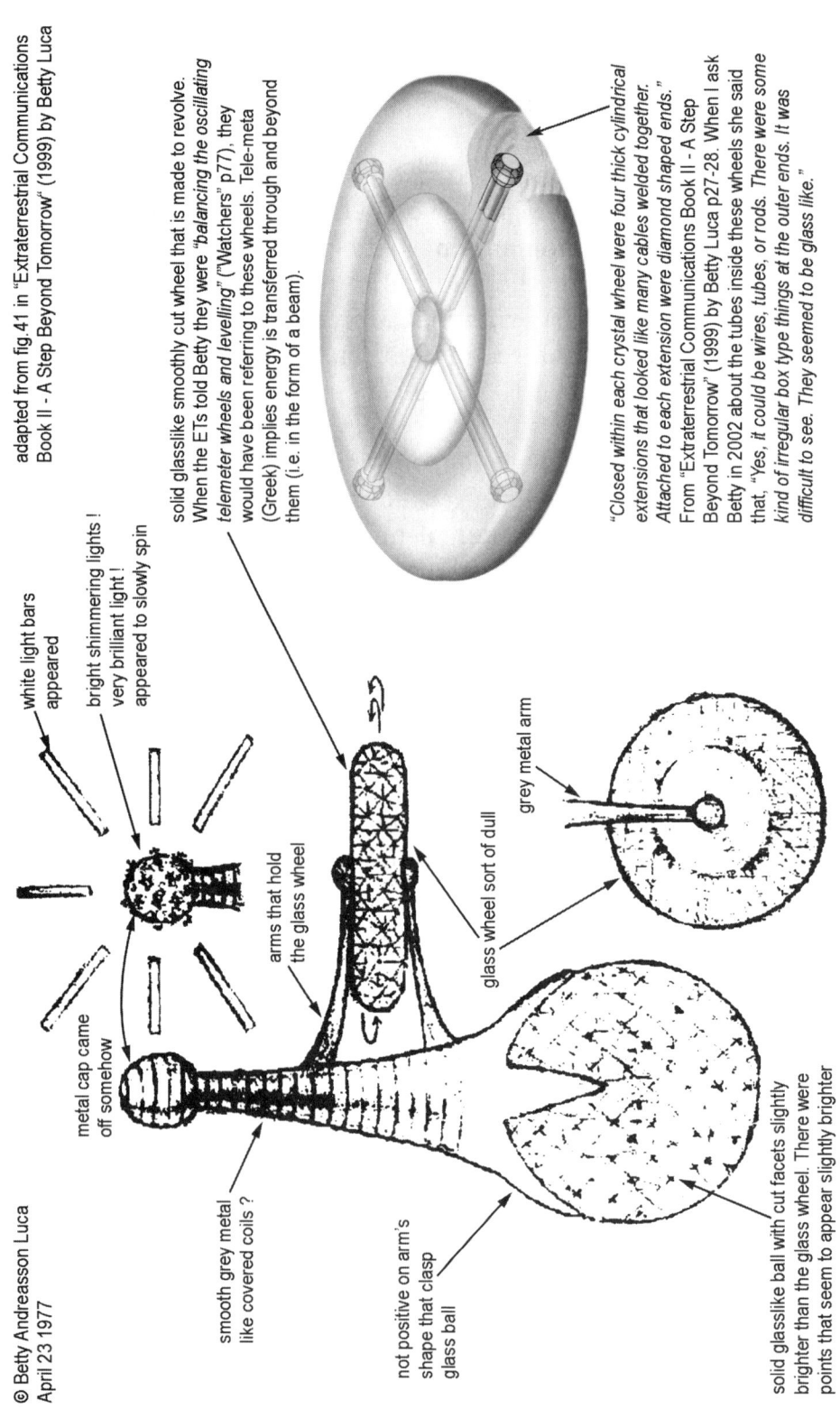

Figure 17. Sphere-sets of top spheres, lower spheres and telemeter wheels

As a result of those spheres feeding into the UFO engine's rotational power, gravitational force and rotational force might then be imparted into the aaUFO craft's adjoining energy systems. And as Hawking has suggested for mini black holes, that power will in proportion be all the more ferocious at this scale than on the astrophysical scale [note: 22]. Indeed, judging from the remarkable scene of these 'almost tangible' bars of intense light spilling out of the four glass spheres of the aaUFO (as seen and drawn by the abductee Betty Luca) with their photons spinning down into the craft's central vortex, this must have been a remarkable sight to behold, one of raw energy under perfect control [note: 23].

2.8 Magnetic Flux Break-Reconnection Ionization (Reconnection Part I)

How then does the aaUFO produce the all-important fuel to make that vortex engine do work, and how powerful can this electro-dynamic engine become? Because straight away there will exist a scientific conundrum, straight away this brand new and novel assumption will suggest that the fuel for this craft will not be liquefied gas, and it will not be a petroleum-based fuel, and it will not be nuclear fusion derived—which thus far are the only fuels presently understood as being able to produce power of any magnitude on this earth. The UFO's fuel has to be able to yield substantial power and it will have to be a fuel that the UFO can either store somewhere, or that it can readily convert from an easily accessible and omnipresent source (of the media outside of it) somehow. The latter statement would seem to be the most preferred because there is no source of energy in space... or is there?

If the reader would take on the task of scanning through the usual reference books for electronics and science, or even asking the academically trained scientists who have been educated the usual curriculums so prevalent on Earth, I would wager that he or she will not be overly encouraged, when looking for mega-powerful energy sources, to venture too far from the ubiquitous lores of electricity power turbines. These are, after-all, the present meat-and-two-veg sources of electric current which feed into the national grids of just about every nation, and which utilize magnetic field *movements* to generate electricity, with these movements being powered by water flow in hydro-electric plants, and steam flow in nuclear or fossil fuel power plants, and more recently through wind flow in wind turbines. All adequate in their own way, but are there more efficient ways to produce electricity that our scientists have not dared to look at?

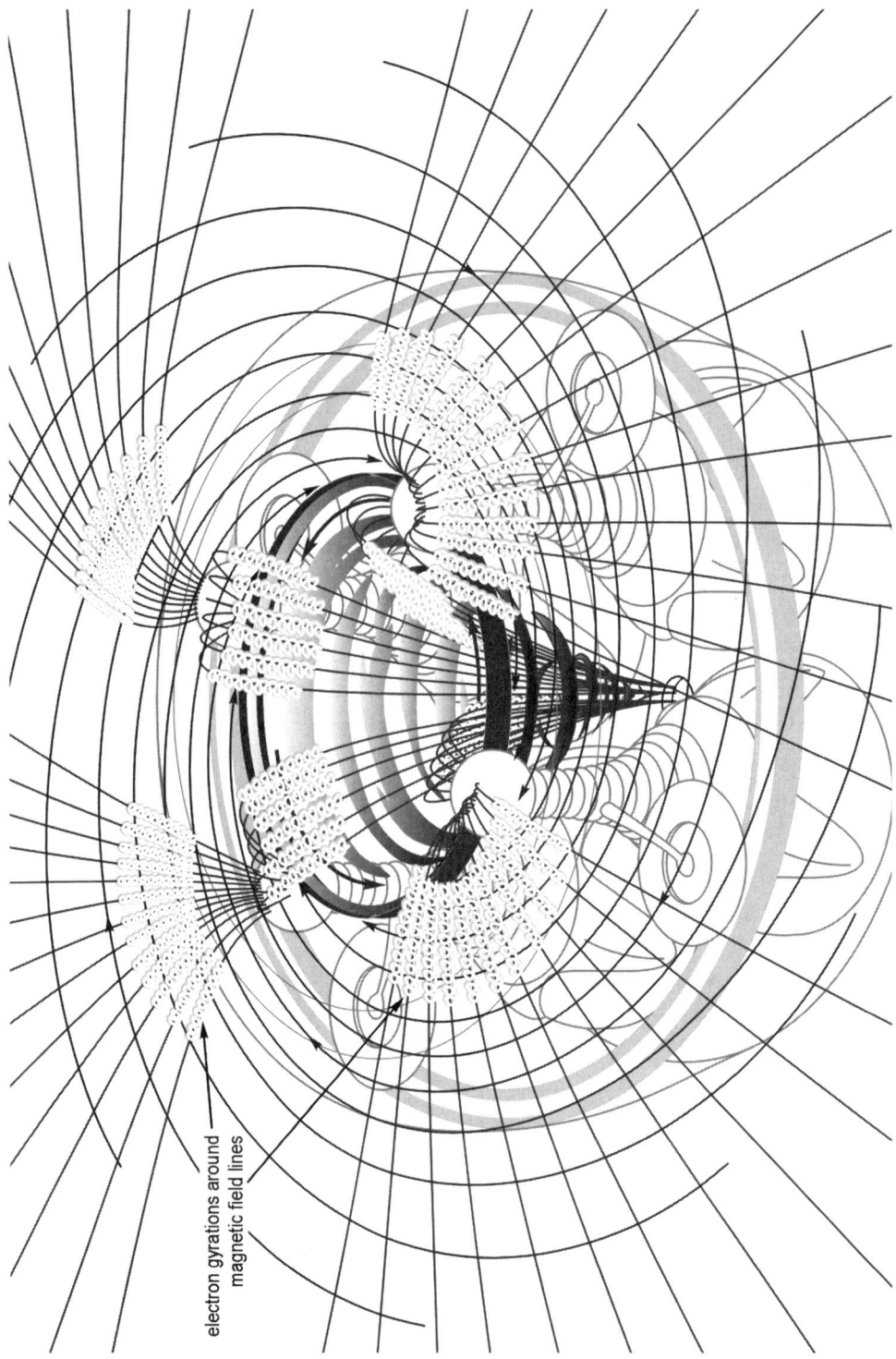

Figure 18. Emissions gyrating around magnetic field lines

However, there is one source of abundant and powerful energy that no one (as yet) has managed to put a price on, and that no one can ascribe a patent number to. Yet it is found in more abundance working with more efficiency than all the power generators in the whole of this world put together. And strangely enough, its principles will not be found anywhere in any text-books of physics, of science, or even of electronics, and neither will it be taught by any teacher in any school curriculum, for in general it is a principle of power generation which doesn't feature in rudimentary electromagnetic theory.

But here it is in astrophysics and its called magnetic field reconnection, and it can be found working in all areas of space at all times, and it is even highly prominent on the surface of the sun and in the sun's solar flares. Its basic working principle is given in the paper of M.A. Ruderman and P.G. Sutherland [see above note: 13] whose authors have found that it works especially well in the rotating fields of accretion disks in space. In their delineation of how it works in a pulsar jet Ruderman and Sutherland point out that it will provide copious productions of electron-positron charged particles, and indeed will produce so much energy as to cause pulsars to continually eject tonnes of surplus particles out into space. These authors also point out, in their explanation of this particle-production principle, that because the magnetic field lines which penetrate a pulsar's jet play a role analogous to that of conducting wires in an electronic circuit, if those magnetic field lines are broken near the pulsar's surface, then the entire voltage-potential-drop (being carried around that magnetic field line) would be developed across that break. Meaning that, where the magnetic flux lines are snapped the electrical voltage that was being carried along the whole of that unbroken line would continue to traverse the gap between the snapped ends of that flux line, and would continue to increase in potential (by self-inductance) were it not for the fact that the virtual-particle field of the vacuum will become unstable (at that gap) and will breakdown to transform those virtual particles into real electrically-charged particles. When that breakdown in the vacuum occurs, says [Ruderman-Sutherland], avalanches of electron-positron pairs will be produced out of that area of space. In space the potential drop threshold for when this effect takes place around the breaking of magnetic flux lines is about 10^{12} volts. The most salient example of this particle producing mechanism can be found, as I say above, around our own sun, where the chromosphere continually produces electrons and positrons by this method and then has to eject out from it around a million tonnes of those charged particles every second [note: 24].

This magnetic field line reconnection process is, as I say, working throughout the whole cosmos, and in active galactic nuclei which are some of the most powerful of all the astrophysical jets, their magnetic field reconnection system is regarded as one of the most efficient producers of charged particles in the galaxy, through which magnetic energy is converted into kinetic and thermal energy by conversions through slow magnetohydrodynamic shock waves. And around accretion disks this shearing-reconnection of their strong magnetic fields produces a dynamo effect (as referring back again to my section 2.4 summary), which gives a rapid amplification of any of its incoming, and smaller electrical fields of charged particles (its seed fields), so that very quickly they will develop into much larger fields, and go on to accelerate particles which will collide with other particles, to produce more particles, and more collisions, which subsequently will lead to avalanche productions of more electron-positron pairs.

In impulsive reconnection mechanisms which also occur in accretion disks, these magnetic field line breaks and reconnections would lead to intense heating, and then anomalous dissipation of additional charged particles in localized regions around those flux lines. With the subsequent production of electron-positron pairs and even of X-ray emissions (especially if shock waves are produced in the vicinity of dense plasma fields) then these particles feed back into the accretion disk to complete the dynamo effect. As the instabilities accumulate in energy and when the electron velocities exceed that of ion-acoustic waves a flurry of fast reconnections occur resulting in explosive outbursts of charged particles [Di Matteo] [note: 25].

In the [Blandford Payne] paper it is explained that the magnetic flux lines will be broken by powerful surges occurring through the shearing rotations of the accretion disk, but that those lines will then be continuously regenerated by the dynamo action of the disk—and when these explosive reconnections take place they will dissipate significant emissions of electrons and positrons, causing flares in and around the accretion disk.

Moreover, from this basic premise, that electron-positron pair production comes from the breaking and reconnection of magnetic field lines, there is much associated research which also indicates that, as well as the production of charged particles there will also be generated substantial bursts of kinetic energy, shock-wave particle accelerations, and also heat conversion processes [Priest-Forbes].

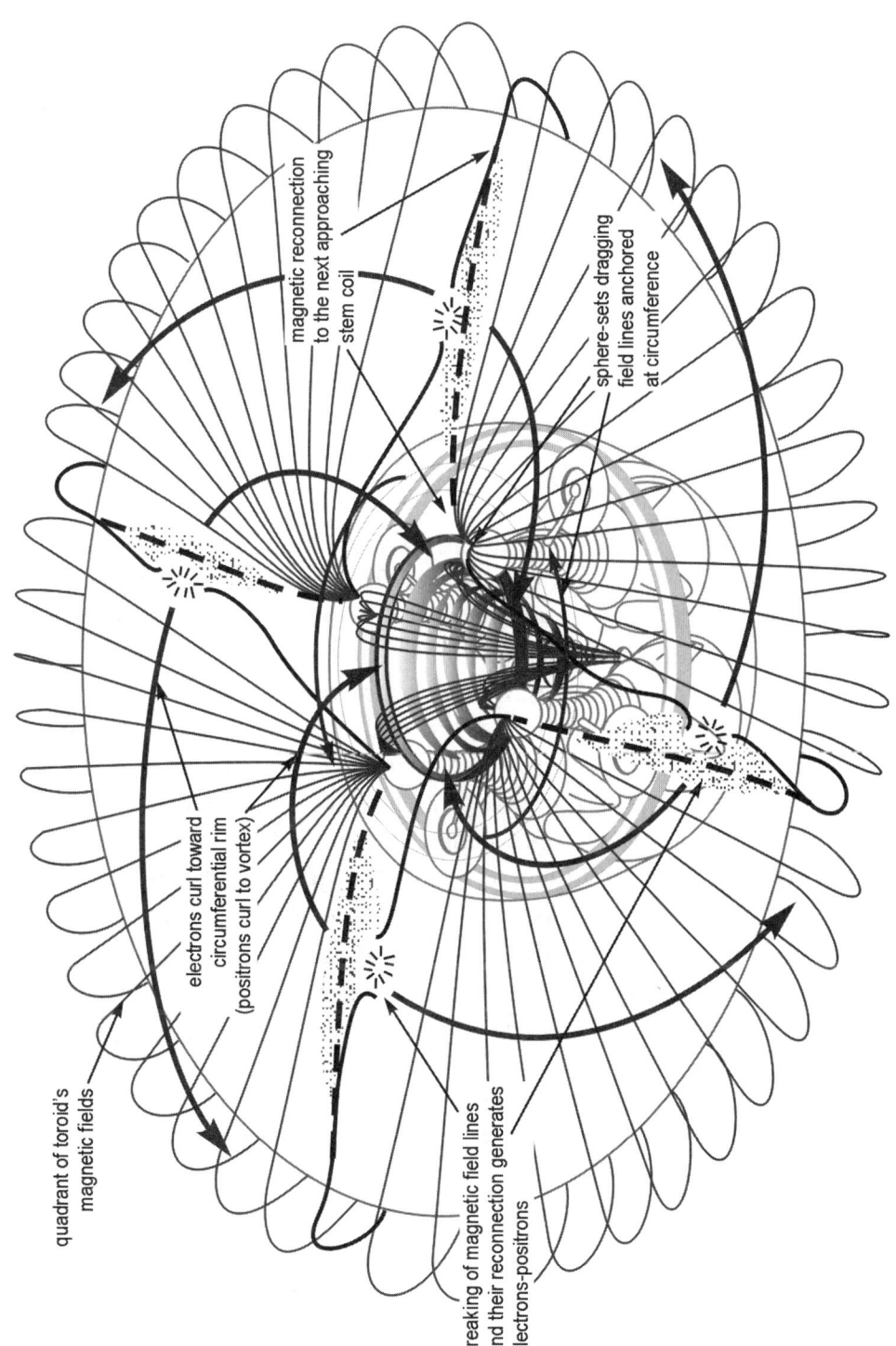

Figure 19. Each of the four sphere-sets grabs a quadrant of the toroid's magnetic flux lines

Where this form of energy production occurs inside the aaUFO and how the electron-positron producing shock waves are developed inside the aaUFO, in regular pulses, will be dealt with in detail throughout the subsequent chapters of this study [note: 26]. But essentially, here is the first abundant supply of fuel which can be used by the aaUFO whether that craft is flying through our sky or whether it is moving through space. For, whether it is attributed to particle dissipation or particle dissociation, the breaking-and-remaking of magnetic field lines produces and then amplifies copious amounts of electrons and positrons from what some have called the 'empty' vacuum of space. And as this is a mechanism by which jets of plasma can be powered for thousands of miles, and in some cases millions of miles through space, then it surely and easily can provide and amplify the significant amount of energy a UFO needs to power its electric circuit and to produce its propulsion.

Notably, this reconnection mechanism was mostly what led to the stressing of the air surrounding the discs of J.R.R. Searl (back in the 1970's), and which provided the copious supplies of charged particles which then became imbued with other kinetic forces, and which subsequently allowed Searl's generating rotors to defy gravity and to rise upward, not only by ionizing the surrounding air but by creating a projected field in the form of a rotating tube-field ahead of and above that disc, which the disc would simply fall upward into. The same reconnection electron-positron producing process is found in Searl's SEG power generation turbine which would also create copious supplies of electrons, which in Searl's turbine would be harvested by inductor coils and fed into an electrical storage circuit (and in an intelligent world such a mechanism would have then powered into a national grid, without the need for fossil fuels or nuclear-power rods) [note: 27].

Further to this mention about John Searl, what made me initially inquire about magnetic flux reconnection and to find out all about it generating electrical energy, was that the basic phenomenon of it is was what caught my eye when I was trying to work out what J.R.R. Searl's SEG electrical generator was doing. Tests had shown that whenever a Searl generator was faithfully constructed and allowed to rotate, huge amounts of electrical energy would be generated, but no one knew then exactly where it was all coming from!

2—Helical Field Tube Technology of Propulsion

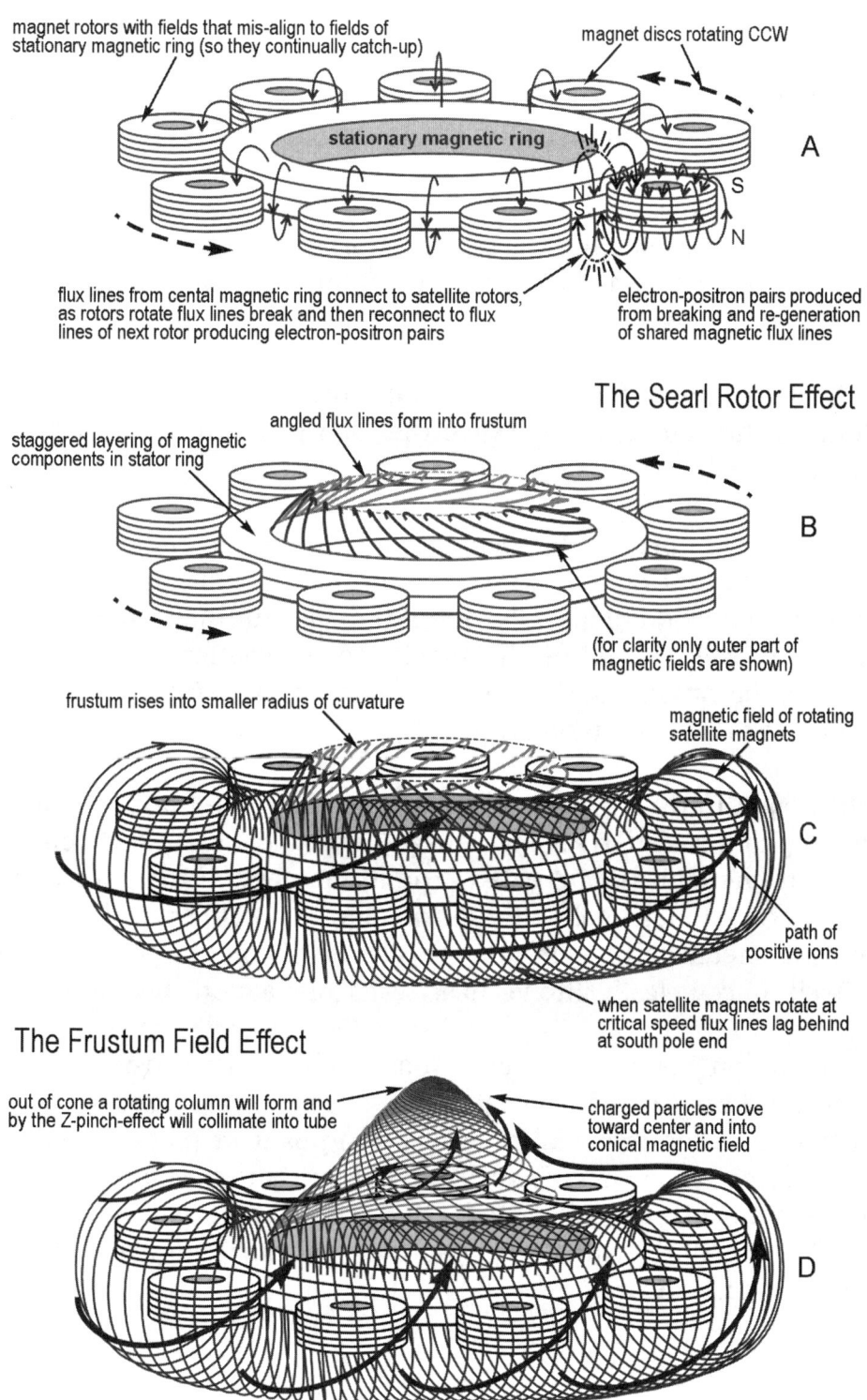

Figure 20. In the Searl rotors a frustum field is first generated

Therefore, regarding the effect of ionization through the making-and-breaking of magnetic flux lines in Searl's rotors, where the satellite magnetic rotors orbit around the stator ring, and in their epicyclic motions the flux lines would establish a shared link between the satellite rotor's magnetic field and the magnetic field of the central stator ring. When those outer rotors rotate around the outside of the stator those shared flux-links will get ripped apart, but then almost immediately they will become reconnected to the NEXT satellite rotor's magnetic flux as it rotates into proximity. This will happen with every rotor and so every rotor will produce electron-positron pairs while it rotates—ad infinitum.

So ALL the magnetic flux lines are being broken and reconnected ALL THE TIME as long as the rotors move. Hence the high generation of charged particles around Searl's rotors!

This reconnection process happens again and again as the twelve or so rotors rotate around the stator ring so the SEG is continually producing copious amounts of electron-positron pairs, quite literally, out of thin air. But then again, looking a little closer into this process, it will be the shock waves produced in this reconnection process that will be producing the electron-positron pairs (and so there will be an oscillatory nature to these emissions).

To add weight to this revelation in energy-generation: Roschin and Godin, two researchers in Russia, reported that in laboratory tests (performed during the 1990's) the extreme ionization created around the version they made of the Searl SEG caused a toroidal magnetic field to form, and this field spread out horizontally from their one meter SEG generator to about 15 meters in diameter, which organized itself concentrically into vertical 'walls' of magnetic flux intensity.

And just so that the reader doesn't loose sight of the astrophysical nature of this phenomena, because that's what it is—a conversion process of the omnipresent virtual energies into real-time electron-positron pairs, exactly the same conversion process which manifests itself out in space around rotating energy fields—I have found a very similar model to Searl's 'current-sheet' fields in a paper by [Kirk-Lyubarsky 2001] which explains the concentric nature of these electric fields in astrophysical pulsars [note: 28].

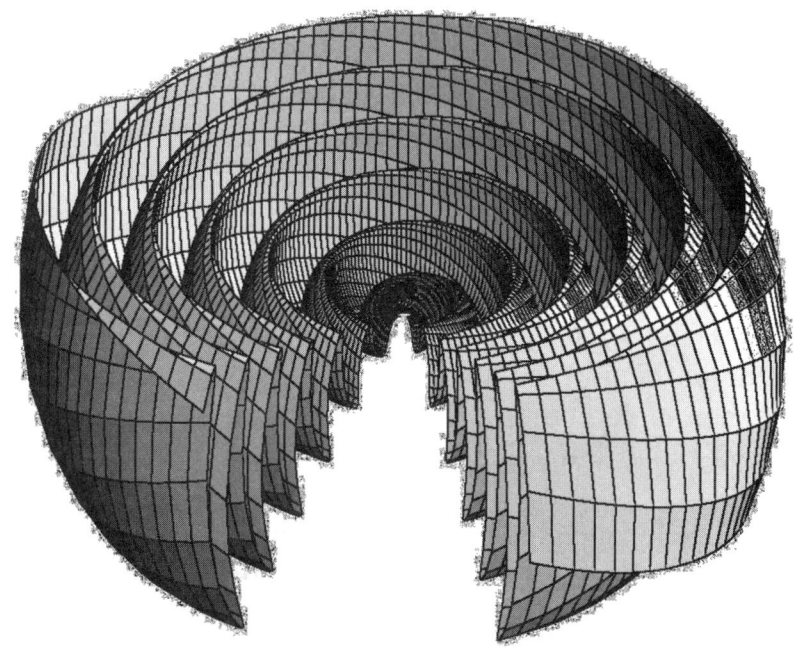

current sheets from the Kirk-Lyubarsky study of magnetic reconnection
(J.G. Kirk, & Y. Lyubarsky, Publ. Astron. Soc. Australia 18 (2001) 415-420)
reproduced with kind permission of John Kirk

Figure 21. Current sheets as a spiral corrugation in an MHD pulsar wind

2.9 The Quest for Stability in the Development of a Filament-tube

The reader will notice that I have only very tentatively associated several black hole mechanisms to the aaUFO. That is because this is only chapter 2, and all I am endeavoring to establish in this early chapter is a generalized link between astrophysics and UFO physics. The previous sections of this chapter will have provided a basic background of energy dynamics which can be expanded upon much more thoroughly in the following chapters. This chapter though is about UFO filament-tubes and astrophysical jets and so that will be what the remaining sections of this chapter will be focusing upon, because there is still a whole new science to find parameters for so that a UFO craft can be able to establish for itself well collimated, structurally stable filament-tubes.

And, whether it is down on paper as a Torsional Alfvén Wave, or a Poynting Vector Flux, or a rotating azimuthal magnetohydrodynamic (MHD) field, or a Lense-Thirring vortical field, the over-riding point of reference throughout this study will be how the dynamic of that rotating energy will be configured as a helical, spiraling, rotating tube of energy. And as all of these hydrodynamic models will, with respect to the rotating engines that have created them, have been formed in the same principle manner as a rotating vortex filament, then those vortical tubes whenever referred to will (as much as possible so as to provide a common string throughout this study) be referred to as the filament-tube. And whatever they have been called in other studies the most important aspect of these filament-tubes will be their stability and straight collimation, for when transposing this technology into the UFO's propulsion system this collimated stability will hold that filament-tube as one long continuous rotating energy field, so that it can then be used to transport pulses of the craft's propulsion-inducing energy. Obviously, understanding fully how these filament-tubes work and what dynamics would be required by the UFO to initiate them, and to maintain them in the air for anything up to a second, will be essential to the process of then understanding precisely what sort of energies might be sent down those filament-tubes that will produce kinetic movement in that UFO.

One of the favored principles behind filament-tube generation can be seen in this image (figure 15) of the torus-cylinder of magnetic flux lines which is created by charged particle momentum wrapping around the primary field lines that are anchored in the rotating core of the accretion disk. The wrapping around will lead to the emergence of a helix-shaped field (filament-tube) that extends along the longitudinal axis of that system to produce field-amplification which will carry and accelerate plasma in jet-like flows, along that same axis, which will be further driven by field-torsion and confined by inward magnetic tension (ie hoop stress).

I clarify this because several research studies offering different viewpoints in the field of astrophysical jet production have proposed in their different models all sorts of different names and terms for their jets, for instance, when Meier [Meier 2004] gives details of his view of how a jet is structured he states that the stability of such a jet is enhanced by a helical magnetic field both inside the well-collimated current driven Poynting flux dominated plasma, and outside of it in what Meier sees as a large electrical current which flows parallel to the main jet flow; and it will be this strong current which will be responsible for generating an intense tightly-wound helical magnetic field around the jet, and that it will be the continued rotation of the entire magnetized plasma about the jet's axis which plays the most important role in that jet's dynamics [note: 29]. This mechanism drives what can be described as a barber's pole torsional Alfvén wave which carries the electromagnetic energy that accelerates the plasma spiraling within it.

Stability of these filament-tubes cannot be over-emphasized. For, as was shown in the image of the Mexico UFO where the video captured the filament-tube initially sent out by that UFO, but which failed to connect with the surrounding air, if the filament-tube collapses then the UFO doesn't go anywhere!

Perhaps, from the fact that a UFO's filament-tube has been shown on film as failing, it would be appropriate to stress here that these filament-tubes will be prone to instability and collapse if the correct parameters are not applied to them continually while they are active. For instance, there can occur instabilities in the filament-tube such that will drive kinks into them that will come from disruptive interplays between pressure and the curvature of magnetic field lines, (ie pressure-driven instabilities resulting from Kelvin-Helmholtz kinks developing out of velocity imbalances in those flows causing wrap-around overlaps). In the early stages of astrophysical jet research when the process of how these filament-tubes were formed was not adequately understood, many theories were proposed to either deal with this kink-overlap problem or to explain this problem in greater detail. This was because, at that time, this was seen as an insurmountable problem, and so because so many researchers took on the task to solve this kink-overlap phenomenon there resulted a whole wealth of theories and information about ways in which these jets can be formed and made stable, and how they can be collimated (and this great wealth of information, of course, will be highly beneficial to the present study).

One of Meier's suggestions, from the Jet Propulsion Laboratory at Cal Tech, is that the most stable type of filament-tube comes from a strongly magnetized, hypersonic (but sub-Alfvén speed), Poynting flux dominated, torsional Alfvén wave. And while Meier does have some good ideas these should be compared with what Spruit says about the configuration of a helical plasma flow, especially where the flow's Alfvén speed should be large compared to the localized sound speed. In this type of jet strong magnetization is again figured to be the most important factor, and examples of such jets are gamma-ray bursts where there is an azimuthal magnetic field inside the jet which will be used to accelerate its particles (the reader can see an azimuthal magnetic field outside the jet in my figure 23 below). Also in these filament-tubes magnetohydrodynamic (MHD) waves can be formed and these in turn can be compressed into MHD shock waves that will then accelerate fields of particles through them (especially if another twisting technique, mentioned by [Karas] is also incorporated into that filament-tube).

2.10 Collimation and Hoop Stress

Collimation is the intrinsic ability of the helical filament-tube to exist as a straight-walled and parallel (or almost parallel) stable cylinder. Hoop stress is an important part of the collimating process, which is defined as being part of the magnetic tension force created by the toroidal magnetic field surrounding that jet, and which collimates and confines the pressure of the plasma into a stable cylinder. In other words, it will be an inward-acting force of tension which extends along the whole length of the filament-tube. This inward tension must be balanced by an outward centrifugal force (and vice versa) all along the whole length of the filament-tube otherwise an instability will form and the tube will become unstable and collapse. Other researchers have coined this collimating mechanism, the sweeping magnetic twist from the Lorentz force mechanism that has been proposed in the [Uchida-Shibata] study, which again, is similar to the model proposed by Meier [Meier 2004] with its torsional Alfvén wave, so as to ensure that the filament-tube is formed to a stable and slender jet shape [note: 30].

It is interesting that while some physicists have dismissed such helical fields as being too unstable to reproduce Spruit, through his axisymmetric-steady calculations, has proved on the contrary that there can be immense stability, and he has outlined the parameters by which excellent collimation of the flow can be maintained in nearly all cases (and this echoes what was said earlier, in the previous section, about the numerous studies carried out on these jets which in the end did prove that there were ways in which stability can be successfully achieved). Spruit explains that many flows become asymptotically parallel to the rotation axis once the filament-tube is established, meaning that the collimation becomes more and more perfect the more it extends (ie until the tube becomes perfectly straight). He notes that where the helical flow is initially formed, as it leaves the accretion disk of the rotating black hole system, the tube may be poorly collimated as it begins its formation process, and that the collimation effect only gathers strength and cohesion when the rotation extends away from the axis and as the filament-tube develops in length [note: 31].

Physicists of the Maryland Centrifugal Experiment (USA) see the phenomenon of hoop stress slightly differently and suggests that the concept of centrifugal confinement is to develop centrifugal forces through a field moving at supersonic rotational speed in order to enhance the force of the magnetic confinement, and that the rotational speed will be determined by an applied electric field which is perpendicular to the confining magnetic field (ie through E x B force)—as is shown in figure 8 (and in figure 32 below). The resulting rotation of the accelerated plasma will be non-uniform resulting in large velocity shear along the length of the filament-tube, which will substantially stabilize that filament-tube as it extends [note: 32].

In another different look toward understanding the phenomenon of collimation, and perhaps controlling it, the paper by [Honda-Honda] proposes that the favored candidate for the collimating force is the actual electron-positron flow flowing out of the accreting disk, which induces a transverse toroidal magnetic field around the jet, and that the screening effect (charge-neutralizing effect) by the electron-positron pairs on their adjacent diffuse envelopes will be what prevents any snakelike distortions from occurring (which is the same thing that Meier was trying to establish with regard to the torsional Alfvén wave central to his theoretical Poynting flux model). Honda then goes on to clarify that it is a stream of excess electrons (the surplus unscreened electrons) which generates the toroidal self-organizing magnetic field, which extends along the whole length of the filament-tube and which contributes a self-pinching mechanism on the electron-positron plasma, and with it a corresponding migration of ions moving radially inward toward the tube's core. In this Honda-Honda model the diameter of the filament tube is adjustable according to the balance of A) the pair-production rate, of B) the return-current strength, and to C) the average particle-density within the filament-tube. Honda's idea, again, is that the magnetic field pressure will balance against the plasma pressure of the jet, although, somewhat differently to majority opinion Honda maintains that the pair-screening effects and return-currents will then expand the bore-diameter of the filament, to the point where such expansion will lead to strong suppression of the instabilities [note: 33].

This idea, as indeed all the above mentioned filament-tube theories, should be compared to that of the Z-pinch effect, which occurs in an electrically charged vortex, or a tornado, and shows the reader the basic form of azimuthal particle rotation (and toroidal-poloidal magnetic field lines) [Thomson].

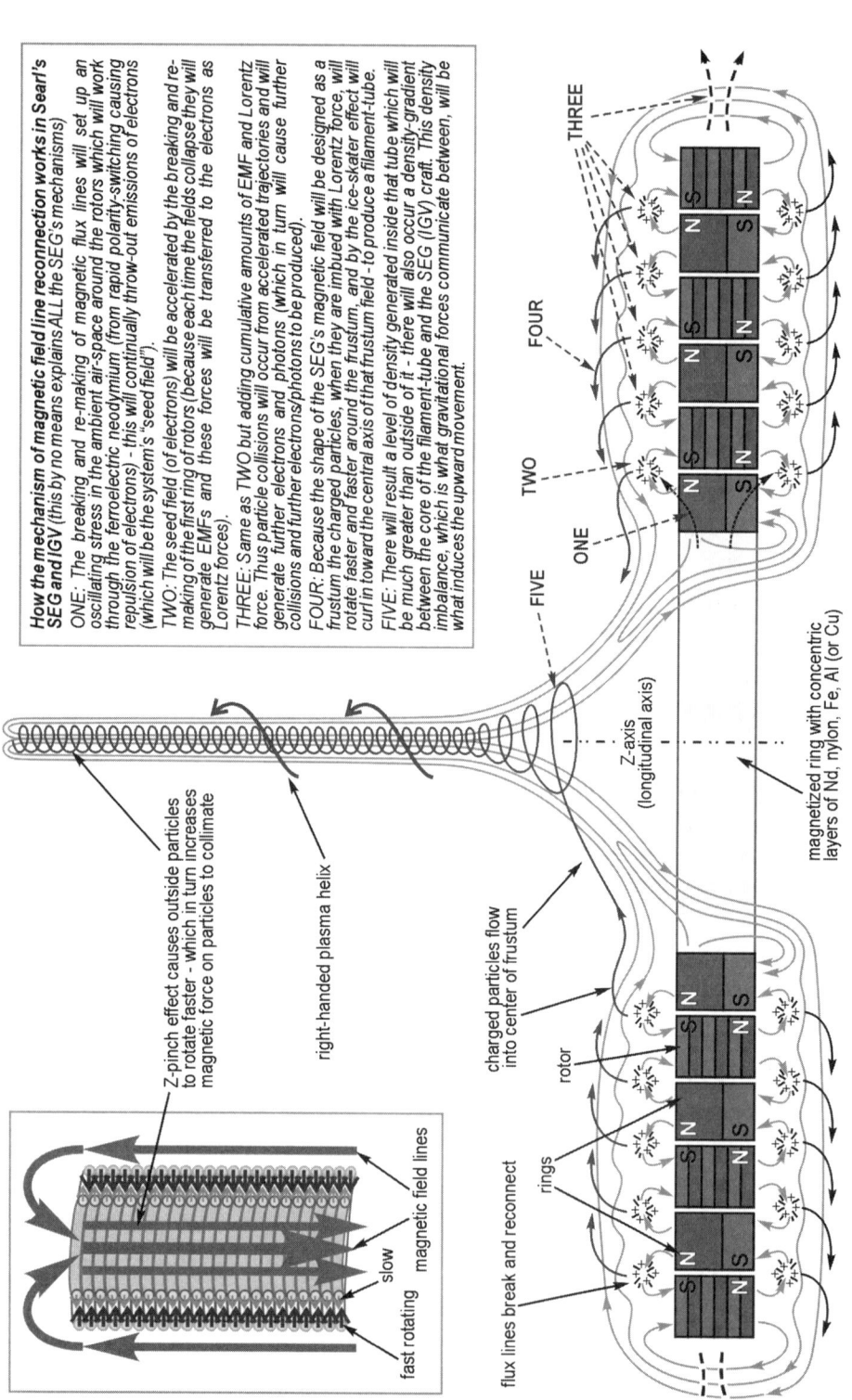

Figure 22. Helical field created by a Searl rotor from charged particles produced through magnetic field breaking and reconnection. Also the Z-pinch effect on a rotating field

Where suppression of instability is mentioned in Spruit [Spruit 1996] it is argued that the necessary ingredient to stabilize a jet is a poloidal magnetic flux anchored in the accretion disk, and the greater the distance from the central axis that anchoring is located the greater the stabilizing effect [Spruit 1999]. And Lebedev expands on this by explaining that out from the initial formation of a wide poloidal field there will form the narrower toroidal field which will surround the filament-tube and give it great stability [Lebedev 2005].

The model of a filament-tube surrounded by a toroidal field is also favored in the [Punsly 2002] model whereby the infinitely long toroid field helps to produce an effect similar to a long cylindrical waveguide surrounded by a solenoidal magnetic field which, as Punsly also says, substantially overcomes the problem of instability.

I did wonder what the reason might be for the darkness of the filament-tube projected by the UFOs over Mexico and captured on video tape, which can be seen in the still-frame of that UFO's filament-tube (see figures 10 and 14 above) [note: 34]. Obviously, the charged particles that make up that filament-tube will have an azimuthal velocity and so it may just be that the rotating energy of that UFO's filament-tube has drawn moisture out of the surrounding air (although when warm air rises up through the middle of that tube that air will cool too), and the rotation has packed that moisture into its tightly coiling layers, which in turn is drawing toward it a sheet of air (and more moisture) all the way up its whole length in a fashion similar to a Lense-Thirring dragging effect which is wrapping layers of air around that tube—and perhaps that's why its darker than the surrounding air.

Further information about the outer shell of a rotating field comes from the [Blandford-Payne] study where its authors say that the best formation of a jet will essentially depend upon the dynamics of the rotating disk that forms it. Intriguingly, these authors suggest that air molecules driven from the inner edge of the rotating disk will have the higher velocity, while molecules driven from the outer parts of that disk will have a much lower velocity, and because of this the filament-tube which results will have a different velocity at each different radius (ie there will occur a frame-dragging effect) and these differential velocities will determine that such a tube will have good stability.

Which, in fact, is the perfect description of how the dome-shaped surface of a UFO will work. Meaning that, in the case of a UFO, with a specifically-domed and circular shape, that curved dome can principally provide the most proficient type of waveguide (or projector) from which to send rotating energies into the air ahead of it—because the apex of its curved dome will promote a higher rotational velocity, while from its perimeter rim there will be projected a lower rotational velocity—so that the inner rotational force from the tube's lead particles will always drag upon the slower particles in the outer part of that filament-tube, and this simple mechanism will give the most stability.

In this paper Blandford and Payne then make the important suggestion that while most of the power will be concentrated within the core of the jet the highest degree of angular momentum and magnetic flux will be carried in the filament-tube's outer wall [note: 35]. Which surely suggests that the greatest degree of entrainment will be inside the filament-tube, rather than outside of it, and also, that if the bore-diameter of that tube can be adjusted, then the power of that entrainment force will be adjustable. Indeed, this is something also mentioned in the Wheeler, Meier, and Wilson study, in that the narrower the jet (ie the tighter the collimation) the faster its flow will be, and so most of the energy would be dissipated into the through-flowing (central) jet rather than into lateral shocks that would compromise the integrity of that filament-tube [Wheeler-Meier-Wilson].

Another advantageous effect that could be utilized through the domed-shape upper shell of a UFO will be 'magnetic drift,' which is caused by the slight variation between the gyroradius and the gyrophase of the charged particles as they rotate around a UFO's upper dome through a non-uniform magnetic field (which as can be seen in the images throughout this book, is quite obviously the case where charged particles are accelerated in a curve around the aaUFO craft's upper dome through the craft's magnetic field lines being formed into a non-uniform configuration above that dome, which ultimately, can be fine-tuned by magnetic pads in that dome).

But this study has already assumed, through the fact that the UFO's upper dome promotes a convergence of magnetic field intensity, that there will be a strong 'drift' of the charged particles that spin around the dome toward that UFO's apex, by what I would call the frustum effect, as will be further explained in sections 7.7 and 10.1 below) [note: 36].

2—Helical Field Tube Technology of Propulsion

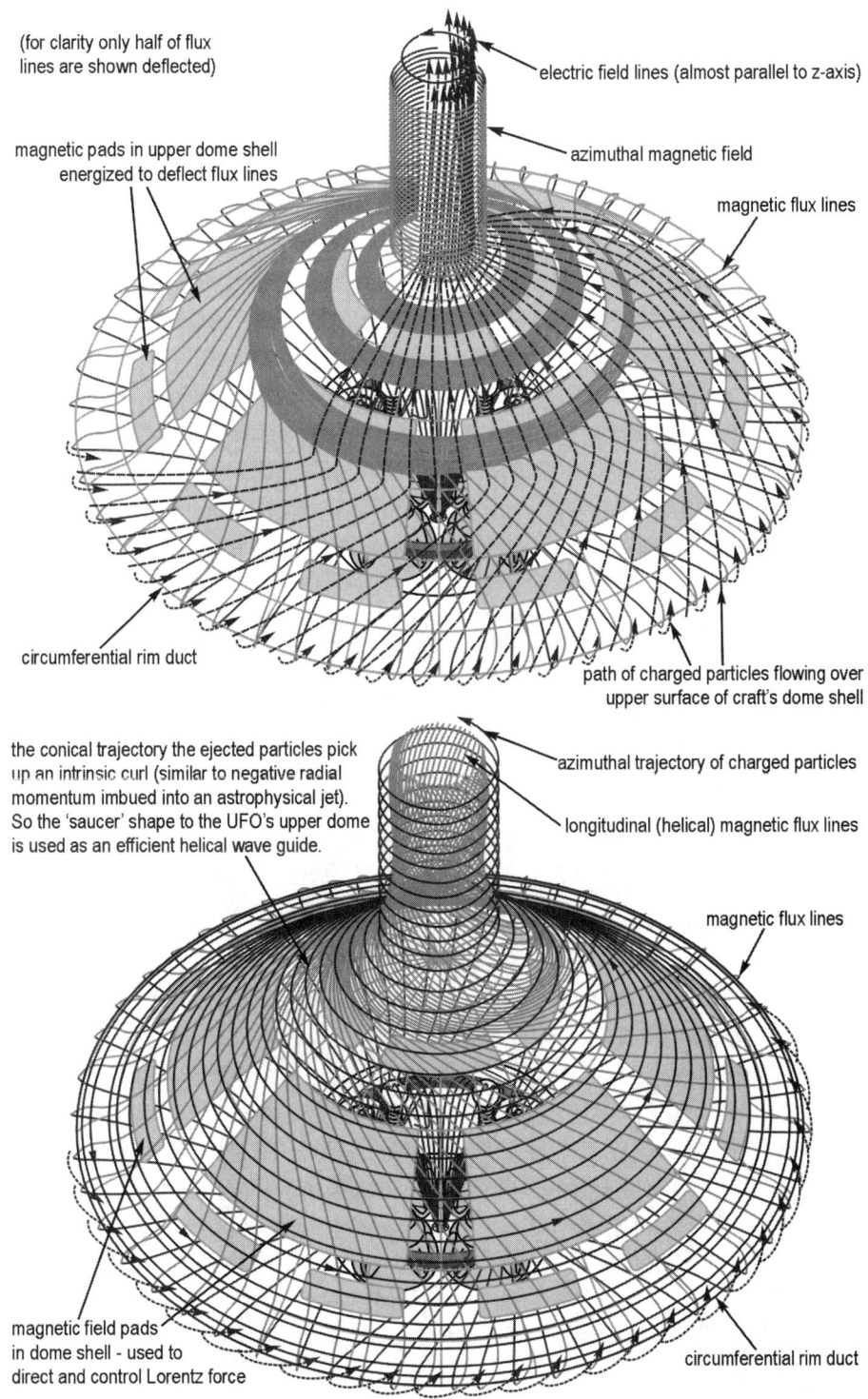

Figure 23. Depending on how its magnetic helical field is configured a domed craft can propel a field of charged particles parallel to its longitudinal axis or in an azimuthal curling trajectory

So that at this early stage of this study, there is already a substantial body of information that can be used to develop propulsion forces which can be extended forward from a UFO into the air or space ahead of it, so that when it does develop its propulsion forces they too will be projected and directed ahead of that craft and with a substantial amount of control.

One point which will be useful to make here, will be that presently it seems to be assumed that such filament-tube jets serve to push energy along them. However, in the above cases, especially in the Blandford and Payne study where the frame-dragging and entrainment will be on the inside of the filament-tube, then surely, there should be expected a degree of tensile drag-force operating inside the filament-tube (rather than a pressure force). What might substantiate this factor of a *tensile drag-force* may possibly be found in a study on the star formation mechanism as proposed by Lery et al, wherein it is suggested that these mechanisms are accompanied by powerful ejection jets from accretion disks [note: 37].

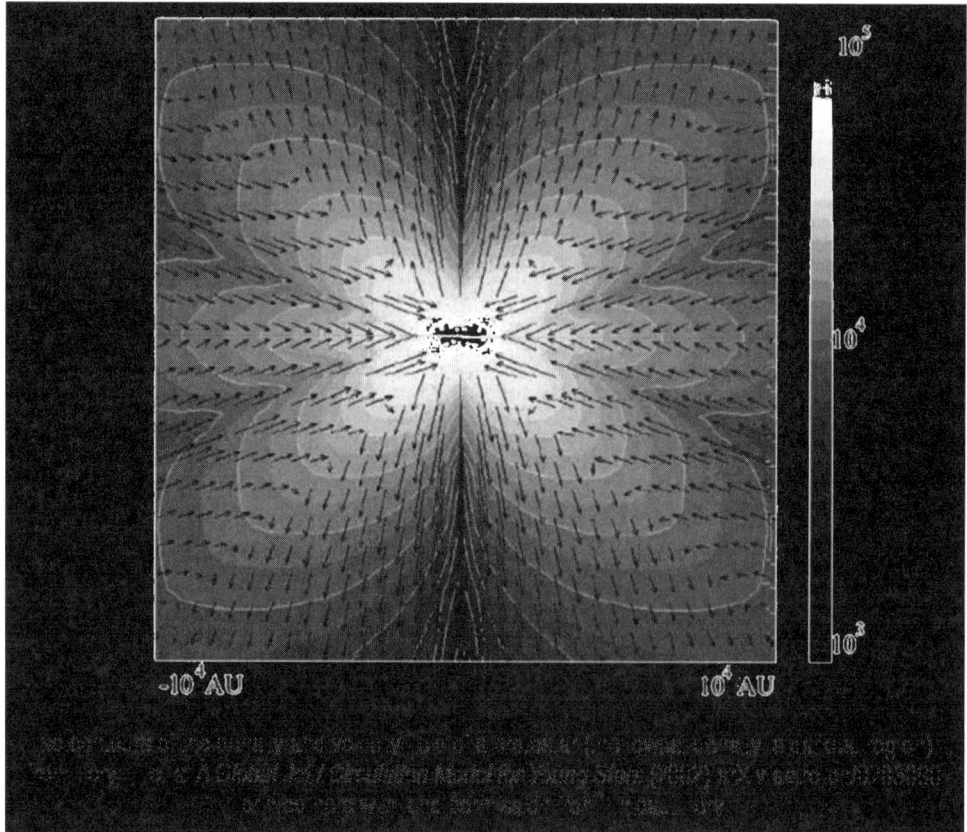

Figure 24. The flow velocity and density fields of a protostar. Is the energy for the jets being pushed into those areas or pulled?

Because then it should be noted that, in this formation process the density, of the protostar's jet around that system's longitudinal axis, will be much lower than the density of the rotating plasma in its disk. Also to be noted in this study, is the fact that the out-flowing co-axial jet which results in such a situation would be collimated cylindrically, again by the pinch effect on its toroidal field, just the same as in the black hole and pulsar systems, and in this particular theory the pinch effect, or hoop stress, will be reinforced by the density-gradient confined along its longitudinal sheath (as can be seen very clearly in this density chart) [Lery 2002]. Consequently, this begs the question—is the jet being pushed out of the protostar-disk system, or is it being pulled…

Intriguingly, if there were a tensile force operating inside a filament-tube then this would facilitate a negative resistance factor for any energies, or wave-packets of energies, being sent down that tube. Furthermore, if that tensile force could then be accumulated, along specific lengths of that tube, which would equate to metastable states of tension inside that tube. Then such a mechanism could be utilized by a UFO craft, to take advantage of a metastable condition whereby vast amounts of tensile force could be stored along the whole length of that filament-tube, which could be held in that state until released by a suitable triggering mechanism. This would be similar to the mechanism that has been dubbed the spring-and-fling mechanism in the [Wheeler-Meier-Wilson] hypothesis, where these authors suggest that from a differential rotation process, the filament-tube's field could be made to coil around (like in a barber-pole curl), so that it would wind into itself very large toroidal strengths, and when this stored energy was released whatever was inside that filament-tube would be projected down that tube at colossal speed. This happens in the case of the protopulsar, where its filament-tube will be anchored at one end to the outer core of its toroid field, while its other end will wrap itself around the axial core and create sufficient amplification through this twisting process that, when triggered, will release a jet-like flow driven by that field torsion's renormalization. With this spring-and-fling it has been conjectured that there will come a critical stage where a simple magnetic switch mechanism can come into play to trigger this high-speed jet-flow. What will be useful to a UFO with this mechanism will be that huge amounts of coiled-up energy could be released merely by activating a magnetic switch through, it should be noted, a significantly lesser force [note: 38].

In space the wrapping of space-time in order to manipulate it (mostly by taking advantage of its inherent renormalizing process which snaps it back to normal again, and in doing so provides substantial force in a much shorter time than it took to warp it up), could be referred to as 'frame-twisting' (not to be mistaken with frame-dragging), in that it would seem to represent one way in which the helical tube, driven by a space vehicle, might attach itself to the geometric and magnetic field lines of space, to coil them into the required torsional tube and thereby 'connect' with the fabric of space-time at a very distant location. This raises the question of how much 'elasticity' does the space-time matrix, or perhaps the interstellar magnetic field lines, have, in order for them to be manipulated by an energy field generated by a UFO craft? And indeed, this question has been raised many times in the search for the dynamics behind some astrophysical jets [note: 39]. But if a way can be found so that the space vehicle's filament-tube after it has been extended from that craft, can be configured so as to 'grab' the distant matrix of space-time, and can then twist into that distant matrix a temporary coiling of its elastic framework, then that distant matrix could be used to store energy which, as in the metastable example given above, could be triggered into releasing all of that stored energy all at once, to thus 'forward-engineer' a path of tensile force between that distant location and that craft—which would allow the craft to be propelled along a force-current without it expending any energy. And if the space-time matrix has not the required elasticity already available in it, then can sufficient elasticity be dialed into the space-time matrix in any way (such as by giving it temporary tension or perhaps a temporary electromagnetic-gravitational force)? On a micro-scale, these same questions appear to have already been answered in the beam-tubes used by the UFO's over Mexico...

2.11 Electrodynamic Anchoring of the Magnetic Field

Coiling up an energy field, which is what a filament-tube is, so as to store energy into it will, of course, depend upon how that filament-tube is anchored. One of the few papers that goes into detail about the anchoring mechanisms of a Kerr black hole out in space is the one by [Punsly-Coroniti 1990b]. The anchor forms a very important role in the whole helical field generating mechanism, for, without an anchor the magnetic field won't be effective in twisting and developing any sort of electromagnetic pressure. In other words, the anchor will work somewhat like a clockwork spring which needs to be anchored before it can do work or store energy. This mechanism of energy storage coupled with a way all that energy can be suddenly released will, of course, be a useful mechanism for a UFO's power drive.

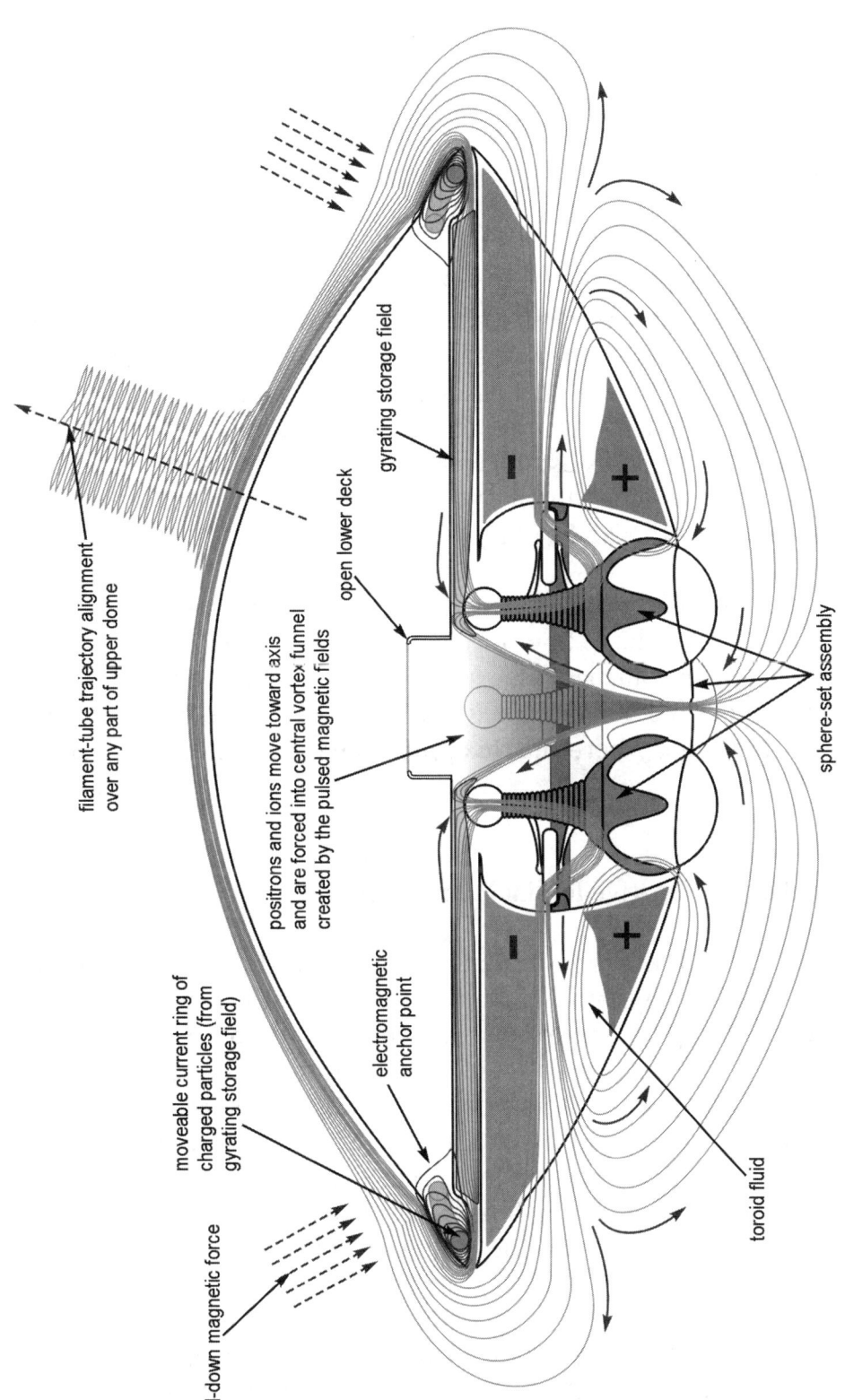

Figure 25. The UFO creates its magnetic anchor inside the craft at the rim

The anchoring region in space, of course, is simply a resistance to rotation (caused by a more powerful form of kinetic motion). And in space this will be achieved mainly where a large accumulation of electrical charges gathers above and below the black hole, and where the electrically charged particles come streaming up from the accretion disk to constantly feed into that accumulation. Then, where large quantities of those charged particles rotate around the magnetic flux lines around the axis of the black hole's rotation, and draw those flux lines in to that longitudinal axis to strengthen and to intensify that magnetic field, the resulting Lorentz force will cause the charged particles to be accelerated around the flux lines and to establish a torquing force in those rotating fields; and this torquing force will form the basis for the system's anchor. Obviously its not a physically-solid anchor but it will set up a holding force between the disk, the black hole, and the interstellar field lines, and the torque will provide a helical form to shape and direct the particles coming off the accretion disk with their surplus kinetic energies (of inertia and angular momentum). Once the jet is established subsequent emissions from the accretion disk will build up the jet's sheath and at the same time maintain the electromagnetic-pressure field at the base of the jet.

The anchor for the aaUFO's magnetic field appears to be located at the circumferential rim where an auxiliary flow of electric current orbits around the craft just inside the craft's upper dome rim (see section 9.6 for more on the aaUFO's anchor), although another possible magnetic field anchor point in the aaUFO might be at the core of its central vortex where the flux lines converge between the large lower spheres.

2.12 Williams and De Felice on Negative Radial Momentum

Somewhat different to all the above approaches regarding filament-tube formation is the Reva K. Williams hypothesis developed from both Roger Penrose's suggestion that it should be possible to extract energy out of a black hole mechanism, and from the Lense-Thirring effect of inertial frame-dragging. With the former development aimed at the black hole's production of high-energy electron-positron particles and hard photons; and the latter development which allows that angular momentum can be imbued into those charged particles, so that by combining these two functions together a flow of particles will ensue that will generate their own inertial frame dragging (or momentum frame-dragging) [note: 40].

One of the reasons why this hypothesis is so different is that it over-rides many of the problems associated with the Blandford-Znajek and Blandford-Payne theories (as referred to above) because, in these two early theories in order for a jet to be formed from the accretion disk's energy there needs to be a very strong magnetic field coupling between both the accretion disk and the black hole's event horizon, but as numerous subsequent studies have shown such a coupling is not certainly described as to how it exists in these two early theories [note: 41]. However, in the Williams hypothesis the gravitational energy-momentum extraction and ejection process occurs independently of any magnetic field, primarily because the rotating particles have developed in them a negative radial momentum force, which gives those rotating particles their own axonpetal (axis-seeking) collimating force (which would be somewhat akin to a self-inducing hoop stress), so that when they leave the system generating them they leave in a vortical trajectory. This astute observation of a rotating particle's intrinsic negative radial momentum force correlates with the negative energy orbital trajectories proposed in the [Leiter-Kafatos] and [Piran-Shaham] studies, and also correlates to the Umstadter CURL effect which sends particles on a curved trajectory (which latter effect is central to the [Maurer-Miller] beam system of craft propulsion [note: 42].

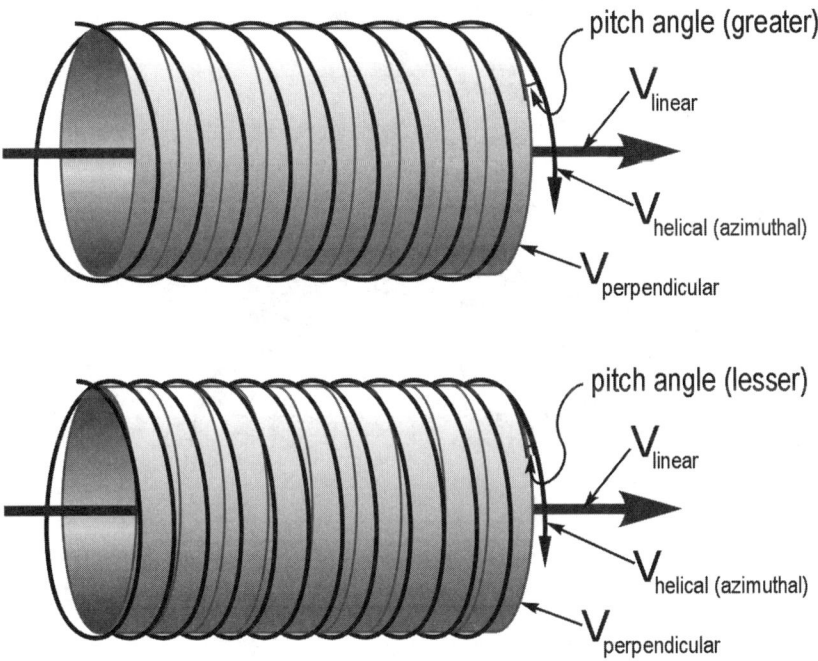

Figure 26. The pitch angle of a helical field

What the frame-dragging does is it establishes its own electric field between that rotating energy-sheath and a vast expanse of air surrounding the UFO. This effectively polarizes the atoms of the surrounding air, causing that expanse of air to become ionized. What the torque of the filament-tube then does is wrap layers, or sheets, of those ionized atoms around the outer wall of that rotating field - and this builds up electric charge in the sheath wall (much like a Meissner sheath effect) - and that's how the electric field gets established between the rotating tube and the surrounding air. This would be the reason for the air around UFOs appearing to be electrically charged, and the reason for magnetically-based instruments in nearby cars and planes going haywire, and for their combustion engines misfiring or stopping altogether - from the fact that the proper current-paths (which should be running through copper wires) in those electronic circuits are being dispersed randomly throughout the ionized air particles surrounding those electronic apparatus, because that ionized air will have just the same conductivity as the copper wires.

within Fibonacci (phi) curves - the leading edge and trailing edge of each wave packet would form a stretched wave-packet as the packet is ejected at the rim. But when they arrive at the apex those wave packets compress and intensify (chirp). At the same time they will trap and draw into the apex their magnetic field lines - so by Lorentz force the wave packets will curl around the apex and shoot off into a helical trajectory

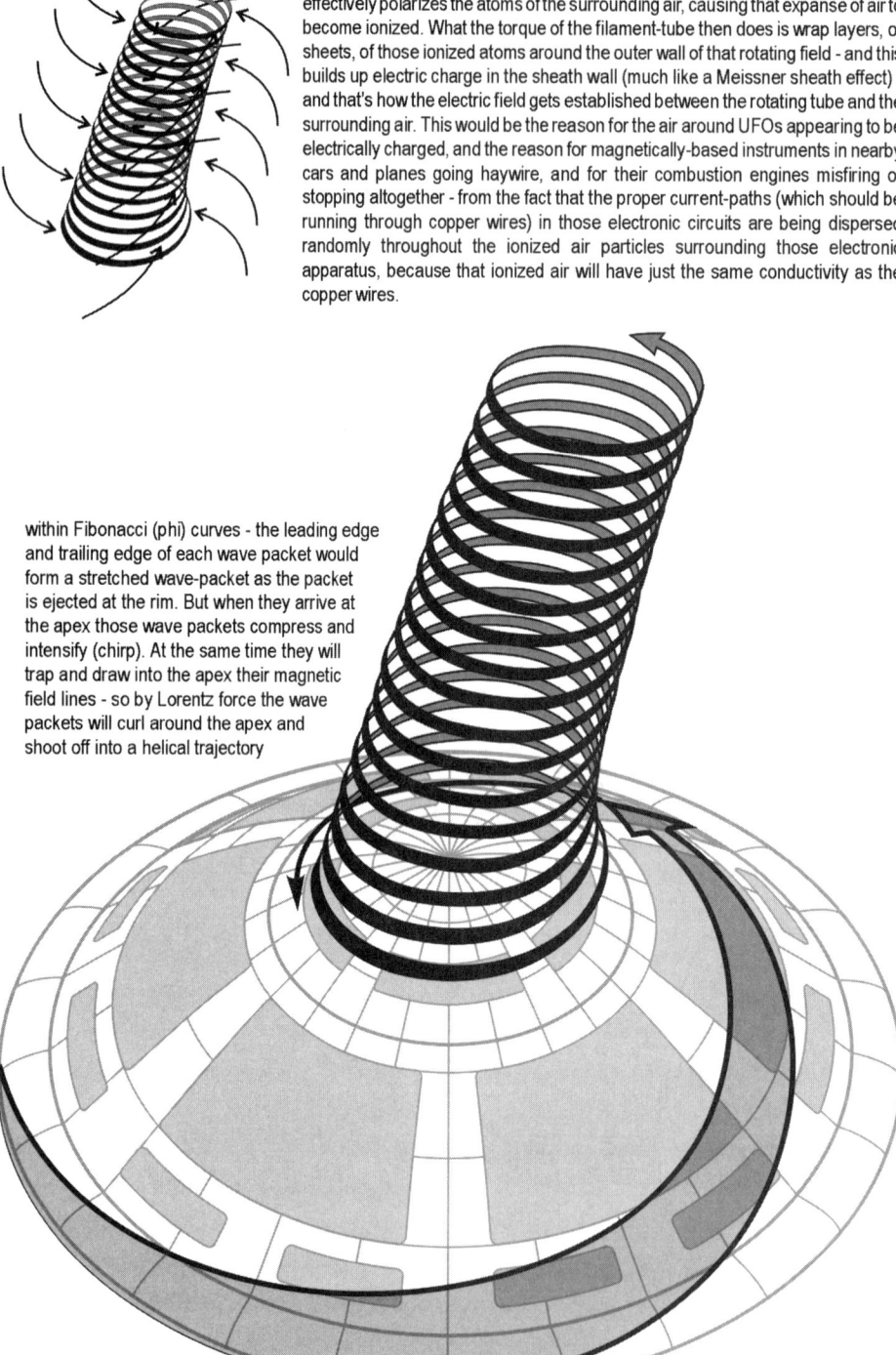

Figure 27. Ejected energy curls around craft's dome in a Fibonacci curve

So therefore in this Williams hypothesis the collimation force which can be used to develop a helical filament-tube is much more substantial and much more stable than in any of the previously mentioned filament-tube mechanisms, essentially from the factor of the space-time frame-dragging effect, or geodesicity, generating the negative radial momentum force intrinsically in each individual particle.

The greatest advantage of all will be that this negative radial momentum can be geometry-induced and therefore it will correspond precisely to a formulated set of geodesic parameters (which have been developed in association with the collimation formulae outlined in the De Felice-Carlotto astrophysical studies, for instance). And so, unlike all the hitherto mentioned MHD types of jet-forming mechanisms which rely on the establishment of a magnetic force alone, with this trajectory mechanism the particles can be given a strong coiling collimation force by setting the intrinsic helical pitch-angle of those rotating particles as they are ejected from, as in this case, the aaUFO's rim, so that when they form into a helical filament-tube field they will automatically self-collimate [note: 43].

And as it will be the helical pitch-angle that will provide the determining factor in the formation of that UFO craft's filament-tube then it becomes very relevant that this helical field should be developed as it leaves that craft around a surface which is domed. Therefore, it follows that when the aaUFO's gyrating energy field of gravitational force is ejected tangentially out of its circumferential duct and it curls over the craft's upper dome, after it then curls around to the apex of that dome then the very curve of that dome will set the pitch-angle and automatically form that ejected energy into a stable well-collimated filament-tube. Then, the propagated wave-packets and the helical filament-tube will be one and the same thing.

3: Holy Grail of Astrophysics: The Black Vortex

Contents:
3.1 Introduction to the Crystal-glass Spheres	83
3.2 Schoolroom Earth or the Universal School of Thought?	86
3.3 Oscillations of the aaUFO's Spheres of Crystal-glass	87
3.4 Plasma: The Electronic State of Matter	95
3.5 Ultra-violet Ionization Mechanisms	95
3.6 Exciton Excitations Around Dielectric Spheres	96
3.7 Electron Fusion and Gyration	97
3.8 Generation and Transfer of Charges from the Toroid Fluid	98
3.9 Formation of the Toroid's Magnetic Field Lines	101
3.10 Toroid's Magnetic Flux Lines Pulled into Shape	105
3.11 Breaking and Reconnection of Flux Lines (Reconnection Pt II)	108
3.12 Top Sphere Emissions Giving off Bars of Light	109
3.13 Gyration of a Cyclotronic Storing Field	112
3.14 Three Intensities of the Magnetic Field to Aid Emissions	115
3.15 Magnetic Funnel Creates a Black Vortex at the Heart of the aaUFO	115
3.16 Vacuous Force—Light Converts to Gravity	118
3.17 Clues Forthcoming in the UFO's Quasi-Black Hole Drive	119
3.18 On the Coherent Amplification from its Black Vortex	124

3.1 Introduction to the Crystal-glass Spheres

I should take this opportunity to remind the reader that as this is the first extensive study to be published showing what areas of universal scientific exploration and physics the extraterrestrials might have initially become involved in prior to their development of workable UFOs, and which has led them to formulate a quite specific series of working principles through which they could begin to construct their UFOs, that in order to do justice to that study I have endeavored to include wherever relevant as many references to existing scientific studies as possible. For this reason this work will proceed very thoughtfully, and my hope will be that its juxtaposition of ample references to existing studies published in well known physics journals and books will be of use to both the casual reader and the interested scientist.

Much of the previous chapter involved the principles of astrophysical dynamics and that realm, of astrophysics rather than mere physics alone, will also feature in some of the following chapters to show where the extraterrestrials have sourced many of their mechanisms from. I fully expect then that even the most scientifically orientated readers of this study will not be wholly familiar with the dynamic principles that have been ascribed to UFOs by those who have witnessed them (and to illustrate this valuable point I was once asked by an ex-NASA engineer how UFOs fly through the sky by traversing what he described seeing as a beam of light extending ahead of that craft, he made enquiries about my work in this field because when he asked his fellow engineers to come up with a technological explanation they told him they hadn't a clue! And nor should they be expected to know, for rocket science works by entirely different principles to those of black holes and quasar jets—which share some of their dynamic principles with UFOs. Incidentally, that technology used by a beam-drive UFO will be understandable to the reader nearer the end of this book). Right now though, I ask the reader for his or her patience and not to expect a UFO's drive system to come flying out toward them from just one or two paragraphs of this study. I will though try and make this study interesting and as I said in my introduction I will not introduce on the unsuspecting reader reams and reams of mathematical formulae and make it more complicated than it needs to be, and instead of the mathematics I will supply diagrams to make it more presentable and understandable. For that's essentially what a breakdown of a UFO's energy dynamics should be—perfectly presentable and understandable —there is no point in treating it like rocket science!

One good place to start with regard to the domed craft that have been observed flying through the skies around the world might be the large spherical orbs that have been observed protruding beyond the hull beneath these craft and which usually feature as a group of four spheres around the center of the craft's lower hull sections. That these quartz crystal or glass spheres are so prevalent in the different designs of what I would term conventional saucer-disc UFOs it suggests that there might be one or two areas of physics that hitherto we on earth have not yet become fully conversant with [note: 1]. Or, it may just be that these spheres comprise a multiplicity of principles which are known about in separate form but which have been brought together and utilized as a singular co-operational ensemble in a hitherto unfamiliar way. Either way, I am confident that such technology when inspected fully may prove to offer this world some extremely useful—and new—energy conversion principles.

At first sight they look rather innate and cumbersome and it must be said that to most people's way of thinking the inclusion of these spheres in an aircraft is completely illogical... four huge one-meter-plus diameter spheres made of heavy glass or quartz being lugged around at the bottom of a UFO, which is essentially an aircraft that should be constructed out of low-weight materials so as to help it defy gravity... Some researchers have ignored them for fear that their wild speculations about how they might function within what sort of mechanism inside a UFO would attract equally wild derision and laughter from those acquainted with the earthly sciences. Quartz itself has been ascribed some pretty wacky attributes by those who have jumped to conclusions about it without properly researching their claims [note: 2]. And so, quartz crystal has seen much ridicule from scientists, particularly since the 1920s when it was jokingly claimed in one scientific journal that quartz would enlarge to four times its size when an electric field was applied to it!

As it will be shown in this chapter (and through the next two chapters) that these spheres are rather complicated acousto-opto-magneto-electro-gravito structures, I would like to point out that while I have discovered most of the functions they perform and can point to certain qualities that might explain how they can be constructed (and with what materials) I should also point out that ET engineering, so I have also discovered, is not quite as straight-forward as the electro-mechanical engineering norm of this planet—in as much as the ET designer's brief favors multi-purposing and multi-functionality. Meaning that while I have found a number of mechanisms operating in these spheres even more functions may be discovered in them later through the course of this on-going research into UFOs.

Obviously, whatever else is discovered about them will alter the components and manufacturing processes involved in engineering and fabricating these spheres. In this respect this is why, in these early chapters, I will refer to them as being made of either glass or quartz or crystal until all their parameters are fully viewed and recognized. So rather than offering the reader a full explanation of them in this one chapter these spheres will shed their secrets through several of the chapters of this book, and once again I appeal to the readers patience in their studies of UFO energy systems, because the only way to fully unravel their complexity is to place these spheres into a surrounding engine, or rather into component mechanisms of that engine, which lays at the heart of a UFO and which needs to exist in order for the craft to develop the necessary energy-fields and power mechanisms that would both generate energy and propel that energy through our atmosphere, and through our part of space.

3.2 Schoolroom Earth or The Universal School of Thought?

One thing the extraterrestrials have done is to recognize what energies are available in space and they seem to have learned one or two tricks about how to collect ambient energies potentially existing in the vacuum of space, and they have learned how to utilize the zero-point fluctuation (ZPF) fields, virtual energy fields, the Higgs-Planck fields [note: 3]. They also seem to know how to manipulate space-time geometry in such a way as to convert that potential energy into useful energies for their craft's propulsion. So ideally, this study needs to look at space from the viewpoint of those who have had to survive in space, or who have learned from its mechanisms how they might subsist in space for long periods of time. Not like our scientists who have had but a few hours now and then to study space as relayed from the innards of small cramped blast vehicles. These ET guys have probably had to exist in space for many years at a time, and have no doubt traveled a wide swathe of it to see how galaxies work, how planetary systems evolve. And probably they have done so since before we on this planet fired our first arrowhead into our first meal. So these ETs have likely been privy to many more insights than us about what space can and can't do and what space energy is all about.

Indeed, having a whole universe to learn from rather than learning from just the achievements of one isolated blue planet, it could rightly be conjectured that the ET physicists may have picked out the very best features of the black holes, quasars, pulsar, and other cosmic systems observable throughout space, and that those ETs have worked out from these non-terrestrial energy dynamics a quite different approach to both producing energy and amplifying energy.

While the comparison between the Andreasson Affair's aaUFO drive system and that of a black hole has already been mentioned tentatively in the previous chapter such a comparison will be gone into in even greater detail the more this study progresses. For, in examining the very mechanisms of the black hole it would appear quite possible that the ETs have not only found the ultimate transmutation processes to use in their UFOs (wherein the photon is exchanged for gravitational force)—but that the ETs have also improved upon that cosmic model. And so, with such a possibility in mind, this study will feature as its central thread the duplication and miniaturization of that most powerful and efficient mechanism of the whole energy-filled universe as the primary engine from which the UFO drive system will be derived...

3.3 Oscillations of the aaUFO's Spheres of Crystal-glass

Anyway, back to the aaUFO spheres. Glass and quartz crystal are dielectric compounds and are semiconductors, this is already known, and this means they can be used as part of an electrical circuit to conduct, to insulate, or be made to oscillate at a resonant frequency or at a number of much higher-octave harmonic frequencies. They can also transduce energy from one type of frequency into another (ie electrical to mechanical and through acoustic waves into phonons). A crystal structure is after all a form of macro-molecule, and as such has an arrangement of atoms, positive holes and negative charge-carriers (electrons) that will perform certain electrochemical and electromechanical functions depending on what precise structuring or composition it is given—particularly if that basic glass or crystal structure has other electro-chemical compounds donated into it during its melt process, or significantly, if that glass is fabricated by the newly developed Sol-Gel process [note: 4]. Interestingly, a detailed inspection of this latter technology will show that the Sol-Gel process would be an excellent way to 'design' the large lower spheres (and even the smaller upper spheres) for the UFO featured in this study, and so this will be further expanded upon throughout the following two chapters.

But the essential quality of the glass or quartz crystal specific to this study is that it has innate within its structure effects which relate to a non-zero renormalizing electrical polarization which occurs above certain frequencies through a particular type of oscillation which can be induced within its lattice structure. In other words, a hitherto un-exploited one-direction polarization effect. With this factor in mind it was interesting that when physicist Kip Thorne was asked to give his scientific viewpoint on the wormhole concept that would feature so prominently in Carl Sagan's *Contact* novel (and movie), that Thorne in his 1988 paper should pick up on a discovery made by one of that story's characters, Eda, whom Carl Sagan had allowed to discover and exploit an energy field which developed an anisotropic stress (a stress that occurs only in one direction so there will result a potential-difference that can be extracted and suitably utilized). Nowadays there are many such effects which can be found in a range of newly discovered materials (and energy fields) which have the potential to one day yield exploitable power; in left-handed neutrinos, in negative-refraction materials (such as photonic crystals), and even in some not so recent discoveries made about dielectrics. Regarding this latter type of effect one of the most intriguing intrinsic properties of a dielectric is that of dielectric absorption for it means that when charges are oscillated through such materials (for instance, resin or perspex) a time-variable differential can exist between the positive and negative charge mobilities when oscillating electrical charges are passed though them [note: 5]. This differential then affects how that dielectric becomes charged within its molecular structure (and how it 'soaks up' one polarity of the oscillating charge more than the other above a critical frequency); and subsequently, how the ambient air surrounding that dielectric also becomes charged. And this anisotropic potential effect in and around a dielectric material can, for instance if it is a glass or a crystal material, be something that can be taken advantage of whenever electrical, magnetic or gravitational forces might be applied to them.

Quartz research nowadays, as we move through the twenty-first century, is thankfully much more respectable than the example I cited above for the 1920's. For instance, in 2000 R.M.L. Baker Jr launched his bid (with US patents 6,417,597granted Jul 9 2002; and 6,784,591 granted Aug 31 2004) to develop several spherical and non-spherical converters of ultra-high frequency gravitational waves (in the Quadrahertz or 10^{15} Hz region), using natural quartz crystal to transform that electromagnetic wave energy into continuous high-voltage power outputs from the Zero-Point Fluctuations (ZPF) in space. Another example of quartz being used in this way can be found in the work project of F.B. Mead and J. Nachamkin whom have jointly delineated a ZPF electromagnetic wave transducer which collects and converts the extremely high frequency incident energy of the homogenous and isotropic ZPF radiation of space, and which through the technology of providing two slightly dissimilar spherical antennas, made of quartz, generates a beat frequency between those two spheres to receive these ZPF energies [note: 6]. In converting one small bandwidth of the extremely high frequency ZPF into a much LOWER frequency of beats those ambient energies omnipresent throughout space can, in theory at least, be collected and subsequently transferred by conventional electronic circuitry, into electrical machines and put to good use.

In both these aforementioned areas of research the emphasis is upon the use of or the development of spherical acoustic waves that interact between the crystal spheres and the omnipresent ambient energies at a harmonic frequency.

The essential beauty of acoustic waves being that they can transport energy, as phonons, through a semiconductor or dielectric material (by the acousto-electric effect), and then deliver that energy out from the 'other end' of that same material by dragging freshly generated photons and electrons (as excitons) into a suitable receiver circuit. Surface acoustic waves (SAWs) while carrying electrons and excitons can, for instance, when suitably amplified, be coupled to microwave-frequency waves to form into wavepackets of energy, and then those wavepackets can be transported through the semiconductor and beyond it by the agent of a stationary magnetic field which curls those electric charges into a helical electromagnetic beam. Similar to this latter type of coupling is the principle behind the curling of electrons through a free electron laser (FEL), or an electron cyclotron maser (ECM), or indeed that of the Graser which activates a curling beam of gravitational radiation in conjunction with a stationary magnetic field. These sorts of energy transferring processes will present themselves in the aaUFO, through the craft's telemeter wheels, for instance (as will be further alluded to in chapters 4, 5, and 6).

With regard to extraterrestrial technology it will be my suggestion throughout this book, that the UFOs central to the *Andreasson Affair* series of books also utilize a number of electro-dynamic arrangements around the dielectric spheres they use, by which a similar conversion is made from the so-called zero-point fluctuations (ZPF), to draw useable energy out of those ultra-high frequency energy fields [note: 7]. These arrangements will form one of many dynamic processes to be brought into action around these sphere-sets in what will be seen as a whole gamut of functions that seem to be operating in concert with each other, and which seem to have been designed to establish acoustic wave fields that will be pervading throughout most of the central part of this UFO.

So the ZPF fluctuations and the energies they bring in are very important to this study because they will eventually be seen as an outside field which coalesces with several other energy mechanisms developed within the inside of this aaUFO [note: 8].

What lends credence to this assumption that UFOs are designed to draw in ZPF derived energy from outside for the craft to utilize in its drive system, and I shall use the aaUFO as a prime example, is that the large lower spheres of the aaUFO are specifically located so that they protrude outside of its metallic upper and lower domes. These lower spheres must therefore be positioned so as to have direct interaction between both the surrounding air (or incident electromagnetic radiation) and the internal electro-dynamic circuit of the craft.

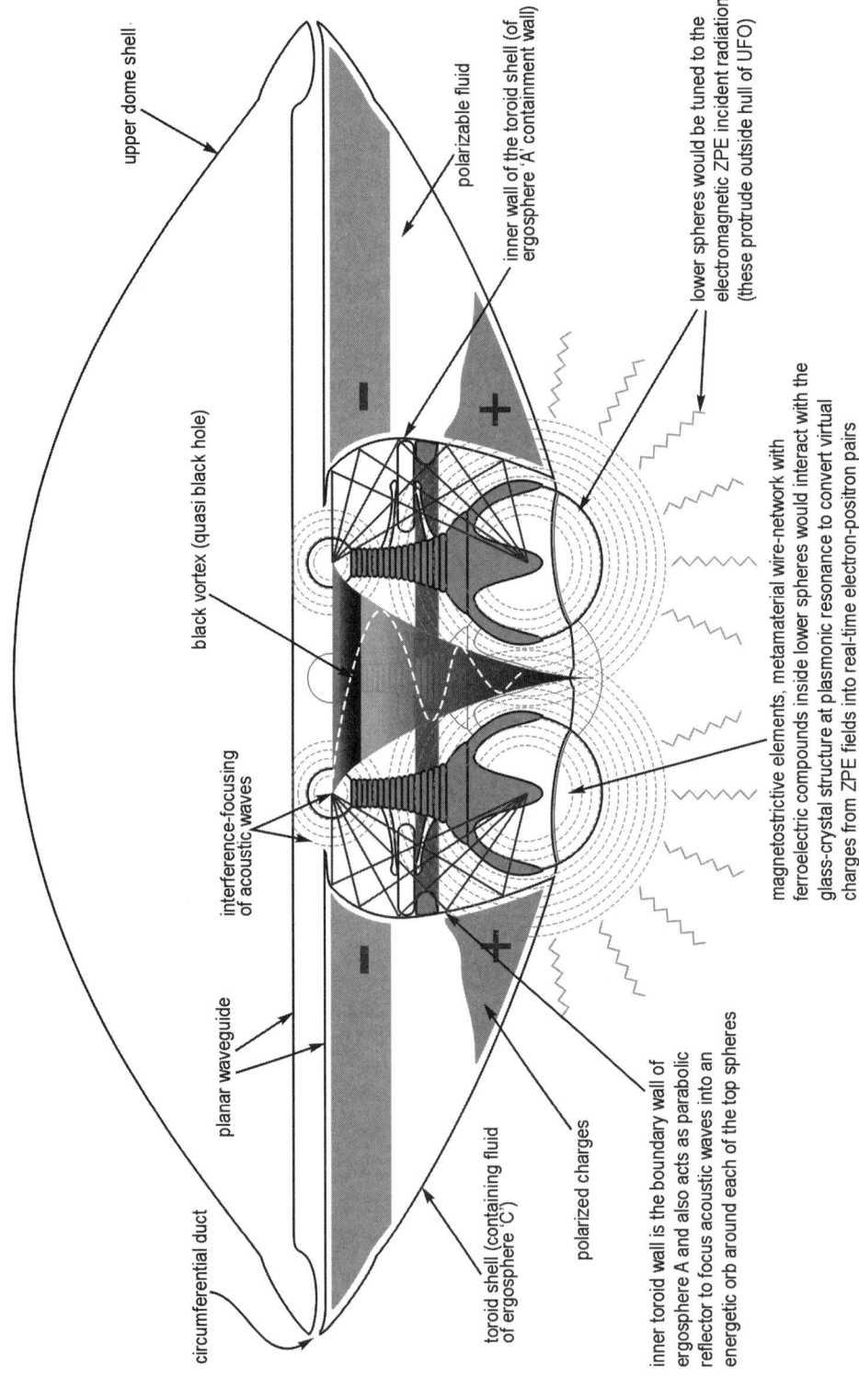

Figure 28. UFO harvests energy from the ZPF field through the acoustic waves around the lower spheres

But rather than allow the ZPF derived inductions to hog all the limelight here I will break off from this fascinating subject and leave it for the later chapters of this study, where its relevance within other UFO energy conversion processes will better fall into place. However, concerning the lower spheres, I would like to point out that these inductions should be expected considering that the craft is predominantly circular and the energy fields which it generates are also circular, and so there will always be an inhomogeneous nature to these fields, and magnetic potential-differences, while they permeate through the lower spheres.

For instance, the four large spheres will have two high intensity non-uniform (inhomogeneous) pulsing magnetic fields that converge at opposite quadrants of each of those spheres; and while each of these large spheres simultaneously revolves about its own axis, and while all four jointly rotate about the central axis of the craft, they should be seen, at the very least, as a configuration of four electrical-charge conductors that are revolving through two separate pulsing magnetic fields. Certainly this latter arrangement alone can be used to initiate piezoelectric oscillations throughout the lower spheres and to create around them a very energetic network of acoustic waves.

So, if a harmonic resonance mechanism can then be engineered between the large lower spheres and the top small spheres with these acoustic waves, then this would be very beneficial for providing several methods by which energies of various kinds could pass from the lower spheres to the upper ones. Energies could pass either through the metal stem-casings (which join both spheres), or they could pass through the tubes or rods that form the coils which connect between the two spheres. But just as conveniently these acoustic waves could oscillate through the open air, by spreading outward toward the concave paraboloid inner wall of the toroid, to be reflected off that wall and to bounce upward into the small top spheres, I say conveniently because by bouncing them off the paraboloid wall, with the wall acting as a reflector, the wall would then focus those acoustic waves, and also transport in them any electric charges accumulating around the toroid wall. Therefore, through these three different routes the top spheres would continually be supplied by, or surrounded by, an oscillating (acoustic) energy field.

This quickly brings us back to the strange appearance of extreme energy excitation seen and described by Betty Luca in the center of a UFO craft she was shown by a group of Greys that specifically wanted her to see how their UFO generated its power (see figure 17 above).

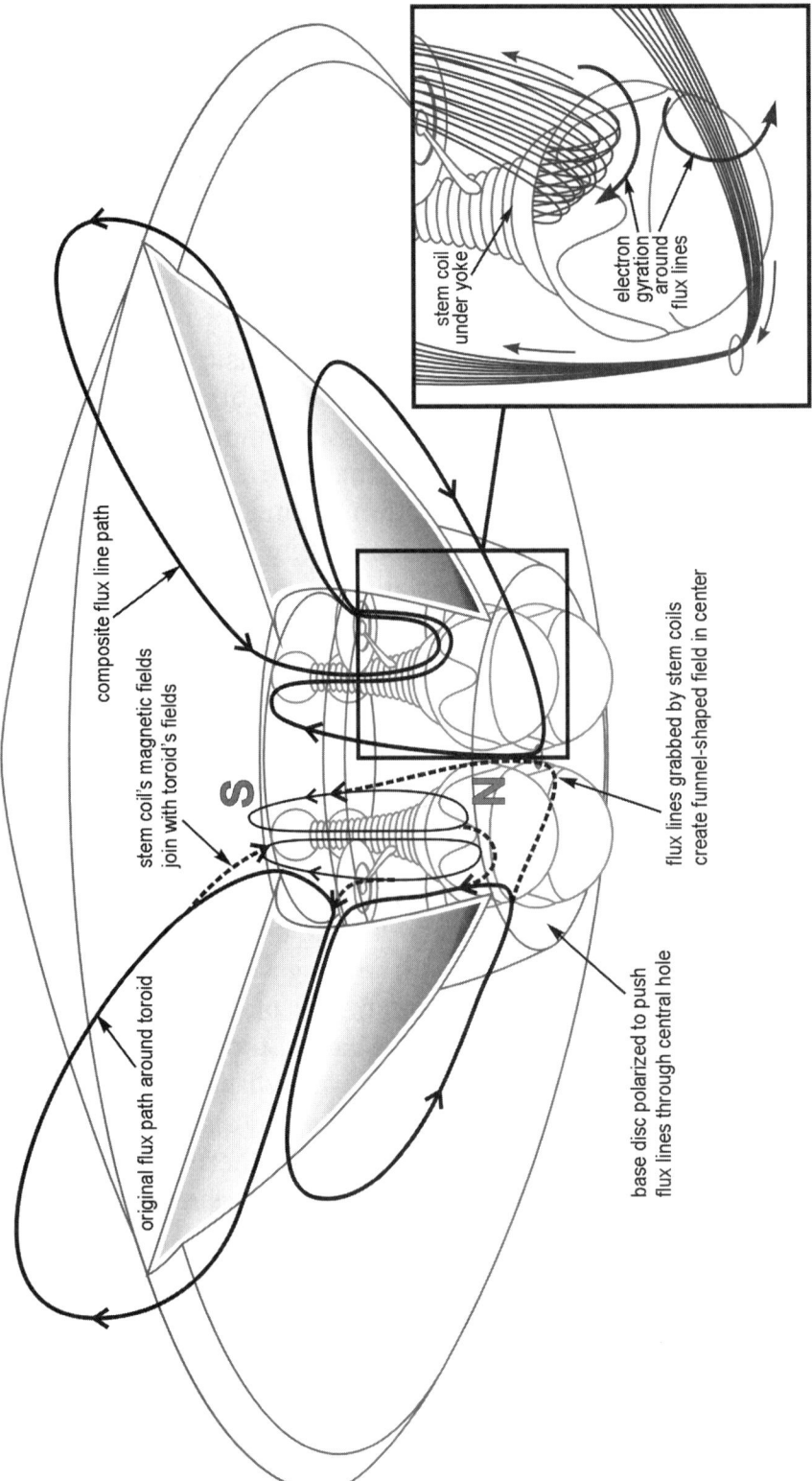

Figure 29. Re-routing of the upper and lower magnetic fields of the toroid forms two intensities around the lower spheres

These are the bright energy bars which hovered around the top spheres and were so bright they shone like an arc welders torch [note: 9]. I've seen a welder's gun and I've used a welder's gun, and I can testify that the blue-white light is so intense that darkened goggles have to be worn otherwise it would damage the retina of the eye [note: 10]). Such a high order of luminescence of photon radiation can be an indication of a number of different effects, of extreme exciton recombination (where the recombination of the exciton's charge-carrier and charge-vacancy leads to the exciton's annihilation and generation of photons), or it could indicate abrupt changes from the excitation states of charged particles, it could be from acute gyro-rotation of the particle's trajectory leading to emissions by curvature radiation, or it could even come from particle fusion.

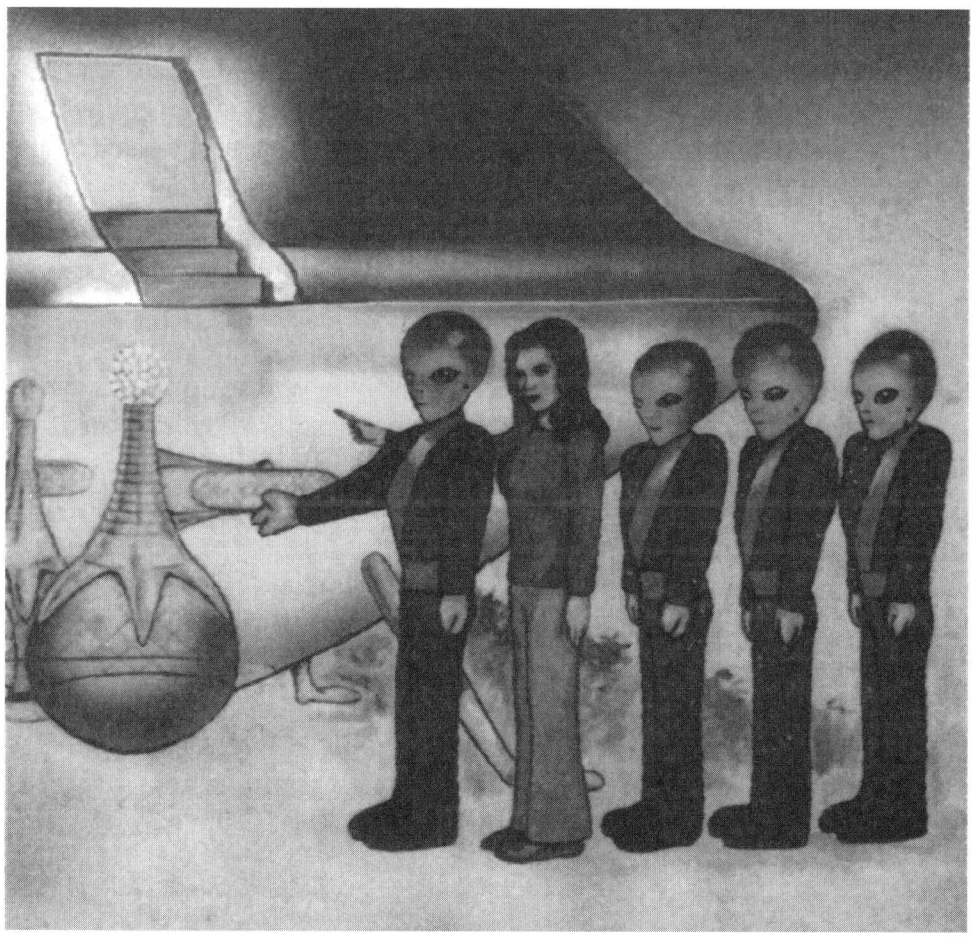

Figure 30. The leading Grey switches the outer shell to transparent mode to show Betty the sphere-sets working inside that UFO (from "Andreasson Affair—Phase Two" front cover)

But to understand how such solid arcs of intense energy can occur firstly let us look at a breakdown of the various elements shown in these energy bars: Firstly, within the principles of static-electric fields there is the isotropic appearance of electric charge field-lines around a sphere, and they follow the well known phenomenon of field-lines connecting to spherical objects when charged electrostatically; such lines indicate that the electrostatic field forces have aligned themselves perpendicular to the surface of the sphere and are non-conducting.

Second, is a phenomenon that might be particularly relevant here, and which has the potential to develop an anisotropic stress in an appropriately designed amplification system, is one that can again be found within the fundamental laws of electrostatics which says that charge density will be inversely proportional to radius of curvature, and which denotes that when two electrically charged spheres of dissimilar volume are interconnected, even though their potentials may be the same, their electrostatic densities (intensities of electric charge) will always be greater for the sphere of smaller curvature: This comes from Coulomb's electrical force law $F = KQ_1 Q_2 / R^2$ (and which was put to good use by Nikola Tesla in his atmospheric charge collector patent US1266175). Which effectively means that most of the electric charge collected or converted at the large lower spheres will automatically drift up to, and accumulate around, the smaller top spheres (because they are inter-connected).

And of course, by running a coil between them inter-connecting them electromagnetically any oscillation established between them will automatically be 180 degrees out of phase. Meaning that the two ends of the coil will be oppositely polarized, and both ends will behave as if harmoniously linked and will be perfectly primed to transfer energy from one to the other (through an alternating current flow). This phenomenon will be important to explain how the upper spheres could be continually replenished with abundant supplies of charged particles over the course of, say, a UFO's journey, from them being continually supplied by the lower spheres (because of the ZPF energies converted through them from the outside media).

In this wise the four glass-crystal sphere-sets should always be seen as structures that *convert* rather than *store* electrical charges (and other forms of energies which will become apparent in later chapters) [note: 11].

3.4 Plasma: The Electronic State of Matter

Obviously, to better understand and duplicate the energy intensities exhibited by these upper and lower spheres the above-mentioned energy fields around them must also be accompanied by a vast amount of charged particles. These charged particles would have to come in great abundance through mechanisms in operation close to the center of the aaUFO's power-drive system [note: 12]. The ones most evident of which are;

- Ultra-violet ionization and photon emission
- Exciton (electron-hole) excitation and photon emission
- Electron excitation and electron-fusion with photon emission
- Ceramic compound excitation and electron emission
- Electron-positron generation from magnetic flux reconnection
- Electron-positron production through the virtual energy fields in space
- Electron-positron production through curvature radiation (rotating fields)
- Comptonized ionization from Penrose pair production
- Graviton-to-photon conversion with electron-positron generation
- Microwave ionization of fluids to induce diffuse charges outside toroid

3.5 Ultra-violet Ionization Mechanisms

Ultra-violet (UV) ionization is a powerful way of supplying abundant amounts of electrons, by the photoelectric effect, and is a system which can readily be initiated within most areas of a UFO craft. Indeed, in the many accounts given by Betty Luca, and throughout the literature reporting abduction phenomena, mention is often made of a strange pink-to-violet illumination operating inside a UFO. So, while at first it would seem that UV ionization will be in the vicinity of UFO drive systems merely to illuminate, it might also exist specifically to contribute to the necessary charged particle environment (perhaps in a way to prime the ambient space into a metastable state in readiness for it to break down and to atomize when triggered electrically). If a metastable state can be maintained then the atoms in the air will themselves become part of the process to provide abundant supplies of charged particles (from the collisions of fast moving particles flowing through them initiating electron cascades or avalanches) [note: 13]. As well as the numerous mentions of this UV illumination throughout the Andreasson literature there are many other clues to be found in the aspects of UFOs, recorded elsewhere in Ufology documentation, to substantiate this point (that the various wavelengths of ultra-violet radiation permeate around a UFO). For instance, the observation that UFO shells are perfectly shiny or have perfectly clean surfaces, is an associable characteristic of UV radiation, from the fact that UV irradiation is a cleaner of metal surfaces (its a method we also use on earth to clean our metal surfaces too). And in some UFO reports where abductees have reported radiation burns on their skin, this radiation burning is a direct result of being exposed to short wavelengths UV frequencies (although, in chapter 11 it will be explained that some of this UV may result from an up-shifting of visible light's frequency).

3.6 Exciton Excitations Around Dielectric Spheres

Another supply of energy inside the aaUFO can come from the production, particularly through dielectric materials, of exciton molecules. Excitons are couplings of charge carriers (electrons) and holes (positive charge vacancies) and are especially advantageous because whereas electrons only exist for a few billionths of a second the exciton couplings have much longer lifetimes.

In recent studies throughout the US, Japan, China, and Russia that have concentrated on the phenomenon of exciton generation and transportation within dielectric crystals and certain silicate glasses, it is suggested that abundant electron and ion production can come through semiconductor crystal materials if their micro-structures are more fully understood and are manipulated on a quantum level (vis. through quantum electrodynamics). These studies have also explained that transportation of electrons can come through the movement of exciton molecules (of excitons coupled with free electrons) moving over crystal surfaces, or just under their surfaces, by them being moved through the agent of acoustic waves, and from the excitons dragging charged particles from within those crystals so that on the outside of them there can form dense plasma clouds. And similarly, from the studies of exciton-polariton generation there can come two lines of theory of relevance here, one that the exciton can be coupled to an electromagnetic wave-particle (as a photon), and the other that the exciton can be coupled to a resonance effect that can occur within an oscillating dielectric substance, meaning that electric charges, when the crystal or glass substance is under acoustic wave or electromagnetic perturbation, can either be transported through that dielectric material, or those electrical charges can form into energetic electric fields fully outside that crystal substance (with the aid of an localized pulsing magnetic field) [note: 14].

Coupled together with the above ionizing properties of semiconductors should also be the process of magnetostriction whereupon such material elements can be used in conjunction with the crystal (or glass) to energize that crystal material upon the application of an energetic magnetic field. Magnetostrictive rods, for instance, inserted into or molded into the crystal's structure would be a good candidate to trigger high-frequency effects within that crystal which would then yield fairly strong charge transferring through piezoelectric and associated mechanisms; and if the air surrounding them is sufficiently primed into a metastable state (as mentioned above) these oscillating effects could lead to their surrounding plasma clouds affecting electron-ion collisions and subsequently electron avalanches around these spheres (most especially around the top spheres) [note: 15].

3.7 Electron Fusion and Gyration

In this area I have been assisted by one particular physicist whom has earnestly addressed the observed factors of UFO propulsion, and has formulated a number of calculations to (a) determine the magnetic flux density and force operating within a UFO-type environment; and (b) to determine the charged particle and magnetic field interactions extending from the craft into the surrounding air (with calculations based upon astrophysical data); and (c) has offered an intriguing theory about rotating magnetic forces acting on charged particles and diamagnetic materials (which, the author has recognized, is particularly transferable to the shield effect utilized in the Bob Lazar UFO, the Onion drive, and the Travis Walton UFO); and (d) has offered the necessary calculations to determine if in a UFO drive system there are sufficient forces generated to initiate electron fusion, proton fusion, and even the transformation of protons into neutrons. The following extract from "The Universal Energies" by Mahmoud E. Yousif (of the University of Nairobi) is included here to give a few insights into the phenomenon of electron fusion in relation to UFO energy systems, which may shed some light on the bars of extreme brightness emanating from these top spheres inside the aaUFO [note: 16].

"In a system where captured charged particles are abundant and energization given by Eq.{16} is continual, orbital charged particles are denoted by n_o, orbits number in one meter along the lines of force is denoted by O_n, therefore the total number of gyrating charged particles in volume of magnetic lines of force is given by

$$N_v = 10^8 \, n_o \, O_n \, l \, B_I \qquad \{19\}$$

{equ.19}

Where, N_v is the number of charged particles gyrating in a specific volume of magnetic lines of force."

"As shown in fig.4, intense B_{EI} given by Eq.{10} decreases radius of gyration, given by Eqs.{12 and 18}, the circumference, and adjacent distances (r_r) between orbital electrons shown in fig.2a, reduced from (a) to (c), therefore production of external magnetic field (ExMF) is at its maximum, substituting Eq.{19} with n_m l in Eq.{10}, hence

$$B_{EE} = 10^8 \left(\gamma_{PS} B_p^2 + \frac{\gamma_{PS} n_o O_n lq^3 B_p^3}{m_e^2 v_c c} \right) T \quad \{20\}$$

{equ.20}

Where, B_{EE} is the maximum ExMF produced by electrons."

"The ExMF needed to give required r_r for Electron-Electron interaction as shown in fig.2, is given by

$$B_E = \frac{\pi m_e v_c}{2 n_o q r_e r_r} T \quad \{25\}$$

{equ.25}

Therefore, distance r_r between adjacent electrons is reduced to Fermi range (10^{-15}), thus enhancing interaction of opposite spinning magnetic fields (SMF), therefore, producing electrons-spinning magnetic force, leading to the electrons fusion."

3.8 Generation and Transfer of Charges from the Toroid Fluid

Solid materials will not be the only sources of charged particles, fluid flowing inside of the toroid shell will also supply electrical charges, and for this reason the toroid has been especially designed to provide three sharp edges to its circular shape. To understand how charges can be transported from the toroid's fluid to the top spheres it should also be borne in mind that acoustic waves will be flowing out of the lower spheres, and while flowing toward the top spheres directly they will also be propagating through the toroid's fluid and mobilizing charges in that dielectric fluid. The acoustic waves propagating through the air and rebounding off the inner wall of the toroid will mobilize and carry with them, on their way to the top spheres, a very large field of positive charges that will continually be replenished on the OUTSIDE of the inner-most point-edge of the toroid (from the copious supplies of negative charges that will accumulate there on the INSIDE of the toroid—see figure 31).

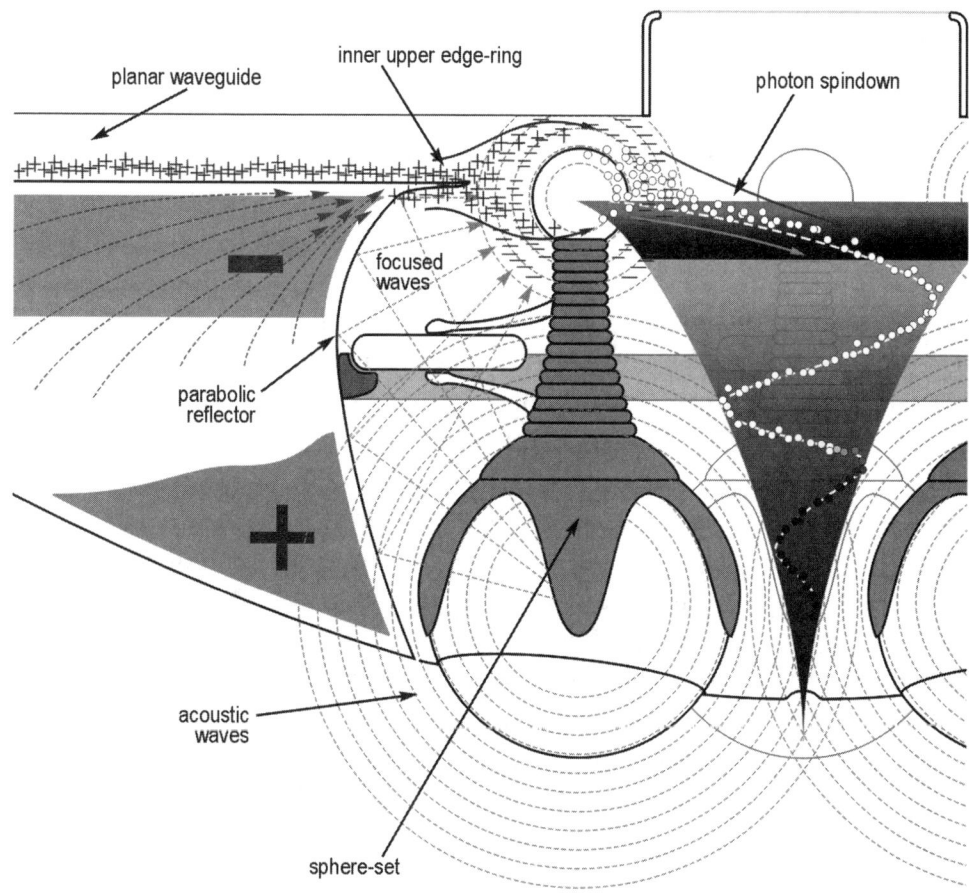

Figure 31. Some of the mechanisms which direct energy toward the top spheres, note the acoustic waves bouncing off the inner wall of the toroid

Both of the upper edges in fact, the outer-most and inner-most point-edge of the upper half of the toroid, seem to have been especially designed by the extraterrestrials so that these two edges will continually be recharged electrically from the inside, as long as there is movement of the polarized fluid which flows inside that toroid shell (this will be gone into in more detail through chapter 6). This will be because, in a polarized fluid of positive and negative charges, where the negative molecules are lighter (less massive) than the positive molecules then the negative molecules will be influenced to a lesser extent by any gravitational forces than those positive ones, and so they will accumulate toward the top of the toroid fluid. And the positive molecules being more massive will more readily accumulate at the toroid's lower inner edge.

The main principle of the toroid fluid being that whatever polarity charges are generated and accumulated on the inside surfaces and edges of the toroid shell, these will be matched by the generation of electric charges of the opposite polarity on the outside surfaces of the toroid shell's walls and faces. This layer of induced charges is known as the diffuse layer and the greatest accumulations of those diffuse ions will be at the sharpest edges of the top half of the toroid (the bottom edge of the toroid will have on the outside a diffuse layer of negative charges). This is a well known by-product of fluids contained in insulated containers, and while electrostatic build-ups do tend to occur and to clog these charges in stationary or slow-moving fluids (because of like-charge-repulsion mechanisms), that problem will be eradicated because this whole fluid system will be constantly circulating around the toroid shell and, as will be detailed in the following chapters, while it circulates it will also be agitated by having microwave pulse beams fired into it.

Therefore, the toroid then becomes another example of the ET's multi-functioning whereby a fluid container actually forms an integral part of the craft's electro-dynamic power circuit. For the toroid is designed to be what is essentially a three-edged shell or three-arced toroid, specifically to mobilize electric charges (rather than it being constructed as a one-piece tube looking like a donut that has a circular cross-section with no charge-transferring edges around it at all). And, once again, this toroid will continually generate an abundance of electrostatic charges within its interior fluid-to-surface and laminae layering (sheath-to-bulk) interfaces [note: 17].

This arrangement between the inner-most edge of the toroid casing and the top spheres could be likened to a cathode-anode arrangement, or a spark gap discharge. But how these discharges interact with the four top spheres which orbit around the craft's central vortex, and how the toroid's fluid will be agitated, and what that fluid is, will be better understood over the next few chapters after certain other parts of the aaUFO are detailed and considered first.

3.9 Formation of the Toroid's Magnetic Field Lines

While the nine or so major candidates for abundant charged particle supply have been mentioned above, these charged particles must all be energized with kinetic energies. And in the aaUFO this is done very purposefully by intensifying, confining, and constantly pulsing its magnetic field flux lines so as to pump up the Lorentz force on those charged particles to a high state of excitation (the Lorentz force is basically where the electrical current will move perpendicularly, at 90 degrees, to the flow direction of the magnetic flux lines.

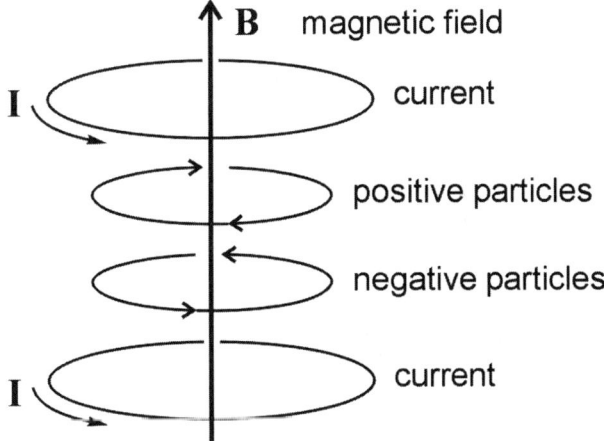

From *Cosmical Electrodynamics* by H. Alfvén & C. Falthammer (1963) p20

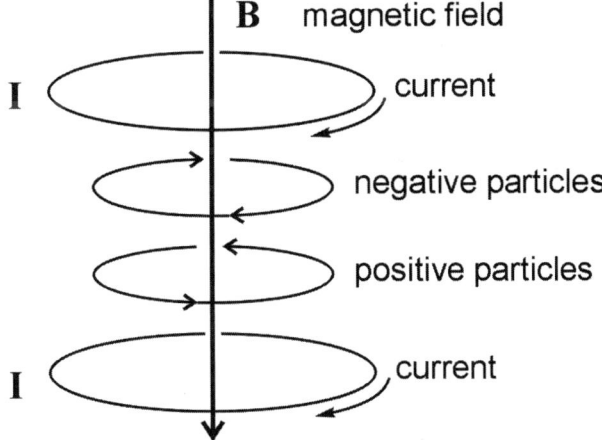

Figure 32. Charged particles will move in certain directions under the influence of fluctuating magnetic field lines. Standard throughout this study will be the circular motions depicted in these charts

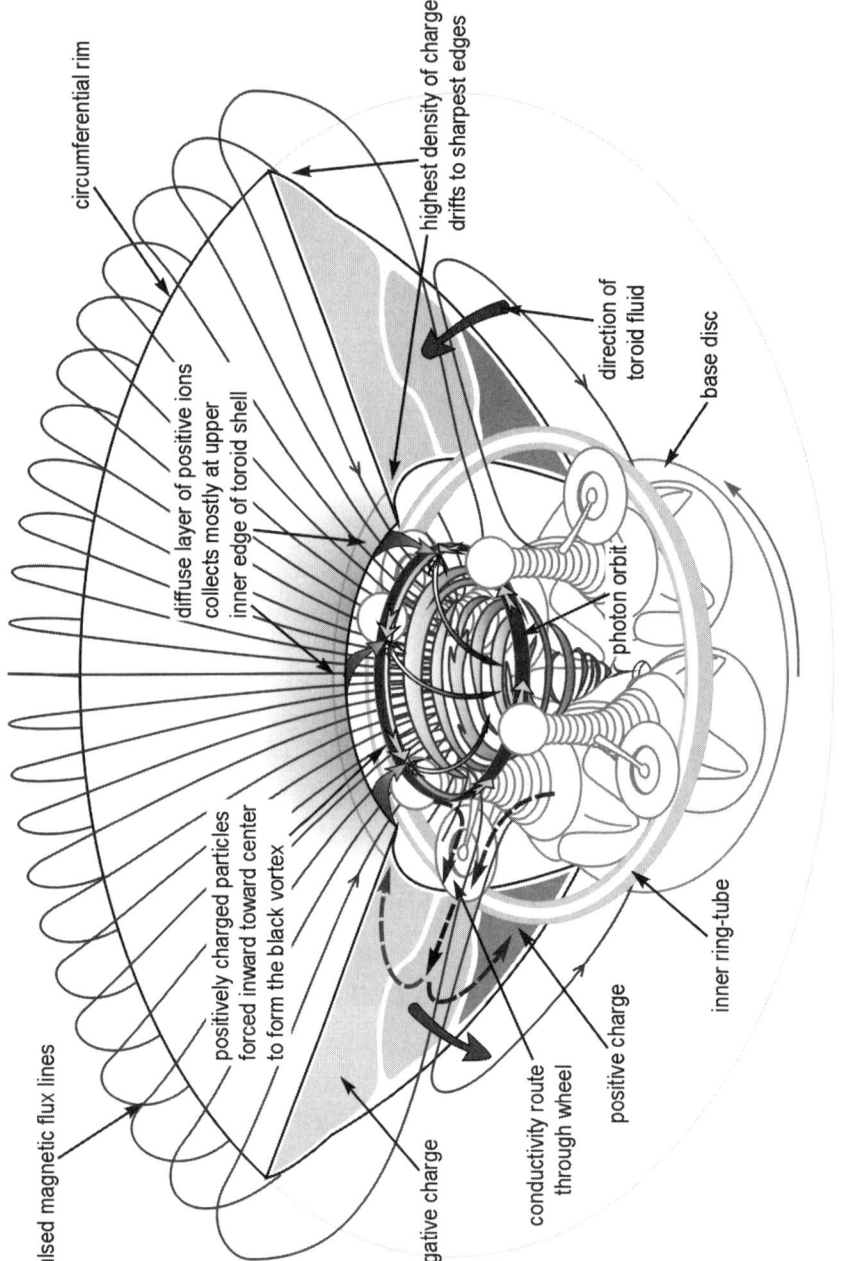

Figure 33. The toroid fluid and the vortex will flow in a counter-clockwise direction, but the positive charges of the diffuse layer will flow in a clockwise direction

Whenever those magnetic flux lines move then an electric current will be induced to flow along a conductor in those flux lines. Similarly, charged particles (as in a plasma) will be induced to move in a direction perpendicular to the motion of the magnetic field lines. This is the basic principle upon which electric motors, dynamos, and of course synchrotron particle accelerators work—see figure 32).

This action starts off with the toroid fluid, because after the electrolytic fluid in the toroid has been made to move through the toroid shell the electrical charges in the fluid will separate (by gravitational and Lorentz forces on a polarized fluid—see section 6.12), and the fluid acts like a current flowing through a conductor, the fluid will generate an electromotive force and establish an orthogonal magnetic field around it. This magnetic field will surround the toroid shell in much the same way as moving electrons generate lines of magnetic flux perpendicular to the current flowing along a wire. And because the toroid is circular those magnetic flux lines will form into an inhomogeneous (non-uniform) field, so that those lines will converge closer together as they approach the inner edge of the toroid (see figure 33).

The movement of the electrolyte fluid within the toroid shell will induce mobility into any charged particles that happen to have accumulated outside the toroid shell. So the diffuse layer of oppositely charged particles will be induced to move in an opposite direction to the toroid fluid (because of their opposite polarity). By these electromotive forces resulting from the toroid's magnetic field, all the negative charged particles (coming from the above mentioned array of ionization processes available to the center of this UFO) will migrate outward toward the outer rim in a counter-clockwise direction, while the positive ions will migrate toward the center to where the vortex is in a clockwise direction. This phenomenon will be supplemented by the fact that the toroid's magnetic field will be non-uniform and converging toward the center, meaning that the actual force of the magnetic field will be stronger toward the inner part of this toroid field than the outer part (causing a spiraling magnetic-drift force on charged particles).

So essentially, when the toroid's fluid moves there will result two flows above and outside the toroid shell, one of negative electrons circulating outward toward the outer rim; and the other flow will be the positive field of ions circulating inward over the inner edge of the toroid and toward the center of the craft to where the sphere-sets are.

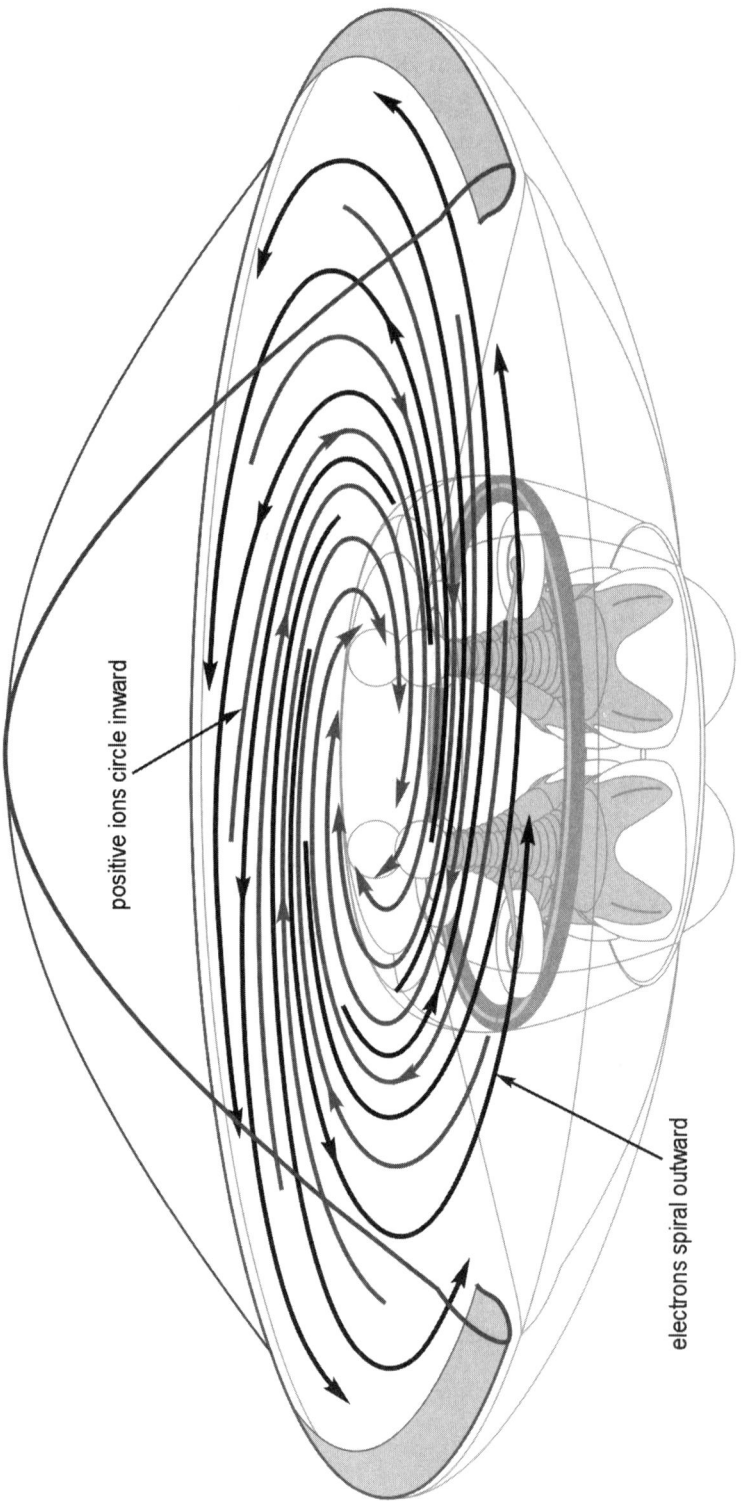

Figure 34. Negative charged particles will flow out from center while positive charges will flow into center

This two-flow mechanism will be what should basically be seen as nothing more than the 'engine started' stage of the aaUFO's drive system, and throughout the following chapters I will be referring back to this basic arrangement and further expanding upon all the mechanisms involved in generating and mobilizing these electrical charges in and around the toroid fluid—and even comparing these gyrating fields to an accretion disk which surrounds an astrophysical black hole. But as I have previously mentioned for chapter two all I will be doing throughout this chapter will be giving a rather generalized unfolding of the aaUFO's drive system.

The aaUFO of course is still parked in its loading bay and its crew are just storing their hand luggage up on the upper decks, because this UFO right now is only just ticking over—idling. With this low rate of energy generation its not likely to go anywhere or do very much. What it will need now is some serious field pumping action to get its numerous mechanisms generating much more energy and bringing all that energy on song... then maybe it can rise up into the air and be off on its travels.

3.10 Toroid's Magnetic Flux Lines Pulled into Shape

Intriguingly, this craft is so well designed that for its pilot to initiate the energize sequence all he or she would need to do is push one solitary button... because the way the electronic system is configured, in the manner of a sequential series of electrodynamic processes, that initial triggering of just one switch would allow an oscillating field to be channeled along an optical fiber transmission line, that would be routed from the upper decks where the control center would be, down to the lower deck area, and to link directly into the aaUFO's inner ring-tube, to operate it. So technically speaking, this UFO would be very simple to operate and build, and even though its maneuverability will be seen to be quite sophisticated the transmission lines it operates by are very few. This factor will obviously mark a huge improvement over existing aerospace aircraft where monumental wiring looms, circuit-boards, switches, relays (and gremlins, etc) are required.

The primary impulse would be transmitted into the inner ring-tube wherein a set of EHD actuators would be triggered so as to initiate a set of magnetic pulses into the magnetic fluid inside that inner ring-tube [note: 18]. This would start to circulate the magnetic fluid of the inner ring by generating a set of small local magnetic-island fields, and when these island-fields move through the ring they will act upon the telemeter wheels so as to allow them to turn. Of course, turning the telemeter wheels also turns the sphere-sets, and as all four sphere-sets are located in the craft's base disc then the whole sphere-set-assembly will turn as one. The sphere-sets being quite frictionless because the only thing that locates them are the telemeter wheels, and where these wheels turn there will be acoustic wave vibrations acting between the wheels and the inner-ring tube.

Once the telemeter wheels are allowed to turn then all the sequential processes in the aaUFO become energized one after the other.

When the telemeter wheels turn, the rods inside them when they line up perpendicular to the toroid wall, they will then make a path of conductivity for the electric current that will run through them and on into the toroid's fluid. In fact, if the reader looks at (figure 33) they will see the route; all the way from the diffuse layers (at the outer surface of the toroid), over the surfaces of the top spheres, through the stem coils, and through the arms, and finally through the wheels along the rods and into the neutral, or zero-potential, mid-zone of the toroid fluid where those electric currents will be dispersed into that fluid.

Furthermore, as this action of the telemeter wheels turning constitutes a conductor moving through a magnetic field (through the toroid's magnetic field) there will be induced in the wheels (and therefore the sphere-set assembly) a Lorentz force (electric motor principle), and hence the sphere-sets will begin to turn with more and more force. This mechanism will be helped by the probability that the wheel-rods will not be aligned simultaneously together but that their alignments will be staggered (and so together they act somewhat like a multiple-phase motor). Although, this is not to say that it will be the primary driving force of these wheels, as will be seen in a later chapter.

But, because the rods only conduct momentarily (when they do align perpendicular to the toroid wall) this means that as the wheels turn the conduction path will be made and then broken, or rather, repeatedly made and broken by all four telemeter wheels, so there will result a pulsing flow of electric current through the stem coils and through the toroid fluid. Then, if the pulsing electric current is passing between the top sphere, and the arms, and wheels, then it will be passing through the stem coils (which ideally will provide the path of least resistance); and so therefore a solenoid-type magnetic field will be generated and established around the stem coils (see figure 29 for instance).

Nothing is specifically designed into this part of the craft's electronic circuit to provide any substantial electromagnetic damping and so this circuit will subsequently generate, by self-inductance, a back-emf voltage that will continually amplify itself and thereby quite voraciously increase the voltage being pumped around this particular electronic circuit. In fact, this will occur almost instantaneously because of the great density of ionized charges already accumulated in this central part of the craft. Because of this there will very quickly come a point when the solenoid-type magnetic fields flowing through the four stem coils will become the most intensified fields in this part of the craft, and certainly much more intense than any other part of the toroid's magnetic field. So these stem coil fields being so strong will then grab the toroid flux lines, so that the upper toroid field will be re-routed through the stem coils and the lower toroid field will re-route through the central base disc. This can be seen in (figure 29 above) although it might be more understandable in an animated diagram [note: 19a].

In essence the upper and lower toroid fields now come under the complete control of the sphere sets—and this is when all the panel lights come on in the pilot's control room. This is where I would consider that the aaUFO's power dynamics will begin to produce the craft's lower range of propulsion forces.

Because, as soon as the inner ring is energized and the sphere-sets begin to rotate and the vast amounts of electromotive force begins to be generated, this primary stage of excitation (as delineated in this chapter between 3.8 to 3.10) will quickly become sufficiently pumped up to initiate what I would term the secondary stage of excitation, of certain (ceramic) material structures so as to initiate charged particle generation and emissions, through the top spheres, that will be sufficient to continually replenish any ejections of energies that will be transferred out of this central engine area which will carry out of the craft its propulsion forces. The secondary stage, which will prove to be the second of FOUR excitation mechanisms centered around the sphere-sets, will be called upon to re-supply copious amounts of variously charged particles, and although the previously mentioned (nine or so) mechanisms of ionization are valid and are still wholly applicable as mechanisms to supply electrons and ions to be energized by the magnetic fields in this craft, I believe those primary mechanisms will only supplement this secondary and entirely more energetic and powerful mechanism through which electrical charges (and other forces) will be produced and amplified to fire up several more of the engine's electrodynamic processes.

3.11 Breaking and Reconnection of Flux Lines (Reconnection Part II)

As is shown in one of the preceding diagrams (see above figure 19) the four sphere-sets will divide the upper toroid field into four 'quadrants' and as the sphere-set assembly rotates the sphere-sets will attempt to pull around with them these four quadrants of flux lines. I say attempt because these quadrants won't fully turn with the sphere-sets, the outer stretches of these four quadrants will be anchored at the rim because of all the other magnetic fields operating there (the anchor which is another astrophysics mechanism derived from a black hole's dynamics and will be further explained in Chapter 9).

Because of this anchoring of the outer stretches of the toroid's upper flux lines the sphere-set assembly essentially converts the toroid's field into a very powerful generator of electricity (electricity power generation corporations please note!). The best way to understand what actually happens in this ingenious mechanism is to consider each flux line individually, for when the sphere-sets turn and pull on the flux lines each one will be stretched until it breaks. But as the electrically polarized fluid (within the toroid) will still be moving around the toroid shell generating a Lorentz force it will demand that the flux lines are quickly re-established. So, as the central assembly rotates and the next of the four sphere-sets approaches that broken flux line it will reconnect itself and also pass through that next sphere-set's stem coil. And of course this whole breaking and reconnection procedure is repeated continually through every single magnetic flux line around the toroid. This can be seen in (figure 19) although it might be more understandable in an animated diagram [note: 19b].

As a result of these flux lines being broken and reconnected with such profusion and with such regularity the toroid's magnetic field becomes a pulsed and highly energetic field, and as every electronics engineer knows, such an alternating or pulsed magnetic field generates large amounts of electromotive force in any nearby conductor. This principle is used to great effect through both the upper and lower spheres which, as can be seen from the diagrams (such as figure 43 below), whereby the pulsing fields are directed through the top crystal-glass spheres and through the upper half of the lower spheres, so that these interactions will initiate separate electrical mechanisms in any materials specifically designed to generate emissions (ie ceramic materials) by these methods (and which will be detailed further in Chapter 4 and 5).

Again, the breaking of magnetic field lines and the re-establishment of those field lines is a known phenomenon in astrophysics (see above in section 2.8), and is one that is known to produce copious amounts of electron-positron pairs out in space nearby to the meeting points (or viscous shear planes) between black holes and their accretion disks, and is a mechanism through which the black hole systems can supply the enormous amounts of charged particles that are needed to form and power their quasar and pulsar jets which extend for millions of miles through space [note: 20]. To understand what happens in the breaking of magnetic field lines and their reconnection is to liken this phenomenon to the breaking of wires conducting an electrical current in an electrical circuit, because if the wire is broken near the power source the entire potential-difference of the circuit could be developed across the gap of that broken wire, and so with this flux line breaking charged particles around the flux lines are instantly accelerated in all directions, hence they collide with other particles in the air and can even trigger electron-positron avalanches around these breaks [note: 21]. Then when the field lines reconnect again a huge amount of charged particles have then to be freshly generated out of the surrounding air, or out of space through the ZPF, so as to accommodate the imbalance in potential difference. The effectiveness of such a mechanism within the aaUFO can be taken a step further than in the black hole environment though, to the degree that such a mechanism can be made to occur in the vicinity of certain charge-donating materials whilst they are embedded into the upper glass-quartz spheres.

3.12 Top Sphere Emissions Giving off Bars of Light

With the above mechanisms of electromagnetic field pumping, and breaking-for-reconnection of flux lines (which will couple to the already present acoustic wave fields focusing into this area), it can be seen that electrons and photons can be copiously supplied within this central area of this UFO craft. But the image of the photon bars supplied by Betty Luca still suggests that there are energy emissions of an even greater caliber coming out of these top spheres and that there is a particular cohesion to them.

For such an image is not showing an indiscriminate dance of photons, it is portraying a quite rigid coordination of locally induced forces that must be coming from an extraordinary intensification of magnetic forces.

But then, what is so highly significant here is that the four top spheres are located at the top of the four stem coils, through which the toroid's four quadrants of pulsing magnetic flux lines will be converging into four of the greatest intensities of the whole of the toroid's magnetic field, and so therefore the greatest empowerments of the Lorentz (spinning) force upon those copious supplies of charged particles will occur where each of those quadrants are directed through the four spheres. And because all of those flux lines will be sequentially pulsed there will be induced extraordinarily powerful Lorentz forces into any subatomic particles moving anywhere near these four spheres.

With this arrangement, it would then seem to be clear that these top spheres should be made not just of quartz-glass alone, but that they should be made of a glass-like substance into which can be embedded certain ceramic compounds which are known to emit electrons or photons when Lorentz stresses are applied to them at particular frequencies of oscillation.

And so, with respect to the images of the sharply defined light bars seen by Betty Luca when the extraterrestrials showed her the workings of these spheres, it should not be too difficult to piece together the other components of this display of power and to understand exactly how this was done.

For while the magnetic flux lines are converging onto the top spheres and the Lorentz forces around them are highly amplified, the resulting emissions bursting out from the ceramics embedded inside these spheres would, like all the other charged particles already mentioned, have quickly been caught into the influence of the intensified magnetic flux lines and immediately trapped into energetic gyrations, by those heightened Lorentz forces, and made to spin tightly around those flux lines. Such close gyrations of these abundant ceramic and other emissions would, in turn, radiate curvature-radiation photons away from them and would form a very vibrant cloud of photons around these spheres.

But because the acoustic wave oscillations emanating out of the large lower spheres will pass through TWO routes to the upper smaller spheres, which will effectively split those oscillations into two slightly different frequencies (where one set of those frequencies are being bounced off the toroid's inner wall and focused into the vicinity of the top spheres, and the other set of acoustic oscillations will be propagating more directly through the stem coils)—AND from the fact that there will be another field of acoustic waves emanating directly out of the top spheres themselves, then there will have then resulted a composite interference pattern from these THREE acoustic fields centered around each of the top spheres (see above figure 28).

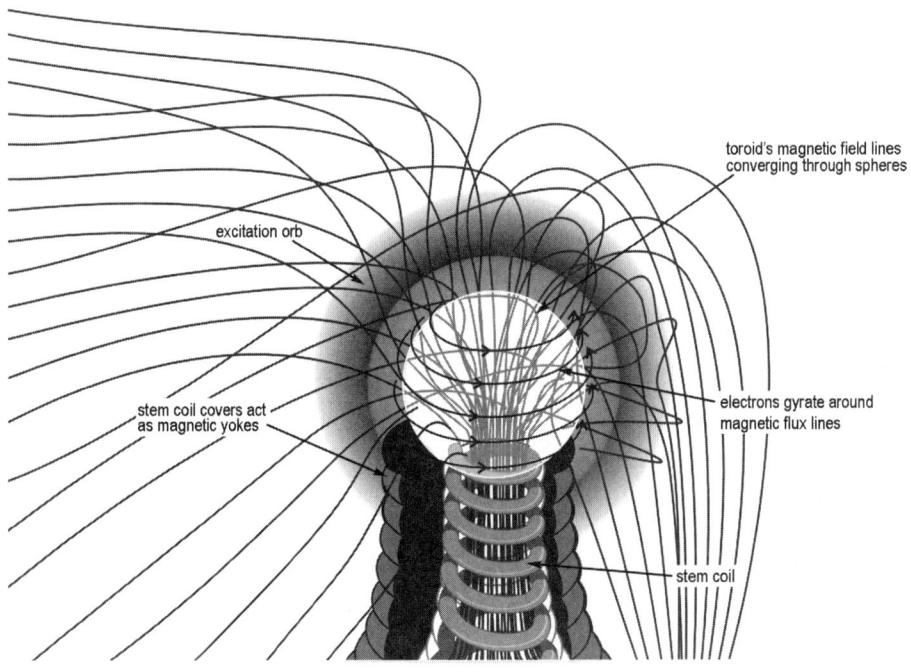

Figure 35. The intensity of magnetic flux lines pulled into stem coils will cause high excitation of charged particles around top spheres

The top sphere's acoustic field would have developed as a spherical zoning around that sphere and would have behaved much in the same way as Bessel spherical waves do as concentric orbs, but then, because of the multiplicity of these frequencies all interfering one upon the other then the interference effect between them all (in similar fashion as to how multiple electromagnetic fields create interference patterns of stripes or ribs) would have resulted in the establishment of several sharply defined hi-low permittivity zones (in a three dimensional pattern) spreading outward from each sphere. These zones would have influenced the charged particles (and subsequently the magnetic flux lines converging into the top spheres) to locate groupings of electrons and positrons to spin in very tight gyrations around the spheres. These would have created very sharply defined bars of light, with the light coming from intense curvature-synchrotron radiation emitted off the electrons and positrons gyrating so tightly around groups of flux lines.

So, the explanation to these energy bars then is that they are a combination of highly excited gyrating electron and positron emissions of photonic radiation, strongly energized by a great intensity of magnetic flux lines converging on these spheres from all different angles, and because the interference pattern from the three acoustic wave fields focussed around each of these spheres would have generated, collected, and shaped all these emissions into high permittivity zones, then these photon-bars would have been very sharply defined [note: 23].

3.13 Gyration of a Cyclotronic Storing Field

One additional point to make about the upper flux lines is that from these diagrams it can be seen that these radial flux lines will generally become angled non-perpendicular to the craft's rim and circumferential duct, and this will cause the negative electron field, while gyrating above the toroid shell, to curl gradually outward toward the outer rim area.

While this field will be gyrating it will be gyrating through the radial planar wave-guide which is essentially the space between the floor deck of the upper part of the UFO, and the top face of the toroid, and this radial planar wave-guide will work just like a giant cyclotron accelerator on the gyrating energies (which will be a rotating field analogous to a black hole's accretion disc).

In the following chapters more will be explained about the amplification properties within this narrow, circular (planar) wave-guide (especially in chapter 9).

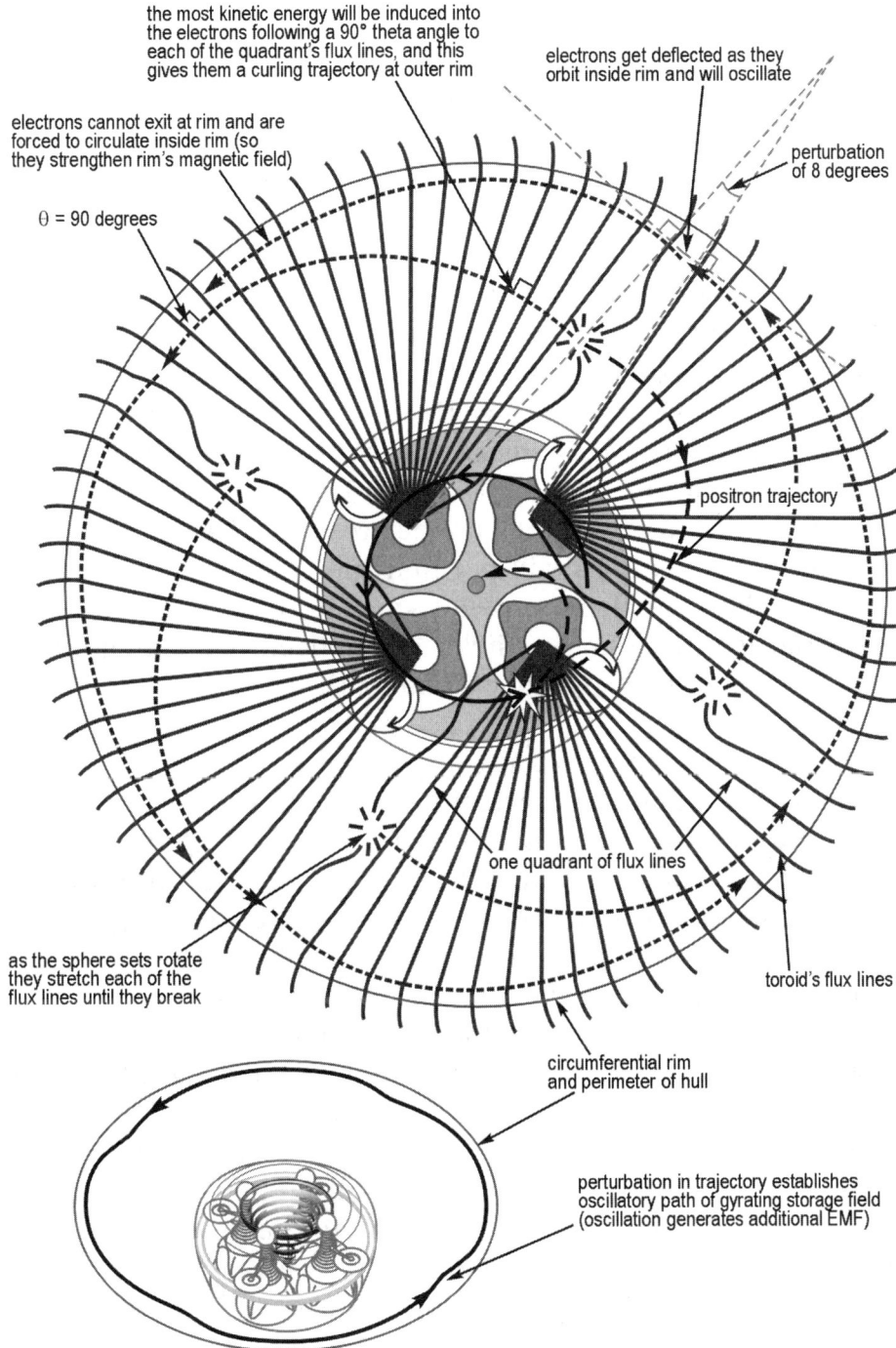

Figure 36. As the sphere-set assembly rotates the magnetic flux lines will break and then reconnect thus propelling any charged particles

But such will be their trajectory over the four quadrants of the toroid's flux lines (see figure 36), making such a shallow angle to the outer rim, that the gyrating particles will not by this Lorentz mechanism alone be directed outward through the duct with any centrifugal force whatsoever. Essentially these particles will be forced only to gyrate at high velocity just inside the rim's duct and become a 'storing energy field' inside the craft.

This would be because, technically, the electrons that get produced by the flux-line breaks will be given the most kinetic energy where their theta-angle to each of the flux lines will be 90 degrees. Looking at the flux-line arrangement (in figure 36) it will be seen that such an arrangement will only give the electrons the CURL trajectory that directs them into a circular orbit just inside the craft's circumferential rim. Although, what can also be seen in this drawing (figure 36), is that these electrons, which will add into the gyrating storage field, will not be accelerated around the craft in a perfectly circular orbit—the flux lines will put four moving perturbations into those orbits and this will oscillate the electron field (which in turn will imbue extra electromotive force into the gyrating storage field, and indeed, will imbue extra electromotive force into the whole electrodynamic circuit of the craft).

In fact, by circulating just inside the rim this oscillating field will bolster the rim area, where the radial planar waveguide contracts into its shallowest depth, because this gyrating field will help to intensify the toroid's and the rim's magnetic fields in that region. Perpetual acceleration of these charged particles will take place because the flux lines will continually be stressed broken and reconnected and generally pulse-oscillated, and because of the magnetic pressure exerted on the accelerated particles which will be proportional to the square of the magnetic flux density (which will be at its greatest close to the top spheres in the craft's center). This will ensure that there will be a continual outflow of electrons from the center toward the outer areas where the flux lines spread over the toroid and become fully intensify at the outer rim—from the forces imbued into them by the curling electrons as mentioned above—and so consequently there will be formed a substantial 'bottleneck' of electromotive and kinetic forces at the rim which will greatly intensify that area's magnetic fields (of course, this complicated arrangement of fields at the rim, which has a number of important energy dynamics enfolded into it will be detailed in Chapters 7, 8, 9, and 10).

From the miniaturization-of-the-astrophysical-model point of view, this gyrating storage field of energy (which is comparable to a black hole's accretion disk), will have the potential to be an even more efficient provider of power than its bigger cousin out in space, considering that it is a gyrating storage field that is fully contained within a purpose-built grouping of ergospheres, wherein energy can be amplified up to specific threshold levels and then released beyond those threshold levels, with the utmost efficiency (which again, will receive more attention in later chapters).

3.14 Three Intensities of the Magnetic Field to Aid Emissions

And this localized intensifying brings us back to a point mentioned at the head of section 3.9, in that one of the most important aspects of the aaUFO drive system is the intensifying of the magnetic fields in localized areas so that the charged particles can be highly energized with kinetic energy. This of course is one of the factors which the designers of the UFO drive system have taken from the black hole model and have greatly improved upon, for the natural environment of the black hole out in space while it utilizes magnetic field bunching that can occur just outside of its event horizon, obviously cannot take advantage of any material walls or structures that localize and direct the particles affected by those magnetic flux lines. But in the aaUFO drive system this is specifically done with surprising effectiveness in three main areas of that UFO, these areas are; ONE at the circumferential duct where the particles move in a cyclotronic motion in readiness before they are forced out of the craft's rim (and then have to be re-accelerated over the upper dome); TWO at the tops of each of the four stem coil assemblies where the top spheres revolve; and the THIRD of these intensified areas is at the center of the craft's base disc where the magnetic field lines of the toroid's lower field converge and form into a funnel shape.

3.15 Magnetic Funnel Creates a Black Vortex at the Heart of the aaUFO

The funneling of those field lines will of course create a vortex or miniature black hole at the heart of the UFO. For electrically polarized particles will be induced to move perpendicular to those magnetic field lines and therefore they will curve into a central gyrating field.

In fact this conical field of gyrating charged particles, formed when the toroid's lower magnetic field flux lines re-route through the base disc [note: 24], will form a continuation of the positive ion field that will already be gyrating toward the top spheres in the center of the craft (see figure 34). The pulse-oscillating (from the breaking and reconnection) of the toroid's flux lines will ensure that the positively charged particles will be accelerated into the center and eventually into the gyromotion that curls towards the central vortex.

Indeed, on closer inspection there may be noticed several more interesting features to this vortex inducing magnetic field: Because, while the intensity of the pulse-oscillated magnetic flux lines will be greatest at the small exit hole in the center of the base disc, which will be the only place where the charged particles can exit— unless they are bounced out by collisions at the top ring of the vortex—then this would mean that as they spin round so as to exit down through the hole at the bottom they would pull inward toward the central axis all of the nearby flux lines and intensify them and generally amplify the rotational force of that 'hole.' If indeed the spinning particles do exit completely, for such will be the high density of flux lines converging together here that there will come a point when the magnetic field density will become too powerful for the spinning particles to continue down through, so then some would be forced to mirror back out of this convergence and would attempt to rise back up out of this vortex, only to be forced down again into it by the Lorentz force of the magnetic flux lines. And so these spinning particles will curl in and out of this hole area again and again in a continuously repeating cycle. Obviously, the energy states of the spinning particles that do pass deeper into this hole would then rise to higher and higher levels of excitation through this continual amplification and generate more electromagnetic Lorentz force into the surrounding field lines, and as this process would forever repeat then these spinning particles would forever be accelerated through a smaller and smaller radius, and have their trajectories turn more and more azimuthal (or perpendicular to the longitudinal axis of that vortex) and become more compacted together. So essentially, this area of the UFO will then possess the highest degree of mass-energy-density and electromechanical force throughout the whole of the craft. And it would be interesting to speculate whether or not those mass-energy particles would eventually become locked into a closed loop of rotation at the base of this vortex [note: 25]. Interestingly, at the hole and just above the base disc, the velocity of those spinning particles would continually increase through relativistic (and perhaps even through superluminal speeds), yet, if they passed through the hole then the magnetic field configuration on the other side of it would be such that those spinning mass particles would then be forced to suddenly expand outwards beneath the craft [note: 26].

This arrangement in the aaUFO is obviously very much like a black hole, but its not, and in order to dissociate this form of spinning drive from being an exact duplication of an astrophysical black hole, I shall coin for this quasi-black hole the term black vortex and give more details about it through chapters 7, 8 and 9.

But with consideration to this new factor, that the aaUFO's drive system has a black vortex at its center, the whole melee of particle generations and emissions that have been pouring into this central area, from all of the above-mentioned ionizing mechanisms can then be perfectly orchestrated into a coordinated spinning-power-drive-system. Consequently, immediately around this black vortex would be established rotating forces that would become the aaUFO's most powerful ionizing mechanism of all, and this unified power would feed back along the magnetic field lines, to trigger the upper sphere's ceramic-emission mechanisms, and back into the break-and-make reconnecting areas above the toroid to where those four epicenters would be producing their ionizing shock waves (as will be further explained in section 9.7), and so possibly there would then occur an oscillating resonance phenomenon between all the different ionizing mechanisms. This would depend on waveguide parameters and dimensions of the surrounding volumes but if enough coordination could inter-connect those mechanisms then there might occur a 'spontaneous ionization' instability whereby such emissions become self-sustaining and continually repeating. If so then the black vortex may well command a supply of energies that are inexhaustible.

3.16 Vacuous Force—Light Converts to Gravity

The reason why it would be interesting to conjecture that at the heart of this UFO there is a black vortex which works like a black hole, is because then as according to the energy dynamics of a black hole, any particle entering into it as a high energy photon mass will be dissociated into an energy-dynamic that imbues a proportional amount of gravitational and angular momentum forces into any adjoining fields of rotational energy nearby it. Which in the case of this aaUFO, will be imbued into its plasma gyrating around the planar wave guide (see above figure 34), and into the polarized fluid rotating around the inside of its toroid shell (and also into the crystal-glass fabrications of the large lower spheres as they revolve about their own axis and orbit so closely around this miniature black hole).

It would have to be seen exactly how all the features of an astrophysical black hole do transfer into a miniature one, but according to the theories of Stephen Hawking, Roger Penrose, and Reva Williams on Black holes there is sufficient evidence to believe that the essentials that this craft needs do transfer [note: 27]. So then it can be assumed that this aaUFO's black vortex will have an event horizon, a marginally bound horizon, and a point of non-communication where intake will be 'lost' to infinity, and THEN, most importantly of all, that it will have a mechanism that converts electrical energy and mass into gravitational force and angular momentum.

Because of this discovery it would then seem that there would be a rhyme-and-reason to the aaUFO's central structure and its outlining structures. Because then it could be expected that surrounding this central black vortex engine there would be found the mechanisms of a gyrating accretion disk, and as mentioned above these are indeed quite noticeable in the aaUFO [note: 28]. And so consequently it is becoming more and more obvious why there should be such a large amount of positive-ion, electron and photon producing mechanisms being employed in this craft, to direct subatomic particles toward the top spheres (even more so than there are mechanisms to extract such particles from them—as will be pointed out throughout chapters 4 and 5).

This would also explain why the top spheres are being used at the meeting point where so much energy is forced to rotate around the mouth of the craft's vortex, which in black hole physics is called the photon orbit, situated just outside the black hole's event horizon. This would also make sense from the point of view of directing the greatest profusion of the craft's pulsing magnetic flux lines into the same area as the rotating photon orbit, because it would mean that those subatomic particles will be given a huge amount of kinetic energy and be forced into a rotational frenzy—just outside the black vortex's event horizon.

In other words, in the middle of this UFO it looks as if the ETs have found a way to develop Lorentz magnetic forces so as to generate a rotating vortex into which light and mass energies are continually pumped in order for that vortex to continually pump back out gravitational and rotational forces—which in the case of a UFO are infinitely more useful forms of energy that can be used in its propulsion mechanisms [note: 29].

3.17 Clues Forthcoming in the UFO's Quasi-Black Hole Drive

For the readers of the book series published about Betty (Andreasson) Luca's experiences in the strangeness of the alien's world this black vortex will be recognized as the cone field that Betty Luca saw and described as spinning like a whirling tornado inside the craft she was taken into and shown by the extraterrestrial Greys (see figure 37). And if the same readers have been intrigued by the strange technical terms used by one or two of the Greys, which have been relayed to us through both Bob and Betty Luca's hypnotic regressions, such as the ET's mention of them purging and lining the cyclonetic trowel in one of their UFOs (from "Watchers" pp76), then they should be pleased to know that throughout the following chapters of this book those phrases will be matched to scientifically recognized dynamics and explained in full.

Certainly, some of those phrases heard by Betty and recalled through her hypnotic regressions have provided invaluable clues to the electrodynamics of the crafts that she had been introduced to.

However, apart from those half-dozen or so phrases and the three or four sketches which only afford a fleeting glance at the central mechanisms to those craft, the avid researcher into UFO physics has had little else presented to them to peak their ingenuity in order that they may formulate in their mind's eye a working blueprint of a UFO. So it is all the more rewarding to recognize those scant details about this UFO's structure within the vastness of the physics detailing the black hole power systems that can be readily seen to exist in space.

And to reiterate the point made at the head of this chapter, now that the comparison has been made it becomes all the more obvious that there should be such a link between extraterrestrial craft design and power systems that pervade space. And while I have already mentioned quite an impressive array of power supplying mechanisms, which in the aaUFO resemble so closely the mechanisms of the astrophysical black hole or pulsar out in space, where energies are drawn out of the ZPF fields of space (through the interactions of just gravitational momentum and magnetic fields), there are many more comparisons to make which can be used to afford a much greater understanding of the aaUFO and other known UFO engines.

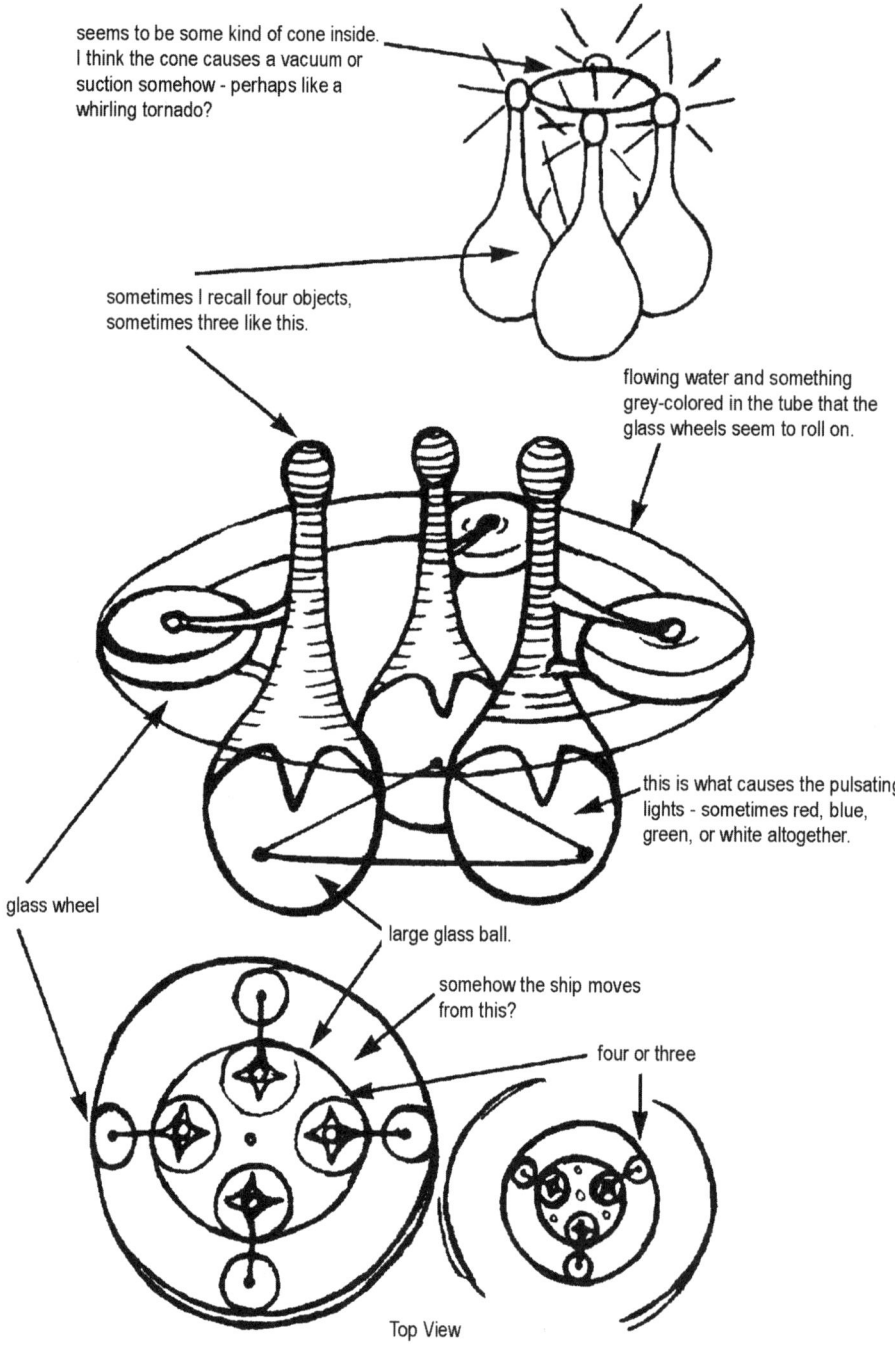

Figure 37. In the middle of the sphere-sets is empowered a whirling vortex (from "Andreasson Affair" fig.8)

For instance, in viewing the aaUFO's drive system it is particularly interesting to see how this craft has, in addition to all the ionizing mechanisms (as alluded to in section 3.4 above), very neatly combined together several of those charged-particle-production mechanisms into tagged-together ones.

One such multiplicity is comprised of curvature radiation (where photons are emitted where charged particles are made to rotate through tight curvatures) at where the black vortex's photon orbit ring rotates; and tagged with this is the [Blandford-Znajek] ionizing coming from the spark gap separation between two counter-rotating energy fields which shear against each other (which in the aaUFO would occur between the CCW black vortex flow and the toroid's diffuse layer of positive ions flowing CW toward the center) [note: 30]. Then meeting together with these at the exact same location is a third mechanism of another ionizing spark gap occurring between the inner upper edge of the toroid shell and the four co-rotating top spheres (and which might be compared to a cathode-to-anode type of arrangement of charge transfer). Then these three tag into the focus-orbs of acoustic waves directed toward the top spheres (as alluded to in section 3.12 above), but in such an energetic and frenzied way, from all the pulsing magnetic flux lines exciting their Lorentz forces, that there then occurs abundant collisions between all these inbound particles against the gyrating particles rotating through the photon orbit. But this melee of multiplicity continues even further; these five mechanisms are then joined by the ZPF energies being routed toward the craft's black vortex from outside the craft through the lower spheres, through the stem coils, and then through the top spheres, which will fire up the sub-atomic emissions from the ceramic materials embedded in the top sphere's glass. So does anyone wonder why I haven't as yet tried to work out the mathematics of all this multi-functioning!

One of the final clues in determining the fundamental chain of processes occurring in this engine came from the fact that while the top spheres are orbiting about the craft's longitudinal axis as part of the sphere-set assembly, each top sphere is individually revolving about its own stem axis. This was clearly observed by Betty Luca when she was shown the internal workings of this aaUFO and the energy emissions around these top spheres [note: 31].

And this revolving of the top spheres will equate to a wholly necessary mechanism, because as the above mentioned multi-tagging all converges upon those top spheres, in order to prevent localized hot-spots developing on their surfaces facing toward the black vortex's rotation they must be made to revolve (see above figure 18). Otherwise, consequent to the quasi black hole's powerful mass-density gradient there will occur a 'tidal' effect around the top spheres in relation to that black hole's gravitational pull. And to explain, a tidal effect is a part of the Hawking theory about black holes which basically observes that a tensile differential will exist between the center-facing side of a mass and its outward facing side (from the gravitational pull being so markedly different between the two faces). And so with regard to these four top spheres because they are situated so close to the rotating forces just outside event horizon of this black vortex there would result, if they didn't revolve, an energy flow streaming off each of them at just one small area of their outer surface. That being the case there would soon develop a hot spot from the excited particles being sucked into the hole and the whole system would become unstable.

Indeed, in principle, the configuration of these four revolving top spheres all orbiting around the miniature black hole in the center of this UFO and shedding huge amounts of photon energy into a spiraling event horizon—with the image of creating a blue glow of ultraviolet and Cherenkov radiation above it—is precisely the same as the image of the 'cannibal' theory of galaxies in the Ostriker-Tremaine hypothesis (which I alluded to in section 2.7) where a central black hole consumes from the ionized energies of its surrounding stars by this tidal effect [note: 32].

But this suggests to me an important point to emphasize about rotating black holes surrounded by an ergosphere and an accretion disk, with regard to what can and what can't be sucked into them [note: 33]. Because, essentially there is no vacuous force directly above the mouth of such a black hole, there is a void directly above the mouth which is not affected by the black hole's gravitational or rotational forces. The main force and flow toward a black hole is lateral and only operates on the basis of horizontal frame-dragging as occurring co-planar to the black hole's equatorial plane. What there is above a black hole is a bulk-effect but it won't be as powerful as the lateral gravitational force. Similarly, in the aaUFO's black vortex the predominant forces will be azimuthal forces not a direct longitudinal force.

This is why an accretion disk accompanies an astrophysical black hole, it forms the main dynamics of entry into the black hole. For instance if the reader looks at the images of microquasar SS 433 which is a binary system in the Aquila region, where the SS 433 star is slowly disintegrating into the black hole (called W50) which accompanies it, then the reader will see what I mean. For subsequent to the star's mass being churned up by the intervening accretion disk, when that stream of mass particles and gases are drawn toward the black hole it will be the interaction between the disk's rotational forces the hole's gravitation (and the stellar magnetic field), which then turns the excess energies of that stream into a bipolar jet which will expel those excess energies into space.

This lateral feeding is one of the reasons why the star of SS 433 cannot be consumed whole in one huge bite by its neighboring black hole. This is especially relevant to the above mentioned act of cannibalism, because the black hole seems to have a very delicate stomach and cannot devour huge chunks of matter, thus all its intake has to be broken down and atomized into subatomic particles and photons first; through, as in the case of the astrophysical black hole, the action of the accretion disk and the ergosphere. And here is the exchange of dynamics that keeps the black hole sustained; before it can digest anything, all matter has to be broken down into digestible masses with electromagnetic and gravitational elements, and it will be what these elements and their rotational forces induce into the black hole's forces (as they then spin down and contribute to the hole) that will maintain the hole's magnetic and gravitational forces and its angular momentum, which in return keeps that black hole running. The whole process is a wholly necessary reciprocal arrangement.

Of course the aaUFO engine doesn't expend nearly so much energy in breaking down matter into photons as does the astrophysical black hole, it mostly eliminates that stage of the process by its designers giving it ample mechanisms that directly produce the highly energetic photons that it needs.

Certainly this aircraft engine seems to have been specifically designed to direct as many atomic and subatomic particles in as many ways as possible toward the top spheres, where those particles are imbued with extra energy from the magnetic fields, and by collisions with other particles flying around in a high state of excitation within the acoustic orbs which surround those spheres; and then when all that excited energy meets with the rotating photon orbit ring, which is basically a hollow cylinder rotating incredibly fast, all that incoming atomic and subatomic energy is dissociated even more: And from then on its the miniature quasi-black hole that completes the digestive process and radiates away energies that are more useful to the UFO. So its the perfect engine for a gravitational manipulation aircraft!

3.18 On the Coherent Amplification from its Black Vortex

Although I have not in this chapter fully answered the question aired at the beginning of it as to what the crystal-glass spheres do (their complexity warrants both of the next two chapters and sporadic mentions in later chapters), what I have shown throughout this chapter with unparalleled success I feel, is that the aaUFO can be compared to the astrophysical black hole. But also, that when its structure has also been compared with the astrophysical systems existing in unconfined space where its accretion disk and ergosphere are only shaped and guided by the immediate bunching or twisting of the magnetic field lines of space, then the UFO clearly shows that while it has been purposely designed to miniaturize the black hole into as compact an engine as possible, it has also been designed to greatly improve upon the efficiency of the astrophysical model.

Because, what these comparisons show most of all in the aaUFO's favor is that its physical energy-containment assemblies have been advantageously designed to provide clear-cut and solid boundaries to all the major rotational fields that associate with the astrophysical model. And in so doing the designers of the aaUFO's ergospheric compartmentalizations have clearly introduced substantial improvements over the original model. For instance, and anyone can observe this fact by looking at the studies of the rotational forces of an astrophysical accretion disk out in space: Because there are no encasing barriers for them they become weaker the farther they spread out from the hole's center (because there is nothing to hold them in open space). But the aaUFO has its storage field gyrating inside a planar waveguide (about 15 cm deep) which extends as a narrow channel outward to the craft's circumferential duct but which contracts down (to approx. 3 cm) with a magnetic contraction at the rim, and so that storage field undergoes confinement amplification by gyrating through a dual constriction, so the aaUFO's version of an accretion disk would appear to be MORE POWERFULLY AMPLIFIED than the astrophysical one.

This study though has only scratched the surface of what has been discovered going on inside a UFO's engine, and to end this chapter one last thought, in the planar waveguide channel because it has such constrictions, as mentioned above, which puts both pneumatic back-pressure and electromotive back-pressure on the energy field gyrating around it, it would mean that as that gyrating energy field nears the circumferential exit duct it would bunch into a more compressed field. This is analogous to parametric amplification where an electromagnetic reactance causes an abrupt slowing of the waves resulting in them bunching up so they gain extra energy before they are finally propagated out of that system [note: 34].

Although somehow, the latter mechanism doesn't quite sound the definitive full tin of biscuits which might be ascribed to a craft that is allegedly capable of accelerating to twenty-thousand-miles-per-hour in a matter of seconds... So perhaps there may be something further to be discovered here in the aaUFO; something that better combines all those available mechanisms and dynamics into a bigger force, which empowers them even more than has been suggested in this chapter.

Because if all these combined forces; of the break-reconnection particle productions, and the ceramic emissions, the acousto-electro-magnetic fields, and the contra-rotating-shear discharges (etc), can all be harmonized in some special way and linked together with the multiple ergospheres, AND with the forces of this engine's black vortex (which has not been coupled directly into the craft's dynamics yet)—by some special oscillating force—then perhaps there could result a grand coherence to this great multitude of energy dynamics, and maybe THEN there would be developed a truly super-inexhaustible drive force for this aaUFO to work the sky with. So, this is to say that there may well be further discoveries to come from these clever little extraterrestrial engineers in the chapters that follow.

4: Ferroelectric Emission Sphere Fabrication

Contents:
4.1 Photon Generator of Gravitational Forces	127
4.2 Electrical Mechanisms Working in Quartz-Glass	129
4.3 Magneto-Electric Oscillations from Magnetostriction Materials	130
4.4 Magnetostrictive and Ferroelectric Ceramics in Spheres	132
4.5 Ferroelectric Emissions and Polarization Reversal Emissions	135
4.6 Colored Luminescence from Within the Spheres	137
4.7 Quartz or Glass, Crystalline or Amorphous?	139
4.8 A Crystal's Acoustoelectric Phonon Wind	141
4.9 Sphere Fabrication with Sol-Gels of Silicons and Lanthanoids	143
4.10 Surface Facets to Enhance Electron and Phonon Exchange	146
4.11 Whirling Zones Inside Focused Acoustic Orbs	148
4.12 Permeability Switching Caps Around Top Spheres	149

4.1 Photon Generator of Gravitational Forces

This chapter will mostly be concerned with the top spheres of the sphere-set assemblies, while the next chapter (chapter 5) will be mostly concerned about the lower spheres of those assemblies, although obviously there will be cross-overs here and there as the basic principle of structure in both types of sphere are very similar, even though their energy outputs are quite dissimilar.

In the previous chapter I mentioned that there was a time when this study had reached the stage where it had become quite obvious that a large amount of positive-ion, electron and photon producing mechanisms were being employed around this UFO's top spheres, and also that numerous mechanisms could be discovered which directed constant supplies of these atomic and subatomic particles toward the top spheres. The puzzle was, why through all these energy-transferring mechanisms was the influx of negative particles complimented with copious amounts of positive ones as well (such as happens at the inner-upper edge of the toroid which is an edge encircling all four of the top spheres specifically shaped so as to propel positive charges toward these top spheres). This is not to be expected in any electronic circuit, its the formula for a short-circuit, positrons and electrons meeting in such a way will lead to the annihilation of both those particles.

Similarly, it wasn't so obvious in the first stages of this study as to WHY there was so many mechanisms which, rather than produce just electrons, they went on to produce excitons and polaritons which actually then go on to convert their initial electric charge into energetic photons. Strictly speaking this would be at odds with a power drive system that was expected to be electrical and which should therefore be assembling mechanisms whereby the end-product should be electrons. For a while I was having empathy for Col. Philip Corso and the engineers that couldn't see what constituted an engine in the Roswell UFO. Just like that Roswell craft this aaUFO didn't conform to customary electromagnetic design. And by applying the usual tools of scientific analysis—through electromagnetic theory, at least—this UFO craft's engine just didn't make any sense at all! Because what use would there be in a power drive system that produces just as much subatomic and photonic energy as electrons and positive ions?

This truly was a brick wall that didn't look as if it was going to give way, there just aren't any principles in electronics to help explain this puzzle. None in physics either. But then, years later, when the hypothesis of the black vortex was floated into these unusual features of this engine's structure THEN SUDDENLY all the lights started flashing and the bells began ringing. It all started to make much more sense. Perfect sense—and the brick wall was no more!

And so all the photon and sub-atomic particle production mechanisms will be of primary importance to this UFO craft: And the strange disposition toward using dielectric materials do then make perfect sense, because these will marry together the acoustic waves and the electrons and the positrons, which will bond together to produce excitons, and these can be transported over the dielectric materials and go on to produce photons. These energy mechanisms are now perfectly understandable as necessary processes that will contribute to the energizing of this UFO's black vortex.

So, here's a brief explanation, in order to generate its much needed gravitational and rotational forces the UFO would be designed around a miniaturized black hole (the black vortex), and that black vortex will have to be fed the right sort of energy to keep it running and powered up. The right sort of energy being mass and magnetic force, both of which for the astrophysical model out in space are converted from subatomic particles and high energy photons and directed into the astrophysical black hole by its rotating accretion disc. The accretion disc's role being to convert electron-positron pairs out of the background radiation of space and to break down all incoming energy into photonic energy, magnetic force and angular momentum. In the aaUFO the pulsed magnetic flux lines will be providing the angular momentum by imbuing Lorentz force on all the charged particles to make them gyrate (as detailed in chapter 3), and it will be the job of the four top spheres to provide the charged particles and conversion process of turning electrons into photonic energy, so that this photonic energy can then be directed into the craft's central black vortex to keep it empowered.

To understand the top spheres fully what needs to be discovered about them is; A) the various processes through which electrons and photons can be generated inside and emitted from these spheres; B) what special materials need to be embedded inside these spheres that will be conducive to the production of electron and photon emissions; C) what Lorentz magnetic and other forces can be manipulated around and through these spheres.

4.2 Electrical Mechanisms Working in Quartz-Glass

In a crystalline material such as glass or quartz crystal there are three basic types of energy-emitting scattering effects that can be triggered to cause electrical polarization and electric field generation; piezoelectric (acoustic mode), polar-optic (optical mode), and deformation-potential (acoustic and optic modes) [note: 1]. The latter type of scattering, as will be gone into in further detail particularly through the next chapter (chapter 5), can also be affected by gravitational forces in special circumstances where those gravitational forces are extremely powerful and when they exist in very close proximity to that crystalline structure. The first of these, the piezoelectric polarization, can be affected by magnetic forces and acoustic forces. But if the top spheres are merely made of glass or quartz then there doesn't appear to be that many ways through which electron and photon emissions can be produced, and so the theory that they might be made from nothing more than these materials is a bit suspect and will need to be looked at more thoroughly—which is what this chapter will be about.

Certainly, if glass or quartz were used on its own then the magnetic energy-triggering fields around these spheres would be rather wasted. And these magnetic energy-triggering fields are impressive, for they comprise of the pulsing magnetic flux lines of the toroid's quadrant fields, the breaking and reconnecting of those same magnetic flux lines, and the intensification of those flux lines into the stem coils through the top spheres. And an additional factor to these converging magnetic fields will be that the faster the central sphere-set assembly rotates then the more this magnetic energy will be intensified. And so, looking at all the mechanisms to be found inside the aaUFO's engine (see figure 18), where there is a grand orchestration to all these very energetic forces and magnetic fields, all converging upon the spheres, and interconnecting into the acoustic force fields surrounding them, and interconnecting with the rotational forces of the black vortex's photon orbit ring, then it would seem that these spheres really do need to be made of something rather special.

4.3 Magneto-Electric Oscillations from Magnetostriction Materials

The most common mechanism to produce ionization or electron emission is the cathode-anode arrangement where the cathode is heated, so that its atoms are excited, and the electrons are drifted toward the anode by an electrical potential difference. But because, at the vicinity of these top spheres, the most abundant forces that are readily available are those associated with the intense magnetic flux lines (which converge through the stem coils) then in this respect it would be advantageous to have embedded into those spheres certain materials which can be magnetically rather than electrically triggered to produce abundance electron emissions.

Alas that is not directly possible with materials such as glass and quartz. But it is possible if the crystal lattice of those materials is doped or embedded with magnetoresistive or magnetostrictive materials. For instance, manganese-based compounds (manganites) can exhibit what is known as colossal-magnetoresistance so that such a compound can become mechanically stressed when magnetic fields are applied to them. Alternatively, magnetostrictive materials can be stretched or compressed by an applied magnetic field, and they can made to oscillate through a magnetically-induced mechanism much in the same way piezoelectric oscillations can be induced in those materials through the application of electrical pulses (indeed, a magnetostrictive rod used in conjunction with quartz is commonly used as a type of oscillator in the electronics industry).

Here then is a possibility for the UFO's spheres, so that the constantly pulsing and highly energizing magnetic forces routed through those top spheres such magnetostrictive inserts used inside them could put the quartz of the spheres into altenating states of strain.

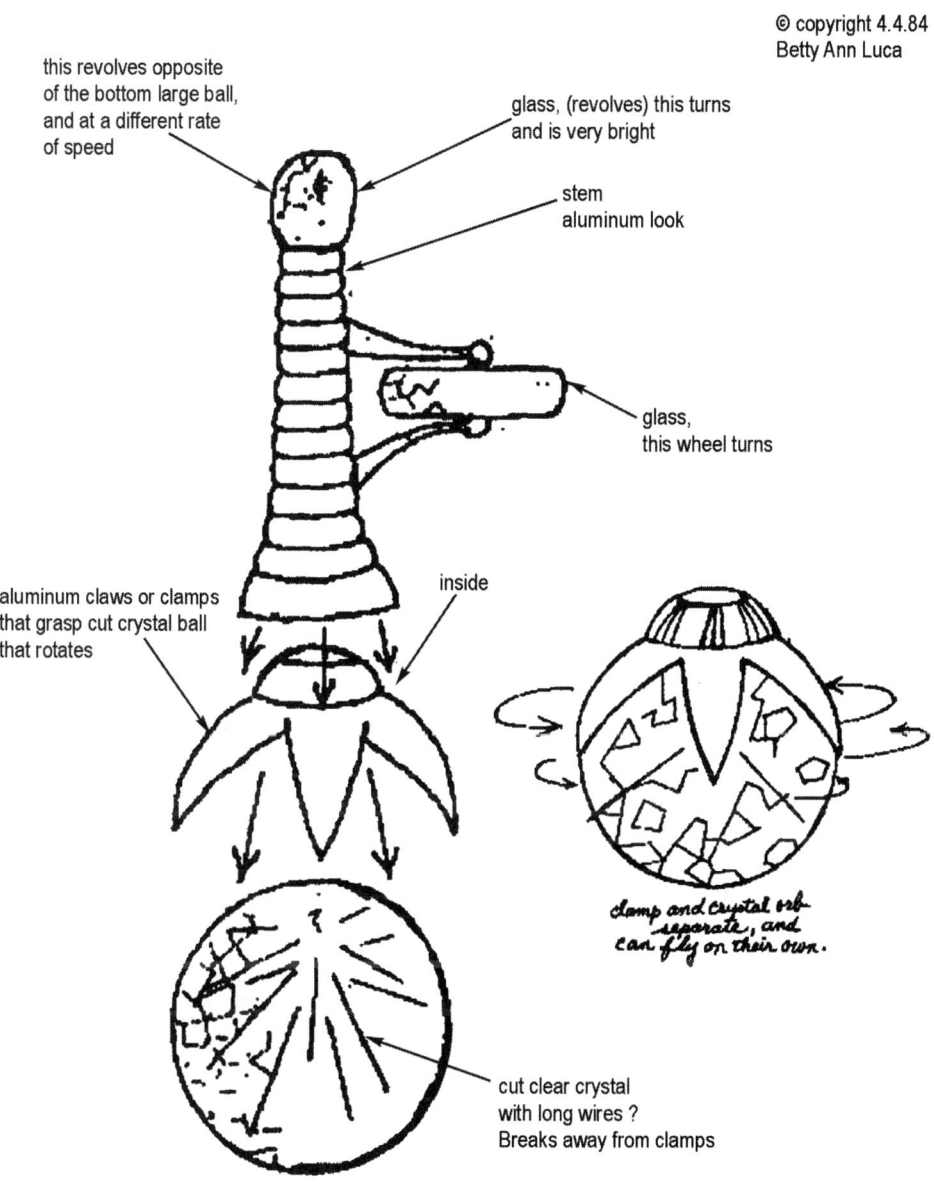

adapted from figs 42 & 43 of
"Extraterrestrial Communications Book II - A Step Beyond Tomorrow" (1999) by Betty Luca

Figure 38. The sphere-set is comprised of a top sphere, a stem coil, a wheel held by conducting arms, a revolving clamp and a lower sphere. In this drawing the lower sphere clearly shows a wire network inside it (from "ASBT-II" drawing 42)

Another feature worth mentioning here might be that when magnetostriction effects occur at fairly low frequencies (of up to 100Khz), which may be well suited to some of the lower harmonics of the acoustic waves coming from the lower spheres in the aaUFO craft's power circuit, the magnetostriction effect will create an audible hum within those material structures, and such a hum has often been reported as being heard coming from a UFO by those who have witnessed them close-by (as for instance the Grangemouth UFO) [note: 2].

Betty Luca did mention that she saw strange wire-like or rod-like structures inside the large lower spheres when the extraterrestrial Greys showed her the internal workings of one of their craft; although having said that I have since found a much better explanation for those structures of wires (as will be explained in chapter 5).

4.4 Magnetostrictive and Ferroelectric Ceramics in Spheres

What these spheres would REALLY benefit from would be ferroelectric materials being embedded inside them. Ferroelectric materials (sometimes known as ceramics) form a special branch of electron-emitting compounds that, when subjected to very rapid electronic field switchings (polarity switchings) will emit from them copious bursts of electrons. And what is significant about this effect is that even many quite common materials will exhibit this ferroelectric electron-emission effect when placed in a rapidly switching electric field [note: 3].

So here is the most feasible solution for getting large amounts of energy out of these top spheres. By combining together the above mentioned attributes of both ferroelectric and magnetostrictive materials, by inserting or doping into the spheres certain ferromagnetic-magnetostrictive compounds (such as nickel, cobalt, or iron, for instance) together with electron-emitting ferroelectric ceramics such as lanthanum, erbium, yttrium, or neodymium (or even elements such as lead zirconate titanate (PNZT) doped with neodymium), then the energetic (and well researched) ferroelectric electron emission phenomenon could be taken advantage of in these top spheres by using the magnetic forces that are directed through them. One set of compounds would work on the other set to produce the electron emissions needed by the UFO (in fact such a combination of ferromagnetic-magnetostrictive compounds was discovered in the so-called 'iron balls' from the burned-out remains of one of these spheres left behind by the Dalnegorsk UFO in 1986, for when these balls were cut open and scanned they were found to be composed of magnesium, iron and nickel, together with lanthanum and cerium—for more on this see below section 5.10).

This will raise a suitable platform from which to progress further, and although its form is still somewhat crude right now there is established a base from which substantial progress can be made to produce a glass-crystal that, with the help of the aaUFO's intense magnetic fields, can generate large amounts of electron and photon power. Because, by using the abundant electromotive forces available around these spheres upon the above-mention magnetostrictive and ferroelectric elements within the spheres, the resulting electron-photon production can then be used for other parts of the aaUFO's electrical circuit—through the black vortex—from the energies radiated away from the black vortex.

But just to back-track slightly, or rather make a quick detour, in many ways this is the same process that happened in the Searl electricity generator (SEG) turbine, whereby the neodymium in Searl's rotors was made to emit copious amounts of electrons through the ferroelectric emission phenomenon (and the high frequency Lamb-Retherford shift effect which it is associated with) being triggered by its alternating magnetic fields [notes: 4a]. Interestingly, with regard to those generators of Searl, he would have extracted those ferroelectric electron emissions by a form of magneto-electric-induction because his SEG rigs were not given any electrical input to start them off running—and yet once they started to rotate they would continue forever. This would occur by the slightly mis-aligned magnetic field lines of the rotors and the stator bucking eachother, thereby generating a kinetic rotational force that was self-sustaining and which resulted in a self-increasing rotational velocity. Once they started to increase in velocity these rotors would almost instantly have developed electric fields of 10^6 volts/cm through the break-reconnection process. But it would have been the resonance effect from the break-remake shock waves rising higher and higher in frequency that would have then triggered the ferroelectric emissions; and as explained by Searl himself, "What you have basically is a solid state device here (see figure on website for [Searl] in bibliography). The electrons are given off from the center element (which is neodymium), and they travel out through the other elements. If the nylon had not been put there, the SEG would act like a laser and one pulse would go out and it would stop, build up, and another pulse would go out. But, with the nylon in there, the nylon acts as a control gate, and that control gate gives you an even flow of electrons throughout the SEG." [note: 4b].

contraction of magnetic strips into magnetic bubbles
(adapted from T.H. O'Dell (1974) *Magnetic Bubbles* as permitted
and acknowledged by Macmillan publishers)

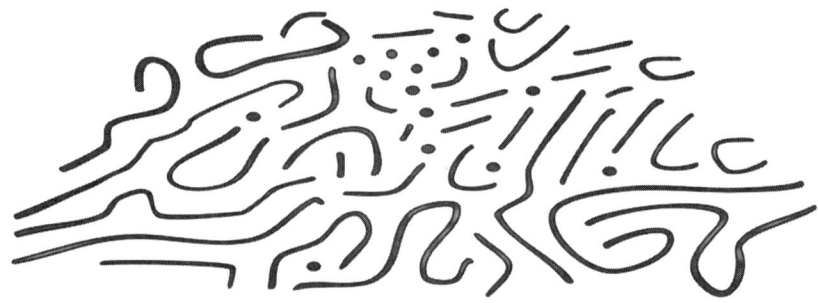

magnetic domains in the process of transition to magnetic bubbles
(adapted from T.H. O'Dell (1974) *Magnetic Bubbles* as permitted
and acknowledged by Macmillan publishers)

Figure 39. On a microscopic scale magnetic flux lines have several stages of existence, magnetic strips can break down into bubble-like domains

Therefore, with respect to the stressing of the ZPF field of virtual energies and the conversion of those virtual energies into actual electron-positron emissions by the SEG, one of the most important parts of the Searl mechanism was the function of the nylon layer in his rollers that acted as both an accumulator and a rectifying gate for the electrons [note: 5]. And it should be noted, nylon is itself another ferroelectric material! Incidentally, recent studies have shown that the ferroelectric emissions from lanthanum are almost twice that of neodymium, and so Searl's design of generator may give even more output if the SEG were to use higher yielding lanthanoids (and it might also benefit from a more effective nylon too). This factor should be seen as food for thought and it may prove to be a beneficial line of inquiry as a complementary mechanism by which greater advantage could be taken from the emission effects in ferroelectric materials as outlined above. In this respect what John

Searl intuitively stumbled upon has proved to be a discovery years ahead of anyone else researching this ferroelectric emission effect. And most intriguingly, the Searl Effect mechanism is light years ahead of the Casimir route, which in my opinion when considering it has yielded so little return, has received a disproportionately large amount of research funding toward, supposedly, developing useable power through the ZPF field in space.

But getting back to the aaUFO's electron-emitting top spheres, this extraterrestrial-derived combination of materials can be compared with a relatively new branch of science which is currently developing lithium-quartz glasses, which are glasses with high ionic conductivity and are used as solid electrolytes for use as power batteries in compact electronic devices. Although obviously, the comparison ends where the latter's power outputs are not suitably amplified by any Lorentz magnetic forces [note: 6].

4.5 Ferroelectric Emissions and Polarization Reversal Emissions

The basic phenomenon of ferroelectric electron emission (FEE) ceramics is where streams of electrons are repulsed out of the surface of that dielectric material, or out of its plasma formation around it, under conditions of fast polarity switching of the electric field being applied to that dielectric. Because, on a molecular scale when the crystalline cells are forced to switch polarity far too rapidly for their normal compensating-neutralizing process to neutralize them back to a state of equilibrium those crystalline cells create multiple layers of electrons which then pile-up on each other (as mentioned above in note 3); this generates an abundant accumulation of non-neutralized high-energy electrons which bulk together and generate a Coulomb repulsion force which ejects those electrons OUT FROM that crystal material and generates substantial electric fields around these materials [note: 7].

Such an effect, for instance, can occur in lead-lanthanum-zirconium-titanate (PLZT) ceramics where electron emissions have been observed as high as one hundred amperes per square centimeter [note: 8]. The best known ferroelectric materials are the ones with ilmenite or perovskite atomic structures because these are most favorably disposed to undergo displacive phase transitions in their cellular structure (and so it is not surprising that they have received the most attention from researchers, and that some of these materials are being developed for use as electronic cathodes and for electron guns).

With such an emission mechanism the FEE material acts less like a semiconductor and more like a saturated charge-capacitor being emptied through a short pulse of high current. In fact, in microelectronics it could be likened to a charged capacitor being discharged through an oscillated thyristor, which is the type of electronic circuit that lasers used to use to trigger their pulses.

Physicists who have been working on these ceramics have found that charge densities for electrons being ejected can be extremely high (in the order of 10^{14} electrons per cm^2 with electric fields rising up to the GV/m range [Gundel 1989]), and in some cases the current densities from ferroelectric emissions have risen to as high as 10^5 A/cm^2. What is significant (in respect to the light bars in the aaUFO that the ETs showed to Betty Luca) is that in many studies these electron ejections can be extremely energetic, having kinetic energies of up to 25 keV [Riege 1994], and possibly even as high as 100 keV [Riege 1998]. But most significantly these ceramics will have emission repetition rates of at least 100 kHz and in some configurations this repetition rate can rise into the MHz frequency range [notes: 9].

But then even further away from classical emission mechanisms, and taking into account the structure of a large network of ferroelectric materials (for instance, embedded inside the four lower spheres which each have a volume of 0.7 mtr^3), what makes FEE emission truly unique is that it does not always require a continually applied external electric field for the emissions to trigger. This is because the emission of electrons in bulk will generate their own electric fields across the cellular domains of that dielectric material, and these generated electric fields will trigger a back-switching reaction which can itself switch the phases of that dielectric material. Which effectively means that those self-generated fields can under certain conditions bounce the cellular polarization back and forth between one polarity and the other. Some ceramics are more susceptible to this back-switching effect than others, for instance certain types of the perovskite and ilmenite compounds can be made to switch polarity and reverse-switch polarity, and obviously keep on switching backward and forward between polarities. A regular frequency of electron emission pulses would then occur from these spontaneous polarization reversal actions within the FEE material itself—which would generate powerful electric fields (see for instance how displacement transitions in ferroelectric emission dielectrics can be triggered by their own self-generated electric fields in [Rosenman] [Kittel] [Scott] and [Yakovlev]).

And looking at this another way, and again, especially concerning the lower dielectric spheres, through this self-induction process, at a molecular level, there will be self-induced from these reversals of phase transitions (that will affect polarity reversals throughout the dielectric material) a corresponding movement of electric charges as phonon transportations of energy, through that solid material; and these energy transportations will move in tune with the reversal switching to set up an alternating phonon resonance effect, or oscillating plasmonic resonance effect, within that dielectric material. So in other words electron charges can be made to oscillate within the dielectric structure of, say, the lower spheres (and this reference to phonon resonance will be further expanded upon in chapter 5).

4.6 Colored Luminescence from Within the Spheres

Obviously, if these ceramics embedded into these glass-crystal spheres can be made to oscillate and emit energy then they will be emitting energy that will have specific coloring. Whatever this colored radiation is it can then be helpful in providing an indication of what the ceramics might be embedded into the dielectrics used by the extraterrestrials. Because the phenomenon of generated color centers is well known to physicists researching crystallography and who work on the different kinds of glass-crystal [note: 10].

This of course, is not to be confused with glass tinting which occurs through a manufacturing process; where it is well known that metallic compounds can be added into glass melts so as to give that glass permanent tints (ie green tinting comes from chromium oxide being added). So specifically, these are color centers being triggered inside the aaUFO's spheres which appear to be coming from energy emissions, and which are temporary glows. And pure and simple they would correspond very favorably toward being ceramic-ferroelectric energy emissions coming from deep within the crystal or glass material.

Betty Luca has mentioned seeing these strange effects within the large lower spheres, from the times when she had to travel in these particular UFOs, or from when she had to stand next to them, saying also that what she noticed was that when these spheres were energized they had small glowing, pulsating zones deep inside them which shone with red, blue, green, and then white light; while on other occasions she even saw long wires or rods embedded into those lower spheres (as in figure 38 above) [note: 11].

In trying to find a scientific explanation to these pulsating energy zones I did find out that sodium silicate glasses, for instance, will luminesce brightly in an electric field, although the glow wouldn't last for much longer than a few microseconds and then it would quickly fade away. But then I discovered, more promisingly, that not only will some materials glow through a range of different colors but also that certain materials will emit colors as a phosphoresce glow when energized, and the materials that undergo phosphoresce emissions glow for several tens of seconds—especially under conditions of photoexcitation, and particularly from being energized by high frequency ultraviolet (UV) or X-ray photon irradiation. So, here at last was another important step forward in the understanding of these upper and lower spheres observed inside a UFO [note: 12]. The phenomenon of photoluminescence has been explained as a quantum confinement effect, although this seems mostly to explain the effect of a blue-shift or a red-shift in such materials [note: 13].

A brief summary of the various colors which can come from various materials would indicate: that quartz can show a brown coloring when irradiated with neutrons, although excitons when recombining in quartz will give off a blue luminescence or a green luminescence if the quartz has been doped with germanium. At very low temperatures (too low for this UFOs energy system) quartz can form very intense luminescence centers when irradiated by a pulsed laser (at temperatures below 180K). Erbium when doped into a crystal can give off green luminescence [note: 14]. In some of the fascinating studies carried out by Russian physicist [Kapustin] it was noticed that when ultrasound frequencies were applied to a chunk of transparent quartz crystal it would cloud over and then visibly glow with fluctuations of different color [note: 15]. And not only color centers, but also colored halos or colored rings have been observed to appear in non-crystalline solids from electron diffraction, neutron diffraction and X-ray diffraction [note: 16]. Interestingly, the perceived colors of yellow, green and blue luminescence from exciton recombination have been recorded in CU_2O (copper oxide) semiconductors, albeit under super-cooled conditions [note: 17]. And finally, zinc-based phosphorus materials used in conjunction with ferroelectrics can also glow with a variety of colors of yellow-orange, red, green and blue [Talbot].

So obviously, the color emissions are a radio frequency (RF) resonance effect occurring in the atoms within the crystal's structure.

There again, if these colored centers are an indication of discrete energy emissions then perhaps within the crystal the phenomenon of particle confinement is being utilized and should be given consideration for further research. This same technology would, incidentally, be quite evident in the optical fiber transmission and power lines that the ETs have been observed to use [note: 18].

4.7 Quartz or Glass, Crystalline or Amorphous?

While its all very well picking the best sorts of emitting ceramics to use in these power spheres, in order to utilize any group of electro-chemical compounds there will be needed a host material to keep all these compounds located and inter-connected and protected (from oxidization). This draws back to the first tentative descriptions of the aaUFO's crystal spheres (in the previous chapters) where it was questioned whether these spheres were made of quartz or of a different kind of crystal or of a special type of glass. Betty Luca's description of them comes from her observing that they looked like a transparent glass-like material which resembled cut-crystal glass with their outside surfaces of these spheres being faceted [note: 19].

Interestingly, it will not so much be quartz crystal that may prove to have the special qualities needed for this task, it may turn out that certain types of glass will have more to offer. The main difference between the two is that quartz is predominantly a crystalline material with a repeating lattice structure, while glass is an amorphous material having no repeating crystalline or long-range-ordered lattice structure, but then glass does however have the necessary atomic structure which can act as a host for charge-donating materials (or ceramics such as the rare-earth lanthanoids which are ferroelectrics). Although, as will be shown in the next chapter (chapter 5) there may yet be another twist in the tale of this special glass.

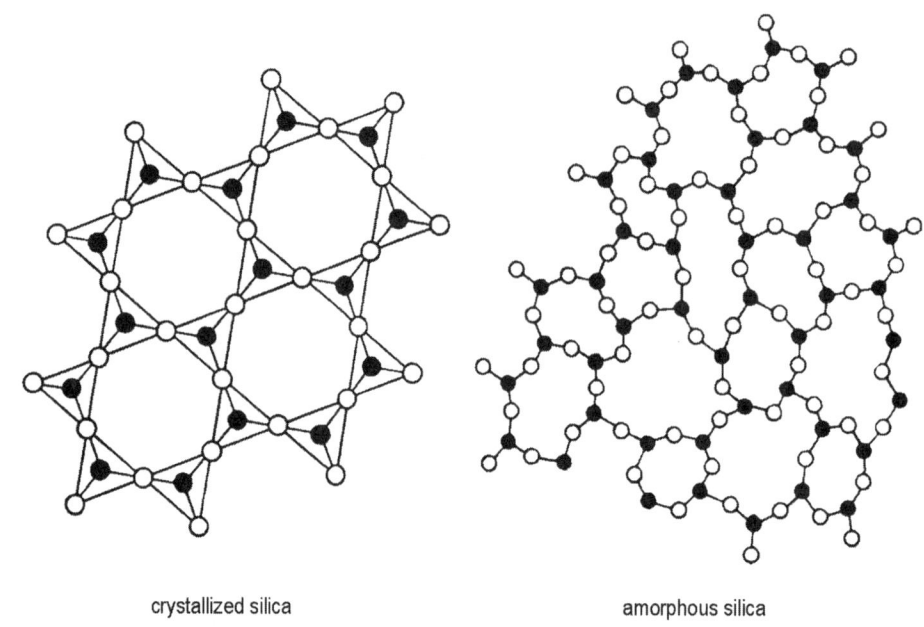

crystallized silica   amorphous silica

Figure 40. Crystal structures of crystallized silica and amorphous silica

But the complex question as to how the aaUFO might generate enough power to fuel its propulsion drive system is now beginning to supply some promising solutions. I hope it will be appreciated how the scant images recalled by Betty Luca and published in all the Ray Fowler books are becoming quite invaluable to this study, and that they will continue to prove invaluable when it comes to progressing any earnest R/D lab work for UFO construction at a later date; as will all the other pieces of information learned from the hundreds of abductees whom have experienced some of the other technological advancements at first hand under the watchful eyes of the extraterrestrials.

Hopefully a way will be found, now that this UFO information is coming out of the closet, to build up a coherent knowledge-base about extraterrestrial sciences, that scientists from all around the world can pool from and develop into as a conditional non-warmongery step forward, and be put to use for the civilian public.

4.8 A Crystal's Acoustoelectric Phonon Wind

I'm not going to use them as proof for the above assumptions, but I will call to the reader's attention a particular group of images that Betty Luca was shown by the ET beings, and which she has published in her ASBT booklets. I will present those images here because they show particularly clearly the carbon-quartz-like lattice structure that the extraterrestrial Greys showed to Betty, which they indicated at the time were structures analogous to the materials they used, but which also, when understood fully would show how the Grey's energy-material-technology worked. With this latter factor in mind what was shown in those holographic-type images was essentially a solid crystalline structure inside of which was contained wind, smoke (or gas), and lightning—all as if in an arrested status. Or, it could be that the ETs were hinting at the fact that some crystalline materials do have latent inside them these very forms of energy—if only their cellular structure can be manipulated in a way that would RELEASE them [note: 20].

My guess is that in these images, which were shown to Betty Luca on three occasions and which she has drawn in diagram form, there are shown what I believe to be the technology of ferroelectric emissions using ferroelectric ceramics embedded into dielectric glass. Because, the main ingredients to the material the extraterrestrials are using would appear to be wind, smoke-gas-plasma, and electricity, in a honeycomb lattice structure.

Indeed, concerning the above drawings what is clearly shown is a honeycomb or tetrahedral, or wurtzite cell structure, which is commonly found in zincblende compounds, or it is the diamond cell structure found in quartz and carbon (and perhaps when it becomes available the solid form of carbon dioxide glass), all these structures are known as the ideal 32 base structures. Depending on what kind of lattice structure that material's cell will have the cells will be made up of cation and anion ions.

The usual way in which electric charge is transported across non-metal material structures is through phonons and plasmons rather then electrons. The mechanisms which do the transporting are called the acoustoelectric effect, and the plasmonic resonance effect. This is because in solid materials the phonon is the equivalent of an electric charge and acts as if it were a gas molecule moving through the lattice spaces of a material, and its influence is transmitted through a material substance as acoustic waves of extremely high frequency (in the order of 10^{12} Hz, which is around the infrared waveband).

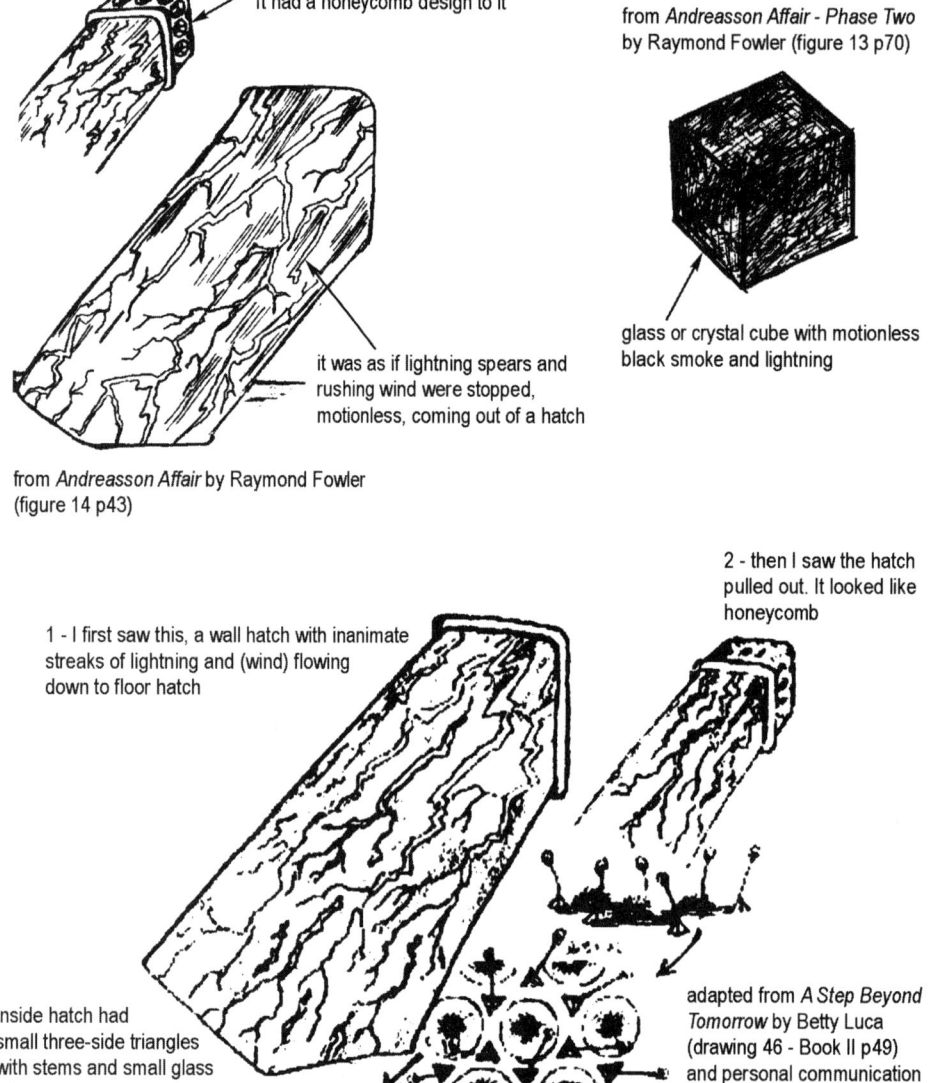

Figure 41. On several occasions the extraterrestrials took time to show Betty a particular facet of their energy storage system

Also within this huge field of research, but involving electromagnetic and ultrasonic excitation of doped glass-quartz, there can be energized near the surfaces of these structures quasi-particles similar to phonons which can then be made to couple together with electron-and-hole pairs and to propagate through the crystal lattice at the surface of the glass or quartz as exciton particles.

Phonons can also interact with electrons alone though, and when a surface acoustic wave (SAW) is propagated around the outside of a crystal, or just under the surface of the crystal, these acoustic waves transfer some of their momentum into the electrons and electron-hole pairs (as exciton molecules) so then the electrons can be mobilized around the dielectric, and so these excitons and exciton molecules (and other charged particles) can be accumulated as an electric plasma field on the outside the crystal (as is also well known that SAWs acting on piezoelectric materials will also generate electric fields in those materials). Subsequent to this energy transfer process the surrounding plasma field can possess what is called an acoustoelectric DC current. And this phenomenon results in a tangible pressure at the face of the crystal known as 'phonon wind,' which is a collective emission of electrons, positive charges and excitons, and these can be made to move a short distance beyond the surface of the crystal, and can be held in the surrounding air just outside the crystal's surface by a resonating acoustic wave field [note: 21].

Is the phonon wind, which is the gas of electrical charges that moves through dielectric materials, the wind-gas-lightning being depicted by the extraterrestrials in their strange analogous representations for their power materials?

4.9 Sphere Fabrication with Sol-Gels of Silicons and Lanthanoids

More to the point, where can the special glass-crystal material needed for the UFO's spheres come from? The best bet would be that such glass would need to be especially designed, and then especially fabricated. Certainly, I feel it would be rather naive to think that the ETs are using in their UFO power drives some odd chunks of quartz crystal they might have found on a beach somewhere on some exotic planet with double shadows! Having said that, one of the PRINCIPLES behind making this special glass-crystal may come from the way crystalline structures are formed and found deep within the core of some distant planet. For the principle behind making such a glass may turn out to be similar to the way diamond is formed deep within this planet, ie at extremely high pressures (more on this in the next chapter).

But as a starting point in the pursuit to duplicate such technology I would suggest that it might be more advisable to think along the lines of a sol-gel type of technology whereby such spheres can be specifically designed and fabricated from a proportioned mixture of specific materials.

If these spheres were to be fabricated to-order they would need to have in them a combination of compounds that can emit electrons, and also compounds that can trigger those emissions which can be induced into action by magnetic fields (which are plentiful especially around the top spheres). The lower spheres might need to have, especially on their lower curve (where they protrude below the craft), just inside their outer surfaces, a predominance of positive vacancies (or holes), and have a structure that would facilitate charge movement from the outside (ie the virtual energy fields that have ZPF fluctuations) into the inner-bulk of the lower sphere's material (more details in following chapters).

I'd tentatively suggest a combination of magnetostrictive elements (such as Terfenol-D) that would generate the magnetic-field-induced oscillations; an abundance of ferroelectric elements (such as erbium, yttrium, lanthanum, or neodymium); and a group of high-mass elements such as lead. For instance, such a group of compounds would then be very similar to the tried-and-tested PLZT formula of ceramics (as mentioned in note 8 above). And then all these compounds would need to be doped into a special sort of glass-crystal host material, for instance, a silicate glass. I'm not a electrochemist, but I would think that there should also be an electro-positive compound included too (made suitable for the outside protrudences), because then a sequence of charge-transferring jumps could be set up around the outside surfaces of the lower spheres in conjunction with the intrinsic piezoelectric mechanisms within those materials, to draw the outside (ZPF) charges in and through these spheres (to where they can be absorbed into the craft's electronic circuit). Finally, whatever would be used as the host material it would also need to protect the ferroelectric compounds from the air, as these ceramic elements would quickly oxidize (see above note 2).

All of the above requirements would suit the sol-gel chemical fabrication and manufacturing process. In sol-gel fabrication the starting materials used in the preparation of the base 'sol' would be inorganic metal salts or metal organic compounds (such as mentioned above). Then (as according to the sol-gel manufacturing process in my references) this compound would be subjected to a series of both hydrolysis and polymeration reactions to form what is called a colloidal suspension (which is the composite sol). After these two processes the composite sol is cast into a mold so that a wet gel can form out of all these components. Then, through further drying and heat-treatment the composite gel would eventually evolve into a dense ceramic-glass material of the required shape and size [note: 22].

If not by the sol-gel method then these energy spheres would ideally need to be loaded with their highly productive electron-emitting compounds (combined with the other elements mentioned above) in a straightforward melt process, such as is used to dope lanthanoid ceramics into a glass-crystal.

Whichever fabrication method can be used to produce these spheres, with the top spheres fabricated in the right way, then the highly energized and pulsing magnetic fields permeating right through these spheres would force the magnetostrictive compounds to pressurize the piezoelectric elements of the (silicate or carbon glass) to create a high voltage electrical (piezoelectric) strain in that glass-crystal, and with such a high voltage strain repeatedly applied to the sphere's cellular structure a resonant frequency would be set up within the structure which would act upon the ferroelectric ceramics (in the spontaneous polarization mode), so that those elements doped into the silicate or carbon glass would trigger the copious emissions of charged particles to be ejected (repulsed) out from the surface of those spheres.

From the way they are used in the aaUFO their immediate environments are already energized, from the power of the central black vortex, from the magnetic fields passing through them, and from the fields of acoustic waves coming up from the lower spheres and focusing onto those top spheres. So then the ejected charged particles streaming out from these top spheres would automatically be grabbed by the Lorentzian forces converging around them (as shown in figures 35 and 42), and they would curl into a highly excited cloud of fast-moving electrons-positrons around those spheres (again, this would be courtesy of the pulsing of the magnetic flux lines converging all around the top spheres).

Such emissions could appear as both a frenzied plasma surrounding the spheres and as multiple ejections of charged particles coming out from every angle of the top sphere's surface—just as Betty's drawing shows—and once these copious emissions of charged particles were being pumped out, even though the electrons would be intensified by all the magnetic field lines around those top spheres the photons given off them, as well as some of those electrons, would either be drawn into the rotating forces of the photon orbit ring which is rotating immediately inside the orbit of the top spheres, or they would be forced to gyrate into the gyrating storage field—which will be gyrating through the craft's planar wave guide, and which will be rotating on the outside of the orbit of those four top spheres.

4.10 Surface Facets to Enhance Electron and Phonon Exchange

In fact, by coupling the above image of a frenzied energy field surrounding the top spheres with the numerous ionizing mechanisms referred to in the previous chapter, which are also expected to be activated in the vicinity of the upper inner edge of the toroid, it will be seen that the energy environment around these spheres will be in an extreme state of excitation. A lot of this particle energy will be grabbed by the magnetic fields converging into the top spheres so that while most-all of these air-born charged particles around the top spheres will be imbued with Lorentz forces, and some of those will be gyrating around the flux lines producing the 'emission bars' that Betty saw—there will be a great amount of emissions from those spheres that will be dragged into the photon orbit ring on their way into the central vortex, and there will be other emissions that will be gyrating very closely around the outer surfaces of those top spheres.

In conjunction with these gyrations around the top sphere surfaces there will be the influence that the two counter-rotating fields (of the one going with the central vortex, and the one going with the storage field) will deliver on this encircling energy around each of the top spheres. And obviously this is a design feature that will greatly amplify the outputs of those spheres if it is understood properly and catered for by anyone considering building these drive systems. Because the counter-rotating field's influence will accelerate the energies already spinning close to the sphere's surfaces to bring their orbits inward into the smallest possible radius around those top spheres.

4—Ferroelectric Emission Sphere Fabrication

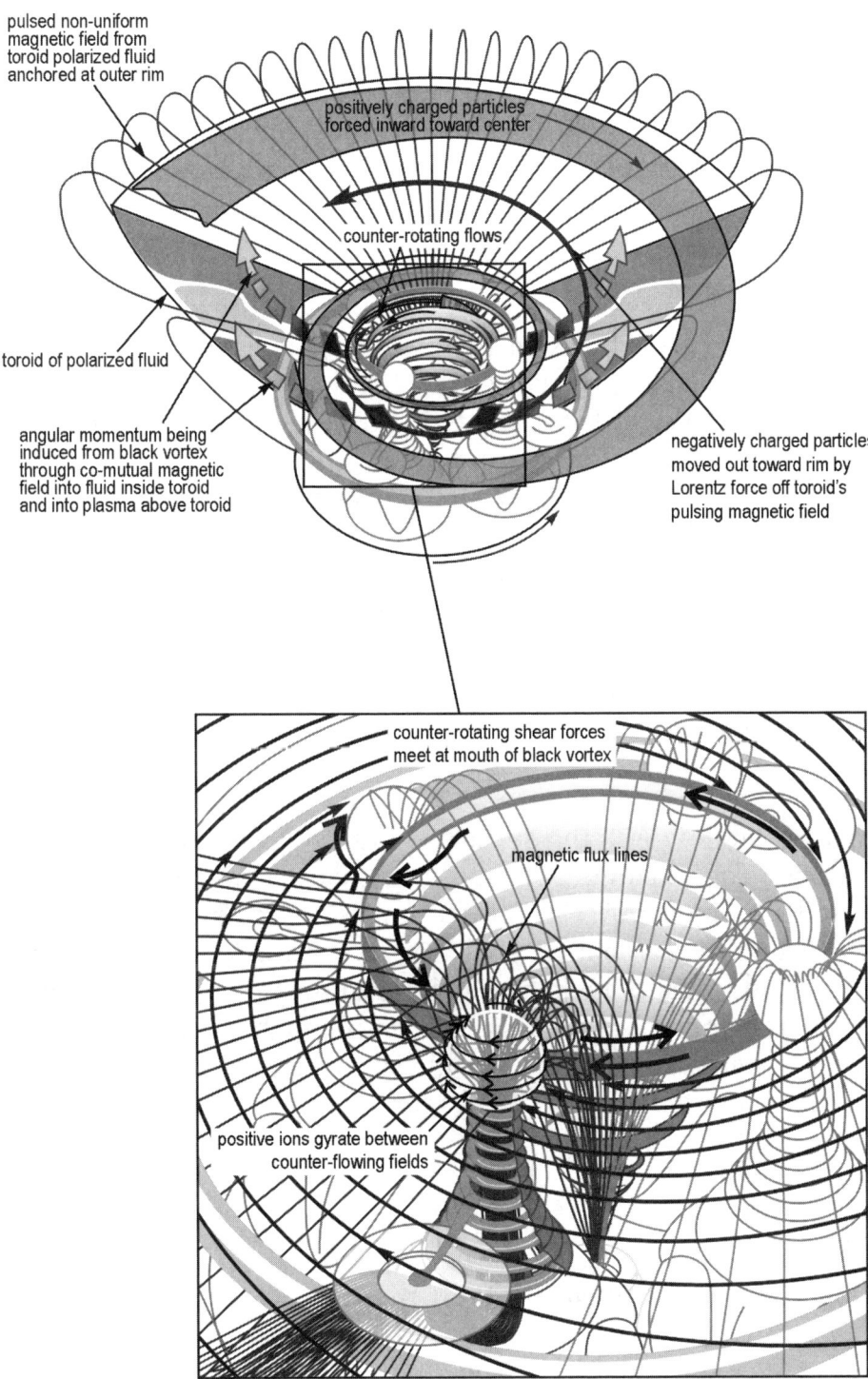

Figure 42. The two counter-rotating flows meet at the photon orbit ring—exactly where the top spheres orbit around the vortex

Meaning that if those surfaces are faceted then when those orbiting charged particles speed up faster and faster while they skim over those facet flats and collide into the facet edge-peaks around those spheres, then this profusion of collisions will generate even more electrons and in the process increase in potential the electric fields around those spheres—and this will equate to yet another energy amplification process.

So here in this amplification process is an additional factor that would compliment the phonon wind emissions that would be excited around these top spheres to mobilize any charged particles in their vicinity. Because, the electrons speeding around the facets of the top spheres would be affected by the flats and edges of those facets, much as if those electrons were passing over a corrugated waveguide (or sawtooth waveguide). In fact some may even tunnel THROUGH these protruding edges, and in doing so connect with the surface excitons and other charged particles which will be gathered just inside the surface of those top spheres. And so at these facets an energy exchange process would occur whereby the electrons and the electron-loaded surface acoustic waves, also moving over the edge-peaks and flats, would interchange momentum with all the other energies operating within the glass-quartz (and trigger, for instance, all the internal ferroelectric emissions). This outside surface spinning particle effect I have dubbed the Ginzburg-Grigoryan effect [note: 23].

This therefore, would be how ALL the surface energies of the four top spheres and their outer plasmas (of excitons, exciton-molecules, electrons, positrons, polaritons, phonons and photons), AND all of the emissions coming out of their ferroelectric inclusions, would be given a very vibrant coalescing boost to form high velocity spinning energy clouds just outside the surfaces of these top spheres. This is why I suggested in the previous chapter that the emission bars may look very powerful around the top spheres, but they were only showing a small percentage of the power that COULD be generated around them (more on this in section 5.8) [note: 24].

4.11 Whirling Zones Inside Focused Acoustic Orbs

At the head of this chapter I did pose the question, why do these spheres produce excitons? Well, the greatest advantage with excitons is that they will have longer lifetime than other particles, which is ideal for these particles to be created inside the surfaces of these spheres and then to be extracted away from the spheres. Eventually these excitons, because they are formed of a negative and a positive particle, will go through the annihilation process to transform ultimately into photons—which of course are the perfect forms of energy with which to feed the black vortex engine of this UFO.

The real beauty of these arrangements will be that the black vortex will consume most all the annihilation photons and sub-atomic particles churned up in the numerous atom-busting mechanisms around the four top spheres. There will also be, where the two counter-rotating charged particle fields meet, a whole range of collision processes where counter-rotating particles will smash into each other to result in Comptonizing ionizations (especially through the Penrose-Williams pair-production processes happening around the photon orbit ring), and those shearing forces will lead to dissociation of molecules into sub-atomic particles, photons, and gravitational binding energy [note: 25]. Curvature ionization, synchrotron radiation, spark-gap ionization will all add into the frenzied production of copious emissions buzzing around these top spheres.

Interestingly, with the acceleration of so much of this sub-atomic and photonic energy around the photon orbit ring it may be possible that some of these high-energy photons and particles that come spinning AWAY from their collisions with the particles of that ring, will pummel right back into these frenzied energy orbs surrounding the top spheres to trigger even more secondary emissions.

4.12 Permeability Switching Caps Around Top Spheres

One last point to be made about the top spheres. When Betty was shown the internal workings of the UFO craft she at first noticed that the top spheres were each covered by a cap, which appeared to be quite separate from the tightly-fitting aluminum (brushed-aluminum-like) stem coil coverings that act as magnetic yokes (which direct the magnetic flux away from the outside of the stem coils to ensure that it intensifies through the middle of those coils—see figure 35) [note: 26].

That these covering caps had to be removed, or dispersed, before the top spheres could be seen to emit their very bright photonic energy bars shows one of the possible ways in which the extraterrestrials were controlling the central vortex, and hence the craft's electrodynamic circuits. That these covering caps suddenly disappeared before Betty's eyes—as if they were dissolved into thin air—may have been indicative of them being made of a special material, such as a metamaterial, with the unusual quality of being able to be switched from opaque to transparent when an oscillating electric field of a specific frequency is applied to them.

The covering caps would have represented the notion that the flux lines could be prohibited from passing through them, so that those flux lines would instead be directed around the outside of the sphere-sets. And if the flux lines couldn't pass through these covering caps then they wouldn't route through the stem coils, and the stem coils would not then be able to grab the toroid's quadrant fields and amplify any of the UFO's electrodynamic power circuits, and there would be no break-and-reconnection of the toroid's flux lines either. So in other words, in their prohibitive configuration these caps would effectively shut down the UFO's power circuits. In fact, the controlled switching of these caps to opaque could even be considered as one of the UFO's power-down mechanisms, or it could be representative of what I would call the 'tick over mode' for this type of UFO, because this action of governing the magnetic flux lines passing through the stem coils would bring most of the top sphere's magnetically triggered emission mechanisms to a steady halt. This in turn would de-intensify the central vortex so that it would power-down considerably and just rotate at its minimum velocity. But then, to get the UFO powered-up from this tick-over mode all four caps would only need to be switched back into transparency, which would permit the magnetic flux lines to again pass through the top spheres and the stem coils, to fire up the emissions, and to energize all the UFO's electrodynamic circuits. Perfect control—merely by switching these four caps between opacity and transparency!

My final suggestion though would be that technically these covering caps could represent a way of providing a graduated switching device for the toroid's flux lines (for controlling how those flux lines route through the stem coils). If the covering caps can be switched through a graduated amount of magnetic permeability then they would allow variable amounts of magnetic force to pass through them and through the stem coils. THEN there would be a way to control the top sphere emissions and the power of the black vortex. So these covering caps, simple though they are, could control the craft's tick-over, the craft's power levels, and provide one of the ways to power-down the craft's main electrodynamic power circuit (the other way being through the control of the inner ring-tube).

But the over-riding consideration will be... how to run a transmission line to these central sphere-sets so as to control this permeability function in those covering caps? And this may prove a little difficult.

So was this scene shown to Betty in this way as an arrangement to be brought to the researcher's attention, to emphasize the fact that there should be a way found to provide an adjustable permittivity-permeability shield for these top spheres? Or was it just showing how an on-off permeability switch would be required? Only this craft's construction will provide the answer.

5: Gravitationally-Polarized Power Spheres

Contents:
5.1 Polarization Changes in Lower Spheres Orbiting Black Vortex	153
5.2 Rotating Forces Enclosed Within UFO's Ergosphere A	161
5.3 Displaced Alignment of Crystalline Lattice Cells	162
5.4 Fitting the Black Vortex into the Craft's Electronic Circuit	167
5.5 Negative Phase Velocity of Metamaterials	170
5.6 Plasmonic Pump from a Thin Wire Lattice Network	174
5.7 Thick Wire and Diamond-lattice Plasmonic Frequencies	176
5.8 Faceted Spheres Inside Microwave Spinning Fields	177
5.9 Summary One: Glass Spheres that Think they are Superconductors	181
5.10 Summary Two: Wire Network in Megaconductor Dielectric Spheres	185

5.1 Polarization Changes in Lower Spheres Orbiting Black Vortex

In chapter 4 it has been established that the small glass-crystal spheres situated directly above the stem coils, as a result of the high density of pulsing magnetic fields passing through them, would operate as powerful emitters of electron and photon energy and would create around them metastable orbs of highly excited electron and photon energy for the black vortex to draw from and consume.

Hopefully, it will be seen in this chapter that the lower spheres will have a quite different series of energy mechanisms to serve and to be affected by compared with those of the top spheres. But even so the fabrication process of the lower spheres should have as its starting point the assumption that these spheres will generally be similar to the top spheres regarding the host glass-crystal material, and that they should also be embedded with magnetoresistance-magnetostrictive materials and ferroelectric ceramics.

But because different clues have been forthcoming regarding what the lower spheres are comprised of there may possibly be offered further into this chapter a quite different format to be added into the fabrication process for them. Indeed, this format may be so different that I feel obliged to offer due warning here, that with the unraveling of this rather new and radical format, there may be given to the reader more than one opportunity to raise an eyebrow (or two even) before the end of this chapter regarding the unusualness of what these lower spheres will be capable of doing.

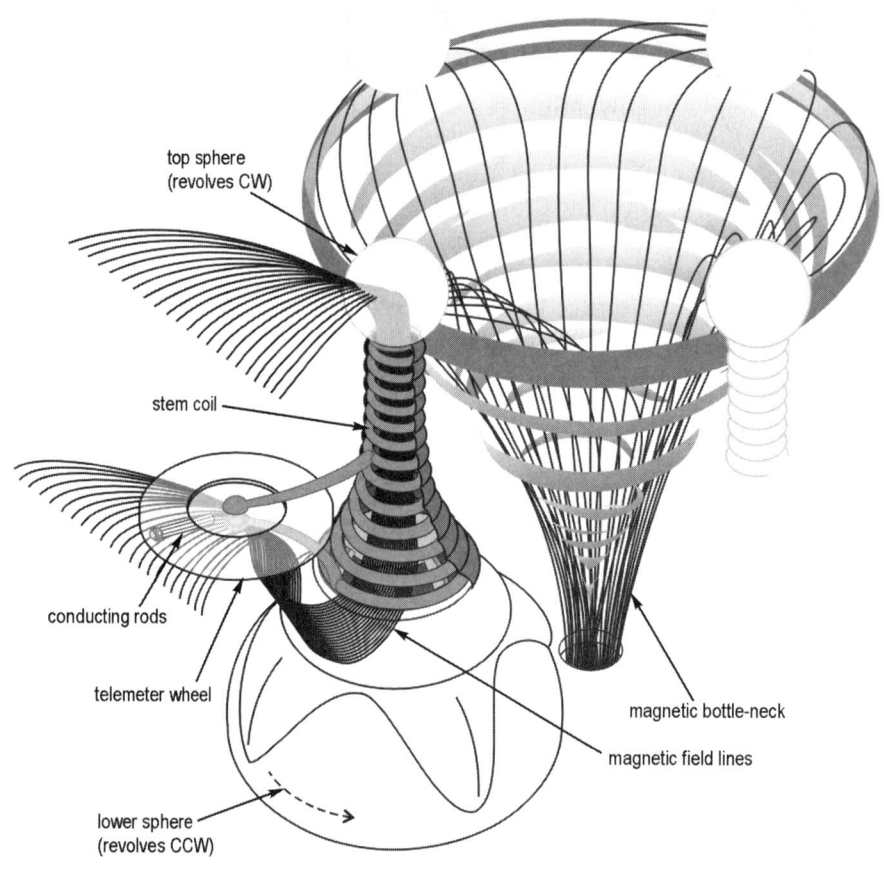

Figure 43. The stem coils direct the craft's magnetic fields through the lower spheres as well as through the top spheres

Unlike the top spheres the larger lower spheres will not end up having the same sort of highly-excited free-electron and photon energy clouds around them. And because they won't be feeding into the black vortex they certainly won't be glowing nearly as brightly as the top spheres will be. The energy environment they will have around them will be quite different from the top spheres. For instance, they will not be sandwiched between two high-energy counter-rotating energy fields. They will however, have immediately next to them, in one direction, the powerful gravitational forces radiating from the central black vortex; and then coming from the other direction, these lower spheres will have the great density of magnetic flux lines that will be passing down through the stem coils, and these will be passing through the tops of the lower spheres. The latter's Lorentz forces will be almost as powerful as those around the top spheres. So therefore the lower spheres will be situated between two very different charge transferring mechanisms, which will induce two very different types of polarization movements within the atomic cells of those lower spheres as they revolve about their own axis and orbit around the black vortex.

But it is the former of these two arrangements that will be the most intriguing because the lower spheres will be tapping some of the gravitational radiation of the black vortex while they orbit around it. Some idea of how this arrangement works can be found in the model for an astrophysical ergosphere, where out in space and inside the ergosphere that surrounds a black hole, a polarization cloud will form around the black hole because of the way the gravitational force radiating away from the black hole will pull on the atomic and molecular structures of the particles forming that surrounding gas cloud: And as that gravitational pull effect will be proportional to the subatomic masses of those particles then it will follow that it will be the more massive nuclei of those atoms that will be pulled away from their normal central alignments within those atoms by the black hole's influence [Bekenstein-Schiffer] [Saa-Schiffer].

The black hole's gravitational forces will pull on the positively charged atomic nuclei more than on the electrons because the electrons, for example in hydrogen atoms, will have 1836 times less mass than their positive nuclei counterparts. And it will be because of this mass imbalance ratio that some astrophysicists have suggested that gravitational influence upon mass will primarily be communicated through this polarization effect, and indeed that it will be mostly through this effect that gravitational forces will be transferred away from the black hole into the rotating energy fields within the ergosphere surrounding it in space. Also, that as a result there will be a bulk-type enhancement to this atomic polarization effect, which will act like a homogenizing force upon all the alignments of every atom's electromagnetic and gravitational-binding forces which will become unbalanced due to that aligning force from that gravitational source.

Therefore, when these bulk polarization motions are initiated within the surrounding gas cloud's atoms and molecules, the resulting force misalignments within each atom will then be accompanied by another atomic force field—which will essentially have to come into play to balance the tendency of the enhanced gravitational pull to rip the positive (and neutral) nuclei away from the electrons. This other inter-atomic force field which would have to develop inside the atom in order to counteract the hole's gravitational pull, to prevent that atom's ionization, would have to be a supplementary one to the atom's electromagnetic force field, and it would have to be an electro-gravitational exchange force which mediates directly with the atom's strong nuclear binding force [note: 1].

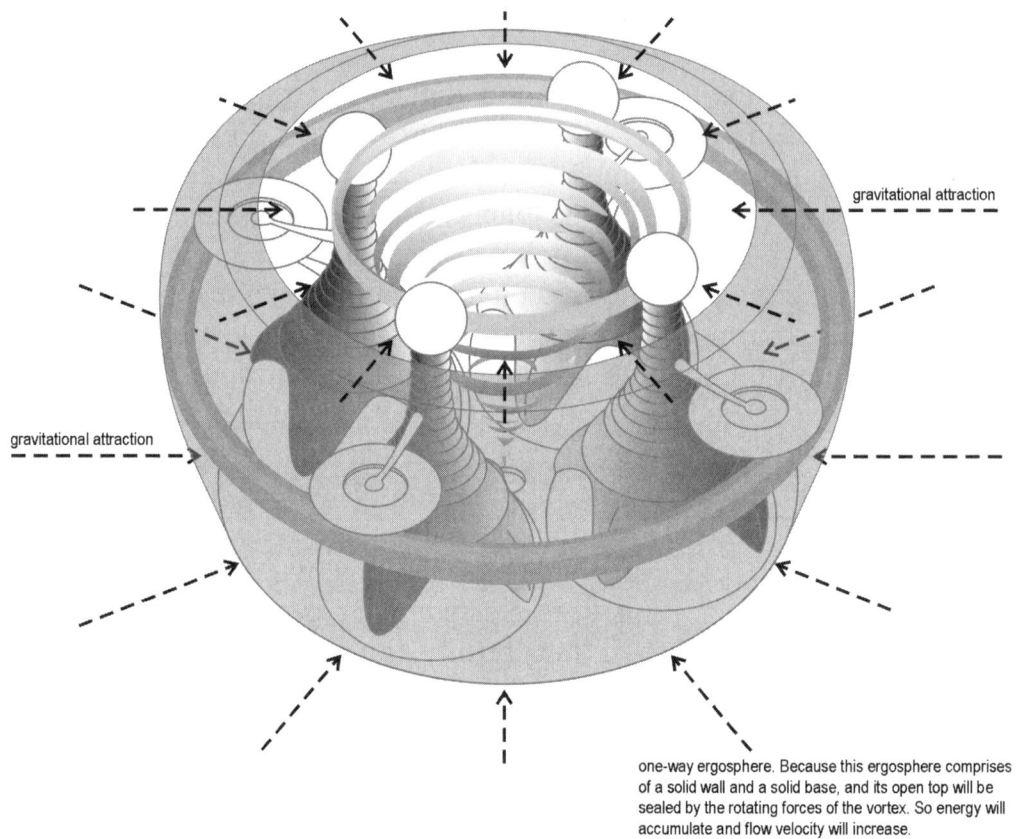

Figure 44. The central ergosphere forms a powerful engine with its central black vortex

Looking at the above example of astrophysical dynamics from the perspective of the aaUFO, that craft's black vortex will produce around it a similar polarization cloud throughout its ergosphere of rotating energy which surrounds its black vortex. And outside that black vortex in the polarization cloud the counter-balancing forces to keep the polarized atoms together will be formed also from the inter-atomic binding-separation forces (which might also be considered as the van der Waals forces) within its atoms and molecules. But there will be a particular difference to this effect within the UFO, because of its smaller size and the way rotating particles inside the aaUFO's ergospheric chamber will have to negotiate around the four sphere-sets (so that their orbits will fluctuate in orbital radius, rather than having a plain circular orbit—as indicated in figure 49 below). This would mean that the misalignment of the electron cloud within each atom will vary and generate a fluctuating resonance in those atoms; and possibly a change to the wave-function of those electrons, from the repetitive changes in the orbital distance between those electrons and the atom's nuclei (as per the term $-e^2/r$). So while this gravitational polarization will take place much like it does around a black hole out in space, in the UFO it will do so at an oscillating frequency [note: 2].

Supplementary to the above assumption of atomic fluctuation coming from the aaUFO's black vortex then the atoms of its polarization cloud, as they orbit around that vortex, will also generate a radial electric field, and although it too would be oscillating, predominantly, it would be more positive toward the aaUFO's vortex and more negative facing away from that vortex. Thus there will be created a concentric electric field with a fluctuating magnetic field (running up through the central area of the UFO) around that craft's black vortex (see figures 21 and 25 above). Also, and as these ergosphere images clearly show, this factor of atomic imbalance will be especially enhanced within the aaUFO's ergospheric cavity because that energy is made to circulate within such a purposefully confined volume. For this ergosphere has a containment wall all around it and so the effect of this atomic polarization will be all the more acutely enhanced. Consequently the craft's variable magnetic field will then become evermore empowered with Lorentz forces, which in exchange, would feed back into the gravitationally-polarized field's generation of its concentric electric fields so that those too will become evermore empowered.

What is also significant, and again, its very evident in these diagrams, is that once energy enters this aaUFO's ergosphere it will be forced to accumulate into the rotating fields surrounding the vortex. And this marks another substantial improvement over the dynamics of the astrophysical black hole (which can only be confined by the magnetic fields of space). This artificial ergosphere's solid walls and its solid rotating base mean that its only 'open' section is at the top—but this open top will be where the mouth of the rotating vortex will be—and so the top will only be open to energy coming into it from above—from the perspective of any energized particles already inside that ergosphere, the top of that whole ergosphere will be sealed-off by the vortex's downward rotating forces preventing anything from leaving. So the greatest aspect of this improvement over the black hole in space will be that while energy can continually be pulled INTO this UFO's ergosphere the only place where energy might LEAVE once it has entered will be through the central base hole! But, and here I reiterate a point already alluded to in chapter 3, in order to progress through that confining base hole that spinning energy will have to become hugely amplified so as to overcome the magnetic mirror effect, from the magnetic fields converging through this hole which will attempt to reverse that spinning energy back up into the body of the vortex. This core section of the black vortex will be analogous to that of the black hole where atomic energies are subjected to intense shearing forces resulting in the release of vast amounts of energy—so, in fact, around this hole (just above the plane of the aaUFO's base disc) would be the perfect place to tap off some of that hugely amplified energy, and this of course, as one might expect, is done very efficiently in several ways by the extraterrestrials (further aspects of this one-way amplified flow will be delineated in chapters 6, 8, 9 and 11).

Without doubt, the kinetic energy of this bulk of rotating particles which is being pumped by all the mechanisms around the mouth of the vortex (particularly by the top sphere emissions) will be enhanced by the containment of that energy within this ergosphere, for as with any rotating body of particles, while the kinetic energy is increasing so will the velocity of those particles also increase. And in general, whenever an object increases its velocity, its relativistic mass also increases (as per e = mc²). So there will be a mass-gain process established inside this ergosphere. And, thinking this effect through a little further, one other effect this mass-increase process would establish would be that there would arise a 'mass-pull-gradient' (gravitational gradient) between this rotating body—of the energy rotating with this vortex and with this ergosphere—and its immediate surroundings. This, purely and simply, is where the aaUFO produces its own localized gravitational field; and this one-way mass-energy accumulation compares favorably to an astrophysical black hole's point of singularity (this will be expanded upon in chapter 10). But even more intriguingly, the mass-gain process in the center of this aaUFO will produce an anomalous effect— one which is certainly not featured in the astrophysical black hole—and that will be the effect of a localized mass-density gradient which will permeate throughout the WHOLE of that domed craft—to which all mass objects will be subjected (which will be expanded upon in chapter 11).

Significantly, the atomic-mass polarizing force from the black vortex and the mass-enhancement upon the energy circulating faster and faster inside ergosphere A will also be felt within the atomic structure of whatever materials are used in the lower spheres. For, just as in the example of the polarization cloud cited above there will result a tendency for the atomic cells in the materials of those spheres to re-align toward the central black vortex, but as these four lower glass-crystal spheres will be revolving and orbiting very close around the black vortex there will result a continually moving and re-alignment effect within the cells of those four sphere's materials.

The heavier positive nuclei of the atoms and ions in the sphere's cells would experience a greater force pulling on them than there would be on the electrons in those cells because of their mass differential, and as mentioned above this (in the atoms of a simple gas) amounts to a 1836:1 weight difference. But within the sphere's glass-crystal that is embedded with high-mass materials the effect would be greatly enhanced because of an even greater mass differential between the more massive positive nuclei and the electrons within those molecular cells. For example, within the heavier lanthanoid series of atoms the mass-differential between one of those atoms and one of its electrons would be almost 300,000:1 and with lead or bismuth this differential ratio would rise to over 400,000:1 (and although this differential comes down when one considers the nucleon cluster as compared to the whole of the electron cloud, ie this ratio would then equate to 4627:1 for bismuth, for example, it is still a very effective differential).

Therefore, within every momentary frame, or snapshot (of the lower sphere's movements as they revolve and orbit around the central vortex), the gravitational pull on those sphere-cells would affect a shifting of the positive nuclei toward the central vortex, and this would be felt inside all those cells as a polarization change because the positively charged nuclei would have their axis displaced relative to the electron orbits. And hence an electrical polarization would result from this gravitational displacement and there would in that snapshot-instant be a potential-difference produced within the crystal cells of that material. This electrical displacement field effect on an atomic scale is only small, but when it happens in bulk in a large body then the effect produced can be substantial. And so generally, whenever these lower spheres are revolving and orbiting around the black vortex, there will be generated a varying displacement of atomic charges in the sphere's material (relative to the vortex's pull and relative to the sphere's revolutions) and so generally there will result a corresponding pumping-up action of the potential-differences (of electric fields) produced within the sphere's structure. In other words, there will be an oscillating electric field produced inside these lower spheres as they revolve and orbit the aaUFO's black vortex [note: 3].

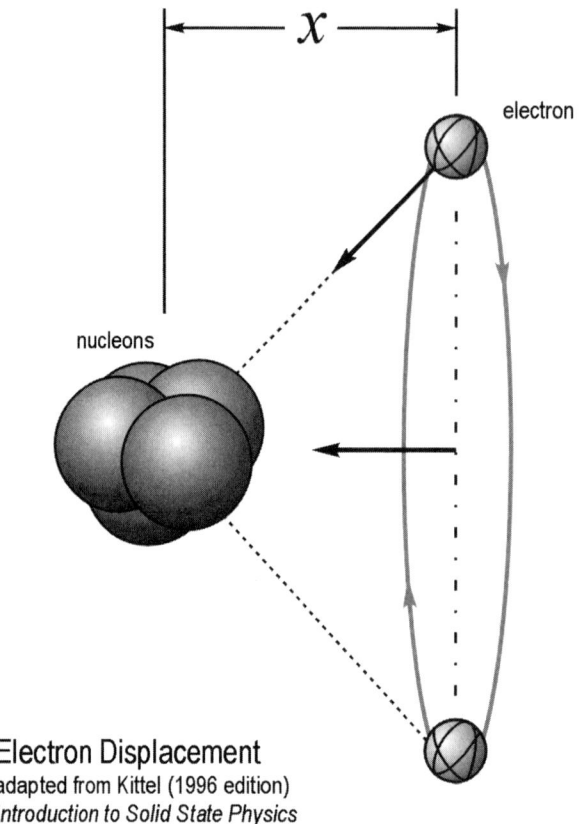

Figure 45. Gravitational forces from the black vortex will displace the more massive nuclei away from the center of the electron orbit

Interestingly, from an electrochemical point of view, this polarization creates within the molecules of the glass-crystal a displacement of the electro-positive cations relative to the electro-negative anions within the glass-crystal's cellular structure, which, most importantly would physically alter the neutralizing positioning of the cations and anions, and undo their equilibrium state-of-coincidence cellular alignments which normally give that material its zero-potential 'ground state' (such a normal 'zero-state' in a material's cellular alignments would physically correspond to the cation-anion alignments being coincident, or in other words in-line and equally-spaced). Therefore, what this gravitational polarization does is move the central plus ion out of that zero-potential neutral state toward the minus ions in that cell, and this move electrically polarizes that cell.

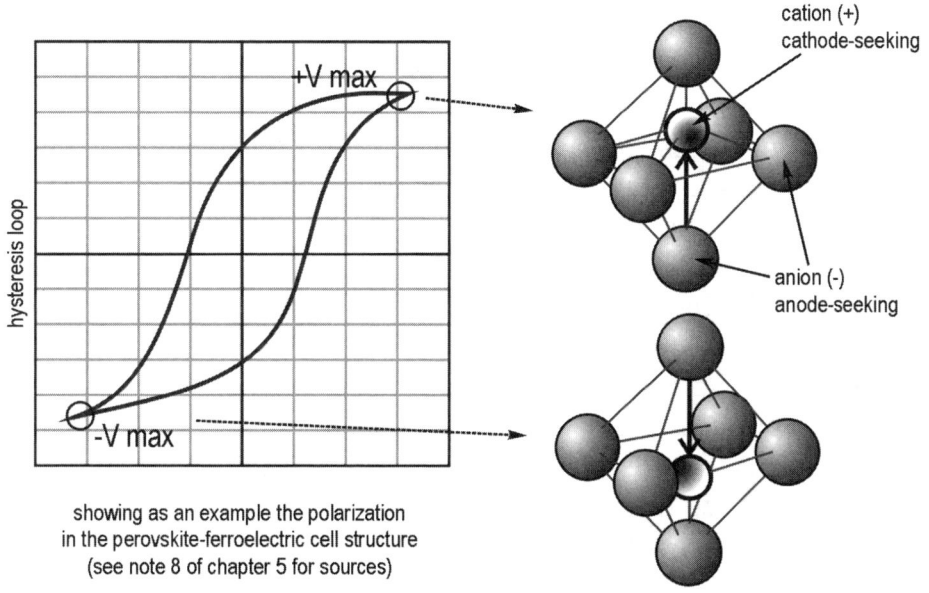

Figure 46. Displacement of positive cations inside the negative oxygen anions

What is even more fascinating though, is that this re-alignment of anions and cations is, I believe, one of the things being indicated in the crystal-honeycomb images given to Betty Luca by the extraterrestrials, as alluded to above in section 4.8 (which was indicated in figure 41 above, and as will be shown in more detail in figure 48 below) [note: 3].

5.2 Rotating Forces Enclosed Within UFO's Ergosphere A

A very prominent feature of this aaUFO engine will be how its gravitational and electromagnetic forces can be increased or decreased in relation to the increase or decrease in rotational speed of the sphere-set assembly, and when these sphere-sets are not rotating at all the craft's power drive can then be seen as being in the 'tick-over' mode. While the actual details of how this rotational velocity is controlled and maintained will be featured in the next chapter, it will be necessary here to assign some figures to a basic rotational velocity for these sphere-sets in order for the reader to imagine how the lower spheres fit into the craft's gravitational and electromagnetic processes.

So in order to set some sort of benchmark I have arbitrarily determined (and these figures can hold for the whole of this study) that the sphere-set assembly will orbit around the black vortex at one-half revolutions per second, and I will assume a slightly higher speed for the revolving speed of the lower spheres (where they revolve about their own axis) of one revolution per second—so as to differentiate their revolving from their orbiting (other figures can be found in chapter 7: note 16).

Going by the descriptions of the top spheres (in ASBT-II p45 drawing 42) which says that the top spheres revolve opposite to the bottom large spheres, and at a different rate of speed, and from the fact that in (ASBT-II p46 drawing 43) the lower spheres are shown revolving with arrows drawn indicating counter-clockwise (as viewed from above). Then I will therefore state that the top spheres rotate clockwise. And again, for ease of calculations, I will assume a speed of 2 revolutions per second for the top spheres. Obviously, the faster the sphere-set assembly rotates the greater will be the UFO's power output because it will mean that the toroid's magnetic fields routed through the top spheres will produce more electron-positron pairs and energize all the more Lorentz force to accelerate the charged particles gyrating inside ergosphere 'B' (of the storage field gyrating inside planar wave guide).

5.3 Displaced Alignment of Crystalline Lattice Cells

Continuing on from section 5.1 and it's brief look at the ergosphere of energy spinning outside the central vortex, another process to consider will be, that while the lower spheres revolve about their own axis and as they orbit the vortex, they will revolve through a volume of fluid (ie of free-electrons, ions and photons) that will be in constant acceleration and which will be affecting momentum-frame-dragging around the central vortex. If the reader imagines a cylinder with a tornado spinning inside it then that's the basic idea, into which should be dropped four equally-spaced sphere-sets (and which in fluid dynamics would equate to four rotating cylinders turning through a rotating fluid), and then into the middle of that rotating fluid would be positioned the craft's black vortex—all tightly nestled together.

But because of their close proximity to the most powerful section of the black vortex these four large lower spheres will experience the greatest gravitational influence. They will also experience the most pronounced tidal effect of differential tension, in a way quite similar to that of the upper spheres (remember, the tidal force effect on the top spheres which hauls energy off the in-facing side with more ferocity than the far side... in section 3.17) but, for the lower spheres this effect will be felt much more acutely on the in-facing, vortex-facing, side of these lower spheres because of their larger diameter: This would equate to a much greater tidal force acting upon the lower spheres than on the smaller upper spheres, from there being a greater angle of divergence (acting between the vortex's core and the large spheres—see figure 47). These will be the gravitational forces through which the atomic structuring inside the lower spheres will be un-neutralized and then polarized.

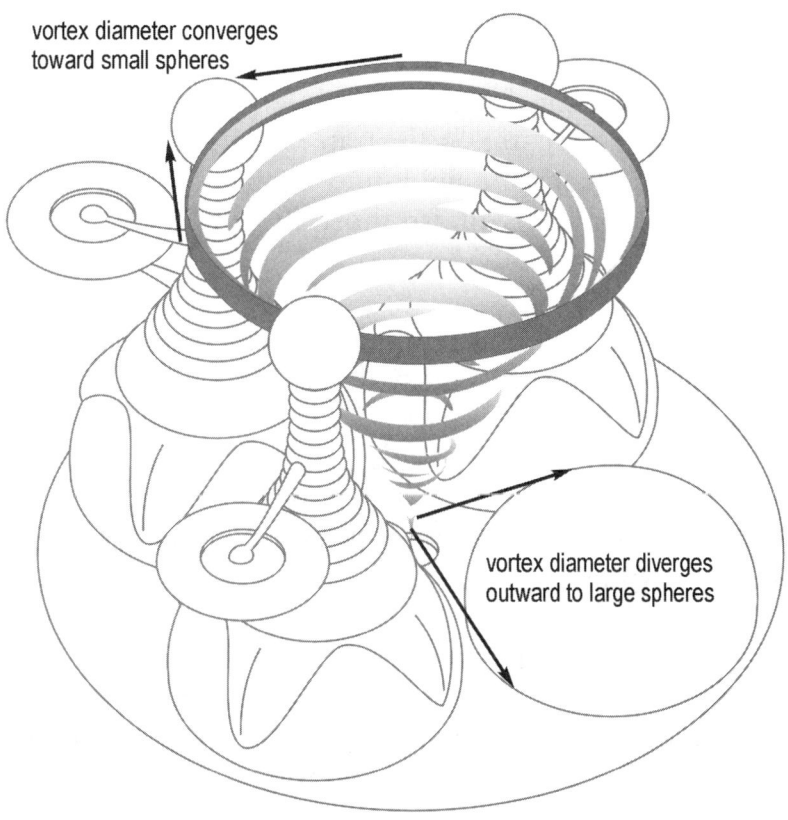

Figure 47. Difference in curvature of influence at top and bottom of black vortex

To offer a better idea as to what this factor of divergence may mean it might be better to compare this with the familiar mechanism of lines of magnetic force, in the knowledge that the actual magnetic attractive force is proportional to the degree by which the magnetic flux lines converge-diverge. Then it might be easier to imagine that if there will be a very localized divergence of the gravitational force lines radiating out from the central vortex, how that diverging gravitational force might act upon the lower spheres. And the reader might, at the same time, compare this divergence toward the lower spheres to the effect upon the top spheres, where the gravitational pull will come from the large diameter photon orbit ring (just above the event horizon), and which shows a marked convergence upon those smaller top spheres; and that from this factor it would follow that the force acting upon these lower spheres will be quite opposite to that acting upon the upper spheres (and if nothing else, this would suggest opposite polarities between the upper and lower spheres).

That the angle of divergence is so wide relative to the whole sphere means that at the small surface area where the sphere is closest to the black vortex there will be a marked variation in the gravitational force induced from the vortex there compared, say, to the force induced at the lower sphere's equator. The craft's magnetic flux density will be all the more greater in the gap between the lower spheres and the vortex too (this is because magnetic field lines cannot enter any sort of quasi-black hole, to pass beyond its event horizon, so those flux lines will tend to bunch up just outside the event horizon).

Similarly, as this will be where the highest intensity of magnetic flux lines will collect, then this will be where the most powerful rotational forces of the whole ergosphere will be concentrated—between the core of the vortex and the inner edge of these spheres. The upshot of all these factors will be that the gravitational polarization effect from the craft's black vortex will be especially localized here and especially strong, and it will be acting most upon a small area of the outer surface of each of these four lower spheres [note: 4].

Together, what these elements amount to is a magnification or intensification of the gravitational pull from the vortex. Therefore, as the spheres revolve round only a small section of its outside surface area will line up momentarily with that gravitational pull, and then as the sphere continues to turn further around that gravitational influence will become reduced on that section (and increased upon the next section as it comes closest to the vortex). Thus by focusing upon a small area of the sphere's crystalline structure the vortex will pull the massive ions of those cells out of their natural alignments—but not all at the same time.

This would provide a possible switching device which, referring back to the electro-positive and electro-negative ion switching process mentioned above, if it can be made to occur as a decisive switching in the crystalline materials, actually within the atomic cells inside these lower spheres, then great advantage can come from this gravitational effect generating a *physical displacement* in those cells which would then affect an electrical-polarity switching process upon each atomic cell inside each of these spheres.

Once the central positive ion has been displaced it will then REMAIN in that displaced position even after the electrical biasing field (or as in this case the gravitational biasing field) is reduced; because of the fact that the ionic displacement will instantaneously generate its own electric field once it moves. And because this localized electric field will be greater than the elastic restoring force, which would attempt to physically push the positive ion back into its neutral position, the displaced ion will remain displaced [note: 5].

Indeed, getting the positive ion to return BACK to its original position will involve just as much effort and a similar sort of field application to that which displaced it initially. For once the positive ion has been displaced in each cell then each cell would require an oppositely-biased electric field to be applied to it before it would move that ion back to the center again [note: 6]. This oppositely-biased electric field is known as the coercive field or coercive voltage (which in ferroelectrics is quite high and in the region of kV/cm).

That these oppositely-biasing fields would be relative to precisely what ceramics are doped into the spheres, would theoretically suggest that there could be a way to design into these spheres provision for a wide variety of power outputs to suit the broadest range of parameters through which these spheres would be able to operate. And so rather than being doped with just one type of ferroelectric ceramic a whole range of operating parameters (including their thermal operating ranges) can be dialed into these spheres and utilized if a variation of ferroelectric ceramics were used within them [note: 7].

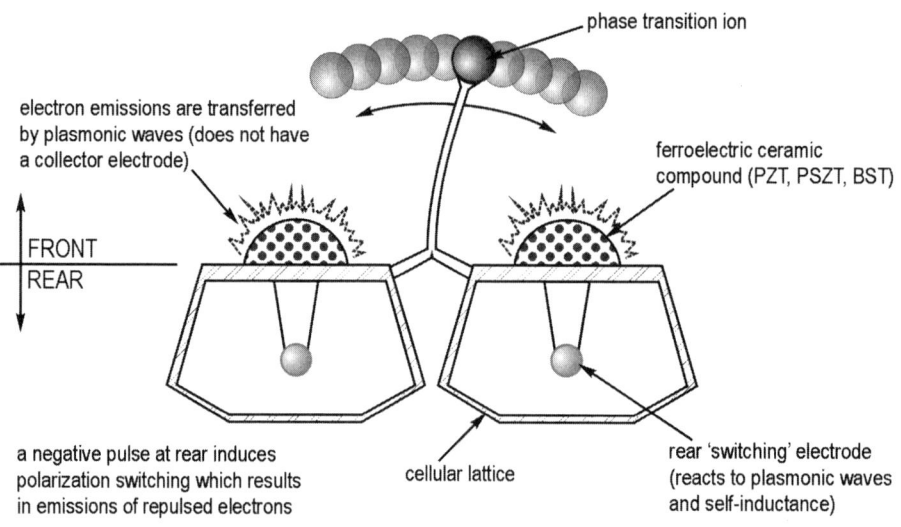

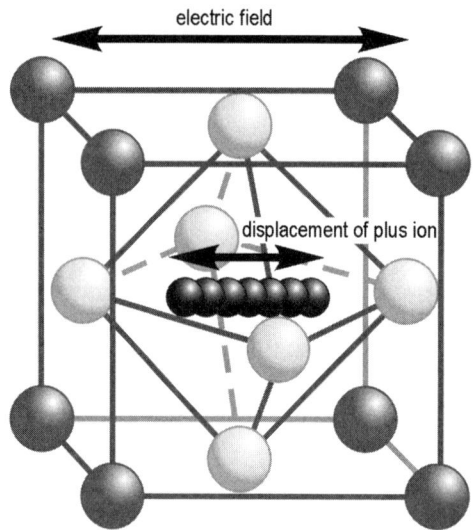

Figure 48. One of the drawings of the extraterrestrial's crystalline power system resembles cation displacement within a cellular structure

Robust switching of the central ion will trigger the most copious ejections of electron emissions from when cellular phase-transitions occur (an enhanced switching effect within ferroelectric ceramics), and will initiate a greater bulk-effect within the material's cellular domains to trigger spontaneous polarizations, and most importantly robust switching will be conducive to polarization reversal switching which is where the ionic phase-configurations repeatedly bounce back and forth from one phase to the other from them being in such a highly excited state, triggering self-induced reversal switching which, as mentioned in section 4.5, takes place when like-charge repulsion forces build up and cause fast-repeating ejections of electron emissions from the cells (there will be more comments about what sort of host material the "glass" of these spheres will need to be made from, in order to help facilitate this required robust switching in section 5.10 below) [note: 8].

Wonderful though this effect is, what has to be borne in mind here is that not the whole of the sphere's structure should be, or need be, fabricated in such a way as to solely to take advantage of this effect alone. Because when all the mechanisms discoverable in these spheres become apparent and are considered together it may well prove advantageous to have several differently configured zones fabricated into such a glass-crystal sphere (each for their own specific purpose, but interconnected so that one emission mechanism would effectively trigger others). In other words, there are still many more mechanisms to look at which might be going on inside these spheres.

5.4 Fitting the Black Vortex into the Craft's Electronic Circuit

Obviously, the copious emissions coming from the spontaneous polarization reversals would be greatly beneficial, for they would transform these lower spheres into power sources—if only these electrically polarizing effects can be made to oscillate backwards and forwards at a cellular level. Before we can look at how that might be achieved there needs to be a step back to see how ANY sort of electrical potential can come out of these spheres.

Looking at what the aaUFO's lower spheres are doing it suggests that a continually moving stress force will be available to act upon the spheres as long as they continue to revolve and orbit around the black vortex. If this stress force were made to pump up and down in strength then by the piezoelectric effect, and by the interfacial effects within the sphere's cells, then those cell's electric charges would have movement induced into them (ie which in a dielectric will be through the movement of phonons) so that they would either move into neighboring cells in other parts of these large spheres and momentarily accumulate in them, or they could be directed into the next part of the UFO's electrical circuit to conduct as electrical current (the next part of the electrical circuit in this case would be the stem coils, on their way to the top spheres).

The question is, how will such accumulations of charge move together in bulk through the glass-crystal material of these spheres from the bottom extremities of those spheres to then conduct into the stem coils? [note: 9]

One of the answers to this question comes from the theoretical model of the electric generator hypothesis of [Blandford-Znajek] which opines that current can be extracted from a black hole's electrodynamics if a conducting mechanism latches itself to both the sharp end and to the blunt end of that black hole, because of the extreme differences in electrical potential established between them.

As can be seen in the top diagram (in figure 49) on the next page, the same potential difference will be established between the top and bottom of the aaUFO's black vortex—so that it will be highly positive at its base and highly negative at the photon orbit ring.

Electronically this would equate to a power source flowing energy through a semiconducting crystal, and then through an inductor coil, through another semiconducting crystal, and then also through an air-gap resistor. In other words through a typical electronics circuit which oscillates (like I say in chapter 1, just like a '555' electronic timer circuit).

So here is the potential for electric charges to be moved out of these spheres because as they would be charged with one electrical polarity (from the bottom of the black vortex) the top spheres would be charged with the opposite polarity (that of the mouth of that black vortex). That the stem coil is the conductor between the lower sphere and the top sphere means that electric current has the potential to flow between them through that coil [note: 1

0].

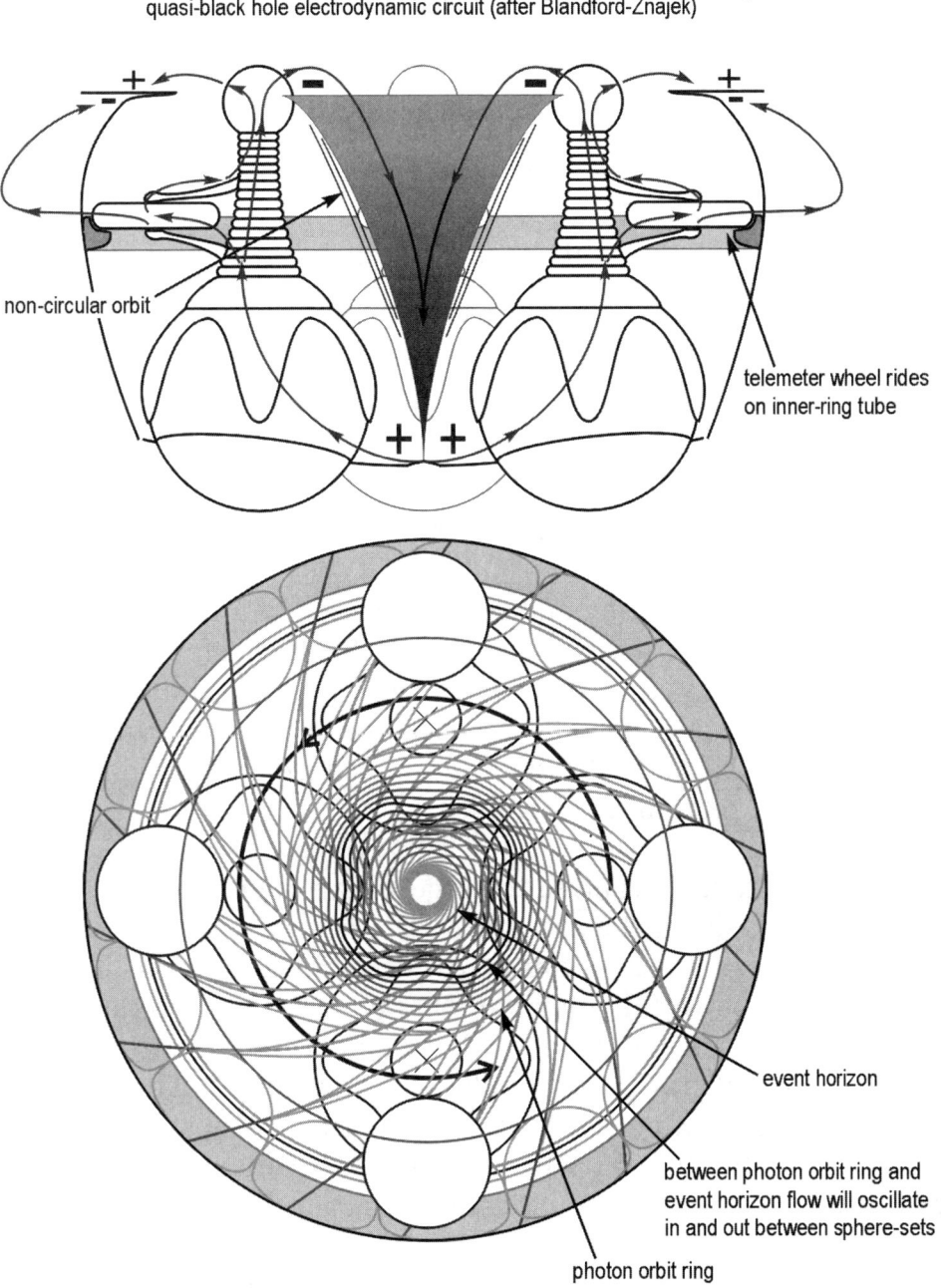

Figure 49. The aaUFO's electrodynamic circuit follows the Blandford-Znajek electric circuit for a black hole. Where the rotating flow negotiates around the four sphere-sets an oscillation will occur in the vortex's mid-section

From a quick look at this circuit what the extremities of the lower spheres need is a current transmission mechanism through which electrical charges can be transferred into the interconnecting rings of the stem coil: What could be inserted into the lower spheres to accommodate this is a spread of metal conducting rods which would somehow connect with the sphere's cellular structure, and these wires could be made to make contact with the stem coils...

Not a bad idea, but then this is not using the genius of the extraterrestrial designer!

5.5 Negative Phase Velocity of Metamaterials

Actually, the un-normalizing and re-establishing of a crystal's atomic structure with electromagnetic or gravitational biasing fields might be established more effectively in another way. And this calls upon a phenomenon known as negative phase velocity which has been noted to occur in some semiconducting dielectric materials but which has also been observed by some astrophysicists to occur immediately outside a rotating black hole.

So it would pay to look again to the ergosphere of energies rotating around the aaUFO's vortex, especially at the area between the lower spheres where the central hole is located in the base disc. At this hole where the vortex runs 'through' the craft is probably the most powerful volume of energy in the whole craft. And it is by no means a fluke of coincidence that the four lower spheres are positioned exactly where the most intense force of gravitational radiation can be induced from the black vortex, at precisely where its rotating energy is most dense and most energetic—for this is also where the magnetic flux lines will be more tightly converged together than at any other location in the whole of this UFO craft. So, ideally, THIS is the prime area to tap-off the highest amounts of energy in a UFO—although, in the aaUFO this is not strictly necessary here, because of the way the co-mutual magnetic field inter-connects all the major electro-dynamic energy fields around the whole craft (as will be shown in figure 71 below). Incidentally though, where the S-4 "Bob Lazar" UFO has a similar arrangement with its quasi-black holes this tapping-off is clearly being done because it is so evident in the way that UFO is structured.

One of the products of negative phase velocity is that in dielectric materials, fluids and fields, electromagnetic and gravitational waves can be bent the 'wrong way' when they attempt to pass through those materials (or through those fluids and fields). Astrophysicists have actually formulated how in the area surrounding a black hole out in space (in its ergosphere) electromagnetic and gravitational waves can be SLOWED down from their normal light-speed velocity to much slower velocities, and that under certain conditions these waves can even bend fully backwards from being too close to a rotating source of gravity [note: 11].

On a somewhat smaller scale this negative phase velocity phenomenon can occur in crystalline materials if the material is naturally conducive to it, by its structure having a negative refraction index (NRI), or if that material is specifically fabricated to exhibit these attributes. Prof. J. B. Pendry suggests that radiated waves when entering NRI materials may get trapped inside that material from a resonance effect that will exist just inside its surface which bounces those waves, once they have entered the material, back and forth by diffraction (ie when a wave is trapped there results a resonance within that material) [note: 12]. This highly technical but fascinating phenomenon, which features just as equally in the aaUFO as it does in the black hole, will be given much more air to breath in chapters 8, 9, and 10 of this study, so I shall give only a brief mention of it here (until it takes the reader to the point I want to introduce next).

In the world of science, negative refraction index materials gained solid recognition around the 1990s, and through its later stages this research progressed into negative phase velocity materials (and then this research, after the late 1990s, went back and brought forward even more effective NRI materials).

To explain why NRI materials research has received so much attention the reader might begin with a look at the studies of Pendry, and observe that much of their special qualities center around the transmission of electron charges through those materials as plasmons, which rather than being the electrons themselves are a collective oscillation of the electron's charges—as happens in a similar way through a phonon—in that the plasmon can transmit electronic charge through the cellular structure of a dielectric material. This plasmon technology then, becomes comparable to electron transmission through metals, although, as will be shown below, this plasmon technology has many advantages over electron-through-metal transmission.

More recent studies into NRI metamaterials were notable for the discovery that the plasmonic resonance (where the flow of electron charges acts like a plasma) could occur at a much lower frequency in these metamaterials than it does in metals (in metals it occurs at UV frequencies). So instead of being hampered by an ultra-high-frequency for this resonance (which would in metals occur in the visible and UV range, as does photoelectric ionization in metals) the great advantage in artificial metamaterials was that this plasmonic resonance could be brought down to the much lower Terahertz frequencies [note: 13].

But when researchers went on to discover more about the parameters for NRI metamaterials they then discovered how the plasmonic frequency could be lowered down even further to GHz frequencies (in the microwave range), and so then metamaterials became a very worthwhile field of research.

At this point the reader might be wondering, what on earth is the point to all this scientific plasmo-babble! And I apologize for all this tech, but I do promise that it will all make sense very soon. Let me say that, the quintessential point to follow here about NRI metamaterials, is that they can be BOTH an insulator AND a mega-conductor of electric charge... and that is something mere metals cannot do.

Because, by this NRI technology a material which for all other intents and purposes is classified as a dielectric, and will have for its base-substance glass or crystal or plastic (etc), can be 'switched on' to conduct electricity—in plasmonic waves—when oscillated above a certain frequency.

So with regard to the aaUFO, if the spheres (upper and lower) can be fabricated in the same way that metamaterials are fabricated, and if an oscillating frequency can be established within them which 'switches on' the sphere's plasmonic oscillations at, say, a particular microwave frequency, then these oscillating electronic charge movements will THEN switch on the all-important atomic cell spontaneous polarization reversal mechanisms within the ferroelectric ceramics embedded into those spheres.

Once the spontaneous polarization reversals are triggered and cause copious emissions of electrons to be ejected out of those ferroelectric cells then these charges would be picked up and transferred in bulk THROUGH the dielectric materials by the plasmonic waves established inside those metamaterial spheres... and from the way the aaUFO is set up those electric charges would be received into the pulsing magnetic flux lines where they enter the lower spheres, just below the stem coils (where the magnetic fields pass through the stem coils), so that then, those charges can pass through the stem coils and pass up into the four top spheres. As a result, the lower spheres can then be seen to perform as OSCILLATING POWER SOURCES that can be switched on or off by the operational frequency of the craft's electrodynamic circuit.

Of great significance for the aaUFO, and indeed, many other types of UFO, is that in NRI metamaterials researches it was eventually discovered, that a significant feature which contributed considerably to lowering of that NRI metamaterial's plasmonic resonance was the inclusion inside it of a network of thin wires.

This immediately brings back the images recalled by Betty Luca through her hypnotic regression sessions (and published in the "Andreasson Affair" books). What was seen by Betty and drawn by her was a long-wire network which was embedded inside the UFO craft's lower spheres (as figure 38), and from the fact that they appeared to Betty as if they converging from the outer extremities of those spheres into a small hub near the top of them, just about where those spheres hook into to the stem coils, then it would suggest that those wires were especially embedded inside the aaUFO's spheres for the same purpose as wire meshes are used inside NRI metamaterials.

This would correlate also with the observation that those UFO spheres when operating were seen to be glowing inside from what appeared to be small luminous color centers, and as previously mentioned (in chapter 4 above), such tell-tale phenomena of color centers are a good indication that they too have energized inside them electron-emitting ferroelectric materials, which are giving off both electrical charges and phosphoresce emissions.

But, without doubt, the fact that the aaUFO's spheres have inclusions of wires in them marks a particularly significant breakthrough in the understanding of the whole field of UFO electrical power sources. And, dare I say it, it may even change the way Science looks at electrical power generators on this planet with regard to fuel-less suppliers of energy for almost any household electrical appliances—and perhaps even with regard to small power units for automobiles!

5.6 Plasmonic Pump from a Thin Wire Lattice Network

While physicists Pendry and Marques noted that by having both negative electrical permittivity and negative magnetic permeability that these materials constitutes them as having negative refraction (a contributive factor in NRI), these physicists' most important observation was that by using a long wire network composed of an infinite number of wires in those dielectric materials, that plasmonic resonance could be achieved in those materials at a much lower frequency than was previously expected [note: 14].

It was then shown that a wire network made of gold-plated tungsten wires could switch on the plasmonic response at a frequency as low as 8 GHz. This means that below 8 GHz the metamaterial is an insulator and above 8 GHz the plasmonic resonance becomes established and the material becomes 'transparent' and transports electrical charge. Then in the plasmonic state the electron charges behave as a collective field rather than individual charges and behaves as a plasma (a plasmon is the quantum of this plasma oscillation) [note: 15]. This was later explained by the fact that these long wire networks were made up of exceptionally thin wire—of an incredibly small diameter of just 20 microns. For it was the thinness of the wire that was found to be crucial to the activation of the plasmonic resonance at that much lower frequency. The thinness of the wire, it seems, reduces the electron density in the metamaterial (because only a small percentage of the material's volume is filled with metal), but the thinner the wire the greater will be the magnetic field induced by the electrons that move along those wires, and so by this factor the effective mass of the electrons is increased out of all proportion [note: 16].

The plasmonic behaviour also comes from the self-inductance (or reversed electromotive force) of the thin-wire network. And so another contributory factor is that by raising the self-inductance (L) the plasma turn-on frequency can be lowered still further, and this can be done both by making the wires even thinner or by putting curling loops into the wire, or both (and it has, with these sorts of modifications, been found that the conduction frequency can come down even further to 6.8 GHz).

But finally, even more recent tests have shown that the surface plasmon resonance of a sphere of wire loops could bring the conductivity switch-on threshold down to just 4.6 GHz —and this is a lot more promising, because for such a technology to be used inside the aaUFO's lower spheres the conduction frequency would need to be as near as possible to my previously established figure for the aaUFO's working frequency of 3-4 GHz (and in the following sections there will be an attempt to see if the lower spheres might indeed be fabricated as a specific sort of glass, and cut with a faceted surface of specific dimensions, in order that they might automatically generate for themselves a resonant frequency of between 3 and 4 GHz—so that seamlessly these two factions can match). And so this is why I am paying so much attention to the technologies of metamaterials, and trying to see if the plasmonic resonance can be lowered down to the frequency that the aaUFO's structuring will resonate to.

By determining that the enhancement of the self-inductance of the wire network would play an important role in lowering the plasmonic resonance down further, it would be very tempting to conjecture that before the thin wire network was constructed, that it might benefit significantly by having those thin wires sleeved inside thin filaments of quartz. And that those quartz-coated wires could additionally be plaited together in bundles of 20 or 30 (thereby offering inductive coilings into that wire) first, before being structured into a repeating network. For such refinements should significantly enhance the amount of self-inductance incorporated into that wire network's electrical circuit, and thus, the collective plasmonic frequency might be brought down even closer to 3-4 GHz. Quartz, as most will know, has inductance and capacitance of its own so it would be an ideal material to both cover the wire and to interface with the host material (the glass) of the sphere. What would also be beneficial about sleeving such thin wire would be that the wire should be considerably more easier to handle during the construction stage of when these spheres were fabricated [note: 17].

Therefore, ideally, the wire for this metamaterial's network would need to be around 1—5 microns in thickness (which compares to human hair which is around 70 microns wide), possibly made of monatomic copper or gold so as to offer little or no electrical resistance [note: 18].

This arrangement may sound quite exotic yet it is precisely the arrangement that has been found at the site of a UFO incident in Russia and subsequently diagnosed in recognized laboratories across Russia. This quartz sleeve hypothesis (of mine) was proposed as a result of piecing together the evidence which came from the burned-out remains of what seems to have been a 'glass-like' sphere operating inside the Dalnegorsk UFO which had over-heated and had partially disintegrated. Thankfully though, there was enough correlative evidence to show that this UFO which made a soft landing on top of a hill had utilised a very similar type of electrical power sphere to that of the Andreasson UFO, because it also showed evidence of having several types of ferroelectric ceramics inside its rather volatile host material—which could well have been a very un-earthly type of glass (and which will be further detailed below).

5.7 Thick Wire and Diamond-lattice Plasmonic Frequencies

As detailed in a paper by Pendry [Pendry 1998] it was proved that by using a long thin wire network made of wire 1 micron thickness there then resulted an anomalous reaction; because by using such thin wire the whole metamaterial structure then acted like a homogeneous dielectric medium, where the resonant frequency wavelength would be derived from the accumulative effect of that thin-wire network—rather than from the actual spacing between its wires. This homogeneous effect brings down the plasmonic activation threshold to a very much lower frequency than the actual lattice dimensions of its wire network would normally oscillate at; so that the plasmonic activation wavelength actually becomes much longer than the lattice spacing in that structure. For instance, in a thick wire array with the wires 5mm apart (so the wavelength would equate to 10 mm) that would represent a frequency of 30 GHz in open air. In a thin-wire array with the wires 5mm apart the plasmonic frequency will come down to 8 GHz, disproportionately lower than the actual lattice spacing of that network [Pendry 1996]. This effect would be even more enhanced when the wires were shaped into a non-uniform (diverging) spread. From this study's point of view, if this homogeneous factor of NRI metamaterials can be improved upon still further and used to reduce the plasmonic activation frequency even lower, then it bodes extremely well for offering an explanation of the designs of wires being used inside the spheres that Betty was shown by the extraterrestrials.

And there are signs that this much lower frequency can be achieved. Because, in one stand-alone study which inadvertently stumbled upon the above mentioned homogeneous effect it did so even though it's authors did not use ultra-thin wires. This meant that the thinness of the wire was not the only crucial parameter! What was so significant about this different approach which improved upon the long-thin-wire theory, was that it used a repeating network of wire which was formed into a hexagonal (graphite-diamond-structure) lattice [note: 19]. In this study it has been suggested that wires in this type of configuration act more like atomic valance bonds and that its geometrical spaces act like atoms, so that the whole structure then performs like a giant molecule. Remarkably, the plasmonic turn-on frequency with this type of diamond-structure lattice was achieved at 6.5 GHz—from this mesh having a wavelength spacing of 2.3 cm—which is all the more remarkable considering that the metal wire used in this mesh had a 1.25 mm square cross-section! So obviously this well designed 3-D structure offers much food for thought, but equally it offers much room for improvement too, because the plasmonic frequency can then be reduced further if that mesh were made of thinner wire (which would raise the self-inductance of the mesh and consequently also increase the effective mass of its electrons). Interestingly, the authors further suggest that high permittivity pellets (ceramic materials) could be placed into that mesh to lower its operational frequency; but I would suggest that the inductance could be improved in a better way, by using high permittivity ceramic pellets alloyed with high permeability iron (as will be referenced and explained in section 5.10 below). Note the similarity of the diamond-structure mesh used in the above study with the quartz-like lattice depicted in the drawings made of the extraterrestrial's crystalline structure (shown in figure 41 above).

5.8 Faceted Spheres Inside Microwave Spinning Fields

As mentioned above in section 5.4 what these spheres need is a way to get them resonating to a precise frequency, so that all these emission mechanisms can switch on. What they need inside them is a way to oscillate backwards and forwards. And the way the extraterrestrials are possibly doing this is incredibly fascinating.

Looking back at the energy environment the lower spheres are in, inside ergosphere A, it will be noticed that all of its energies will be, a) rotating, and b) following a curved trajectory. In fact, inside ergosphere A there will be no straight-line path for any of its energies to flow along.

And again, from the fact that ergosphere A is so confined there will be great advantage to be gained from the Lense-Thirring frame-dragging going on between the rotating body of the central vortex and the side walls of this ergosphere.

Because, while I did mention above that as energy will be continually pouring into this circulating field, and that it will forever accumulate there because the only provision for that energy's exit would be through the base-disc's central hole, there is another way it can exit. It can exit through the lower spheres: it can pass through the lower spheres and then out through the wheels and into the adjoining toroid shell's fluid; or it can pass through the lower spheres, up through the stem coils to the top spheres, and then be emitted out of the top spheres back into the black vortex, wherein it will either then become more densified or its equivalent energies will be radiated away.

But, what will be needed in order to do this, to bring the lower spheres into the black vortex's electronic circuit and to allow energy to flow through them, will be that the plasmonic resonance of those spheres will have to be energized first so that these lower spheres become charge-transferring spheres.

Surprisingly, what will facilitate the paths of conduction out of this ergosphere, through the lower spheres, will be the faceting cut into the sphere's surfaces.

In order to see why, the ergosphere these lower spheres are in should be viewed very much as it was in section 5.1 above, as if being inside a whirling tornado. And because all those whirling energies will be flowing in curved trajectories, the energy spinning inside ergosphere A will then curve around the sphere-sets in the same way an eddy current becomes established in a fast-flowing fluid and flows around the outside of curved objects. That rotating energy will flow around the four sphere-sets in the opposite direction to that of the main vortex. So, while the vortex will flow counter-clockwise (as seen from above) the eddy currents around each of the sphere-sets will all flow clockwise. This clockwise flow will hug tightly around the lower spheres and be powered by the vortex.

But instead of travelling over a smooth surface that fast rotating energy will have to circulate around, and negotiate its way over, all the ridges and flats of the sphere's facets. With a little imagination the reader may visualize this curved flow reverberating around the outside of the spheres. And from a waveguide point of view this reverberating action would be equivalent to spreading the flat facets of that curved material out in a straight line and having that energy flow up and down over its hills and valleys; and this would then be equivalent to seeing an energy beam passing over a saw-tooth wave guide.

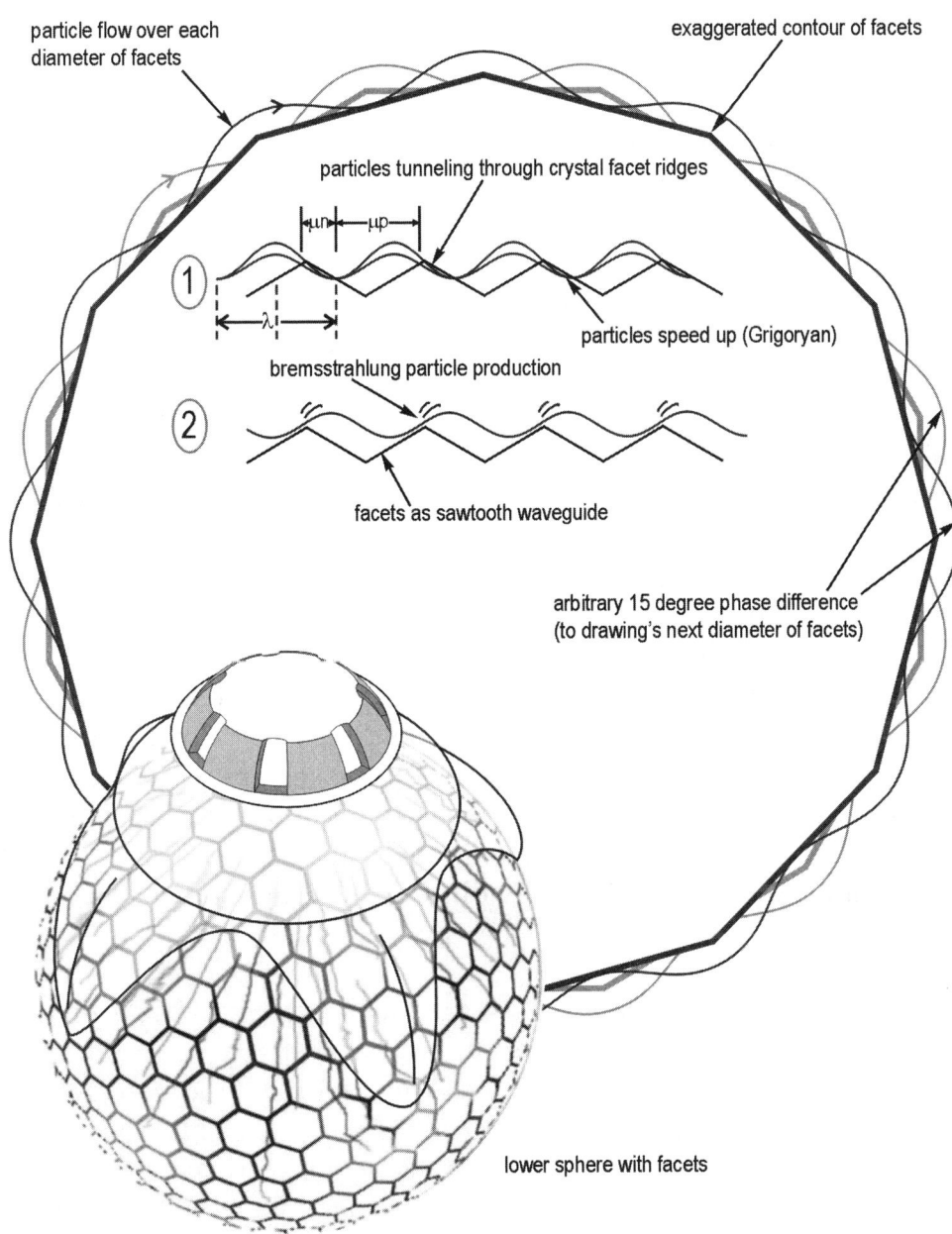

Figure 50. Charged particles that will be forced to rotate around the outside of these spheres will move in an undulating trajectory. At different diameters their undulations will be varied and so will set up a harmonic resonance

So, the faceting around the lower (and the upper) spheres would act as a waveguide for the charged particles speeding around them; and whatever particles (ions around the lower spheres and electrons around the top spheres) were circulating around them they would resonate to the frequency equivalent to a beam being guided over a straight-line saw-tooth waveguide with similar proportions. In fact, this novel waveguide mechanism will oscillate that flowing energy to a precise frequency proportional to the dimensions of the facets around those spheres [note: 20].

For example, if the dimensions of the facets can be roughly estimated for now (there is no equivalent research on faceted waveguides as yet so its an area that will need fresh investigation, particularly to see if it develops a homegeneous resonance from there being so many adjoining facets), with a ballpark estimated width set to 5cm—which would equate to a wavelength of 10cm—then this would mean that a field of particles traversing this curved waveguide would oscillate at around 3 GHz.

Such a concept would correspond firstly, to a phenomenon noted in a paper by [Grigoryan 2005a] who was investigating how relativistic electrons interact with a spherical dielectric when those charged particles rotate around it (and partially inside) in an equatorial orbit: And secondly, it will correspond to an external beam effect proposed by Russian physicist Ginzburg where charged particles are passed over the surface of a dielectric to generate microwaves [note: 21]. Therefore, as these UFO facets will generate oscillations in the microwave region in much the same way as these two aforementioned studies, I will dub this novel microwave oscillation by faceted waveguides around a dielectric sphere the Ginzburg-Grigoryan external beam effect.

The exact formulation for the frequency this effect will give will most probably be affected not by just one circumference of facets around the sphere, but by the whole multiplicity of facets over all the different circumferences of that sphere (which would turn it into a multiple-plane waveguide). And so this is why I say above, that this multiplicity of faceting may result in a homogeneous resonance effect; so therefore the facet width may end up needing to be less than 5cm for the top spheres and more than 5cm for the lower spheres to accommodate for this factor.

Ideally though, this action of the ergosphere's rotational energies spinning around the facets will be assumed to initiate oscillations around the four lower spheres in the 3 GHz to 4 GHz region. But in order to secure that range of frequencies this Ginzburg-Grigoryan external beam effect must also have factored into it the gravitational radiation and angular momentum coming from the black vortex, particularly for when the sphere-sets begin to rotate (because then these 'anti-vortices' around the sphere-sets will increase in velocity from the black vortex's increased power). Looking at this another way, this will provide another of the craft's on-off power switches to switch on the plasmonic activation for the wire networks inside the spheres. Because, this mechanism of microwaves oscillating around the lower spheres should be designed so that the reverberating oscillations will be held below the plasmonic frequency for the ferroelectrics inside them while the sphere-set assembly is held stationary, but then, as soon as the sphere-set assembly is allowed to rotate and a pre-established threshold would be surpassed (when the ergosphere's rotations rise to a certain velocity), then these Ginzburg-Grigoryan microwave oscillations over the facets would rise above a critical frequency and they would then trigger the plasmonic switch-on to occur within these spheres. This automatic switching effect would conform to the design function I mentioned earlier (in chapter 3), whereby the best way to design a UFO's power drive is to use appropriate semiconducting materials and structures—into which energy thresholds can be set to switch on at appropriate stages through the energy development process of this engine—that way fewer transmission lines are needed (and also there will be less thinking for the pilot to do).

5.9 Summary One: Glass Spheres that Think they are Superconductors

Also towards the head of this chapter I pointed out that on the other side of the lower spheres, opposite to the central vortex, on the out-facing (away facing) side of each of the lower spheres, there will be a sharply converging high density of magnetic field lines, into which, and through which, these lower sphere will revolve. And because these magnetic field lines coming from the toroid will be pulsing and highly energized they will imbue a large amount of Lorentz force into any body of electrons, or electric charges, moving through them. As per $e = mc^2$ the result will be that those electrons will take on additional mass.

The following hypothesis may be seen as a way to provide the necessary cell-switching that would counter-act against the vortex's gravitational force switching on the transiting ions inside the sphere's cells. Because, for spontaneous reversal switching to occur in the ferroelectric materials embedded into the spheres, the central ions have to be directly switched out of one polarized alignment and then forcibly switched into the other polarized alignment, and this will have to be done in a continually repeating to-and-fro action.

Switching can occur at the two places where magnetic field intensity will be at its strongest, and as above, this will occur as the lower sphere revolves and the atomic cells within them momentarily become aligned adjacent to the central vortex. This switching might also occur at the away-side of the sphere, where the electrons in the cells become more massively charged due to their induced Lorentz force there. The increase in effective-mass imbued upon both the electrons and the protons from the magnetic fields on this away-side of the lower spheres will be disproportionately greater for the electrons than for the protons (from the electron's extra momentum) (see note 16 above). Thus on the away side of the sphere, away from the vortex, these more massive and more energized electrons will orbit around the atomic nucleus at a greater velocity, and therefore push the positive and neutral nuclei back into their central alignments within those atoms. This will initiate that back-switching of the sphere's cells.

As the spheres continue to revolve around the far side of the lower sphere's turning circle and then return back toward the central vortex, out of the magnetic flux, this re-alignment through the energized electron cloud would cease and so the electrons would loose their effective mass, so when that cell comes up to the vortex again (and as the facets line up to the gravitational radiation) the transiting positive ions get displaced back to their previous alignment. This switching continues for every facet as it lines up adjacent to the vortex and then turns away from it, and moves into the magnetic field intensified on the opposite side. What intensifies this switching further will be that as the cells approach the vortex so they will also pass through the most intensive section of the magnetic field that surrounds the vortex, so again their effective masses become substantially increased either side of when the facets pass in and out of perfect alignment with the vortex, causing the atomic alignments to jitter back-and-forth momentarily. This causes the gravitational displacement switching to occur all the more suddenly and robustly for the positive ion. And because the switching occurs as each small facet in turn aligns to the black vortex the lower spheres only need to revolve-rotate slowly to get all the facets to sequentially switch and to set up a resonance effect.

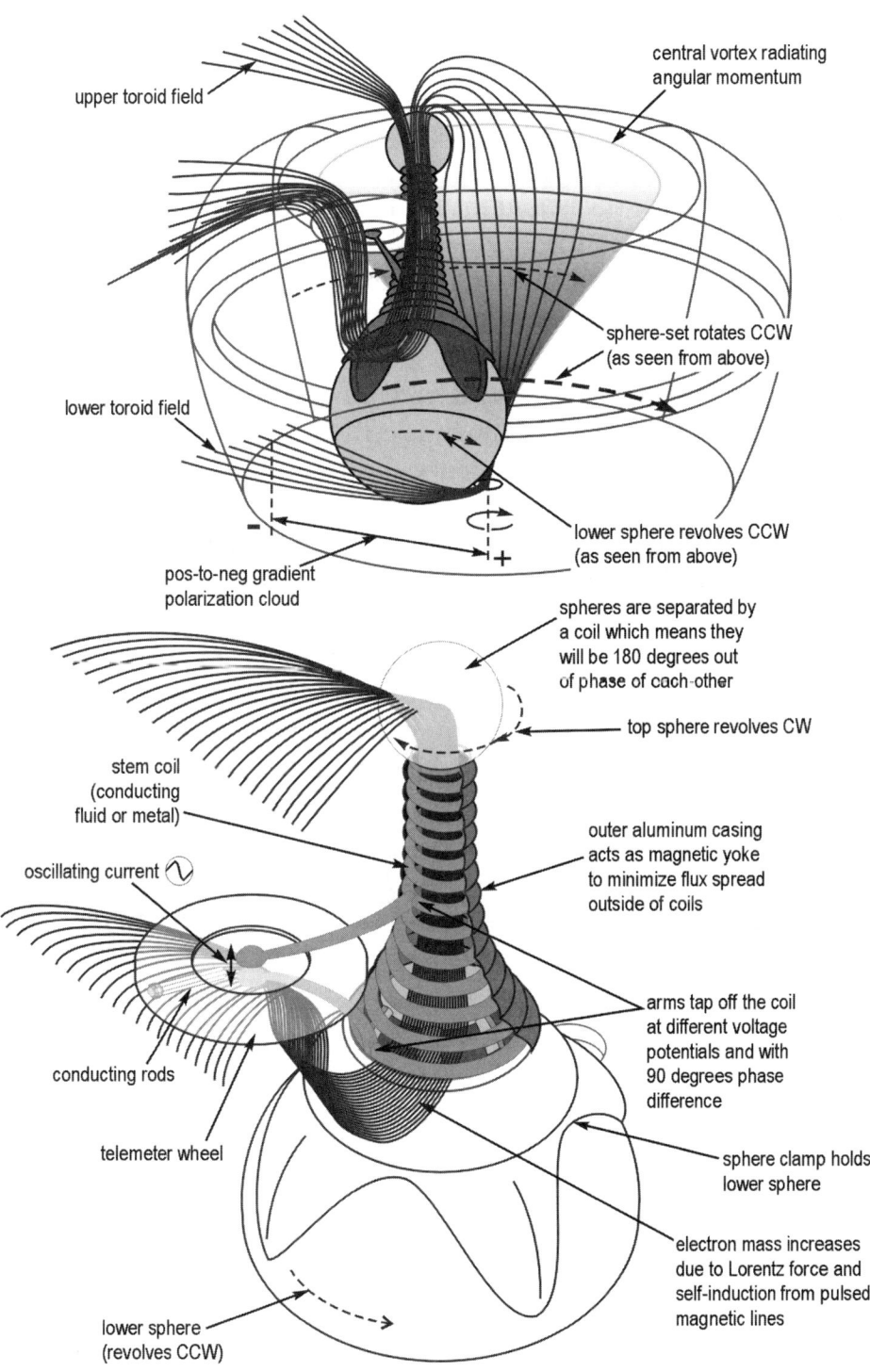

Figure 51. Spheres revolving into intense magnetic field at bottom of stem coils will increase the effective mass of electrons in cells

Just to track-back to summarize all these factors in the lower spheres; in summary, the essential ingredient for these spheres will be:

1) a non-uniform wire lattice mesh
2) artificial dielectric host material (glass)
3) cellular structure of atomic mass components to be gravitationally polarized
4) plasmonic pump frequency threshold (at microwave oscillation)
5) provision for dynamic magnetic field amplification on cellular electrons
6) spontaneous polarization reversal ferroelectric inclusions
7) ZPF one-way (rectifier) gate mechanism (for outside surface cells)

Once the spheres are fabricated (as mentioned above, through the sol-gel process) they will just need to be oscillated above a critical frequency for them to deliver electrical power.

This is to say that while quartz isn't a superconductor (and nor will it ever be), a specially fabricated glass can be hoodwinked into behaving like a megaconductor when it is given a wire-mesh lattice, some ferroelectric ceramics... and just a little quartz. Because then this glass will become a plasmonic medium acting like an insulator below a certain frequency and then like a resonating plasma-pump above that frequency. So by anybody's standard such a composite material can be recognized as a much more efficient accumulator-conductor than any metal-wound generator.

5.10 Summary Two: Wire Network in Megaconductor Dielectric Spheres

Verification of the above hypothesis may be found in the extensively researched and highly informative reports prepared for a Russian UFO investigator about what was discovered in the discarded debris left behind from when a UFO had seemingly ignited one of its glass-crystal spheres while in mid-air, whilst flying into the earth's atmosphere over Russia, and which caused the craft to make an emergency landing in the Primorye Territory of Russia to affect a repair. For on January 29, 1986, at 19.55 local time, near a town called Dalnegorsk, near the far-Eastern coast of Russia towards the Sea of Japan, a UFO was seen by several witnesses to be in trouble and flying on a horizontal trajectory at an altitude of about 800 meters at an estimated speed of 15 meters per second (35 mph) with a crimson flame shooting out of it. Then suddenly, according to the extensive reports prepared by Valeri Dvuzhilny, upon nearing the Izvesrkovaya mountain close-by to Dalnegorsk the UFO was seen to alter its trajectory and to undergo a steady and apparently controlled landing on top of the mountain, where the craft remained for approximately two hours [note: 22].

Interestingly, during the first half-hour that the craft was seen to be on the mountain (also known as Hill 611), witnesses watching it from their houses in the town said that the stricken craft tried to rise back into the air no less than six times, and that each time it rose it almost immediately lowered itself down to the ground again (these witnesses noting also that the craft's surrounding energy field visually intensified whenever the craft did try to rise up, and diminished as it lowered back down to the ground). After these six aborted attempts to take-off the UFO appeared to remain inactive for a further one-and-a-half hours on top of the mountain, although during that time it still had surrounding it a very bright energy field. Indeed, before the UFO did finally remove itself from the mountain and it disappeared into the night sky, those who watched the whole event later remarked that during the whole two hours that it was on the mountain a bright sphere of glowing energy could clearly be seen to fully surround the UFO, and that out of this glowing energy there were occasionally produced a number of high-intensity flashes of light that the witnesses compared to the flashes of electronic arc-welding...

When Dvuzhilny, who in previous years had worked with Russian (CCCP) military personnel and had also worked with the Russian Anti-Aircraft Defence (AAD) on radar detection of UFOs (particularly in this area), was informed about the incident by some of Dalnegorsk's schoolboys, who thought it may have been a meteorite come down, he was intrigued and immediately wanted to discover for himself whether or not the incident involved a UFO, because he knew full well that meteorites do not burn or ignite other materials when they hit the ground!

His suspicions were confirmed when, some two days later, he was led to the landing site by the Dalnegorsk schoolboys, for on the cold winter's day when he went up the mountain he was surprised to discover a four-meter wide area completely devoid of any snow—whereas all around the rest of the mountain and all along his route up there had been a covering of snow fully two-feet (50 cm) deep. Obviously he found that the UFO had gone, but what Dvuzhilny did find remaining all around the four-meter clearing were unmistakable signs that something had been there very recently, and that whatever it was had generated a tremendous amount of heat while it was there! For he found that many of the rocks around the cleared area had been cracked as if exposed to extremely high temperatures (subsequent investigation of the magnetic field signatures still remaining in the siliceous rocks would also reveal the extremely high power levels the UFO must have generated while it ascended and descended six times over different areas of the rocks). Dvuzhilny found too that many of the rocks were hazed-over with a chemical film, and that over a wide area spits of metal had splayed out and were aligned in an outward direction (the subsequent investigation would establish that most of these spits were of lead). At the edge of the cleared ground on some of the trees, strangely, the bark had been separated from the wood of the tree but it was not burned or blackened, merely did it have a yellow coloring—indicating that the energy field inside the orb that surrounded the UFO, while the UFO was on the mountain, was completely devoid of oxygen and was possibly vacuous in character.

As a result of this initial investigation numerous objects of metal and other non-metal materials were collected by Dvuzhilny for further analysis; these included burned-out remains of a strange carbon-glass-like material (within which were preserved the remains of a wire mesh structure), the numerous spits of metals (such as lead, bismuth, tin and cadmium), and a large amount of small 'iron-rare-earth' balls (these were found to contain many elements from the lanthanoid rare-earth series) [note: 23].

The Russian-language websites which cover this report from Valeri Dvuzhilny say a lot more about the UFO's landing site than the Western websites, and in particular that in the months and years that followed both the materials found at the site, and the mountain landing site itself, underwent a very comprehensive array of scientific investigations under the personal direction of Dr. Dvuzhilny, who is a pharmacological researcher. Additionally, when other scientists, in 1988, visited the top of Dalnegorsk's mountain to investigate this landing site using their extensive range of geological and geophysical instruments, they began to discover many phenomena equally as important as the original pieces of UFO debris (although strangely, these phenomena were not mentioned in any of the English-language websites I visited for Dalnegorsk) [note: 24]. For instance, while they were doing geophysics over the affected ground in 1988 these scientists discovered that six places within the small circular clearing still had unusually high amounts of residual magnetism that could be detected in the surrounding rocks (as a mark of their thoroughness, they were even able to estimate how much power in Gigawatts the UFO must have generated in order to have affected the magnetic signatures that their instruments detected). And one of these scientists said that judging by the instrument readings he had taken, whatever the object was at the center of such powerful emissions it must have had its own independent power-generating source, and that such a power source would had to have operated in the SHF frequency range.

SHF frequency range? This golden nugget of information is highly significant—because it correlates perfectly with the metamaterial data I have mentioned above, and even more significantly it correlates perfectly to the operational frequencies I have arrived at, from entirely different criteria, for the Andreasson UFO and even for a number of other UFOs. This is because, the SHF frequency range is the electromagnetic frequency range between 3 GHz and 30 GHz, with wavelengths from 10cm to 1cm (ie it covers the microwave range of frequencies that I've been working with throughout this aaUFO study).

This microwave frequency discovery at Dalnegorsk was also perfectly correlated to the observation that when the branches of the bushes nearby to where the UFO landed were analysed the inner layer of wood showed signs of being boiled from the inside out; and this microwave signature also correlated to the results of a rather unusual investigation carried out by Dr. Dvuzhilny on biological samples dug out of the ground at the site; whereby, in the subsequent laboratory report produced it was indicated that earth-wasp larvae, which had been discovered seven centimeters below the ground-surface at the site, had been charred as if by a very intense heating mechanism, even though the soil taken from the surface-layer directly above them was completely free of any such heating effect! And again, such a heating mechanism would be conducive to the area being radiated by microwave energies; because, as many an astute cook will know, a similar principle occurs in microwave food-processing, which forms the principle by which food is cooked, whereby it is the heating of the water-content within the food that cooks the food not the action of the microwaves on the material itself. So similarly, this would be why the wasp-grubs dug up at the site would have been charred (from the inside out) and not the soil above and around them. While this is an unusual discovery it certainly adds weight to the diagnosis that the Dalnegorsk UFO's power drive operated at a microwave frequency of between 3 GHz and 30 GHz. And so this whole field of diagnostics carried out at the Dalnegorsk site is especially welcome to the study of UFOs because it helps to confirm the operational frequency range of several other UFOs [note: 25].

Obviously, the Dalnegorsk UFO craft flying through our atmosphere that night was in trouble and some parts of it seemed to be malfunctioning or overheating, that can be ascertained from the sightings of it flying through the air with its surrounding energy orb punctured by a short crimson flame (possibly hot plasma). The melted debris of carbon and glass found near the trees at this Dalnegorsk site, the discolored bark of the trees, the magnetic signatures precisely impregnated into the rocks at six separate locations, and the spreads of metal that had melted and sprayed out over the immediate area, all confirm that some sort of structure had undergone disintegration from intense heat (which was estimated by the Russian scientists to have risen to well above 3500 degrees C); and the later magnetometer readings also confirm that a very powerful generation of electromagnetic energy took place around those rocks.

Indeed, the reader might here be reminded of the Roswell UFO, which in 1947 also suffered an ignition of its gases and/or combustive materials, which led to that craft's surrounding energy orb being engulfed by a fiery plasma as a result of that UFO being struck by lightning, so that it scorched the earth and sand where it initially came down and skimmed the ground (before it rose back into the air and finally crash-landed several miles further on, killing three of its alien crew). What was so different between these two events, though, was that at Dalnegorsk there were actual eyewitnesses watching this UFO, watching it land, and watching it until it took off. And whatever was found to remain from the ignited plasma and combustive materials after that Dalnegorsk UFO departed was retrieved by non-military persons—and painstakingly analysed in scientific laboratories—and there was enough of that UFO's debris (which remains today) that can tell us a very significant story [note: 26].

Of course, as happens in England and America the reports about this UFO were hushed up or diluted from different standpoints, just like the UFOs which were seen and photographed clearing the radioactive fall-out from Chernobyl's nuclear power plant accident back in the 1980's: The Western press-agencies turned their backs on the alarming facts about Chernobyl as if those events never happened! Likewise for Dalnegorsk, the first spokesperson for the Russian authorities, once they got to hear of this Dalnegorsk UFO, announced that what happened on the Dalnegorsk mountain was a geological stress-field which hovered over the area and absorbed these metallic and mineral materials into it… but then because that report was deemed too farcical another Russian academic presentation wholly dismissed the Russian authority's geological explanation and announced that what happened at Dalnegorsk was a scientific experiment…which went wrong [note: 27]. Governments may be very good at shouting loudly above the voices of individuals when these sorts of things happen, but they can't erase people's memories of them happening… as has been the case with the Roswell UFO!

It is not difficult to see how the Dalnegorsk UFO data would have had a blanket put over it because it represents, in many respects, our misplaced servitude to our military hierarchies who clearly don't possess as much intelligence as we think they should have (this has clearly been the case with the military's denial of UFOs in the past). But then, I look at UFO technologies in a different way, because, if researched correctly these UFO technologies would represent a huge array of scientific advancements for any COMMERCIAL ENERGY SECTOR (as shall be explained below). This, of course, would generate huge profits for this technology's first developer anywhere in the world... In Russia, Dvuzhilny approached no less than 19 scientific research institutes (from Leningrad, Novosibirsk, Vladivostok, Kiev, Khabarovsk, Irkutsk, Moscow, and Kamchatka) for scientific analysis of the UFO's remnants he had discovered. In all 18 investigators and 8 physicists were employed on a very broadly-based program of scientific analysis and discovery, so that Dvuzhilny could formulate a comprehensive report on whatever that UFO's power generation might be, and also on the wholly unconventional materials which the UFO had discarded and were discovered by him on Hill 611. These scientific studies encompassed a wide diversity of disciplines from chemistry to geophysics to mineralogy to metallurgy.

So what did all these scientific investigations carried out by those Russian academics show? Well, I believe, quite conclusively, that all the disparate studies carried out on the Dalnegorsk UFO remnants correlate together toward an unprecedented degree of clarity. Because they quite clearly point to a high-yield power source which used metamaterial structuring and ferroelectric electron emitters—and the magnetic readings made at the site where the UFO struggled to take off indicate that the UFO was developing electrical energy in the multi-Gigawatt power range!

Piecing together these findings I would suggest that the burning something that the stricken UFO dumped so unceremoniously on top of Dalnegorsk's mountain in 1986, was originally a specially manufactured glass sphere which was being used by that UFO as one of its power sources (and if that UFO was similar to the other UFOs which have been observed having three or four of these types of glass sphere, then it would have been possible for a UFO to 'limp home,' possibly to another larger craft, with whatever other spheres it still had intact—although, it might also be suggested that the crew of that UFO organised a crude manufacturing process and affected some sort of on-site repair to their craft before they did leave). Either way, I will include below some of the findings and one-by-one I will relate them each to known metamaterial and ferroelectric technologies, to show why I believe these technologies were being used by that UFO.

My reasoning which points to this conclusion is that central to the Dalnegorsk UFO investigation there were discovered substantial chunks of thin-wire mesh structure, which, although it was embedded inside a glass material which had obviously suffered extremely high temperatures, because this mesh was so well preserved inside that melted carbon-glass, and because there was so much of it encapsulated inside that 'glass preserve,' it provided the Russian researcher, Dvuzhilny, with substantial opportunity to dispatch numerous samples of it to several laboratories for specialized analysis. What the carbon-glass preservation was also good for was that it preserved (more-or-less) the very lattice structure that the thin-wire network was originally formed into, and although these remnants now show a somewhat amorphous lattice some sections of it do indicate that it might have originally been formed as a repeating structure, which would have been like a cubic lattice (certainly, from the electron-microscope photographs I have seen of it I would conjecture that it might have originally been very similar to the cubic mesh as depicted in Pendry's 1998 physics paper, for instance).

Furthermore, Dvuzhilny's report also speaks of iron and rare-earth pellets (with diameters of 2 mm to 8 mm, the size of ball-bearings) taken from that UFO site which had an extremely complicated composition—of being made up of lanthanoids, fluorine and argon gases, and in particular that they had high concentrations of barium and strontium (the latter two elements are both highly ferroelectric and highly piezoelectric, as has been noted in chapter 4's note 8). So straightaway, there is a strong suggestion that these extraterrestrials might be using electrochemical processes to produce their high levels of power, and more than a hint that they were using different combinations of electron-emitting lanthanoid ceramics [note: 28].

It is interesting that the extraterrestrials were using iron and nickel in these ball-pellets too, because iron and nickel both have a high magnetic permeability, and in electronic circuits iron is used inside an induction coil to increase the inductance of that electrical circuit (in electronics inductance is proportional to the ratio between electric current and magnetic flux, which in turn is proportional to magnetic permeability ($\Box$): Indeed, three of the elements found in the debris left behind by that Dalnegorsk UFO, iron, nickel, and silicon, all have very high magnetic permeabilities: So, having iron-nickel in these pellets would have contributed additional inductance to that structure and improved the way the electron charges would have oscillated inside the whole metamaterial structure (this follows along the same lines as that mentioned above for the 3D metamaterial structures in sections 5.6 and 5.7, for instance, and particularly because inductance becomes a very important factor in metamaterials in that it is a contributory factor which can lower the plasma frequency of that metamaterial's thin-wire network from UV-visible light frequencies down to GHz frequencies) [note: 29].

The list of trace-elements found dispersed inside the melted carbon-glass mesh included iron (Fe), molybdenum (Mo), germanium (Ge), and tungsten (W), all of which are electronegative; zirconium (Zr), cerium (Ce), neodymium (Nd), lanthanum (La), and praseodymium (Pr), all of which are electropositive. What is significant about this list is that it quite clearly includes, once again, many exotic elements from the lanthanoid (rare-earth) series (which are in the transition elements group of ferroelectrics).

In fact, when the large amounts of lead found at the site are also taken into consideration the above group of elements are very much the ones which an electrochemist would expect to find as groupings in ferroelectric ceramics, particularly in the lead-based perovskite compounds. The lead found in the UFO's debris, of course, is also a high-mass metal and would be ideal for the gravitational-to-electrical polarizations known to occur within UFOs. Such ceramic compounds though are used very specifically for electron emissions, and with respect to them possibly being used as a power-source for a UFO, this would strongly suggest that the extraterrestrials were using this mix of compounds primarily in a ferroelectric spontaneous-polarization configuration. And, along with the iron, nickel, and cobalt materials found, this would seem to echo the same type of magnetostrictive technique which can be used to oscillate and to spontaneously polarize any ferroelectric materials that are embedded inside such glass spheres (as I've postulated would occur in the Andreasson UFO, as previously mentioned in sections 4.3 and 4.4 above).

Interesting though these different elements are, as far as the academic investigations went, it was without doubt the thin-wire network preserved inside the carbon-glass which generated the most interest at the various research facilities throughout Russia, as samples of it underwent a whole range of scientific tests to establish exactly what it might be. Particularly as there were found a few analogous effects for when this wire was heated to the highest of temperatures: For instance, it would withstand temperatures of 2800 degrees C without deforming its shape (merely did it just glow red), even though, essentially, the wire was found to be primarily composed of gold, silver, and nickel [note: 30].

Once the mesh samples were investigated in the Russian labs it was found that inside them the ultra-thin wires of 'gold' (gold-silver-nickel alloy), while they were structured as single strands of wire, each of them were encased inside an artificially-fabricated quartz sleeve or filament, and these quartz filaments (with the wire inside them) would then be braided-twisted together in bunches of 20-30, presumably so that, as a braided multi-strand bunch they would, again, have a higher degree of self-inductance (and capacitance). The gold-alloy wire itself would have a thickness of just 7 microns, and the sleeve around the gold wire would have an outside diameter of 17 microns (remember a human hair is 70 microns in thickness) [see note: 30].

These braided multi-strand bunches of quartz-filamented thin-gold-wires were then structured into what looks, as I say above, very much like a cubic or quasi-cubic lattice network, and they would have then, judging by the blackened carbon-glass which had melted and then solidified around them, have originally been embedded into a larger structure made from a special glass material—most probably with the glass manufactured into some form of sphere (this conjecture is based upon the way the glass-spheres in the Andreasson UFO were manufactured).

As the reader will recall from the previous sections of this chapter, with artificial metals, which is basically what metamaterials are, they will be designed and manufactured in such a way that they will act like an insulator below a particular frequency of oscillation, and then they will act like a emitter-generator above that particular frequency. This is quite a new concept, and of course, while conventional metals are wonderful, and conventional wiring looms are also wonderful, they will ONLY transmit electrical energy when they are physically connected, and switched to conduct between a generator of power and an electrical load.

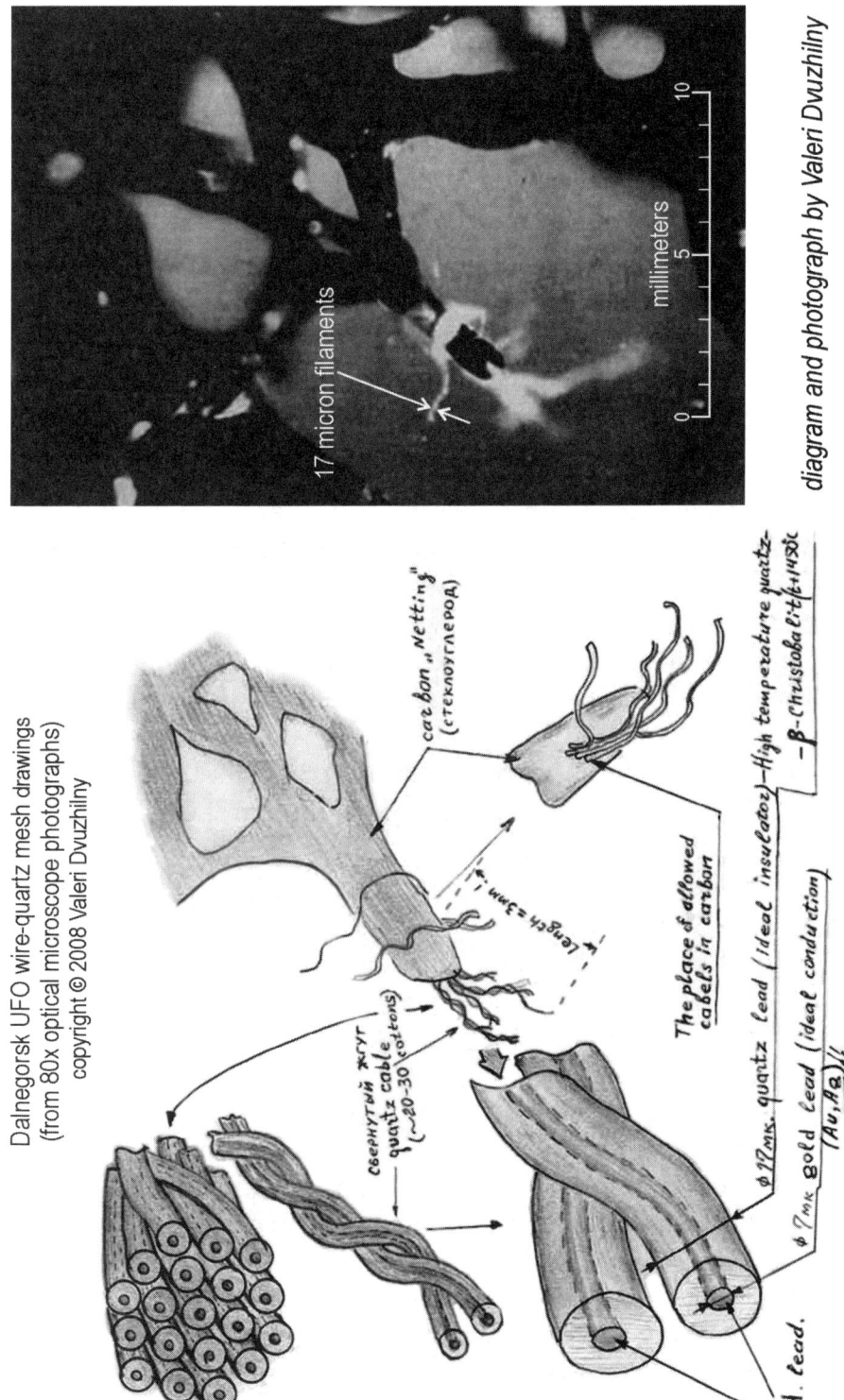

Figure 52. Low-resistance gold wire encased inside quartz filaments braided together in bunches of 20 to 30 filaments were found to be formed into a mesh network encapsulated inside a carbon-glass material

A metamaterial structure embedded with ferroelectric ceramic donors is in itself a generator capable of emitting huge amounts of electrons, and it will be an insulated device and dormant until it is oscillated above a particular frequency, above which frequency these metamaterial structures automatically switch themselves on to become plasmonic generators and transmitters of electron charges. What is so revolutionary about them (which might be especially interesting for the commercial sector of this planet) is that they are a power source—and a wiring loom—all in one. And they will also have the cost-cutting advantage of having in-built into them an ON–OFF switch!

From the electron-microscope photographs of the mesh that were kindly sent to me by Valeri Dvuzhilny, it appears that the original lattice-period of the extraterrestrial's mesh-network would have been approximately 4 or 5 mm. Now this is very interesting indeed, because it means these braided strands of the ET's can be compared directly to the thin-wire structures featured in Pendry's studies on metamaterials (which were spaced 5 mm apart in a quasi-cubic lattice [Pendry 1996] [Pendry 1998]). The reader will recall, from my mention of it in section 5.7, that Pendry's mesh when it was manufactured and oscillated established a benchmark plasmonic switch-on frequency of 8 GHz, and so these studies can provide a good starting point from which further developments can be made.

With this in mind though it should be pointed out that the ET's mesh is considerably more refined than the meshes used in those early researches by Pendry. The fact that the ET's mesh comprises of multi-stranded gold wires sleeved by quartz which are braid-twisted in bunches of twenty or thirty, will before anything else is considered add significantly to the inductance and capacitance of those wires. Therefore, it would be expected that the final plasmonic frequency that the extraterrestrials have designed into their metamaterials will be considerably lower than 8 GHz. Indubitably, with all the additional refinements discovered it can only be conjectured what that final frequency might turn out to be, but bringing into the equation; A) the above-mentioned research on metamaterials regarding the homogeneous resonance effect (mentioned above in sections 5.6 and 5.7) which lowers down the plasmonic frequency; and B) the use of such specialized amorphous mesh-structuring as used by the ETs being quite similar to the structure used in the [Sievenpiper-Sickmiller-Yablonovitch] researches; and C) the ET's use of ultra-thin 7-micron gold wire; and D) the high-capacitance and induction enhancements of the quartz sleevings and multiple braidings of those wires; and most importantly E) the inclusions between those wires of large amounts of small iron-rare-earth high-permittivity pellets (some elements of which would also increase the magnetic flux influencing those wires)... then certainly it should be scientifically possible for the Dalnegorsk mesh-glass-sphere to have operated inside that UFO with a resonant plasmonic frequency as low as 3 to 4 GHz.

And so therefore, just as I requested at the beginning of section 5.4 of this study, here might well be the mechanical solution for how to control and maintain the resonant frequency of oscillation within these spheres, and here is the material environment where the ferroelectric elements can be located so that they would be subjected to their cellular polarization reversals—which would force them to emit copious amounts of electron charges. So thanks to these discoveries in Russia the question as to how the extraterrestrials develop electrical power inside these huge glass-dielectric spheres is finally answered. Such spheres would have supplied the UFO craft using them with an almost limitless reservoir of electrical power; and these spheres could also have been used as power-pumps which would have converted even more electrical energy from the surrounding ZPF field outside of those UFO craft (this facility will be further explained at the end of chapter 10). Being so independent, then naturally, those UFO's would have carte-blanche to travel into or out of our atmosphere without auxiliary fuel supplies and with complete stealth [note: 31].

As an aside, this presents somewhat of a quandary, I know the operational environment inside a UFO, particularly inside the aaUFO, would be especially conducive to the subjection of microwave pulses upon these metamaterial spheres (as I've shown in sections 5.1 and 5.8 above), but if these metamaterial structures can be manufactured to operate at such low microwave frequencies, then this technology will suit earth's commercial sector just as much as it will suit UFO research, because it could just as easily be developed as a prototype for universal household-size electricity generators (which would have ZERO carbon-emissions, of course). Indeed, scientifically speaking, this could be the moment when the extraterrestrial water of the holy Greyl could be turned into wine—for everyone on this planet—because such technology could replace the nuclear and coal-fired power station generators of electricity! So, I have to say, did the scientists in the West miss-out on this important scientific discovery solely because the American and British news-media people deigned not to report the scientific facts about UFOs? Perhaps that secrecy has built into it a double-edged sword.

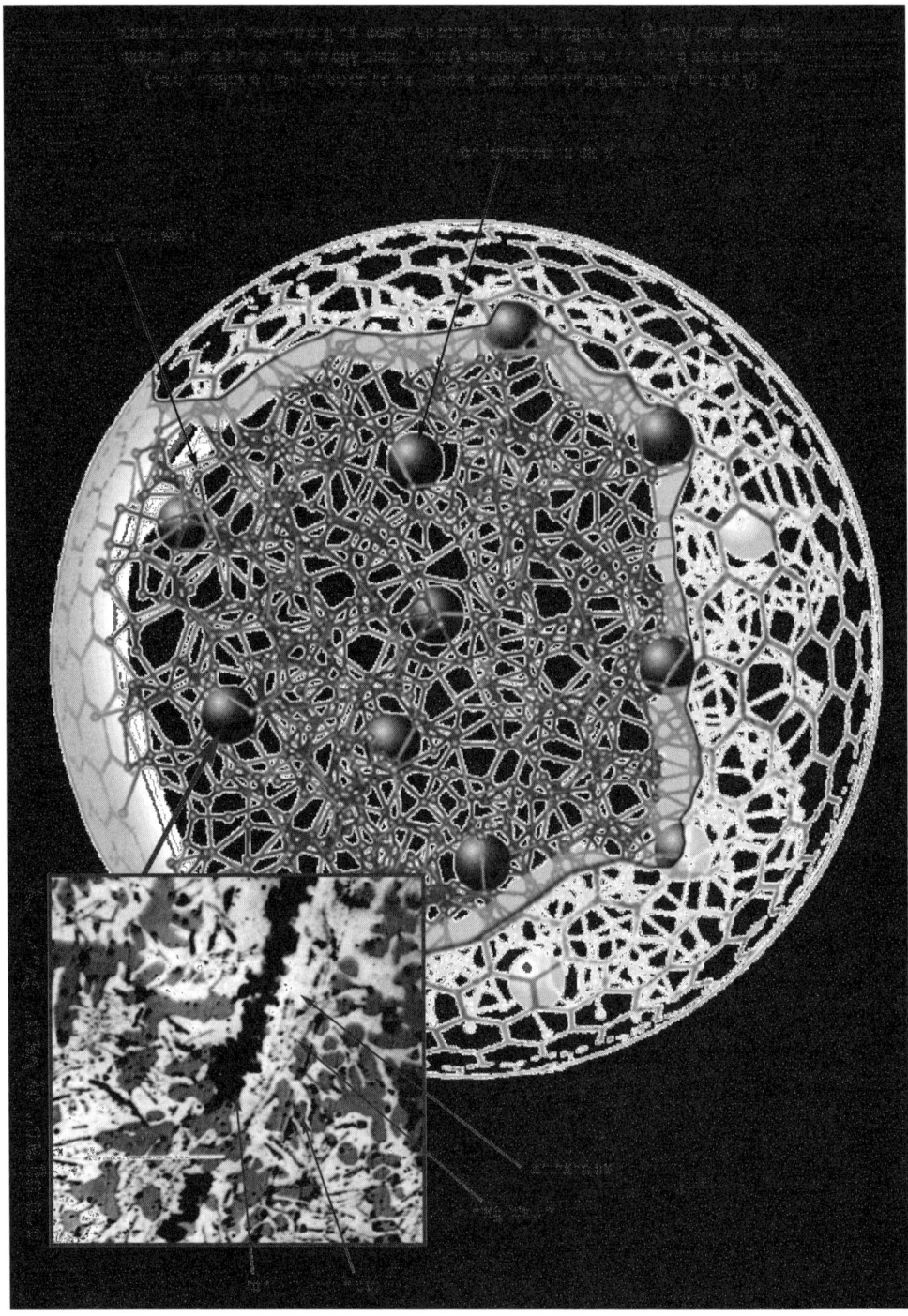

Figure 53. From the detailed analysis of the artifacts left behind by the Dalnegorsk UFO, and from the observations made about the Andreasson UFO, it has been possible to reconstruct the basic design of the power spheres used by these UFOs

Anyway, going back to the UFO artifacts found at Dalnegorsk, what was even more intriguing about the blackened material recovered there was that the material which surrounded the special wire mesh was a carbon-based material which had quite clearly melted. Because, that the heat caused that glassy structure to transform into a material pitted with holes this surely singles it out to be a carbon-based glass material—not silica-based [note: 32].

This then raises the intriguing question, especially as the carbon-glass when analyzed was recognized as having an almost beta-cristobalite structure, was this strange material used by the ETs an entirely new form of glass? Similar in some ways to carbon-dioxide glass? Presently (in 2008), we know of it as a-carbonia but we don't yet manufacture this new type of glass on this planet (even though an Italian-led group of scientists are presently working on trying to produce this new type of carbon-glass) [note: 33]. Little is known about this special new material other than it is formed at extremely high-pressures and at extremely high-temperatures, but it is hoped that this new type of glass when the secrets of its successful manufacture can be found will have a much stiffer structure than normal glass and be much more rigid like diamond (which, certainly, would switch polarity more rigidly to better suit the ferroelectric spontaneous switching mechanism).

But the added refinement with the Dalnegorsk UFO's glass though is that it has had all of its oxygen (and dioxide) atoms removed from it, so that essentially it has a purely carbonized structure. Obviously this makes the Dalnegorsk glass completely different to any other material known on this earth, and it could either have been specifically manufactured in this unique and special way by the ETs, or the oxygen could have been removed from it during the time when the UFO's vacuous energy field surrounded it on the mountain-top, and when the intense heat around that quartz-glass sphere would have caused the glass to melt and spread into the wire mesh.

So, in conclusion, in my hypothesis I would see the original Dalnegorsk UFO's power units as being a spherical fabrication of a carbon glass (structured like carbon-dioxide glass), into which would have been embedded a thin-gold-wire network (sleeved with high inductance quartz filaments). This wire network would have been oscillated at around 3 to 4 GHz—perhaps like the Andreasson UFO as a direct result of the high energy plasma rotating around the UFO's central vortex interacting with the faceted surfaces of these spheres—or through electrostrictive and magnetostrictive elements embedded inside these spheres which can be triggered by external magnetic fields to oscillate at microwave frequencies. Then, once the wire mesh was oscillated up to its own resonant frequency, the iron-rare-earth balls (alloyed with ferroelectric-lanthanoid ceramic elements inside them) inter-dispersed inside the gaps in the thin-wire network (see figure 53), would have been oscillated on a molecular level, and all their anion-cation alignments would have undergone rapid polarity-switching, and as a result of these rapid switchings every one of those ferroelectric molecules would have been forced to accumulate, and then by like-charge repulsion, to emit electron charges (for the reasons explained in section 5.3 above). These electron charges would have flooded out of each sphere's wire network at the rate of four-thousand-million times per second, and would have been transmitted in plasmonic waves through the UFO's dielectric spheres to their surfaces. Then, at the surface of those spheres external magnetic fields, or magnetic inductor coils, would have drawn those electron charges out of the spheres and would have converted them into electromotive-force-electrons to be used by the UFO for power (which is precisely what conversion the stem-coils are used for in the Andreasson UFOs, as will be explained in section 6.3 below, of course). Again, what supports this hypothesis very powerfully is that at Dalnegorsk the intensity of the magnetic fields imbued into and still preserved into the rocks surrounding that UFO site, when measured by geophysicists, was calculated to have been applied by that power source at a constant rate of 2.5 Oersteds with variable peaks of over 1000 Oersteds for when the UFO tried to take-off, and it was from these magnetic readings that the Ukraine geophysicists could conclude that the UFO must have had a multi-Gigawatt power source [note: 34].

Why these glass spheres over-heat and can occasionally ignite is not a great mystery if they were part of an engine similar to the Andreasson UFO. For as I explain in section 3.17 on the tidal effect, the gravitational pull from the central vortex (in the Andreasson UFO) can be so strong that these energy supplying spheres will have to continually revolve around their source of gravitational pull, especially when that vortex is running at full revs. If for some reason these glass spheres get jammed and can't revolve then the gravitational source will create a hot-spot in the ferroelectrics inside them [note: 35]. So if they have been made of carbon dioxide type of glass and they get so hot that the glass ignites then they will burn all the more ferociously and disintegrate, on a molecular scale from the inside outward. Precisely as seems to have happened in the case of the Dalnegorsk UFO.

6: The Craft's Electric Circuit

Contents:
6.1 Radial Momentum Induced into the Telemeter Wheels	199
6.2 Pathways of Conductivity for an Electrical Circuit	203
6.3 Electrical Energy from the Black Vortex	204
6.4 Array of Tube-guides with Crystal-like Firing Lenses	205
6.5 Orthogonal Magnetic Field Turns Beams Helical	209
6.6 Pumping up the Microwave Beams	210
6.7 Firing Through the Toroid Wall	212
6.8 The UFO Takes on Water to Breakdown into Electronic Fluid	214
6.9 Pulse Firing into the Toroid to Ionize its Fluid	215
6.10 Fluid Polarized by Black Vortex at Inner Rim	218
6.11 Inner Ring-tube Controlling Mechanism	226
6.12 Active Governing Control Through the Inner Ring-tube	227
6.13 The Soft Polymer-plastic of the Inner Ring-tube	229
6.14 Adjustability of the Toroid Fluid's Permittivity	233

6.1 Radial Momentum Induced into the Telemeter Wheels

This chapter will be about what fluid is used in the toroid, how it is propelled through the toroid shell, how this fluid is electrically polarized to establish the craft's electrical circuit, and how it then generates electrical energies to both feed back into the top spheres, and to feed into the gyrating storage fields rotating above the toroid. This chapter will also propose how the aaUFO's power can be governed.

In the most basic terms, the aaUFO's toroid shell establishes containment for three separate rotating energy fields, its enclosing surfaces physically establishes two of the craft's ergospheres; ergosphere A inside the toroid's inner vertical wall, and ergosphere C within which the fluid of the toroid is rotated. The top face of the toroid also provides the lower of the two planar walls that constitute the planar waveguide (wherein ergosphere B is gyrated). So fundamentally it is a structure which contains and is surrounded by a range of different fields of energy all rotating at different speeds. In astrophysical studies the definition for an ergosphere, where it surrounds a black hole, is a cylindrical field which contains within it energies that are in constant motion, similarly that same definition applies to all three of the aaUFO's internal ergospheres (and to its external ones).

200 Anti-Gravity Propulsion Dynamics

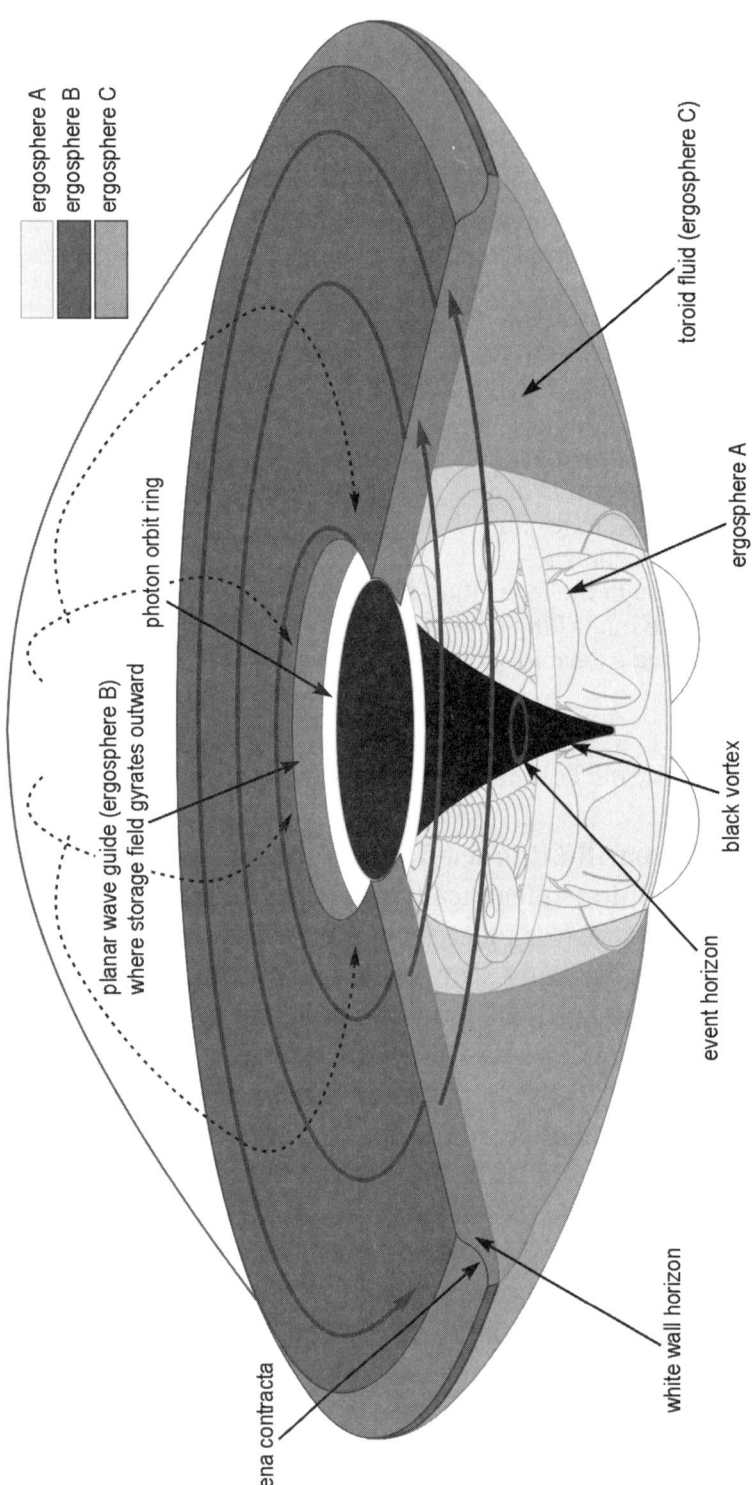

Figure 54. The lower section of the aaUFO is made up of three different compartments wherein three (or four counting the vortex) fields of energy rotate at different velocities

Once the magnetic fields (which run from the four top spheres to the central hole in the base disc), are en-placed and the charged particles rotate into that magnetic funnel to establish the rotating vortex at the center of ergosphere A, and the sub-atomic energies and photon emissions begin pouring into that vortex from the top spheres into this ergosphere to continually power it up, there will be established a fierce turbine of rotating and ever-increasing power.

Some of the rotational force coming from the black vortex will establish a torque force in the sphere-set assembly directly. And some of that torque force will come from the rotation of the telemeter wheels, although mostly the rotation of the telemeter wheels will be used as a method of governing the velocity of the central sphere-set assembly, to slow it down from rotating too fast.

However, when the sphere-set assembly has to be moved from stationary and the central vortex is merely ticking over, in order to get that assembly rotating another torque force will also be required, and this will be used until the central vortex comes up to operational levels and the Lense-Thirring rotating forces within the central ergosphere come up to speed.

This secondary turning force, from the telemeter wheels, will come from the angular momentum advected outward from the central vortex. As shown in figure 55 the curve or concavity of the inner toroid wall will be an area which will induce a Berry's vortical phase motion around the outside of the rotating field which encircles the vortex [note: 1]. These 'eddy currents' will rotate oppositely to the main vortex direction, and the angular momentum from them will be what will induce the rotational torque into each of these four telemeter wheels and which will be powerful enough to turn the whole sphere-set assembly comfortably through the lower power levels. This will be because as they will rotate through a smaller rotational diameter than the main vortex, even though the vortex will be only partially energized, those eddy currents will have induced into them a rotational velocity, and torque, that will be a proportionally greater than the rotational velocity of the central vortex. This factor comes from fluid dynamics where the outlining eddy currents will rotate faster than the central rotating field in order to match the same overall amount of angular momentum. And the torque advantage also comes from the fact that where the wall of the ergosphere is most curved (in the vicinity of the inner ring-tube and the wheels) there will be the greatest amount of kinetic force available to turn these wheels.

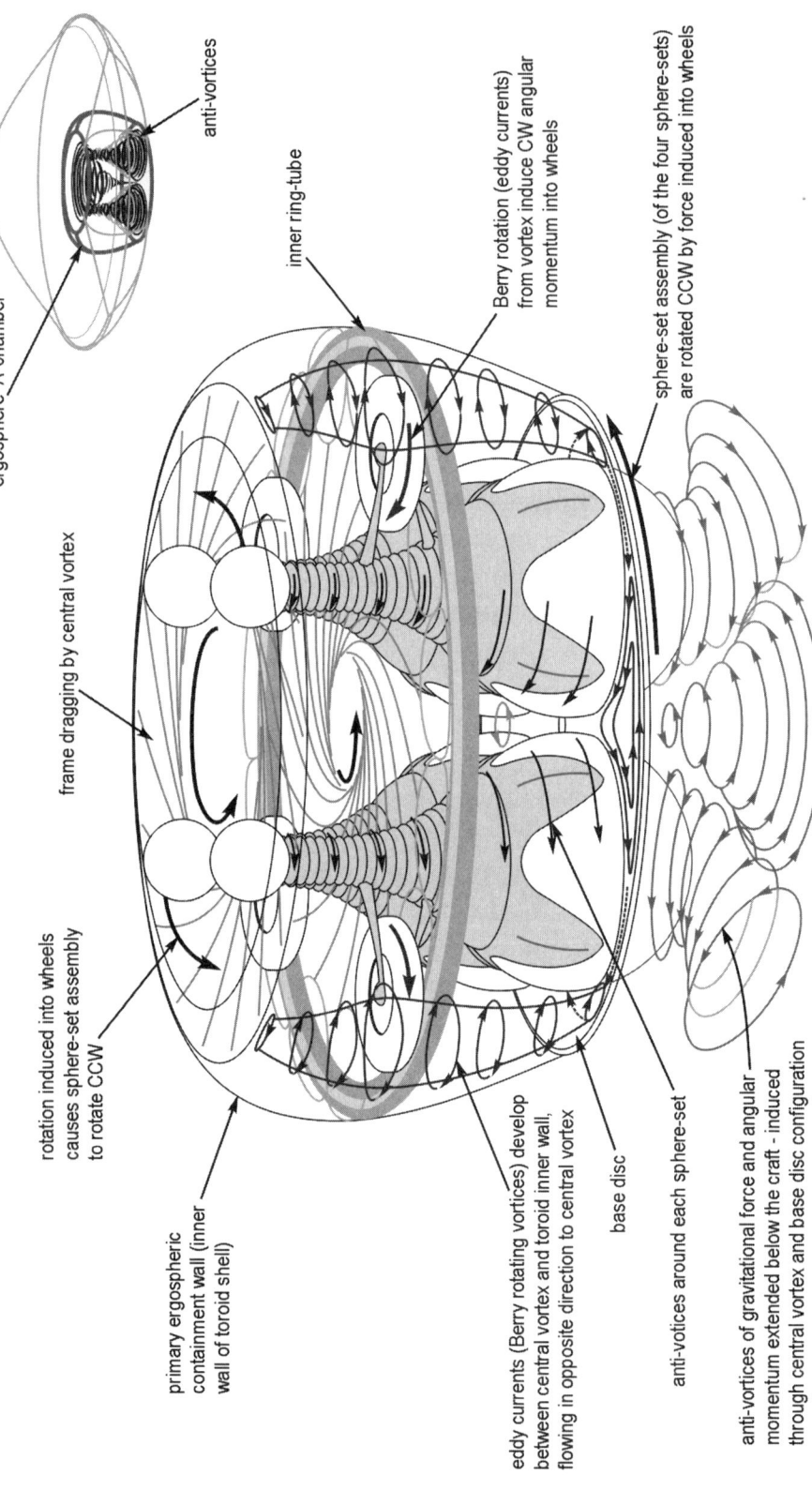

Figure 55. Ergosphere 'A' is a very compact space where energy will be constantly drawn into so that its rotational velocity will ever-increase

This factor will allow, of course, that when the main vortex does power up and rotates faster and faster and drives its energy in toward the center (with its negative radial momentum), then the amount of angular velocity imbued into these outward vortical elements will proportionally be much greater also (and these will have a positive radial momentum) [note: 2].

Together, both the torque from the Lense-Thirring energies following around the vortex, and the eddy currents rotating the telemeter wheels will provide a smooth and continual force to keep the central sphere-set assembly moving.

6.2 Pathways of Conductivity for an Electrical Circuit

While the main electro-dynamic circuit of this craft's black vortex has been delineated in the previous chapter, the circuit of particular interest here will be the electrical circuit-within-a-circuit that mainly comprises the four stem coils (running between the upper and lower spheres), the four sets of arms, and the four telemeter wheels. The path of conductivity of immediate interest to this section of the electrical circuit is that which runs from the four lower spheres up through the clamps into the stem coils, from the stem coils into the arms, through the telemeter wheel (through the tubes inside these wheels), and then through to the toroid's moving fluid.

Because of the way the black vortex fits into the electrical circuit of the stem coils (see figure 49) there will always be a conductivity path running through the stem coils, however, this is not necessarily the case with the path of conductivity into the toroid's fluid.

A potential-difference will always exist between the power-output coming away from the lower spheres and the mid-section of the toroid (because it will be the most neutral section mid-way between the upper fluid's negative charges and the lower fluid's positive charges, and so it will assume a 'relative zero voltage zone' between the two polarities throughout the toroid and with respect to all other areas throughout the lower section of the craft), this central area also corresponds to the central channeling of the toroid's magnetic fields and so electric charge will always be moved into this part of the toroid.

However, it will only move into this area if a path for that conductivity is established for it to flow along [note: 3]. But, as it will be obvious to the astute observer, the path of conductivity is not always established for that flow to take place because of how the telemeter wheels are designed, from them having four conducting tubes embedded into them which will have to line up with the toroid wall in order to establish that full conductivity path between the lower spheres and the toroid's fluid (which is why these wheels turn on the toroid wall). This means that, rather than being designed to conduct continuously, by the inclusion of these tubes into this particular circuit the idea will be to deliver short pulses of energy through the rotation of these wheels. That this will occur only when the telemeter wheels line up their tubes perpendicular to the toroid's wall, means that the sphere-set assembly has to be allowed to rotate continually for this to happen.

So, it can be seen that this action contributes to the UFO's engine powering-up mechanism, and when this circuit engages and the assembly rotates then the faster it rotates the faster will be the pulsings delivered through these wheel tubes into the toroid's fluid. To set the scene more fully, this will occur at all four separate locations around the toroid through all four telemeter wheels turning and aligning their conducting tubes with the toroid's wall (and the fluid on the other side of that wall).

Roughly speaking, if the assembly rotates at 0.5 Hz (one-half revolution per second) these wheels will fire bursts of microwave oscillating energies—coming from the lower spheres—at a repetition rate of between 9 Hz and 38 Hz (between 9 and 38 times a second). The reason WHY these beams will need to be fired into the toroid fluid will become clearer toward the end of this chapter and through subsequent chapters [note: 4].

6.3 Electrical Energy from the Black Vortex

There are several possibilities for what sort of energy will be radiated through these hollow tubes set into the telemeter wheels. The set-up is intriguing, and when they align perpendicular to the toroid wall there is also the chance that they may even interact with the magnetic field of the inner ring-tube that the wheels turn on (but, I don't want to get too complicated here so I shall leave that nugget for later). What the tubes could draw through them and could tap off the stem coils is a static DC electrical field, but in this UFO's electrical circuits DC current (direct current) will not be much use at all anywhere so that's unlikely. They might, for instance, transmit the gravitational radiation coming from the bottom of the vortex resonating through the lower spheres as waves, but I somehow think that unlikely also. The best fit scenario would be that these hollow tubes transmit an oscillating electric current which they will draw through the arms (tapping off the stem-coils), and this current will come from the direction of the vortex, but more specifically through the lower sphere's plasmonic resonances, which will be electric plasmon charges which will be converted into electric current when they transit from the metamaterial lower spheres and pass into the magnetic fields, through their Lorentz forces, into the stem coils. Then through the stem coils they will pass as oscillating current into the arms—and while arriving at the wheel axle will enter into a hollow cavity there, where they will be distributed into the hollow tubes when they line up. So eventually, what will then be propagated through the wheel's hollow tubes will be an oscillating electric current.

Electronically this would make perfect sense, for it can readily be seen that while there is a potential-difference between the lower and upper spheres there will also exist a 180-degree phase-difference between those two spheres, because of the fact that there is a conducting-coil running between them (as shown in figure 51), and this would automatically make the upper sphere to lower sphere circuit an alternating circuit. This coil may not be a conventional copper-wire coil (it may be a fluid or plasma conductor, for instance) but nevertheless it will be there to do the same job as a metal coil in that it will generate a solenoidal magnetic field around that coil. The tapping principle here will be the same as that for a self-tapping transformer, and so providing the lower arm taps into the stem-coil at its lower end, and the upper arm taps that coil half-way up it, then this coil arrangement will give a voltage-difference equal to half the voltage being established between the upper spheres and the lower spheres, and the electric current should have a phase-difference of 90 degrees. So, in other words, there will be an oscillating current running between these two arms, just as there is an oscillating current coming out of the lower spheres (via the craft's co-mutual magnetic field).

The big question will be though, what will be the final frequency of that alternating electric current converted from the lower sphere's plasmonic resonance? Obviously it would be beneficial to tap the same microwave oscillations that are being generated at the lower spheres, which have been estimated as possibly having a frequency of 3-4 GHz. How will this frequency of microwave energy be of use through this next stage of the aaUFO's electrodynamic circuit though?

6.4 Array of Tube-guides with Crystal-like Firing Lenses

Another engineering factor that separates the extraterrestrial designer from the terrestrial designer is the former's preference for non-metal materials. For everyday electrical appliances on earth metal is assumed to be the most useful of materials to use for its strength and for its electrical conductivity. But as was shown in the previous chapter metal is not the superior material when it comes to its usefulness in transferring oscillating electronic circuits—metamaterials and semiconductors are. For this same reason the telemeter wheels would also be made predominantly of various types of semiconducting materials (and even its tubes may be more useful made of such materials).

These wheels have been described as having an outside surface which is smooth (not faceted like the spheres). This fits in with the fact that they won't have any specific electrically charged energies rotating outside their surfaces to interact with, but mostly the aspect of smoothness will be necessary because they will have to turn upon the soft surface of the inner ring-tube whenever the sphere-set assembly rotates (as will be detailed below). Therefore, it would follow that the smoother they are the better.

Betty Luca has described seeing these glass wheels as having inside them, "four thick cylindrical extensions that looked like many cables welded together" (ASBT-II p27). For my working hypothesis I have taken this description to mean many tubes welded together. They could be wires, possibly bunches of thick wire of, say, aluminum or copper so that they would do the same job as Leitz wire to take advantage of the skin effect when oscillations are of an ultra-high frequency (from the fact that ultra-high frequency pulses only traverse a conductor at skin level), thus if the conducting channels are then made up of a multiple group of thick wires or tubes this would be to increase the surface area of those conductors to improve the overall strength of transmission. This would put the current oscillations being used by the ETs firmly in the MHz range of frequencies, because that's the range the skin effect works best in [note: 5] [and see note: 22].

What they might also be though are hollow wires specially made from very thin hollow cylindrical metal (of around 1 mm thickness), for recent research has discovered that such wires when used as waveguides for high-GHz frequencies propagate plasmons along the surface of those wires with little attenuation [Deibel] [note: 6].

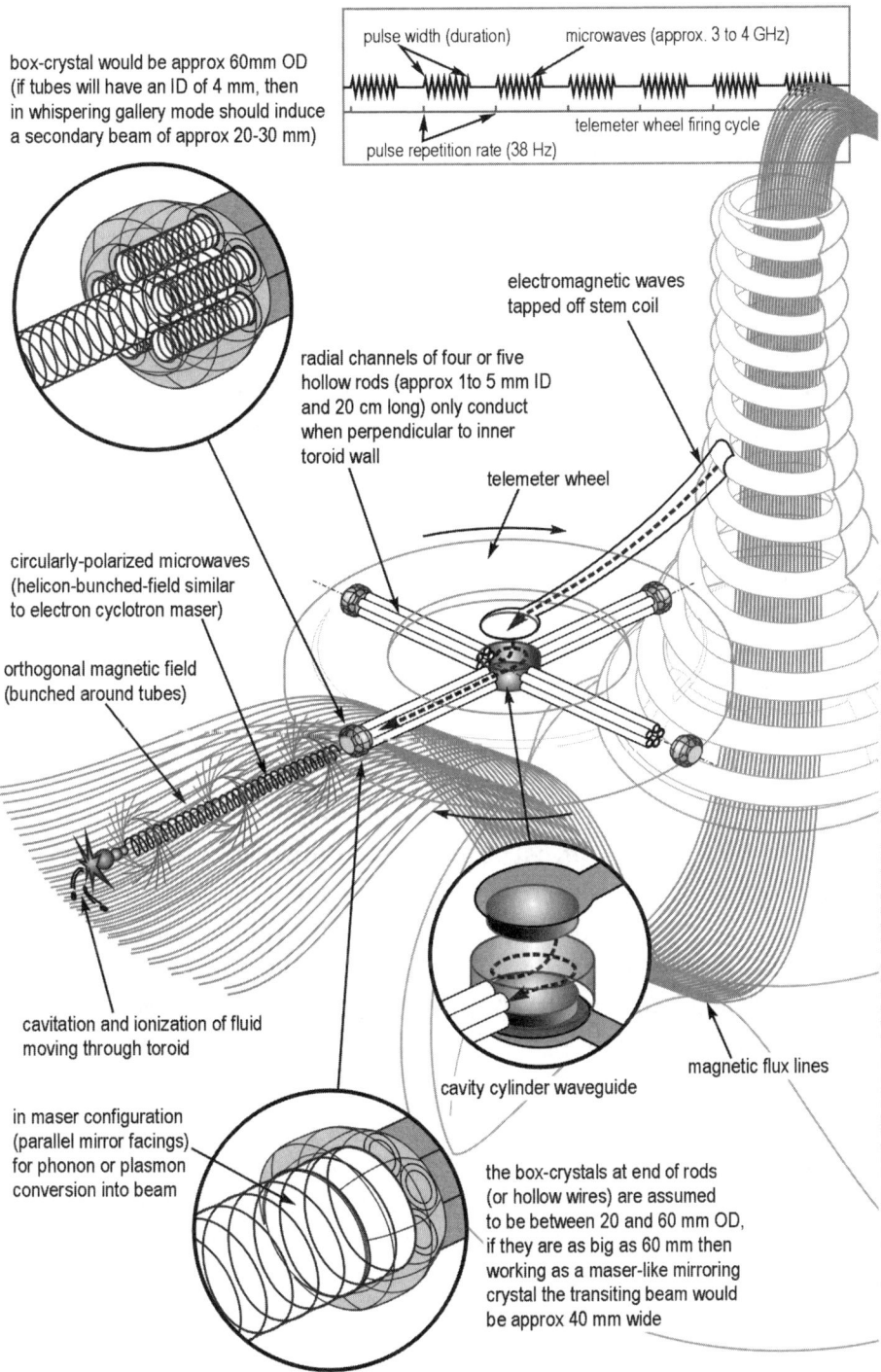

Figure 56. As the conducting tubes align with the toroid wall they will also align parallel to the toroid's magnetic field lines, this will cause the oscillating energies to turn in a helical path

This latter field of study would strengthen the view that tubes would be better than wires, so I would still favor these conductors being bunches of hollow tubes rather than the thick wires. The main point about these conductors being tubes is that only as hollow tubes will they transmit energies at microwave frequencies (which, as will be shown through the following sections will prove the more useful range of frequencies to use here) whereas wires cannot transmit microwaves.

The other point about these rods or tubes (or hollow wires) as conducting channels, is that each of these radial channels is terminated by a cylindrical crystal which looks as if it would work on the same principles as a maser crystal [note: 7]. Indeed, because these channels are terminated with such box-shaped crystals the whole complexion of what energies these channels might transmit changes somewhat, and it disposes them even more toward being microwave tubes, or possibly quasi-optical waveguides (or even acousto-optical waveguides) [note: 8].

Strictly speaking an optical waveguide might suggest that the energy to be delivered will be an intense light-photon energy (such as with a laser beam) and this might be difficult to justify because there is no photon power-pumping amplifier to accompany these hollow tubes and their crystal ends, so that theory is only just balancing upon the window sill.

That the energy beams propagated out of these wheels have to pass through the toroid wall suggests that the energy must undergo an intermediary conversion process, possibly into phonons or into plasmons. One way then of transferring energy through the material wall of the toroid might be to treat these tubes as acousto-optical waveguides (as alluded to above) which would produce a conversion process which would involve exciton generation and then exciton transfer: where these glass-crystal wheels would be irradiated with SAWs (surface acoustic waves), and so there would be exciton activity just below the outer surfaces of these glass-crystal wheels [note: 9]. If however, the microwave beam was to couple to the acoustic phonons, through the crystal cylinder then that beam may be used to drag phonons if they are weakly coupled to it through the material interfaces (maser-quartz-wall) of the toroid, especially if the beam is made to gyrate through the crystal in such a way as to bunch the electrons inside the maser-like crystal cylinder. The presence of the orthogonal magnetic field would imbue more energy into the electrons that would compensate for the slight attenuation of the SAWs transportation [note: 10]. This being the case then the crystals at the end of the tubes would facilitate an electromagnetic-to-phonon conversion process to then send the energies through the material wall as phonons (and reconvert them back into electromagnetic waves on the inside of the toroid, through the co-mutual magnetic field that passes through both the tubes and the toroid fluid) and this process would certainly provide an interesting form of one-way transport for this undulating field. This might seem to be a workable idea.

Another possibility might be that instead of the plasmonic resonating charges from the lower spheres being converted into electromagnetic energy and passing through the wheel tubes as electromagnetic waves, they might be transferable directly into these tubes, as plasmonic waves, and made to pass through into the toroid fluid as plasmon charges.

6.5 Orthogonal Magnetic Field Turns Beams Helical

Interestingly, the way that the toroid's magnetic field runs through the toroid's wall on its way into the stem coils will constitute an orthogonal magnetic field as it passes through the telemeter wheels, meaning that whenever the conducting tubes line up perpendicular to the toroid wall the magnetic flux lines will mostly run parallel with the conducting tubes (as shown in the figures 51 and 56 above).

This will be an additional factor which will help determine what the final energy configuration will be for the energies passing through these tubes, for it is well known that in the presence of a stationary magnetic field electrons propagating through that field will be turned into a helical path due to the Lorentz (v x B) force (whereby the electron rotational frequency will be proportional to that applied magnetic field) and so this configuration could then circularly polarize the beams of electrons running through these tubes.

That being the case, if the electrons are following a helical path, they can also then be bunched up. This electron bunching would provide a power amplification process, and this might be what the crystal cylinders at the ends of the tubes are specifically employed to enhance by reflecting or mirroring them backward-and-forward, to amplify them into a more powerful beam.

But an oscillating electromagnetic field which runs through a stationary magnetic field and turns that electromagnetic field into a helical beam ALSO forms another useful conversion process: Indeed, such an orthogonal stationary magnetic field configuration is featured in the theoretical conversion of electromagnetic (spin-1) waves into gravitational (spin-2) waves. Because, according to Zeldovich (following on from Gertsenshtein) electromagnetic waves can be transformed into gravitational waves if the electromagnetic waves are propagated through a static orthogonal magnetic field—relative to the action of the stress-tensor being established between these two fields. This might be useful technology here, but if the ETs have this conversion in mind then they are also adding to these two field combinations, an extra contribution of angular momentum, and also, they are possibly introducing a mass-density gradient (between the black vortex and the toroid fluid).

But considering that there are other much more simple ways of transferring gravitational radiation from this black vortex to other parts of this aaUFO then this conversion mechanism, impressive though it sounds, will be seen to be an unnecessary one. [note: 11]. Its just interesting that physicists have been working on a similar conversion process where electromagnetic waves, in theory, can be converted into gravitational waves [note: 12].

Such a magnetic field set-up is equally typical of a free electron laser (FEL), but with the intensity of the static magnetic field being so high then the set-up is more like a Graser, similar to a free electron laser but which runs electromagnetic waves through a static magnetic field and theoretically converts them to gravitational waves, or visa versa, by circularly-polarizing them into a helical wave when those energies are propagated through hollow tubes (as undulator tubes of approx 3mm I.D.) and so theoretically these gravitons may be what are being transmitted (or converted) through the tubes and into the maser-like crystals terminating those tubes [note: 13].

Otherwise, the set-up resembles an electron cyclotron maser (ECM), and in such a mechanism electrons can be phase-bunched into the gyrating helical trajectory of the beam in such a way as to accelerate them [note: 14].

However, the above theories are mostly rather redundant from the fact that in the aaUFO's configuration of magnetic flux lines, the ones coming down from the upper toroid will be pulsing and the flux lines coming up from the toroid's lower magnetic field will also be pulsing (from them all being stretched and broken by the stem coils). So, while the aaUFO does have orthogonal fields routed through all four of the telemeter wheels they will not be stationary or static. Indeed, it should be noted that the aaUFO does not have anywhere in its electromagnetic circuit any stationary magnetic fields.

6.6 Pumping up the Microwave Beams

So many possibilities, but what has to be remembered is that this configuration in the aaUFO is very simple, and it doesn't work by having loads of pre-amplification circuits and high-voltage triggerings that will be dimming its lights: Particularly as the sum of its components are; microwave energy, central hub-cavity, hollow (metal) tubes, orthogonal (pulsing) magnetic field, crystal maser-like cylinder—all encased inside a glass-crystal wheel. And the energy has to be delivered through a dielectric material wall.

Looking at this configuration from another angle might be a better idea. And it might be better to take a closer look at what is available in this particular electromechanical circuit.

In each wheel there are four perpendicular radial channels each approximately 20cm long that spread out from the wheel axle. The wheel axle I am presuming is some sort of cylindrical cavity which will be used to shape the microwave oscillations, to prepare them for the conducting channels (the details of these weren't seen and so they could have been plain empty cavities or have had crystal waveguides inside them). The four perpendicular channels would have been made up of four or five hollow tubes (or wires) welded together. Frequency wise, if they were around 3mm I.D. that could equate to a 20 to 100 GHz microwave frequency, and if smaller such as the 1 mm hollow wire example mentioned above they could work around 50 to 150 GHz although I suspect they might eventually be found to have internal diameters closer to 4mm—5mm, which I would tentatively equate to 15 to 30 Ghz) [note: 15]. And with that conjecture I would also see that whatever will be the microwave frequency generated by these tubes, this frequency will greatly help in the determination of what sort of fluid will be contained inside the toroid shell. These two factors will therefore need to be worked on in tandem.

That Betty saw these conducting tubes terminated by crystals which were 'box-shaped' inside these glass-crystal wheels offers an indication that the most predominant factor to them is that both their inner and outer faces are parallel. This suggests quite strongly that they might be used to mirror-amplify whatever beam goes through them—just like a maser crystal mirrors its energy beam to amplify it, so as to bounce that beam back and forth between its mirrored faces to pump up the beam's amplitude. In the laser process such a crystal is used to intensify its input beam through chirped pulse amplification, where a crystal of neodymium-glass or titanium-sapphire is used (these materials having exceptionally good energy storage qualities) [Mourou-Umstadter 1992]. Otherwise, there may be a population inversion mechanism used through these box crystals, and again, the crystal would be 'pumped' with more and more energy so as to excite the atoms coming through it to a point of excitation where most of the atoms rise up into a metastable excited state, and then when additionally triggered those atoms all-together dump suddenly to their ground state, stimulating large numbers of photon emissions; which through a laser causes the energy to exit the crystal as a powerful beam. In this aaUFO configuration the principle will probably be the same, but working at a much lower frequency of electromagnetic wave energy.

Going back to the point made above about the set-up resembling a gyrotron assembly or an electron cyclotron maser assembly; where a beam of microwave electrons are sent down a tube and parallel to that tube is a very strong orthogonal magnetic field—so that the beam becomes circularly polarized and forms a helical path through that tube. There are very obvious similarities here to those two configurations, however, if these conductors inside this UFO's wheels are tubes, why group them into four or five tubes? Here, it may prove interesting to speculate that if there were at least four hollow tubes, then they could deliver four rotating microwave beams through to the box-shaped crystal at the end of those tubes.

This could set up a whispering-gallery type of flow within the box-shaped crystal, at the center of which would be induced another secondary rotating beam. This secondary beam would have a precise frequency, which would be automatically pre-set by the dimensions of the multi-tube assembly (and the configuration of the box crystal) and this frequency would be the most desired frequency to send onward through to the toroid. Such a secondary beam would be greatly amplified through the crystal by that beam mirroring backward and forward between the crystal's front and rear parallel mirrored faces; and that beam would be greatly amplified by there being, as in this case, four or five rotating beams all coalescing together to induce the one central beam within the box-shaped crystal. This follows a particular type of whispering-gallery electron-cyclotron-resonance maser-oscillator which uses this very same type of set-up in order to pump up its output beam to a much greater level of power [note: 16]. This power amplification process would depend upon how the microwave cavity at the telemeter wheel axle distributed the electromagnetic energy into the tubes, but, if one beam can be directed down that channel why not have four or five hollow tubes to direct four or five beams down the same channel all at the same time. A factor which will greatly help this amplification configuration will be the strong magnetic field that the wheels will be operating in continually (see figure 56).

So the quintessential reasoning behind using this gyrotron-like whispering-gallery configuration would be to produce a beam several orders of magnitude greater than through a conventional microwave tube system [note: 17]. Obviously experimentation would be needed to establish precisely what the frequencies might be, but from the arbitrary scaling of my drawings the box-shaped crystal would look to have an outside diameter of about 60mm, while the tubes would have inside diameters approximately 4-5mm (but possibly as large as 10mm), and so the induced secondary beam in the middle should be between 10mm—30mm in diameter which would equate to a microwave frequency of somewhere between 3 to 30 GHz.

For the box-shaped crystal terminating the conducting tubes there are many crystal material combinations which exist for this type of task, all giving a large range of oscillations to choose from, and so going back to the comment made above, the oscillation frequency of the microwave tubes and these terminal crystals will need to correspond to the ionization frequency of whatever fluid is being used in the toroid, which, in the case of the aaUFO, will most likely be a hydrogen-based liquid such as water or seawater [note: 18].

6.7 Firing Through the Toroid Wall

Electrons being pumped directly through a wall? Well yes, if you consider that an electron is after-all something that enjoys particle-wave duality. Electrons can simultaneously be a wave or a particle (or even both at the same time). If one considers that an electron is made up of photons, and photons are well established as being the quanta of the electromagnetic field. This was demonstrated in the beam of electrons sent through a crystal in the Davisson-Germer experiment back in 1927, and in the cathode rays of electrons that Phillipp Lenard (in 1906) discovered would pass through materials such as aluminum.

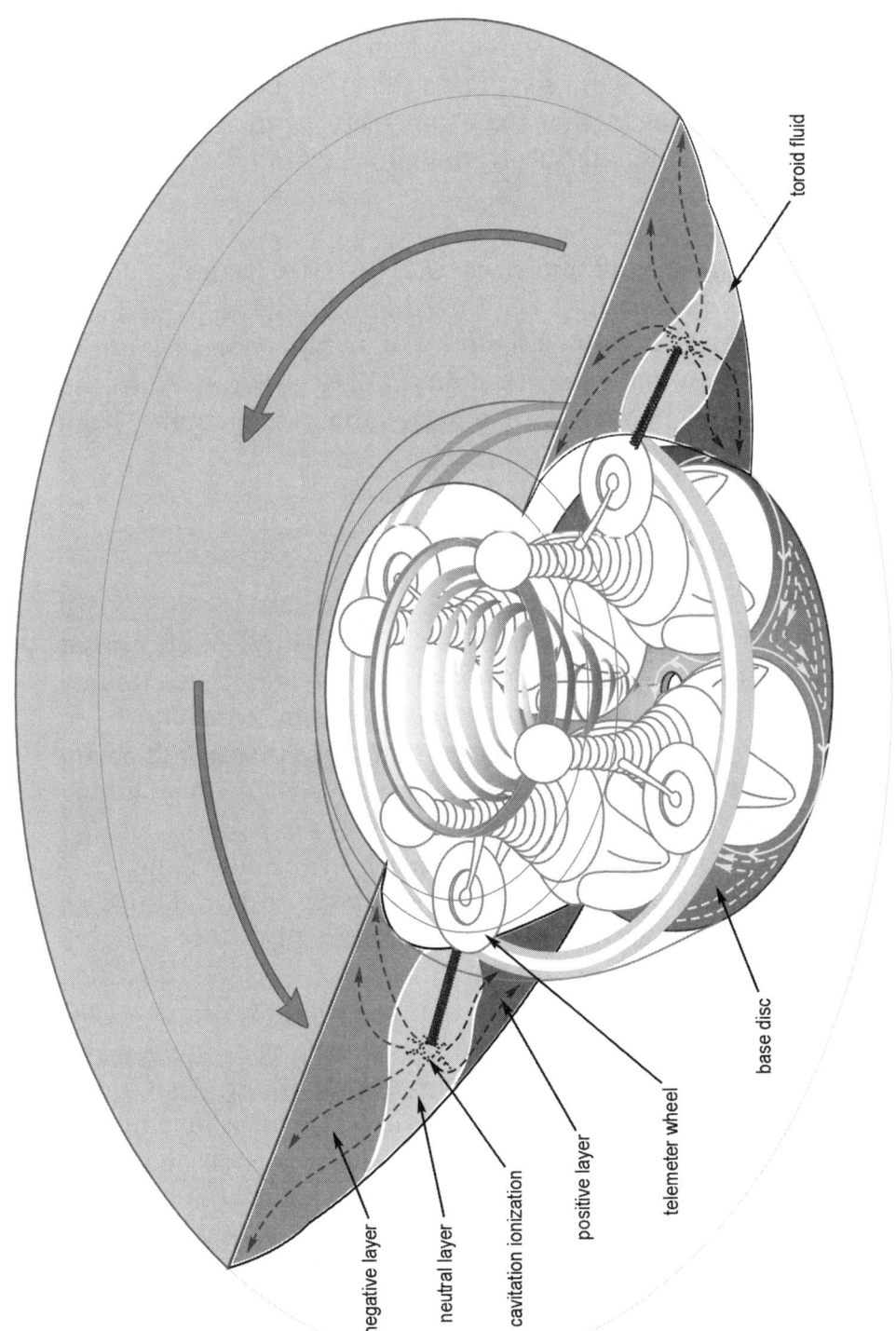

Figure 57. The telemeter wheels fire ionizing beams into the fluid moving around the toroid

Perhaps also, these box-crystals at the ends of these tubes are there to convert the electron beams into electron waves in readiness to tunnel through the toroid wall [note: 19].

To reiterate what has gone before though, while exciton and electron transportations through the toroid wall are feasible, I think when all the clues are assembled, and there are yet still more to be delivered further into this study, what is most likely is that a plasmonic resonance will be set up in the crystal cylinders fed by the microwave tubes and so the charges (rather than the electrons themselves) will be transported through the glass-crystal materials through to the fluid flowing inside the toroid.

On the other side of the toroid wall interface the fluid (ie water) will be circulating through the toroid shell (see figure 57 of toroid flow) moving around as a toroid field by the same mechanism that rotates the fluid of an accretion disk out in space, through the agency of the co-mutual magnetic field running between the toroid and the black vortex, so that it will be imbued with angular momentum coming from the craft's black vortex.

6.8 The UFO Takes on Water to Breakdown into Electronic Fluid

In the case of the aaUFO there is nowhere in Betty Luca's recollection of that particular UFO's structure any indication whatsoever of what might be contained inside the toroid shell. So, this factor has been adjudged partly by what has been deemed the most appropriate fluid for this whole electro-dynamic system, and partly from the evidence which has showed that the extraterrestrials were observed taking on-board water prior to them performing a mechano-electrical ionizing process on that water. Certainly, it would seem highly likely that the aaUFO's toroid will be filled with an ionizable fluid, and that this fluid will be a hydrogen-based liquid such as fresh water or seawater so that it can be easily replenished [note: 20].

To lend credence to the water theory is the fact that several observers of UFO activity other than Betty Luca have reported seeing UFOs hovering over the sea and over lakes apparently drawing up into them huge quantities of water. That those UFOs were also seen dumping some water back suggests that whatever water they do use in their craft becomes depleted in some way. Betty Luca herself mentions watching this phenomenon, and she does actually mention seeing sparks of electricity flying all over the place when the process was taking place [note: 21]. But then, just like physicist Dr. Hal Puthoff often says of the zero-point-fluctuations that there are huge amounts of energy just waiting to be extracted from but a thimble of the ZPF in space, so too is there is a huge amount of energy encapsulated in simple water—if you know how to extract it.

The collected water that fills the aaUFO's toroid though will be of little use if it cannot be ionized and polarized into separate positive and negative components. But if it can be ionized and separated then the water will contribute as a major component in the craft's electrical circuit. For this to happen in water this will mean separating the negative hydroxyl ions (OH^{minus}) from the positive hydroxonium ions ($H3O^{plus}$), and my studies have indicated that under certain conditions water can perform this task very well and can be expected to have a very good rate of structural diffusion of its component ions in order for this polarization to occur [note: 22]. Although, if it were possible to bond a heavier chemical element to the positive hydroxonium ions then this polarization would occur all the more efficiently, I feel (this relates to the hypothesis delineated in section 6.10 below).

Another factor with water is that when separated the positive oxygen molecules will have an almost 1000 times higher magnetic susceptibility than the water's negative hydrogen molecules. This will point to a significant, and exploitable, differential when it comes to the routing of the magnetic field lines through the middle of the toroid, which is also where the telemeter wheel beams will be firing into the toroid's water.

6.9 Pulse Firing into the Toroid to Ionize its Fluid

I have already hypothesized that the fluid inside the toroid should be ionized and polarized (sections 3.8 and 3.9). This can be done through the beams being fired into the toroid by the telemeter wheels. When the short pulses of microwave energy are fired through the toroid wall from the four telemeter wheels into this moving fluid then these beams will continually ionize small pockets of that fluid, possibly through a process similar to how microwave radiation can Comptonize-ionize atoms to knock electrons off the fluid's atoms and molecules. And instead of those ionized particles recombining almost instantly (which is what normally happens in stationary fluids) the fact that the fluid will be fast moving through the toroid will greatly hamper that recombination and aid the polarization process.

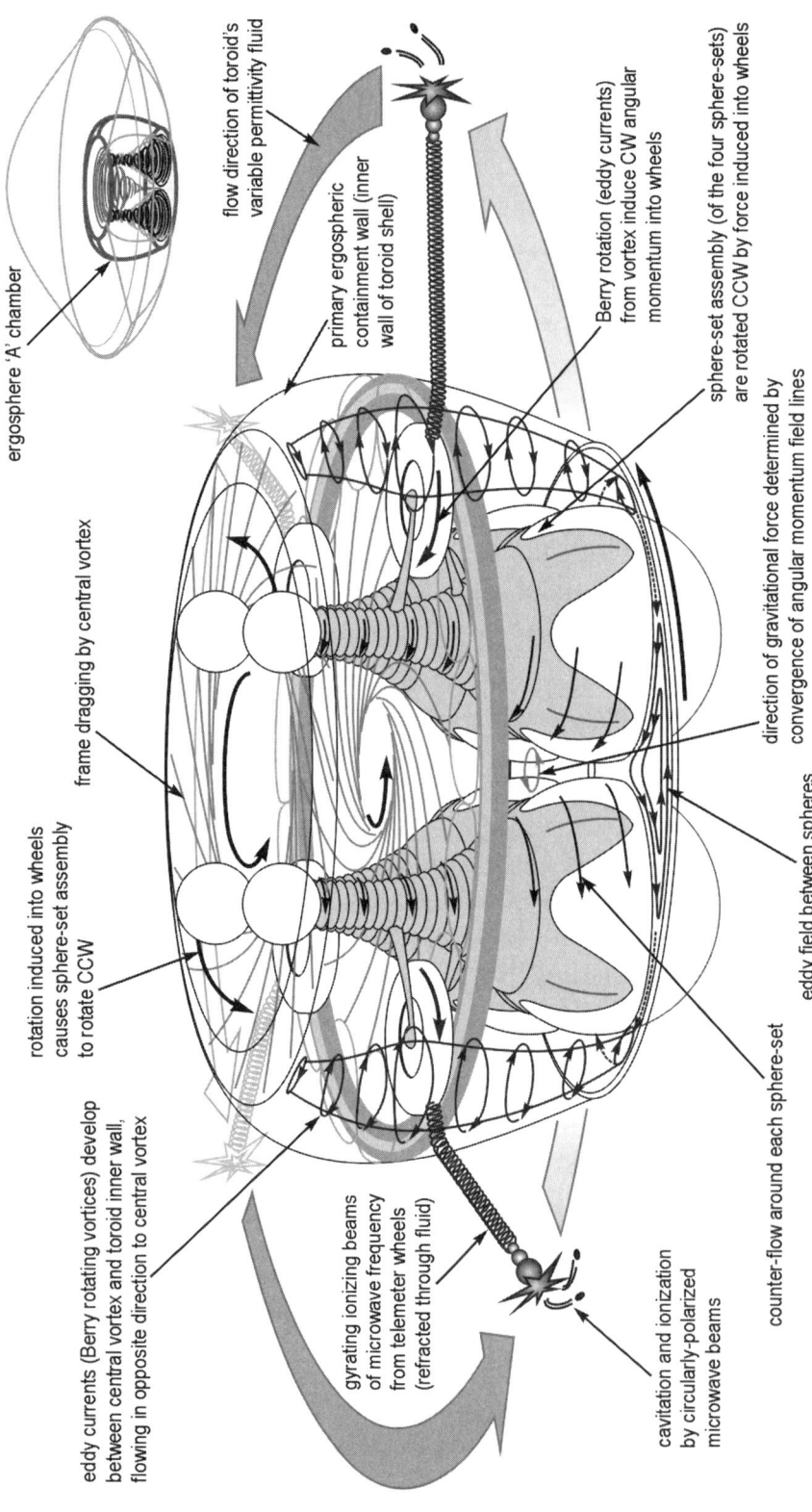

Figure 58. The sphere-set assembly will rotate by the force of the black vortex to fire pulses into fluid

The type of process that would be useful here would be one similar to a water ionization process initiated by a laser beam; where this sort of process (which was developed to carry out studies on seawater) initiates an electrical breakdown of the surrounding water molecules when the water is perturbed by the laser's beam. In this process the laser beam cavitates a small pocket of the water, and as inside these small pockets the air is much hotter than the surrounding water the pockets tends to expand outward, creating a gas-filled bubble which goes on to generate an acoustic shock wave, which then further expands through the water [note: 23].

But this same ionizing effect should also be possible at microwave frequencies by the helical beams being fired through the telemeter wheels (and at the same time this will have a heating effect on the water). And in the heating of the toroid's water, here is another mechanism through which that water can be propelled even faster around the toroid shell. For where the beams hit the fluid in the center of the toroid (see figures 57 and 58), if their microwave frequency is high enough they will heat up and cavitate small pockets of it (just like in the case above with the lasers), and the water immediately surrounding the cavitation pockets will expand and become less dense than the bulk fluid moving through the toroid, so what will happen is that the resulting density-imbalances (in the four locations where these beams fire into the water) will cause the water to move faster. The knock-on effect of moving the water through the toroid shell faster will be that its ionized-polarized molecules will generate a more powerful magnetic field around the outside of the toroid shell, and this magnetic field, while it is the same one that feeds into the craft's black vortex, will amplify the Lorentz forces inside the central ergosphere and it will imbue extra force into the rotational torque of the black vortex [note: 24].

Significantly, through this mechanism, as the intensification of the beams being fired into the toroid increases, so would the velocity of the toroid's fluid also increase and the more that fluid would empower the craft's electrodynamic forces. This would be one way in which the aaUFO's electrodynamic power could be increased and decreased, because to feed more Lorentz force into the craft's black vortex would increase its rotational velocity and its demand of subatomic and photonic energy (and then more gravitational force and more angular momentum would be fed into the gyrating storage field, and its other ergospheres).

Preliminary calculations for the beam firing repetition rate, and the reader will be reminded that these will only be based on a sphere-set assembly rotation speed of just one-half revolution per second (0.5 Hz), suggest that pulses (containing the 3-30 GHz microwaves) can be fired through these telemeter wheel tubes at a pulse repetition rate of 9.5 Hz (or every 105.14 milliseconds—per wheel) with a pulse duration of 11.68 milliseconds. This is quite a conservative estimate though and there is a lot of leeway in these figures for when the sphere-set assembly powers up and rotates faster (although, while the pulse rate does increase so the pulse duration becomes smaller and smaller, and so there will, theoretically, be a maximum working pulse-rate that will allow that ionization to be sustained inside the toroid, before this method becomes ineffectual, and this will depend on how short a pulse can still be used to cavitate the water) [note: 25].

In this regard then, it may possibly be that the intention is to both ionize AND to propel (and to heat) that fluid, and thus-wise provide the means to; ionize (and polarize the toroid's fluid so as to separate the negative charges from the heavier positive particles), and to accelerate its motion through the toroid shell, which to some extent would turn the actual structure of the toroid into a giant charging capacitor and a Lorentz-force generator, which would amplify the rest of the craft's electrical energies (and at the same time provide greater stability to them—by the water flowing in only one direction, which will be a point returned to in section 9.1).

However, as far as the water inside the toroid is concerned, that fluid will be polarized (and it would continually be ionized through these pulsings) and the electrical charges that its separate polarities generate will accumulate into fields at opposite areas inside the circular toroid shell, and these charges will help generate outside of that shell the induced electrical fields (the diffuse layers of oppositely charged ions) that will add into the energy fields gyrating inside the planar waveguide of this UFO. By viewing this action from further away, notice here that there will be a recycling circuit of those energies, for where these negative charges accumulating inside the toroid subsequently induce a diffuse layer of positive ions on the outside, the fact that these positive ions will be joining in with the gyrating flow of positive charges curling into the center of the UFO, into the black vortex, means that a flow-path will always be established for that energy to return back into the craft's vortex—and then, of course, these will eventually be returned back into the toroid because the vortex will always empower all the moving-fluid flows by imbuing them with rotational force. So energy dissipation losses should be kept to a bare minimum.

6.10 Fluid Polarized by Black Vortex at Inner Rim

Water and seawater does suggest a lot of weight to be carried in this flying craft... but then what is weight? Weight is determined solely by whatever force the local gravitational field imbues upon that mass material. The notion of weight is only apparent because of this planet's gravitational field. But this toroid's fluid will not be predominantly affected by the down-flowing gravitational field of earth, it will be affected by the STRONGEST source of gravitational force in its vicinity—which will always be the craft's own black vortex. The toroid's fluid will become imbued with the horizontal gravitational attractive force emanating radially from that black vortex, so this UFO's fluid won't have anything that can be regarded as weight!

This is why I would suggest the toroid should have its water as 'heavy' as it can be, and that it should have mixed into its water some even heavier elements which could bond with the positive hydroxonium (H3O+) ions of the water (elements such as manganese, potassium, or sodium, for instance). Because from the process of ionization which will allow the water's positive ions to separate from the negative ions, this process will allow the more massive positive nuclei to be attracted toward the lower-inward edge of the toroid because of the greater amount of gravitational pull being focused into that edge from the central vortex. By this same gravitational mechanism the negative ions of the water (being lighter and less massive) would be left to migrate to the upper edges of the toroid, following freely with the kinetic flow of the water (corresponding, also, with the Lorentz polarizing force on a moving fluid) [note: 26].

The gravitational polarizing action upon the toroid's fluid will correspond very closely with what is going on inside the central ergosphere (ergosphere A) where, as the reader will recall from the previous chapter, there will be inertial frame dragging and a polarization cloud of charged particles established around the vortex wherein the protons and the neutrons will be aligned toward the black vortex by gravitational force, and will be displaced away from their normal central alignments relative to their circulating electrons (as above section 5.1).

And, what this molecular polarization effect also shows is a way in which the gravitational radiation force can be even more efficiently extended through to the adjoining materials of the outlining parts of that UFO (as is shown in figure 59).

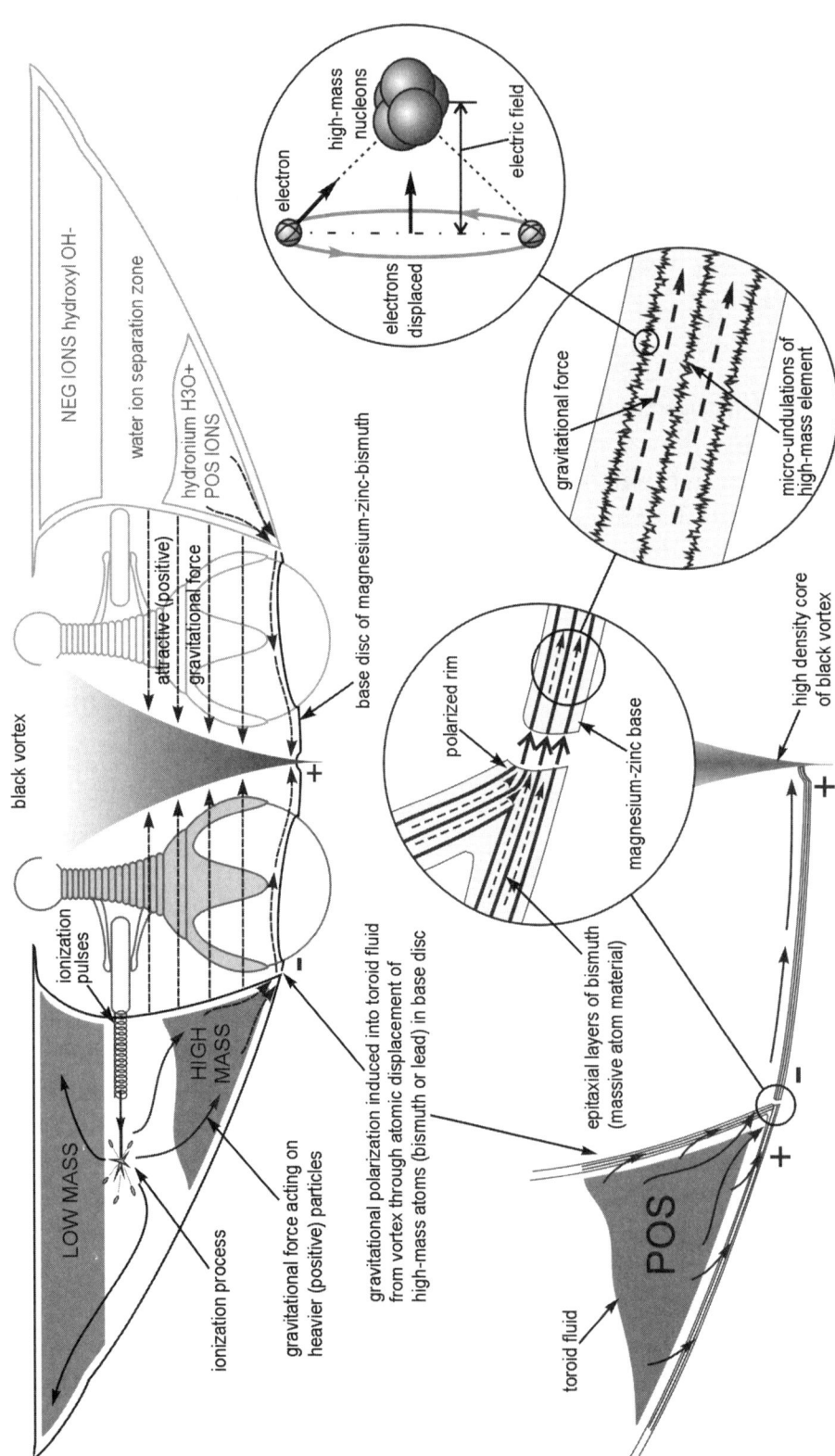

Figure 59. Gravitational forces will be induced upon the heavier atomic nuclei of the fluid's molecular structure, through the heavy atoms that make up the base disc and the toroid casing

Because, at the lower end of this particular UFO's vortex, where the gravitational force will be at its strongest, a transference of that gravitational force can occur—even more effectively than through the polarization cloud process—by using the actual molecular structure of the base disc, if the base disc has fabricated within it high-atomic-mass metallic compounds which can be gravitationally displaced. This gravitational force transference can then be amplified, through constriction (by using the smallness of cross-section for those compounds), to beyond the disc's perimeter to gravitationally polarize other materials or fluids which are then inter-connected to it. Which specifically, inside the aaUFO, would occur where the base disc abuts up to the lower inner rim of the toroid shell where the inner corner of the toroid shell presents a sharp edge to that base disc. This would ensure that at this inner rim edge there would occur the greatest exchange of gravitational force between the black vortex and the toroid shell, and consequently, the fluid circulating inside the toroid at this lower inner rim will have transferred into it a much greater amount of gravitational influence than will be transferred into that fluid at any other part of the craft. This would guarantee that the more massive positive ions of the toroid's fluid would be drawn in toward that inner edge and separated from the negative ions.

The correlations, poignantly, for this transferring mechanism of molecular polarization in UFO structuring can be found in both the UFO that Bob Lazar worked on at Area 51, and in the UFO that was retrieved from Roswell. I will give a more comprehensive explanation of this mechanism, and for how it fits into the Area 51 UFO, in the supplement chapter 12 of this book. However, here, because it is illustrated so graphically in one of the samples of strange metals retained from the Roswell UFO crash debris, I shall detail that particular piece from Roswell, so that I can build up a background knowledge for the reader ahead of the more technical stuff to come.

This particular piece of debris was apparently one of many collected by the townspeople of Roswell in 1947 following the discovery of that UFO's crash site, and while 99.9% of artifacts had to be handed over to the military, allegedly under threat of death, this small piece was secretly stashed away until 1996, when it arrived under mysterious circumstances in the mail at the offices of a US radio station.

It was a very light piece of metal just a few inches long, quite strong and similar to aluminum but very brittle. All the edges were jagged and showed breaks, suggesting that it would have been part of a larger construction, and from the fact that it was a shattered fragment it can be conjectured that whatever the larger component was, that it was originally part of, that it must have taken the brunt of the impact force of when that UFO hit the ground for such small pieces like this to be shattered off it.

When physicists were given this metal fragment to analyze they reported back that it was made from an alternate layering of magnesium-zinc and bismuth, with the bismuth layers (of 1-4 micron thickness) being sandwiched between several thicker layers (of 100-200 microns) of a magnesium-zinc alloy. Linda Moulton Howe, who carried on the research to find out more about this fragment, was told at the time by several metallurgical labs that they had never seen anything like it before and that no one had any idea why it would have been structured in such a manner—the bismuth layers were so thin between the magnesium-zinc layers that they could only have been put in place as ultra-thin epitaxial coatings upon each subsequent layering of magnesium (epitaxial layering is used in semiconductor manufacturing to produce a very thin layer between different compounds). Subsequent to further electrochemical tests performed on this sample, it was eventually theorized that such a material would be useful working as an electronic fuel cell. And this fuel cell attribute is certainly one possibility, but I think such a theory only identifies one of its incidental attributes, there are a lot more factors about this fragment which I think are considerably more explanatory toward to its primary functions.

According to the lab reports that Howe got back the 'bismuth' in this fragment had an anomalous molecular structure but it did show an atomic mass of 208, which is consistent with earth's bismuth, and which is a high-mass material. Bismuth itself is not a strong metal and for it to be useful it would need to be structured into a metal with greater strength, which of course magnesium does have when it is alloyed with zinc, hence the lamination in the Roswell sample is perfectly understandable.

Leaving the Roswell craft for just a few moments, if such a metallic compound were to be used in the aaUFO it would ideally feature in the construction of that UFO's base disc, because the bismuth's high atomic mass would be very well suited to transfer the gravitational influence of its black vortex through that high-mass molecular structure to the other parts of that UFO (as shown in figure 59). The gravitational influence of the central vortex would be able to 'conduct' through the massive positive ions of that metal by the displacement polarization of its individual atomic cells. In fact, it would suit the aaUFO to also have most of (but not all of) the under-surface of its toroid shell fabricated with such a metallic compound, so that the craft's gravitational influence could be focused and intensified and then transferred into the fluid contained inside its toroid so as to polarize its molecules (as explained above). Interestingly, what enhances this polarization theory for the aaUFO is that one of the anomalies noted in the Roswell metal laboratory reports, was that in this bismuth-magnesium-zinc (Bi-Mg-Zn) sample the rate of emission of Mg positive (+) ions was over 60 times greater than the rate from the purest magnesium standard known on earth, which goes to show that it was the positive-ion configuration—the higher atomic-mass nuclei configuration—that the extraterrestrials were interested in exploiting for this particular part of their craft [note: 27].

In this world the use of magnesium-zinc alloys has for a long time been quite commonplace (particularly in the manufacturing of car engines) for their lightness and strength. And a look at the chemical periodic tables shows that magnesium is the most electropositive of all structural metals, while bismuth is one of the most electronegative. Mainstream researches into such bismuth-magnesium compounds per se are not that extensive although for compounds with one of these two in them numerous studies have been carried out and published. Interestingly, for a similar magnesium alloy, for instance lead-magnesium-niobate (PMN), the dielectric constant has been found to be as high as 30,000 K [note: 28]. And if a similarly high dielectric constant can be transposed into the magnesium alloy layers of the aaUFO's base disc then this would be very beneficial for the transference of the atomic displacement electric field, from the gravitational polarization influence of the central vortex, through the aaUFO's base disc to adjoining materials: because the gravitational influence could be channeled extremely efficiently through these ultra-thin layers of bismuth running between that base disc's insulating layers of magnesium-zinc with negligible losses.

Likewise, in the Roswell UFO's Bi-Mg-Zn configuration the intention would probably have been to transmit, and at the same time to intensify (by using the thinnest possible layers of bismuth), the gravitational influence of its central vortex outward through that craft's base disc; while the magnesium-zinc strata of that base disc (if its layers also possessed a high dielectric constant) would have determined that the original Roswell component would channel its gravitational and its cellular electronic charges with great efficiency to wherever they needed to go (ie that craft's polarized fluid). Looking at it another way, the Roswell UFO's Bi-Mg-Zn configuration would work not like a conventional (electrode-to-electrode) *longitudinal* capacitor—but like a *transversal* one, because the horizontal constriction of the bismuth would have ensured that the gravitational polarization would only be transmitted radially and horizontally.

For either UFO this would mean that all around the perimeter edge of the base disc, but not anywhere over the upper or lower planar faces of that disc, the gravitational force would be almost as intense as it was at the central hole where the highest mass-density of the black vortex exists.

As mentioned above, this influence will transfer into the toroid's fluid, but what it will also do is help to generate some of the UFO's power. For when the UFO powers-up and the base disc is rotated, the gravitational influence that extends out from the perimeter of the base disc will drag the magnetic flux lines that run through the gap between the toroid and the base disc (these will be the flux lines that don't route through the stem coils, of course). And again, as happens above the toroid, when those flux lines get stretched and break (and subsequently have to reconnect) huge amounts of electron-positron pairs will be produced, from the stressing of the surrounding air (or surrounding space) through the field-reconnection mechanism.

And so this would be one of the ways whereby energies produced on the OUTSIDE of the UFO can be harvested for use INSIDE that UFO (more on this energy harvesting in section 9.7 and in figure 74, and also in section 10.11).

There are some additional factors presented in this Roswell sample worth mentioning but these will be gone into in further detail in section 6.13 below.

Mainly though, the most important thing to consider is that this ingenious mechanism designed by the extraterrestrials effectively polarizes the toroid's fluid, and in so doing it ALSO establishes the craft's electrical circuit: By establishing a negatively-charged top (of negative hydroxyl ions) and a positively-charged bottom (with positive hydroxonium ions); and a high-permittivity non-conducting region in-between; and in so doing the toroid extends the electro-dynamic circuit of the black vortex (and ergosphere A) outward to include the toroid fluid (as ergosphere C): And as I have mentioned above, because the toroid fluid will have a high permittivity and because it will be flowing it will act just like a macro-capacitor and provide substantial damping to the UFO's electro-dynamic circuits to keep them stabilized.

However, because of the need for the toroid's polarization to be continually maintained (by the wheels continually firing their ionizing beams into that fluid), this would suggest that the toroid's water would be exhaustible, and that at some stage the water will become 'ionized-out' and depleted on a molecular level. And so in this respect the water it uses will be a component that at appropriate time intervals would need to be replenished with fresh supplies [note: 29].

This would correlate with the evidence supplied in the "Andreasson Affair" books regarding a process that was carried out on one of the UFO's in the UFO-by-the-lake scene observed by Betty Luca (see this chapter's note 21, and also section 7.4 below) which must have been a water-replenishment process. And judging by that grand lesson in UFO electro-dynamics, as set up by the Greys in that UFO-by-the-lake episode and which was performed specifically for Betty Luca and those other abductees present to observe and take note of, it would seem that those ETs were showing them (and showing the reader, of course) that appropriate consideration should be given to a number of dangers inherent in that particular re-fueling procedure for a UFO. For instance, in that colorful episode one of its primary lessons would seem to have been, do not allow the central vortex to power-down while this replenishment process is taking place—otherwise the UFO will extinguish all of its power generation; and another lesson would have been, do not cross the polarized water-transfer hoses during this electro-mechanical process—because they are liable to short-circuit! And a further piece of instruction that the Greys might have been showing these abductees was that once the lower inner rim is hyperpolarized (in response to the gravitational pull of the central vortex) there needs to be implemented a method by which it remains polarized throughout this whole water re-fueling procedure—otherwise the UFO's electronic circuit becomes un-established [note: 30].

Incidentally, it now becomes quite clear, from that UFO-by-the-lake episode (and after the resulting transversal shock incident was explained to Betty by the ETs), that the aaUFO's lower inner rim which butts up to the base disc is what those extraterrestrials were referring to as the depolarized rim, because, where the water-replenishing incident led to a short-circuiting of the toroid's fluid and the whole of the UFOs electrical circuit, because the positive charges had neutralized inside the toroid and it could not then retain its magnetic field, so the craft's central black vortex also expired—and once the central vortex expired that UFO became decommissioned. Hence, this was the reason why its energy fields had to be fired up externally by another UFO craft —because it had lost its ability to generate its own power (as shown in figure 63 below).

So, to summarize then how the aaUFO's electro-dynamic circuit works: When the microwave oscillations of the lower sphere energies reach the required frequency they trigger the plasma mobilizing frequency of the wire network and ceramic dopants in the spheres, these dopants, of mostly ferroelectric lanthanoids, will emit huge amounts of electron charges in plasmonic waves which ripple through the spheres and become transferred (as plasmon charges which convert into electromagnetic charges through the toroid's magnetic field), through the commutator ring and the craft's magnetic field into the stem coils. These electromagnetic charges conduct up to the top spheres via the stem coils, where they are converted through annihilation with positive charges into energetic photons, some of which (at the photon orbit ring) get dragged into the black vortex, which gives the vortex more mass-density and rotational velocity, and which enables it to oscillate and to polarize the lower spheres. As a result of the establishment of this primary electro-dynamic flow, some of those electromagnetic charges are tapped off half-way up the stem coils by the telemeter wheel arms (by each arm collecting an electromagnetic potential which is 90 degrees phase-shifted between them), this electromagnetic potential is then made to oscillate through those arms and into the glass wheels and through the hollow tubes running within those glass wheels, to produce a pulsed beam which punches into the toroid fluid as helical microwaves which ionize that water-fluid. The faster these beams are fired the faster the toroid's fluid will flow (and the more empowered the craft's vortex will become). The water in the toroid becomes polarized by the gravitational forces of the UFO's black vortex to provide a negatively charged upper section and a positively charged lower section, separated by a high-permittivity dielectric layer between them in the middle of that toroid, just like a very large capacitor... Simple!

Obviously, the polarization of the toroid fluid inaugurates, establishes, and stabilizes, the magnetic fields surrounding the toroid—which feeds and nourishes the craft's black vortex and all the other electro-dynamic components operating in this

UFO, and again, this corresponds perfectly with the Lorentz forces operating within the toroid's moving fluid.

That is if the reader at this stage of the study has got the gist of the magnetic fields of this aaUFO, and is conversant with the basic routes of the aaUFO's magnetic flux lines, if not then the reader can have a brief recap by referring back to figure 12.

6.11 Inner Ring-tube Controlling Mechanism

Another highly efficient piece of designing by the extraterrestrials ensures that whatever forces are being imbued into the pulsing magnetic fields inside the toroid, where those magnetic flux lines emerge through the center of the toroid, they will the exert greatest amount of Lorentz force upon the telemeter wheels because of the wheels positioning. Briefly, in these cross-section figures it can be seen that the magnetic flux lines of the upper and lower toroid fields very purposefully run parallel to the horizontal axis of the telemeter wheels held by the sphere set assemblies. There is a refinement to this alignment though, for in this figure (figure 12) it also shows that when the inner ring tube's magnetic field is energized it pulls the toroid's lower field up above that ring-tube—and this is to ensure that those magnetic flux lines traverse right through the telemeter wheels (and through the arms that hold them) perpendicular to the wheel's horizontal axis. Another view of this field can be seen in figure 60.

Earlier into my researches it was difficult to understand the exact purpose of the inner ring-tube (see figure 25) from the initial descriptions of the aaUFO that would take Betty Luca on her nocturnal visits to the alien's world. But from the details Betty forwarded in her ASBT booklets [note: 31] and in personal communications it eventually became clear that inside this ring must have been a fluid formulated to energize a strong and mobile magnetic field localized just outside that ring around the outside of the sphere-set assembly. When I initially did some digging around at my local university library I found that there are indeed such things as magnetic fluids and the behavior of these fluids is very different from what one might expect from a solid magnetic material, in that the magnetic fluid comprises moving islands of opposite polarity magnetic poles that can move freely around, or, as for instance inside the aaUFO's inner ring-tube, they can move freely along that tube. Also, that fluid's magnetic pole-islands can be spread out to weaken their intensity, or bunched together to strengthen their magnetic intensity [note: 32].

The wonderful thing nowadays, as mentioned in chapter 3, is that polymer science has become so advanced in the twenty-first century that now any gel can be designed and imbued with either electrical or magnetic functions, as is the case with the most recently introduced ferro-gels (magnetic field sensitive polymers such as polyvinyl alcohol hydrogel filled with magnetite particles) [note: 33]. Betty actually describes this inner ring as a clear soft plastic-like donut filled with grey fluid-like gel [note: 34].

And too, there are technologies now which exist that allow fluids to be moved electronically; by using electrohydrodynamic (EHD) actuators to initiate electro-osmosis motions in fluids (based on electric double-layer flow which produces electro-kinetic transport).

That the EHD system could have an array of dielectric actuators located around the inner ring-tube, which could be controlled by a transmission line (that would possibly involve fiber-optic channels rather than electric cables) this would mean that the magnetic fluid flowing inside this ring-tube could be efficiently controlled all the way from the UFO's control center by using just one transmission line [note: 35]. This would satisfy another proviso I have made with my breakdown of this aaUFO, in that the transmission lines that actually link between the pilot and the 'engine' will be kept to a very bare minimum for constructional simplicity. For this reason I hope the astute among you reading this study will have observed that so far there is still only ONE OPTICAL TRANSMISSION LINE required. And if the design parameters are well thought out beforehand (regarding semi-conductor thresholds) then that will be all that will be necessary for a basic start-fly-upward-return-to-ground functioning of the aaUFO—and so how's that for Vorsprung durch Technik..!

6.12 Active Governing Control Through the Inner Ring-tube

Looking more closely at magnetic fields created in magnetic fluids; what they do is create high and low zones (or islands) alternately throughout the fluid. Therefore, by controlling the flow of such a fluid's magnetic islands inside the inner ring-tube it would allow those islands in the magnetic fluid to act as dynamic flux-movers. More specifically, flux-movers or flux-governors, which could be made to accelerate or brake the movement of any electromagnetic component that moves within their influence.

Such would be the action on the telemeter wheels as they rotate above this ring-tube. While the ring's magnetic fluids are not that powerful in themselves, especially while the islands are allowed to roam freely, they will though be strong enough to re-route the much more powerful magnetic flux lines flowing through the toroid (as shown in figure 12 which shows how the toroid's magnetic fields are routed over and around the inner ring-tubes). And then, once the sphere-sets get up and running and powers up all those flux lines, the inner ring-tube's flux lines will then also be imbued with greater magnetic force.

Indeed, and this is another important point to make about this inner ring, in that, within the craft's magnetic field dynamics there is evident a mutual-induction process that couples this inner magnetic ring-tube with the craft's outer rim magnetic field, and which in turn also couples to the central black vortex (because this inductive-coupling is through the co-mutual magnetic field of the toroid). This will be good for stability and good for amplification (it would be analogous to having a series of transformers with multiple windings inter-connecting their cores with variously positioned taps from which could be drawn different amounts of power). Moreover, whatever Lorentz force is generated at the vortex (and at the photon orbit ring) this will induce a proportional amount of Lorentz force at the craft's outer rim at the toroid's outer perimeter, and vice versa.

But furthermore, whatever force will be generated at the outer rim will be multiplied significantly at the photon orbit ring and around the central vortex because of their smaller radii… But then, the inner ring-tube being positioned right in the middle between both of these electro-dynamic force generators, it will gain the greatest amplification from BOTH of them. And so, when the whole UFO engine powers up and comes up to operational energy levels, the magnetic forces that will be COUPLED INTO the inner ring will be significant—and the amount of Lorentz force EMPOWERED FROM this inner ring-tube will also be significant.

This will mean that with its magnetic force the inner ring fluid can significantly control the toroid's fields, and particularly how these flux lines will run through the telemeter wheels on their way into the bottom of each of the four stem coils. The inner ring can either bunch up the toroid's flux lines as those lines run through the telemeter wheels and into the stems to intensify them—or it can spread them out to weaken them.

This mechanism will correspond to a control mechanism over the amount of Lorentz force amplification that can be exercised on the electron charges coming out of the lower spheres through the plasmonic resonance. Hence, to a large degree the power output of the whole UFO craft can be controlled merely by controlling how the inner ring's fluid is configured.

From an electrodynamic power-drive point of view, this part of the craft's engine is such that once it engages and the sphere-sets begin to rotate, the craft's power would increase exponentially to its maximum output if the sphere-set were not braked. So therefore in order to be able to control the UFO its electrodynamic power-drive will require a method of active governing to control its output (rather than active amplification control). In this way the inner ring-tube has to provide a way of braking the rotations of the telemeter wheels (which of course will govern the speed of the sphere-set assembly which controls the craft) [note: 36]. This could be done quite easily, because, where the toroid's flux lines flow through the wheels and run parallel to the microwave tubes in those wheels, those flux lines could be made to converge with a wide convergence or converge with a narrow convergence. To do this those flux lines would be bunched-up behind the telemeter wheel, or bunched-up in-front of those wheels; in this way the wheels could be slowed or accelerated, and this will be how the wheels and the sphere-set assembly rotation can be controlled [note: 37].

The trade-off from this braking of the sphere-set assembly is that there will be a compensating effect around each of the sphere-sets, in particular that the anti-votices around each of the sphere-sets (as shown in figure 55) will speed up to the degree the assembly is prevented from rotating along with the central vortex. In fact these anti-vortices will help conserve energy in the main vortex [Lin 2002] [Visser 1998a]. But also, that these anti-vortices will be encouraged to flow around each sphere-set (and orbit the main black vortex with those sphere-sets) it suggests that the lower spheres are perfectly positioned so as to capture whatever gravitational and electromagnetic energies are induced into these anti-vortices from the central vortex (possibly through the mechanism described in section 5.8). And possibly, a separate mechanism could even be incorporated to drive these anti-votices while the central vortex is made to go through changes in its configuration in order to use its gravitational forces externally (as will be alluded to in section 11.4).

6.13 The Soft Polymer-plastic of the Inner Ring-tube

There will be a good reason for the inner ring-tube casing being manufactured from a soft polymer-plastic. So as not to dampen the oscillating frequency being generated within the sphere-sets.

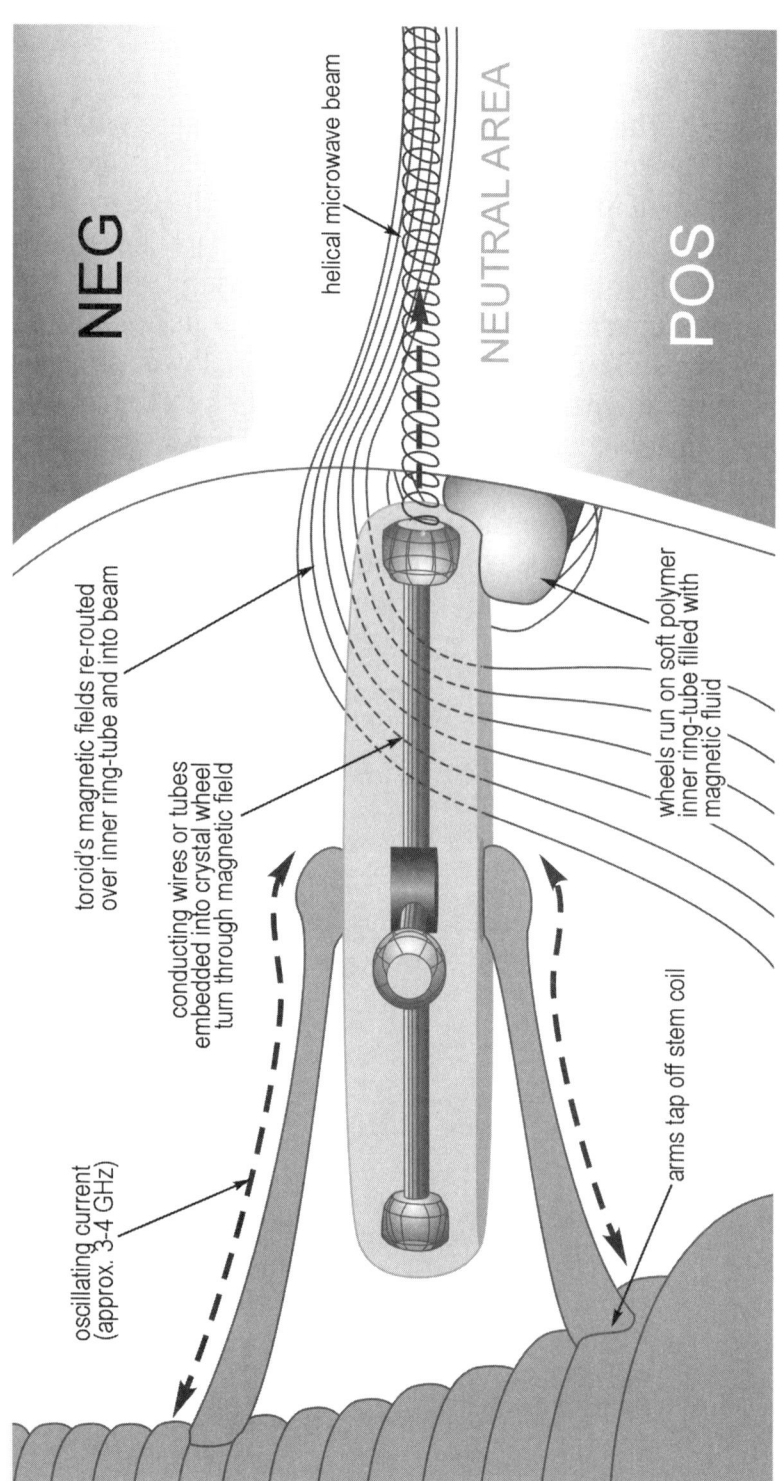

Figure 60. Microwave energies pass through the conducting tubes and become amplified in the maser-like crystals before being fired through the toroid wall into the toroid's fluid

The rotating sphere-set assembly is the only major component in the craft's engine which must be allowed to rotate, it is also the only major component that must be allowed to resonate unhindered at its own particular frequency—which will be determined by the sphere's plasmonic resonances and the acoustic resonances coming from those spheres. But the toroid shell (which is also the lower hull) has no need to oscillate and so it will not be resonating at anywhere near the same frequency as the central assembly (strictly speaking it shouldn't be oscillating at all). So the central assembly's oscillating components must be isolated in a special way from the non-oscillating toroid shell.

Because, if the telemeter wheels butted-up to the solid walls of the toroid, or, as more to the point, to a solid casing around the inner ring-tube, then those glass wheels would vibrate themselves to pieces in a very short space of time—and so would the inner ring-tube AND the central part of the toroid—they'd all vibrate against eachother and eventually malfunction. In order to prevent this from happening the inner ring-tube has to be made of a soft yielding material—to form a frequency-bridge between the two structures. In a way this corresponds, in electronics, to the isolation of a resonating circuit; and so by isolating the central sphere-set assembly in the aaUFO that assembly will be able to build up its own harmonic resonance and develop an extremely high Q-factor [note: 38]. Furthermore, what will help maintain that high Q-factor in the central assembly will be how the base disc is manufactured, because whatever materials are used in its manufacture collectively they will have to resonate to a frequency that will either match, or harmonize with, the plasmonic resonant frequency of the glass spheres, otherwise a mis-matching of frequencies will dampen down that operational frequency of the spheres and the Q-factor of the whole assembly will be detrimentally impaired. In fact, regarding the central assembly, every single part of it should probably be considered as being an integral component of one whole tuned-circuit, and they should all be manufactured accordingly.

To this end, for the base disc a combination such as magnesium-zinc-bismuth would be preferable because it could be manufactured to resonate to a particular frequency with respect to the craft's pulsing magnetic fields that will be applied around it [note: 39]. Again, an interesting corollary can be found in the Roswell sample, because magnesium-alloys are known to be highly electrostrictive, which means that they will produce electro-kinetic movements when an electric field is applied to them (which is an effect which works in a similar fashion to piezoelectric materials). In the Roswell sample, when it was being tested in the labs a small piece of it was subjected to an alternating AC field superimposed by a million-volt electrostatic field, which caused it to jump sideways very energetically (confirming quite spectacularly that it was indeed an electrostrictive compound) [note: 40]. So obviously a material such as the Roswell compound being part of a circular disc would readily oscillate with that UFO's magnetic and electric fields pulsing or alternating through it.

Incidentally, yet another refinement in the Roswell sample, of bismuth-magnesium-zinc, was that its epitaxial layers were undulated microscopically. And this undulation would also make a lot of sense with respect to where that base disc is located, right next to the black vortex which will be radiating a gravitational force: Because if the molecular displacement, from the black vortex, which will be happening at a particular frequency, produces an interrupted electric field within the structure of that base disc which expands radially through that structure (as aided by the pulsing magnetic fields), then that radial electric field expanding through those undulations of bismuth would be forced to oscillate at a pre-determinable frequency (so, in other words, the dimensions of the micro-undulations would help to govern the resonant frequency of that base disc). This micro-undulation theme will be continued later in this book's supplement where its influence within this UFO, and other UFOs, will be more keenly appreciated.

In conclusion then, because the base disc would be better constructed from materials which can be made to oscillate somewhere near the resonant frequency of the lower spheres (probably within a fairly narrow bandwidth to harmonize with those spheres), then by using the gravitational radiation coming from the black vortex, and the magnetic fields pulsing around it, a radial electric field will be generated—and this radial electric field will be made to oscillate through the very thin layers of high-mass bismuth. The two main purposes for this will be a) to transmit the gravitational force from the black vortex to the toroid fluid, and b) to establish and maintain the high Q-factor essential for the central assembly's resonating frequency. Although, what this oscillating base disc will also do is c) help bunch the toroid's flux lines toward the central hole, and d) it would also create a rippling effect around each of the lower spheres (which may be how they are made to revolve), and e) when turned the base disc will pull on the magnetic flux lines around the lower inner edge of the toroid to stretch and break them to produce copious amounts of electron-positron pairs.

Actually the reader might marvel at just how un-complicated the sphere-set assembly is located in the middle of the toroid's central ergosphere—if the reader considers that this assembly is the whole UFO's engine. Yet its base disc has no contact with the toroid shell (there is a small air-gap all around the base disc), and the ONLY contact the central sphere-set assembly will have with the toroid shell will be where the telemeter wheels turn upon the inner ring-tube. So it really is that simple!

6.14 Adjustability of the Toroid Fluid's Permittivity

Finally, from whatever pulses that are fired into the toroid's fluid, through the telemeter wheels, there will be produced in that fluid a proportional change to its dielectric permittivity [note: 41]. The toroid though, as seen as ergosphere C, will be needed to perform a few more special tasks in more electrodynamic and gravitational processes, yet to be delineated through the following chapters of this book, and so the actual make-up of the toroid fluid still cannot be fully determined as yet, for the permittivity and the flow velocity of the toroid's fluid will consequently become crucial factors for these other mechanisms which will be operating through the aaUFO's power drive system—which will employ ergosphere C especially to slow down the velocities of the electromagnetic waves and the gravitational waves coming out of the vortex and the central ergosphere).

But then, if it was thought to be simple up to this point then from here on in this UFO gets a lot more complicated...

7: Black Vortex of Immense (Silent) Power

Contents:
7.1 Intake and Outflow of an Astrophysical Hole in Space	235
7.2 How the aaUFO's Black Vortex Inhales and Exhales	245
7.3 Artificial Engineering of a Functional Black Vortex	243
7.4 Real-time Engineering of a Functional Black Vortex	244
7.5 Lining up the Sphere-sets to the Central Vortex	249
7.6 Seeing a Complete Range of Frequencies Inside the UFO Craft	252
7.7 Gyrating Storage Field Feeds Rim Field (Current Ring)	253
7.8 The Gully Paradox of a Small Gap Leading to Bigger Things	258

7.1 Intake and Outflow of an Astrophysical Hole in Space

The black vortex in the center of this aaUFO is not a naturally-miniaturized black hole, it is an enforced miniaturized quasi-black hole. It is specifically engineered through the craft's magnetic fields which form a magnetic funnel, at the bottom of which is a magnetic bottle-neck. But surrounding that magnetic funnel is a very compact ergosphere of rotating energy producing frame-dragging and rotational turbulence. And as mentioned in section 5.1, because of the magnetic bottle-neck and the solid base disc at the bottom of this ergosphere which together will effectively close-off the bottom of ergosphere A, there will result a one-way energy accumulation process whereby whatever energy gets pulled into this vortex won't easily be able to escape from it. This energy accumulation will affect a densifying process on those energies being thrust down into the vortex from the top spheres at the photon orbit ring, and this continual in-pouring of energy will also affect an increase in the rotational velocity of that field [note: 1].

Where this same effect occurs out in space in an astrophysical black hole its ergosphere of rotating energies will rhythmically expand outward to accommodate its continual influx of energy (and when the ergosphere's magnetic field periodically tightens-up), or it will rotate faster and faster until it expands its frame-dragging forces as an exhaust flow along its longitudinal axis, to form a pulsar or quasar jet to thrust some of that excess energy away into space (this can be seen in the technical breakdowns for the M87 and NGC 6251 jets, and it can be seen in the density chart for a black hole, in the [Lery 2002] study for instance).

In comparison, outside the aaUFO's black vortex its ergosphere cannot expand outward because of its solid containment walls, so by design its energies will become more dense and have to rotate faster and faster, and that will be so that its central ergosphere can accommodate and compensate for its continual intake of energy. And so just like a black hole the aaUFO's vortex will eventually be driven with an angular velocity that will have increased beyond the velocity of light (c), where light waves will become trapped inside that spinning vortex [note: 2]. This light-trapping event will certainly occur toward the core of this craft's vortex where it spins the fastest through an ever-decreasing diameter above the hole in the base disc.

Where its actual horizontal event horizon might occur could be thought of as a particular diameter of rotation inside that black vortex, but with regard to the full compliment of rotating forces driven faster than that event horizon then, in the aaUFO's case, its vortex's 3-dimentional event horizon boundary would have the shape of an elongated cone.

However, the way in which energy is constantly being drawn into the black vortex refers back to something I mentioned in section 3.17, in that the vacuous forces of a black hole and quasi-black hole don't happen in the sort of straight-line-path that would see a mass-particle 'drop down' through the mouth of that quasi-black hole. That mass particle would have to enter by the path of the rotating field which determines that quasi-black hole [note: 3]. Because of this factor, that all energy has to rotate into this black vortex through the photon orbit ring, it will be observed that the photon orbit ring will command a significant role within this UFO craft's energy dynamics.

With this factor in mind it will be seen that the particles being caught in the gravitational pull of that black vortex will come mostly from the positive ions in the storage field gyrating over the toroid's upper edges, and from the electrons and photons in the orbs of energies spinning around top spheres; and it should be seen that while some of that ionic, sub-atomic and photon energy will readily blend into the black vortex flow most of it will initially be moving in the wrong direction or it will be moving at a much slower speed than the rotational speed of that vortex (this was shown in figure 42 above).

Therefore, the vortex's photon orbit ring should be seen overall as having an average rotational velocity of a little less than that of the event horizon, so that dynamically it would follow that the photon orbit ring will be outside of (and above) the event horizon [note: 4]. And herein stands the very mechanism which breaks down matter, for if the matter-energy entering into the rotating field of a quasi-black hole doesn't all arrive at the same speed (and the same direction) as that field then particle collisions will occur, and these particle collisions will be what breaks down mass components into their constituent subatomic-quanta energies [note: 5]. These collisions at the photon orbit ring will involve Penrose pair-production, and possibly Comptonized scattering, although, as these particles are drawn down beyond the event horizon, through the vortex's increasing shear forces there will occur atomic dissociation of particles into pions, muons, gamma rays, and then into raw energy.

The former energy producing mechanism was first alluded to when Roger Penrose observed this differential in rotational velocity around a black hole back in 1969 through his paper on Gravitational Collapse, wherein Penrose proposed that as such rotating differentials exist around a black hole then there should be a substantial amount of energy generation going on just outside the event horizon [note: 6]. Penrose, in this superbly illustrated paper, proposed that it should be possible to extract energy from the rotating mass-energy anywhere between the event horizon, where the rotational force will be extremely high, and the stationary limit, where the rotational force only begins to take effect.

Reva Williams (in 1995) then took this observation that permits rotational energy to be extracted from a rotating black hole and explained it even further: Williams suggested that a particle, say particle (p_1) coming from infinity while falling into the ergosphere of a black hole, if before it passes over the event horizon of that black hole it collides with a bound particle (p_2) which in this instance would be a particle bound to the rotational forces of the photon orbit ring, and as a result of that collision the scattering particle (p_2) is knocked backwards into a retrograde orbit, then that (p_2) particle has effectively given up its energy to the incoming (p_1) particle before it then spins down with a retrograde energy (negative energy) into the rotating black hole. While that (p_2) particle, through this Penrose-Williams hypothesis, feeds into the black hole the (p_1) particle that was incoming, after its collision at the photon orbit ring, is then rocketed out of the system imbued with a *huge increase to its energy*, which it has collected from its collision with that high energy particle rotating in the black hole. So in this mechanism the black hole transforms into both a consumer of energy and an amplifier of energy. This is the secret to the aaUFO's drive mechanism.

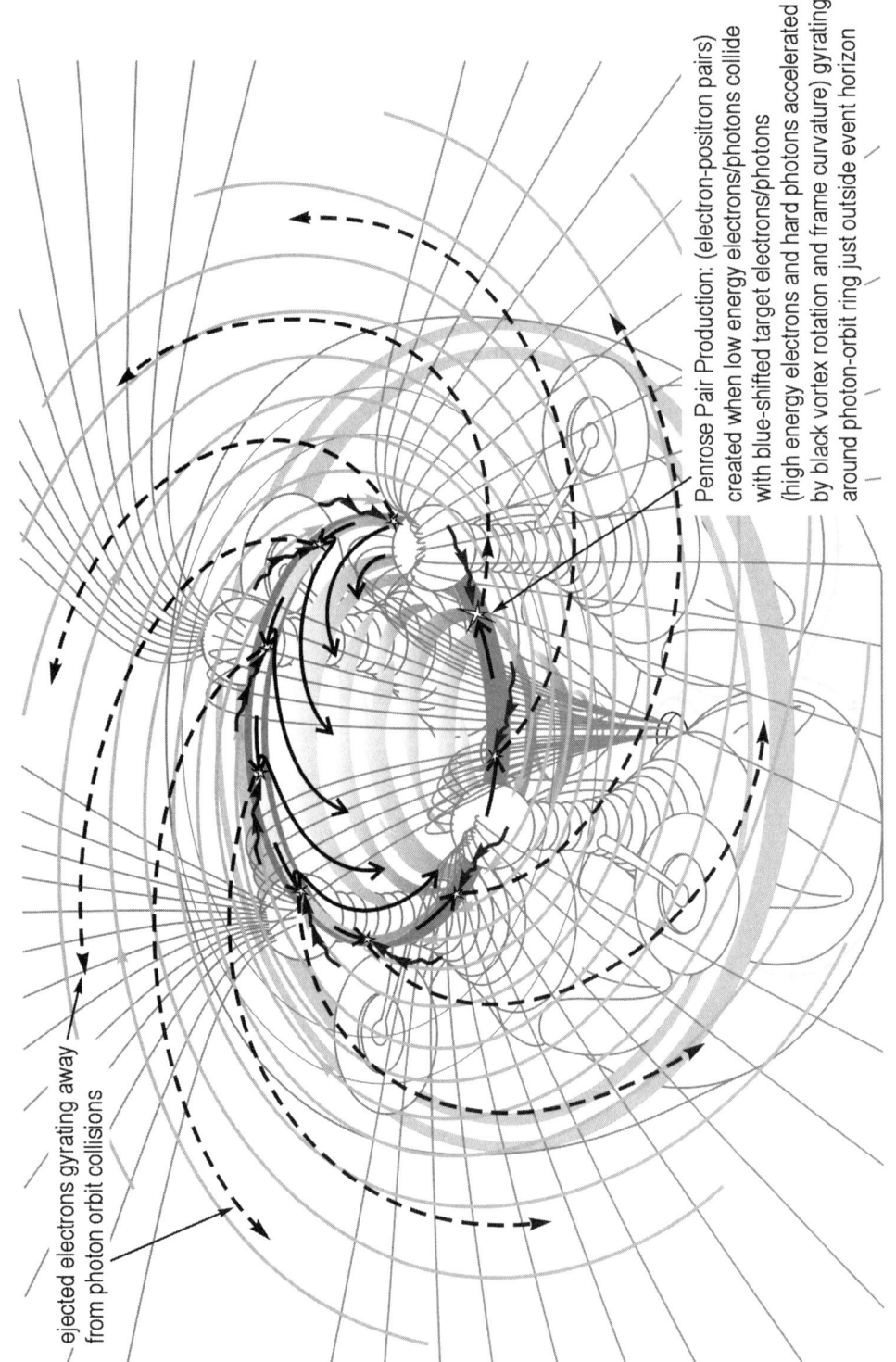

Figure 61. At the photon orbit ring the rotational forces will be so energetic that the incoming positive particles will collide through Penrose-Williams collisions and will be ejected forcibly away toward the outer rim of the aaUFO

By translating this effect into the aaUFO at its photon orbit ring it can be seen that the incoming particle-ion field rotating into the center of the craft will also collide with the rotating particles (the Penrose-Williams' bound particles) of the aaUFO's black vortex, and while half that collision energy will feed into the black vortex the other half will be blasting off in amplified gyrations out toward the perimeter rim of that UFO, and these amplified gyrations will feed into the storage field of charged particles gyrating within the UFO's planar wave-guide. So here is a case of astrophysics putting forth a principle by which a UFO's energy dynamics can be explained, and equally, a case where the UFO's structure is putting forth the evidence to show how the extraterrestrials are improving on that basic astrophysical principle.

And to borrow again from the Penrose-Williams hypothesis (and from Williams' mathematics), these ejected particles from the aaUFO's vortex will subsequently have MORE MASS-ENERGY than the sum of those particles prior to their collisions! This would be because the out-rocketing particles will take with them the Lorentz rotational force that the scattering particles (p_2) originally had, and they will also take with them the gravitational force that was pulling on those (p_2) particles while they were bound into that black vortex (and by this premise it can be envisaged how the output of this 'engine' will rise proportionally to the degree the velocity of that vortex is increased).

It would also follow that after these ejected particles energetically and forcibly gyrate away from the black vortex (now in the same direction as that vortex, ie, counter-clockwise as seen from above) that they will be imbued with a particular gravitational-Lorentz force that would dictate that those particles would escape from that aaUFO in a particular vortical trajectory (ie, they will be imbued with a particular degree of CURL force, but more on that in Chapter 10) [note: 7].

The trade-off, if it can be called that, will be that the aaUFO's rotating black vortex will lose energy in the form of gravitational force and Lorentz force in this process. But as Williams shows in her calculations for astrophysical black hole systems, those losses will be compensated for handsomely through the secondary stage of this energy-ejection process, whereby the relativistic particles being ejected out of this system will on their way out initiate significant particle collisions with existing particle accumulations (in the top sphere orbs and in the gyrating storage field), and therefore will cause further electron-positron pairs to be produced, which, as they occur in the aaUFO, will then provide substantially more fuel to be fed back into the black vortex, and to go through the very same mechanism as that delineated above, so that the process becomes self-perpetuating [note: 8].

Returning to the Penrose-Williams hypothesis, this process was proposed as the possible source mechanism for the copious amounts of energy required for the hugely powerful active galactic nuclei (AGN) jets and for the jets like M87 as previously mentioned (in section 2.5). And in the context of correlating even more knowledge about this UFO's mechanisms I would add that, for the most part, this process also follows along the lines of the [Punsly-Coroniti 1989] study (see figure 16 above) and the [Leiter-Kafatos] explanation of how the rotating forces of a black hole produces high energy protons, gamma-rays, X-rays, and electron-positron pairs, all mainly coming from Compton and inverse-Compton scattering, or through the synchrotron and curvature radiation emissions occurring within its ergospheres and accretion disk.

Obviously, the UFO is a micro-version of the astrophysical in size, but as far as rotational velocities are concerned the two systems will both run at relativistic velocities and therefore will fundamentally be the same. Particle-collision forces will also be very similar, and so the kinetic energies expected in the UFO will be highly comparable to the data given for astrophysical black holes—and Reva Williams has calculated an ejected-particle-energy rating of up to 4 GeV for each of these photon orbit collisions (which compares to NASA's calculation for its new antiproton-proton annihilation reaction producing 1.876 GeV).

Further, what is especially interesting about the Williams treatment of the Penrose-Compton scattering effects (at the photon orbit ring where this translates into the aaUFO's black vortex), is that many types of low energy photons, for instance low energy UV rays, which have already been noted as being prominent in this area of the aaUFO (see the list in section 3.5), will be boosted up to hard X-ray and gamma-ray power levels, from those UV photons being drawn into the UFO's photon orbit ring and colliding with the rotating vortex and being ejected away from that photon orbit ring and being imbued with extra energy. And again, when these then burst through the energy orbs surrounding the top spheres and collide with their air-born particles those collisions will lead to more electron-positron pairs being produced from those secondary collisions [note: 9].

7.2 How the aaUFO's Black Vortex Inhales and Exhales

As has been mentioned above, while the aaUFO is designed to have confinement wall structurings which surround its ergospheres of rotating energies, by placing the ever-energetic photon orbit ring into the black vortex's electrodynamic circuit, and then immediately surrounding it with the inner-edge of the toroid shell this ensures that the whole of the craft's lower-section will be proportionally energized outward from this central ergosphere through to the surrounding ergospheres, by the agent of those purposely shaped structures (see figure 44 above). Furthermore, under normal running conditions this whole electrodynamic circuit will not be prone to short-circuiting because the path of conductivity will be routed through the lower spheres and those spheres will only conduct through the interrupted conductivity of the plasmonic oscillations. Intriguingly, by this very conductivity arrangement it would suggest that the craft's black vortex would oscillate-radiate to the same frequency as the lower spheres. This assumption would certainly befit the design-prowess so far evident in this craft, and obviously, this is a refinement with huge implications, this will give the whole electrodynamic circuit a resonance or coherence factor, and as I mentioned at the end of the previous chapter, there should be an interesting Q-factor to this circuit which would raise its overall power efficiency to a very high level, not only electrically but also gravitationally. For this would mean that the gravitational waves being radiated away from this UFO's black vortex will also subtend to that same working frequency, or at least to a harmonic of it. Because of this resonance effect (in relation to the vortex's gravitational radiation) it could then be assumed that the quintessential effect of the gravitational polarization cloud, that the black vortex uses to communicate its forces through to the whole of the lower section of this craft, will also resonate to that same shared frequency—and this would provide a control mechanism for the gravitational forces used throughout the craft's structure (more on this though in a later study, I feel).

Certainly, it would be interesting to do some number-crunching to compare the electro-mechanical efficiency of the extraterrestrial's aaUFO with that of the astrophysical black hole, and I believe there would be found a greater electrodynamic efficiency here than in any of the presently known energy-providing processes on earth, including nuclear fusion). Especially because, as a result of my researches, as shown above, so many very important areas have now been found in UFOs that have not yet been taken up by the astrophysicists in their black hole studies, but which should perhaps be further examined in relation to energy-transformation mechanisms that are wholly reproducible on a micro-scale. The study of UFOs has been a very broad-based learning process, about energy-generation just as much as it has been about flying aircraft, and to this end I feel that further consideration should be given to the principles of the UFO where it mimics the astrophysical accretion disk's coupling to its black hole, and the unique part that gravitational attraction plays in inducing extractable energy out of that disk-hole coupling: Not just for the creation of a flying aircraft, but also for energy-generation principles, to be used in a commercial enterprise which would deliver abundant electrical energy on a public-utility scale. There is a HUGE potential here for the development of new forms of abundant energy production. The reader might be interested to know that in astrophysics the estimates of a black hole's efficiency as a system that turns mass into energy are exceptionally good, ranging from 6% for a non-rotating hole to a 32% efficiency for a rotating hole—while nuclear fusion slumbers along behind them woefully at less than 1% efficient... With suitable modifications, such as those featured in UFO electro-mechanics, black hole dynamos could be turned into extremely useful energy generators, this is an especially worthwhile thought when one considers that the electrical induction field of an astrophysical black hole can accelerate charged particles up to energies of 10^{18} GeV [note: 10]. To progress this notion further I would propose further study into these hitherto ignored astrophysical principles, and I would further propose that the best studies addressing the issue of macro-to-micro scaling are those collectively found in the Williams-

Penrose, Punsly-Coroniti, De Felice-Carlotto model theories on energy extraction from a black hole... although for the miniaturization of a quasi-black hole I'd have to opine that no study is better than the present one! And no model of miniaturized black hole dynamics is better than the one presented by the extraterrestrials in this aaUFO—in view of the fact that it takes advantage of no less than three energy-churning ergospheres entrained inside three purpose-built confinement cylinders.

7.3 Artificial Engineering of a Functional Black Vortex

Referring back to chapter 3 where I tentatively questioned whether the vortex generated at the center of the aaUFO could actually become a black hole, I did cite Stephen Hawking's theory on the miniaturizing of black holes, but Hawking's is not the only study to propose this idea [note: 11].

Constructing a black hole in a laboratory environment is presently a much awaited scientific development, but it is showing promise with William Unruh leading several experiments to understand black hole dynamics within a laboratory environment. The present analog to the black hole is termed the draining bathtub, obviously referring to the similarity in certain arrangements of fluid flow. Also in the laboratory [Dimopoulis-Landsberg] have created an electrodynamic mechanism which produces one black hole every second (and which collapses at a rate of one every second also). Miniaturized astrophysical jet production is also being pursued and powerful jets are being produced successfully in the laboratory by using conical wire arrays (of around 10mm to 30mm in length) [note: 12].

An ardent supporter of the miniaturization of black hole dynamics, Leonhardt, explains in many of his studies how a rotating fluid vortex can mimic a black hole, and particularly, that one of the benefits of the vortex that is transferable from the macro-scale black hole, is the development of an acoustic horizon, where the velocity of its fluid flow can exceed the local speed of sound so that sound waves are unable to propagate away from it. And, the same physicist establishes that the essential attribute of an optical black hole that needs to be established within an artificial vortex is that it will have to generate a strongly-falling pressure at its core; and another proviso for the artificial creation of a black hole is that it will have to have radial motion that leads to its core (ie an inward radial component)—which of course the aaUFO vortex does have. In fact, the aaUFO's black vortex scores on both counts, for both these criteria are evident in its black vortex from the magnetic flux lines being drawn in toward the hole in the UFO's base disc, with their increasing Lorentz force upon all charged particles making those particles spin faster toward the vortex's axis. No mention is made of an artificial vortex needing to create a point of singularity. The emphasis seems to be on it only needing to be a rotating shearing field which either traps sound waves (sonic black hole) or which traps light waves (optical black hole), or both [note: 13].

Importantly, in the UFO, which has been featured throughout this study, it was actually noted (see figure 62) that in the central core of its rotating energy vortex there appeared to be rotating zone of blackness. From this observation, of the inner blackness to that vortical energy field, it can then be assumed that both the miniature versions of a sonic black hole *and that of an optical black hole* were being developed deep in the center of that UFO's inner ergosphere.

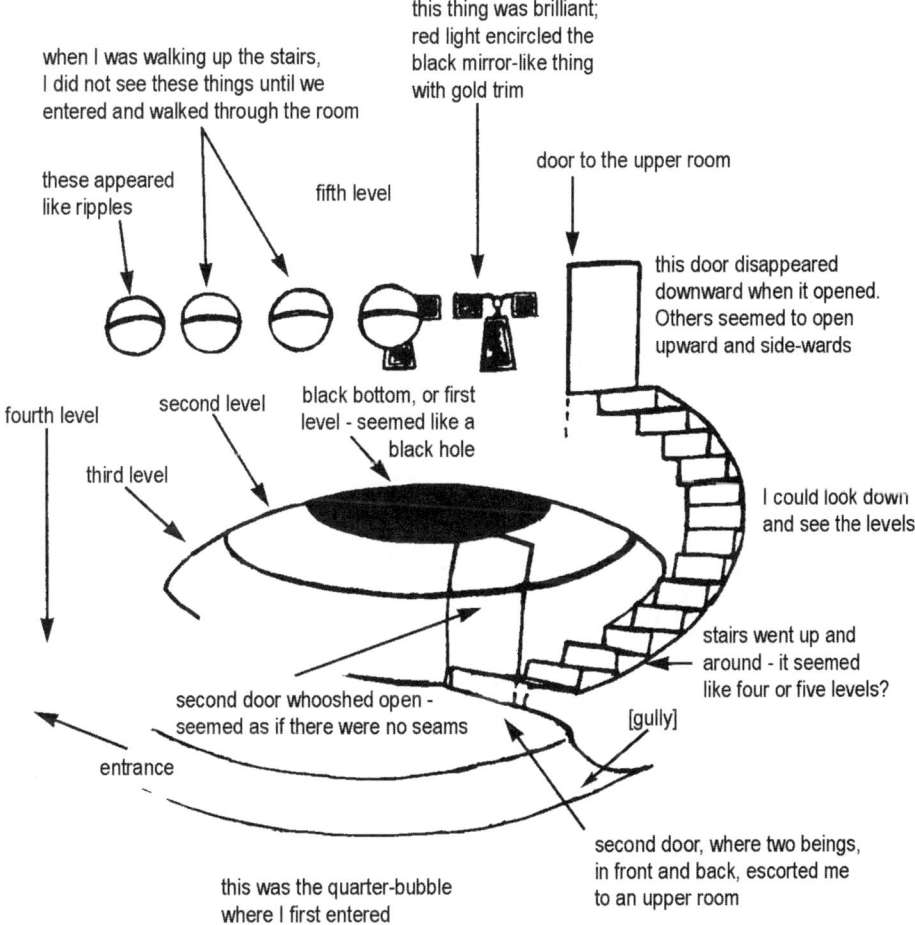

Figure 62. Interior decks of the aaUFO craft (from "Andreasson Affair" fig.11)

The circumstances by which Betty Luca observed that quasi-black hole with its gyrating fields INSIDE the UFO craft (which she was escorted into during her 1967 abduction) were unusual for many reasons, but none more so for the fact that the top of that black vortex was open to the upper decks of that UFO craft (that this was so will provide a stunning new revelation about black hole dynamics that has hitherto never been recognized by any physicist on earth, and which I will be expanding upon in chapter 11). But what was also important about what was observed in that particular abduction experience, which will provide much food for thought in the present discussion on macro-micro scaling, was that these rotating energy fields were being successfully transferred from one UFO to another. This was observed to happen through rotating forces generated by the Greys on the OUTSIDE of their UFOs, as will be alluded to below.

7.4 Real-time Engineering of a Functional Black Vortex

It should be said now in retrospect, after looking through all of Ray Fowler's study and the hypnotherapy notes for Betty Luca's abduction experiences, that there was a substantial amount of information being given to Betty Luca by the extraterrestrials through her abduction experiences about the dynamics of these UFO-type quasi-black holes. Such information was un-correlated at the time of its recall, of course, and it had to be correlated and focused upon (eventually), by matching it up with recognized scientific principles, which in this case has turned out to be the known principles of astrophysical science. It has taken a considerable amount of effort to make this discovery but the hard work has paid off for it will be through this scientific correlation that anyone now can recognize and de-mystify even more of the UFO-extraterrestrial peculiarities that have been observed and remembered by others, and it will be useful as a database if and when more UFOs present themselves in the future. For this study here though, these insights given us by the extraterrestrial engineers can tell us an enormous amount about the black hole miniaturization process.

For instance, I did for a long while assume that the craft's central vortex had to be established anew with a particular 'start-up' procedure before power-up and take-off could commence in this aaUFO. But on realizing that the black vortex, instead, would have to be in operation continually, and that at such times when the UFO was landed and stationary its vortex would simply be powered-down into some sort of tick-over mode, then the pre-assumed start-up procedure became redundant. For, I had realized that the UFO's vortex would have to be running continually after I'd studied the description Betty gave of the UFO which had stopped by the lake near her home back in 1967, and which had malfunctioned so comprehensively that its whole electronic circuit had failed and it became grounded and inoperable. And really, it was because of Betty being given the opportunity to witness this event and her description of how this UFO came to be short-circuited, and the method by which that craft's engine had to be re-energized by another UFO so that it could get back up into the air, that I finally understood how the whole electrodynamic process in these UFOs works.

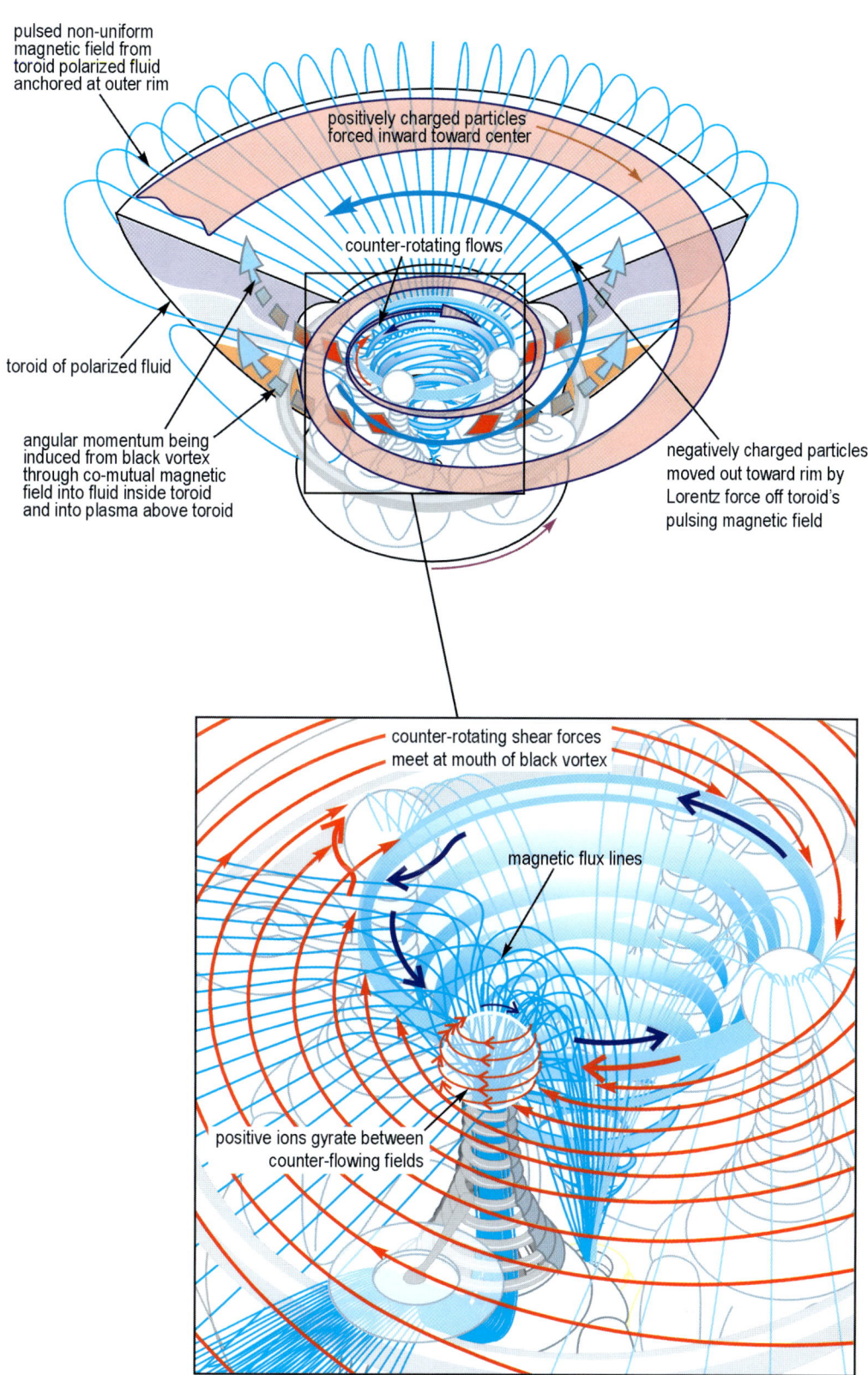

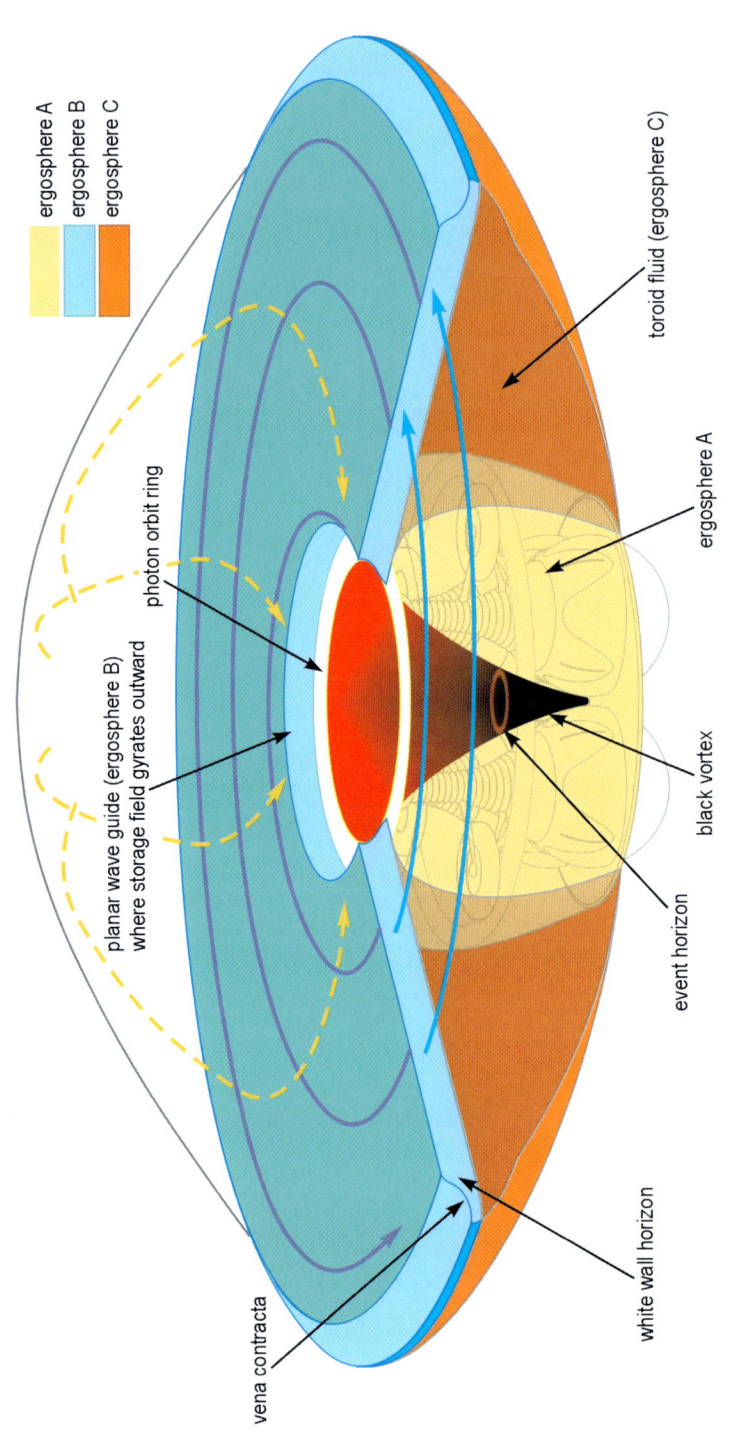

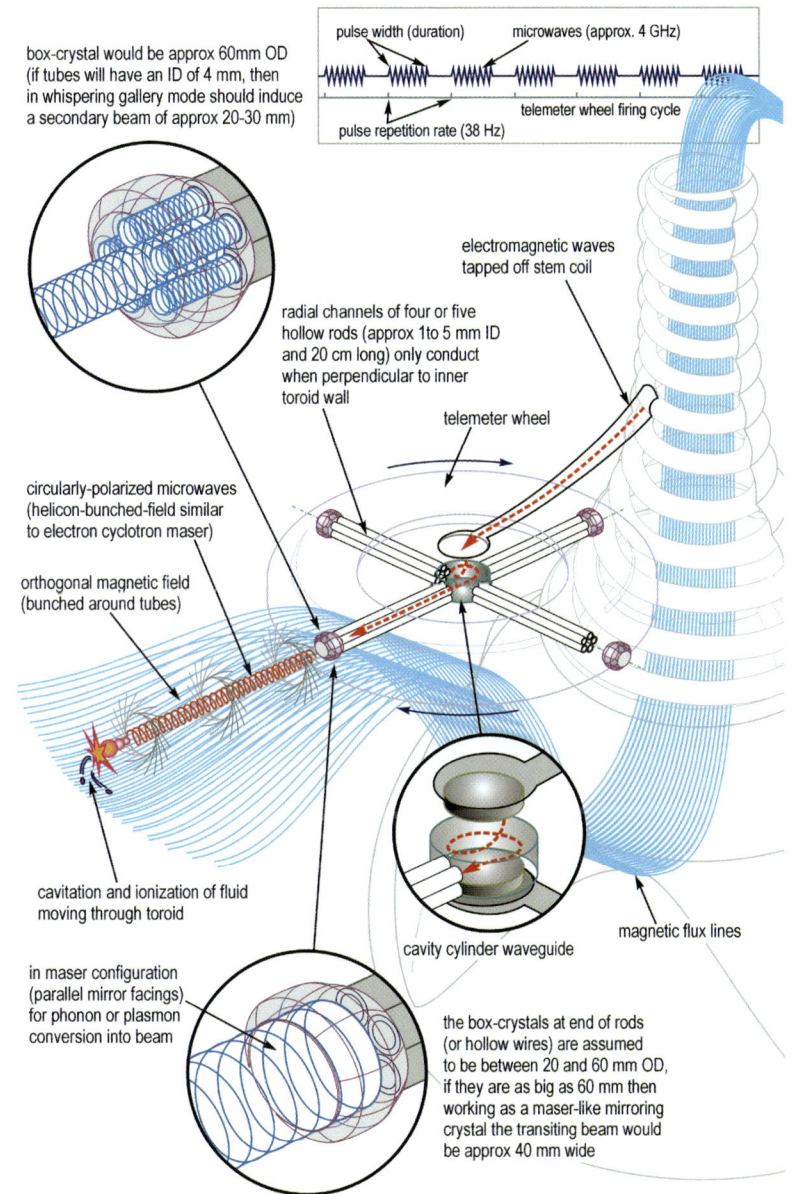

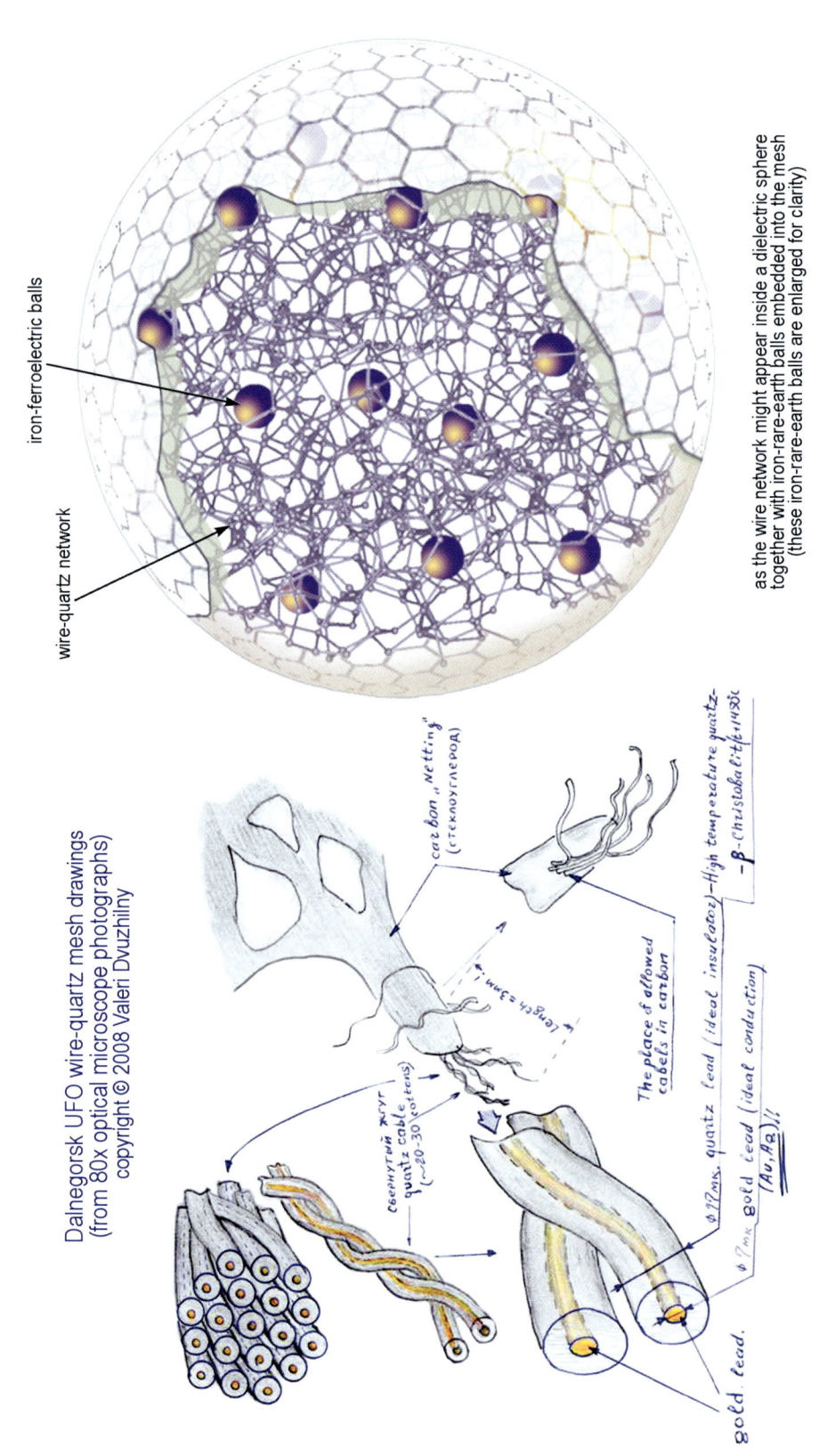

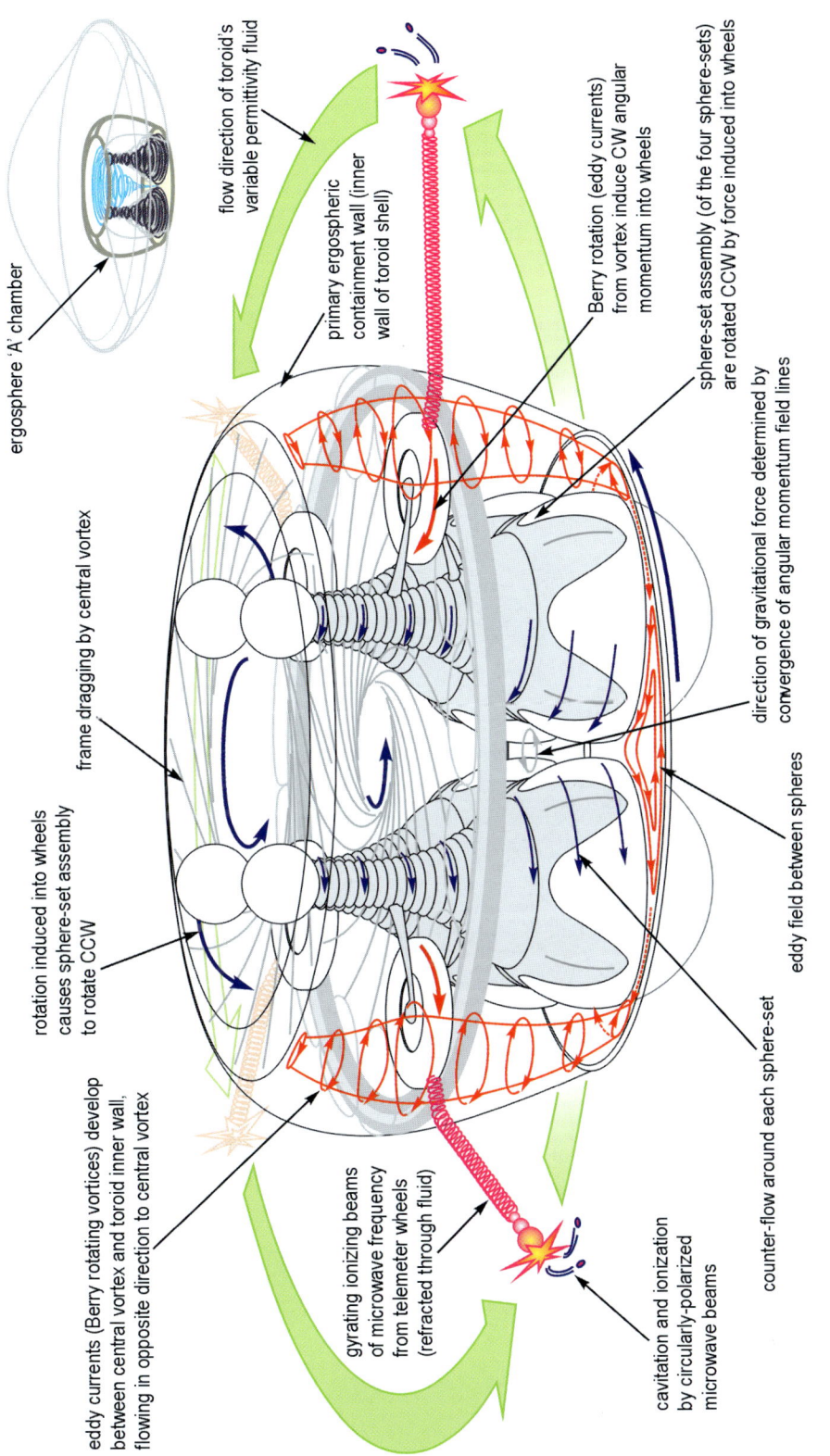

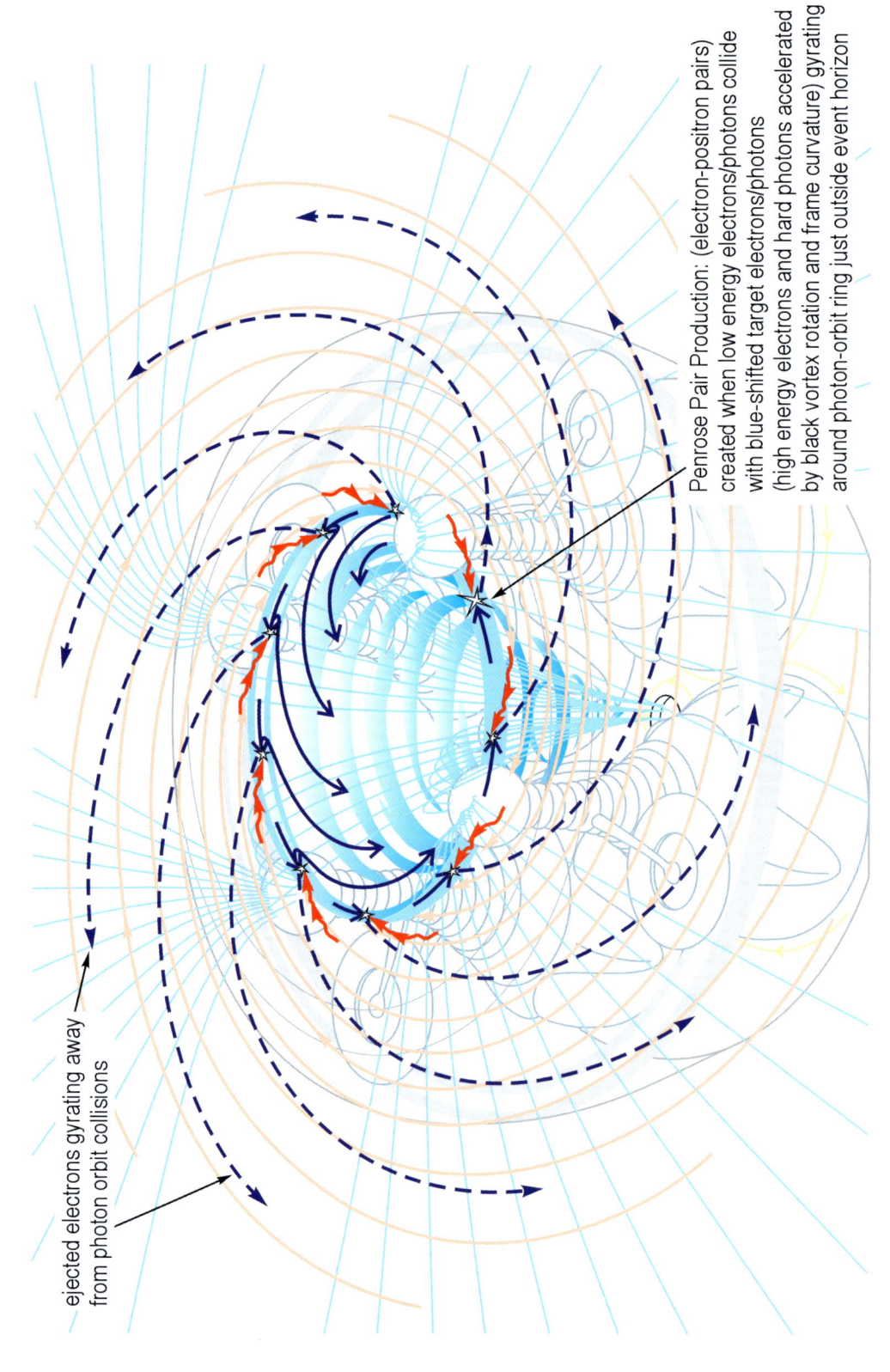

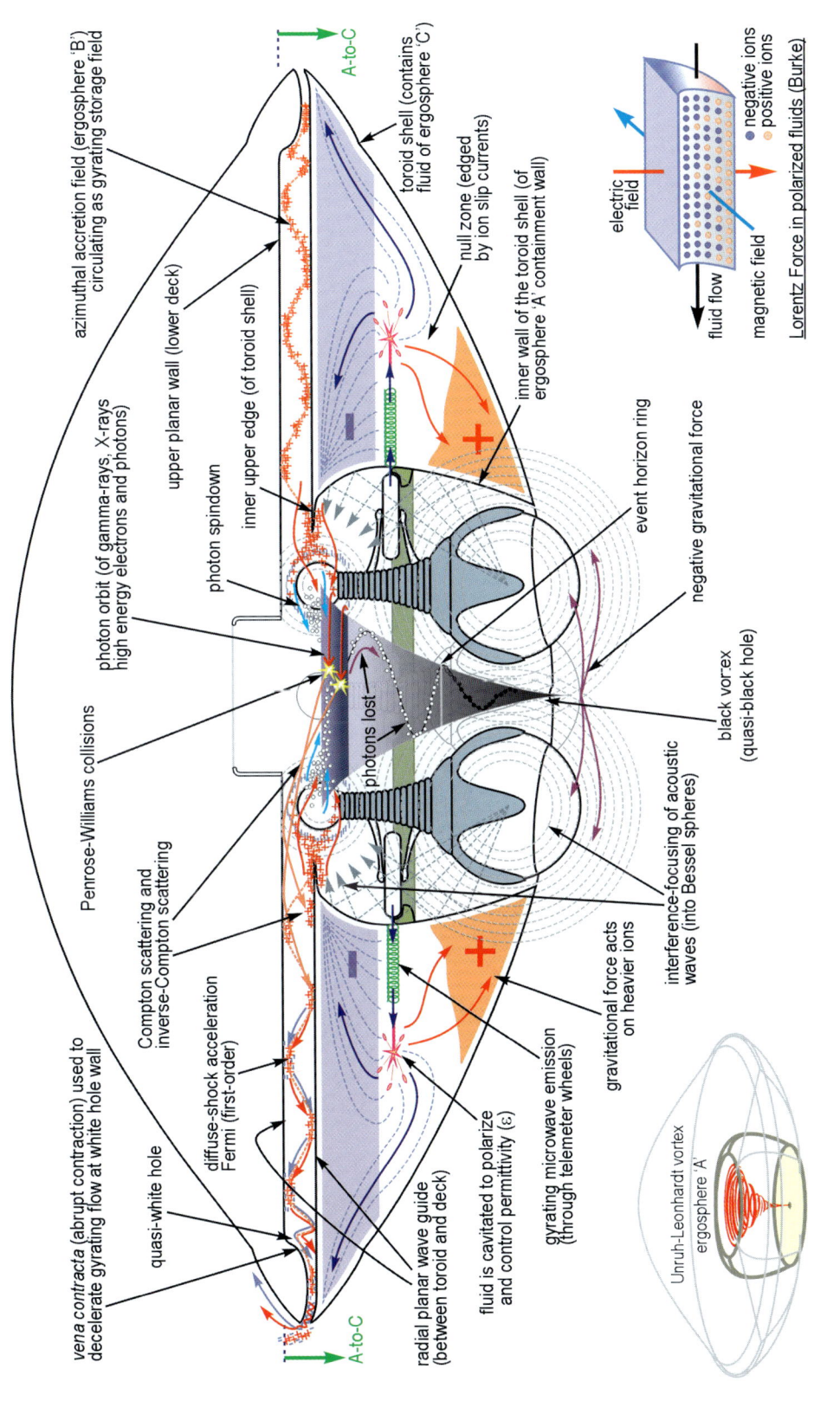

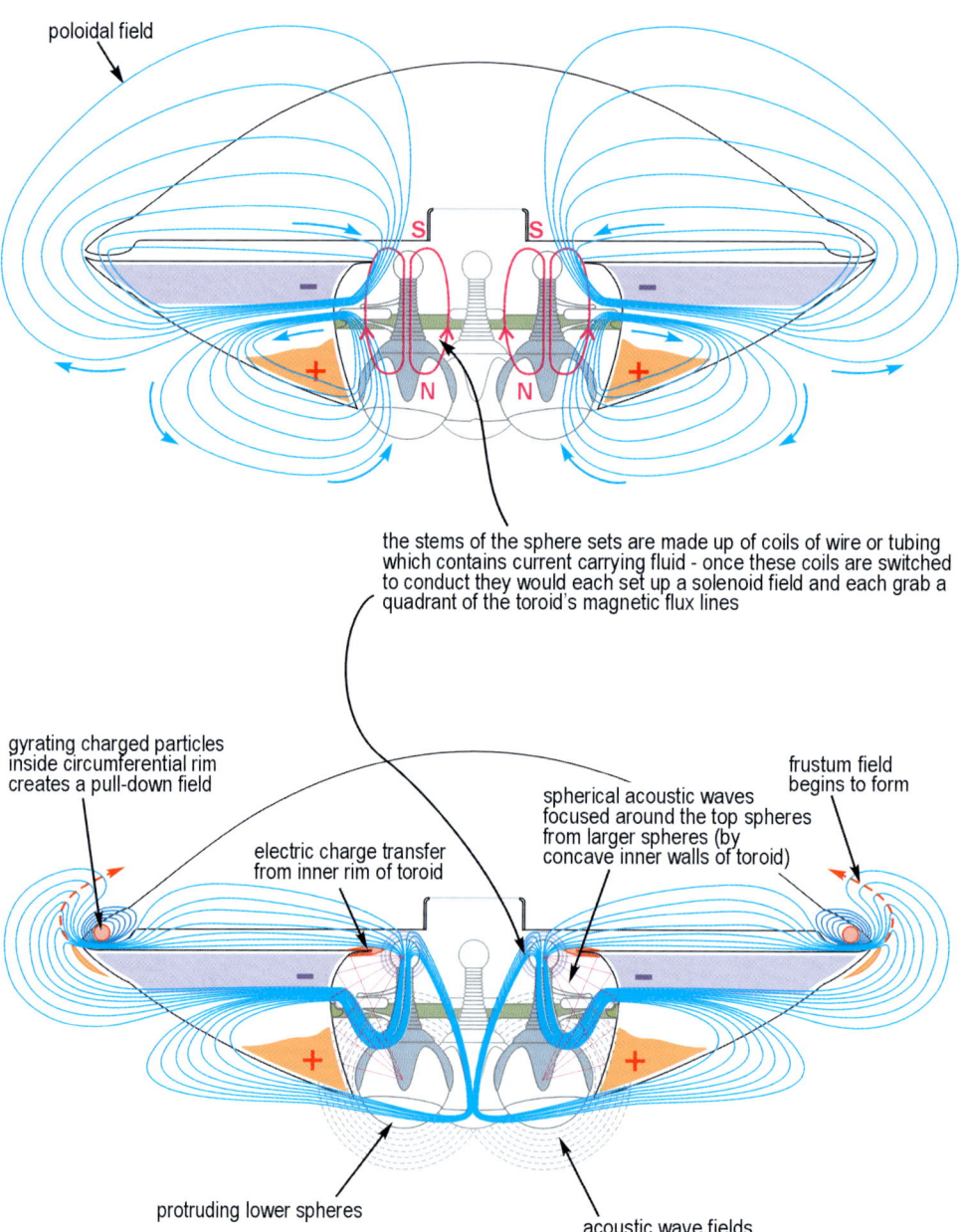

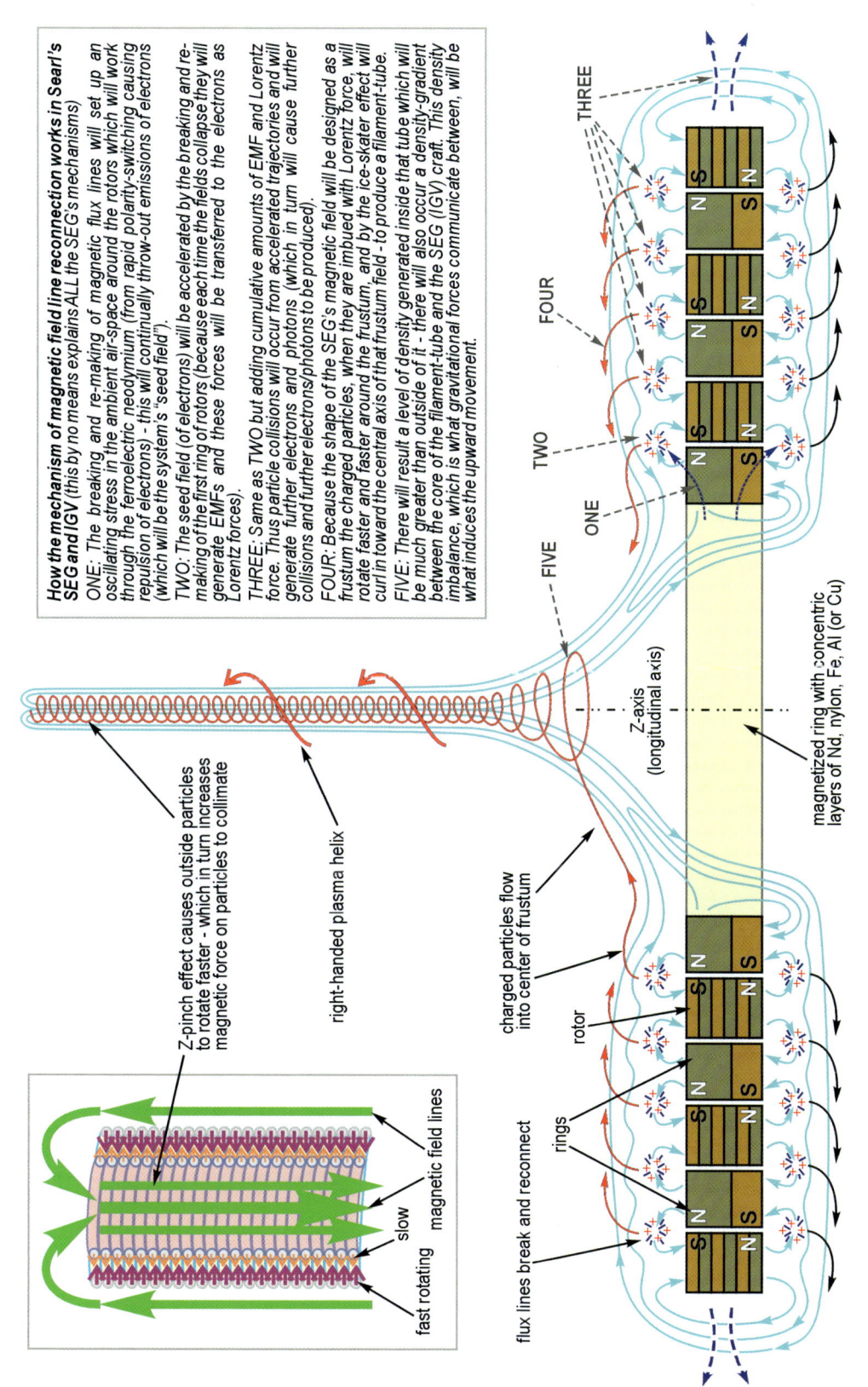

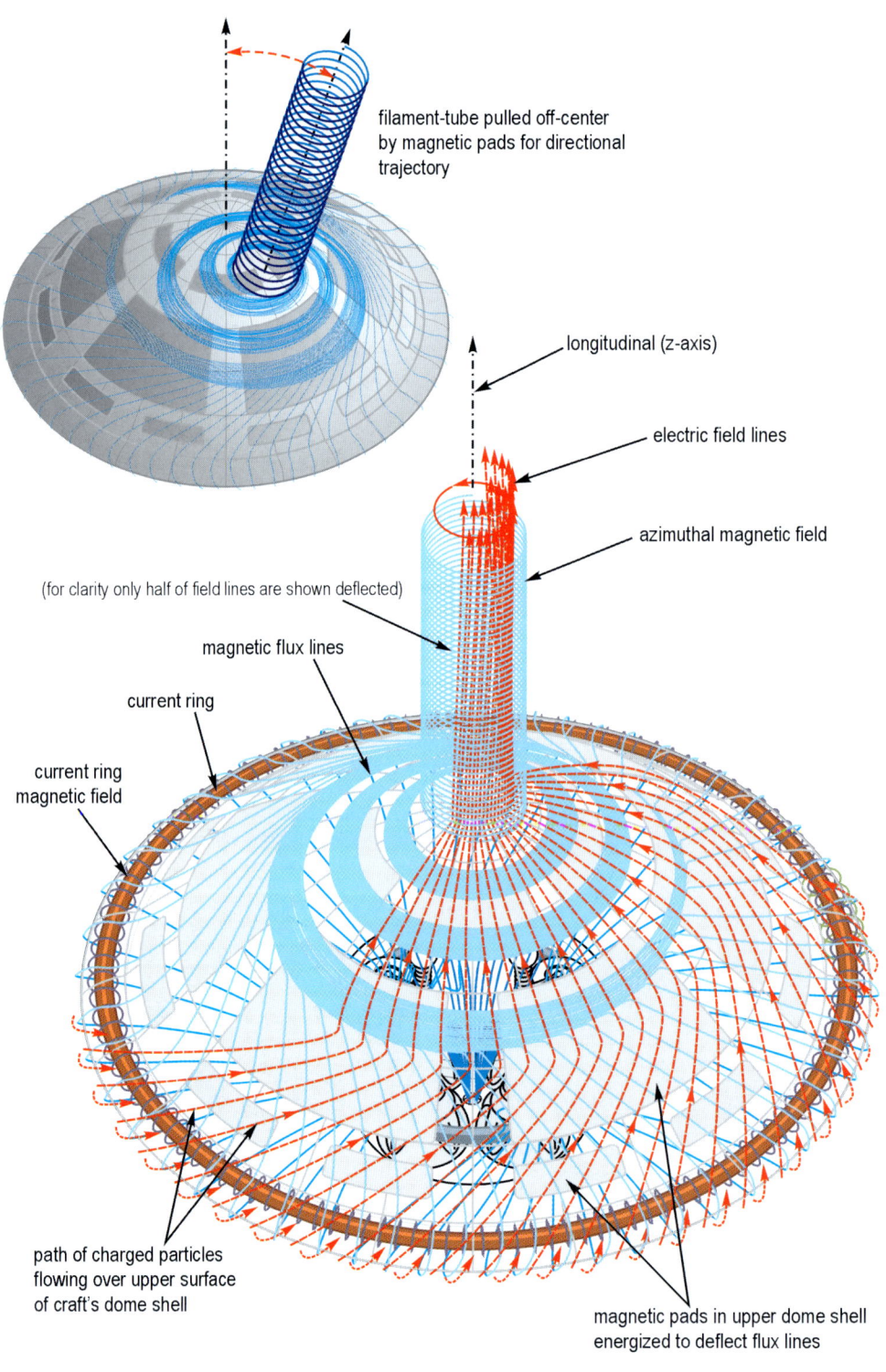

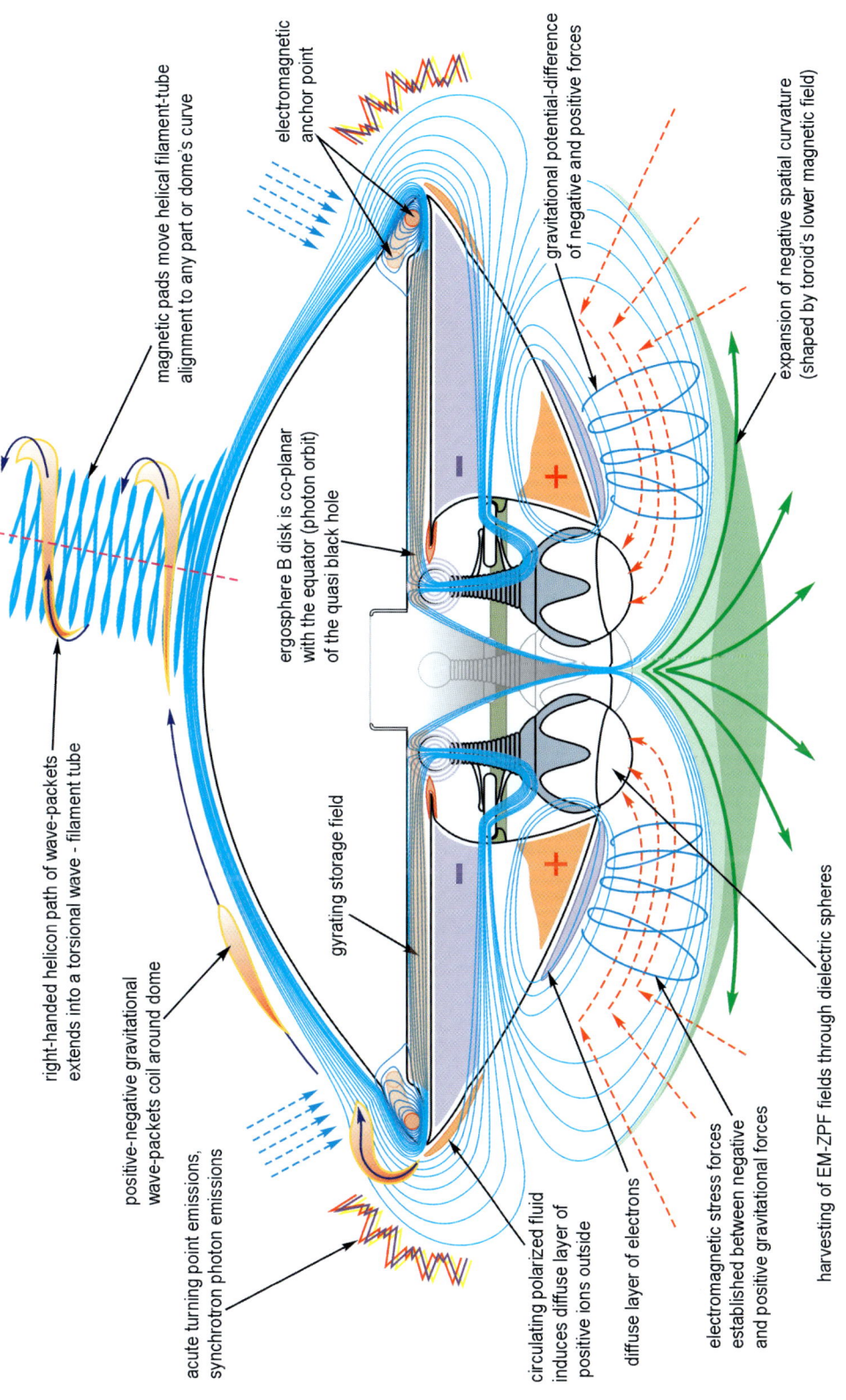

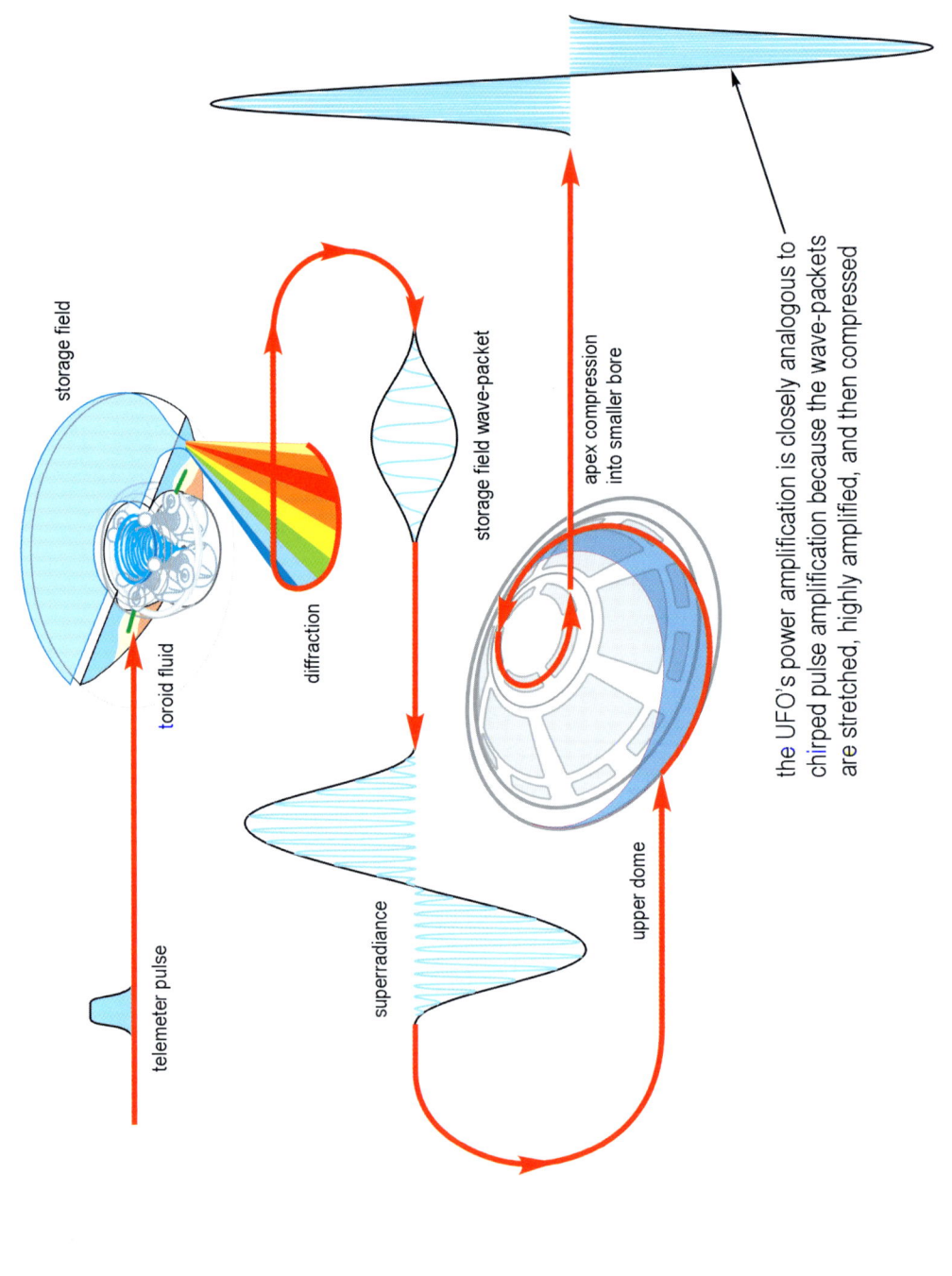

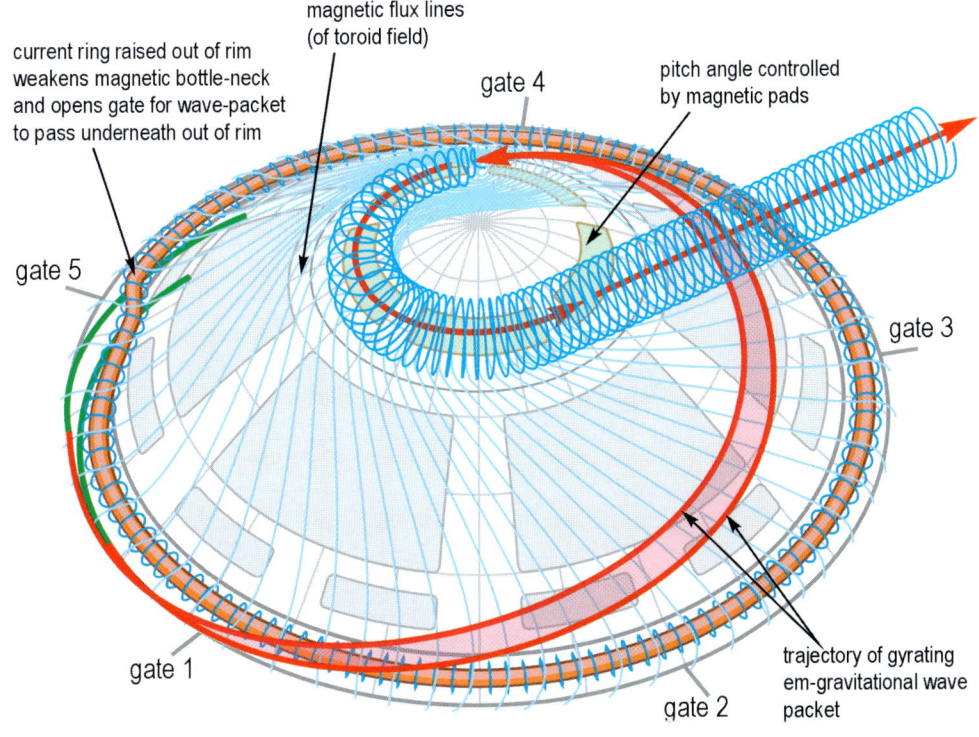
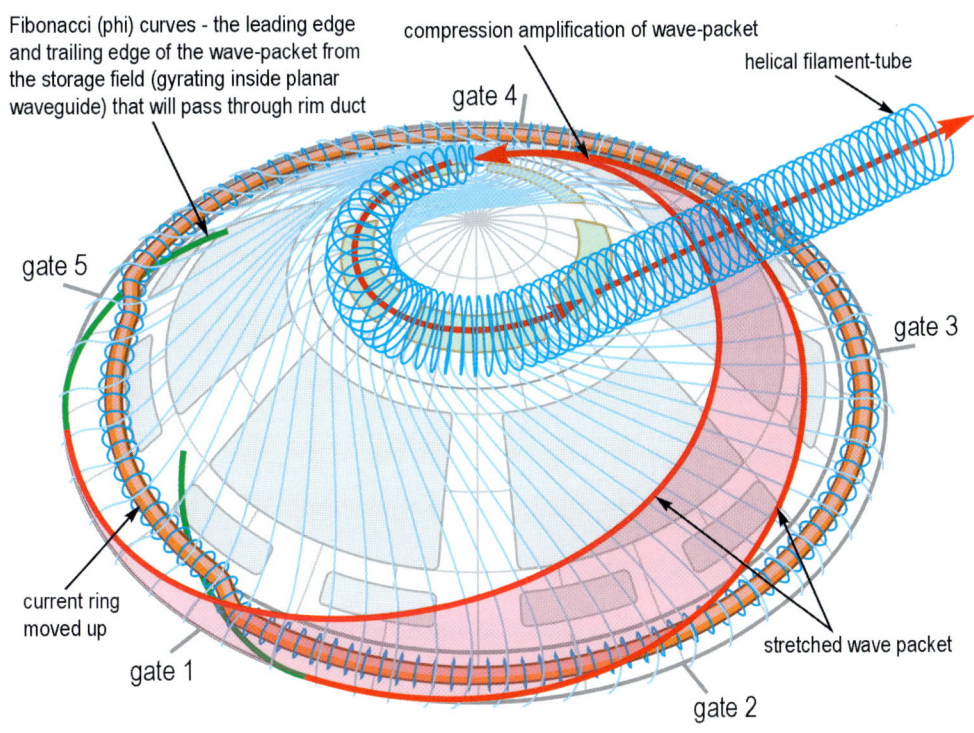

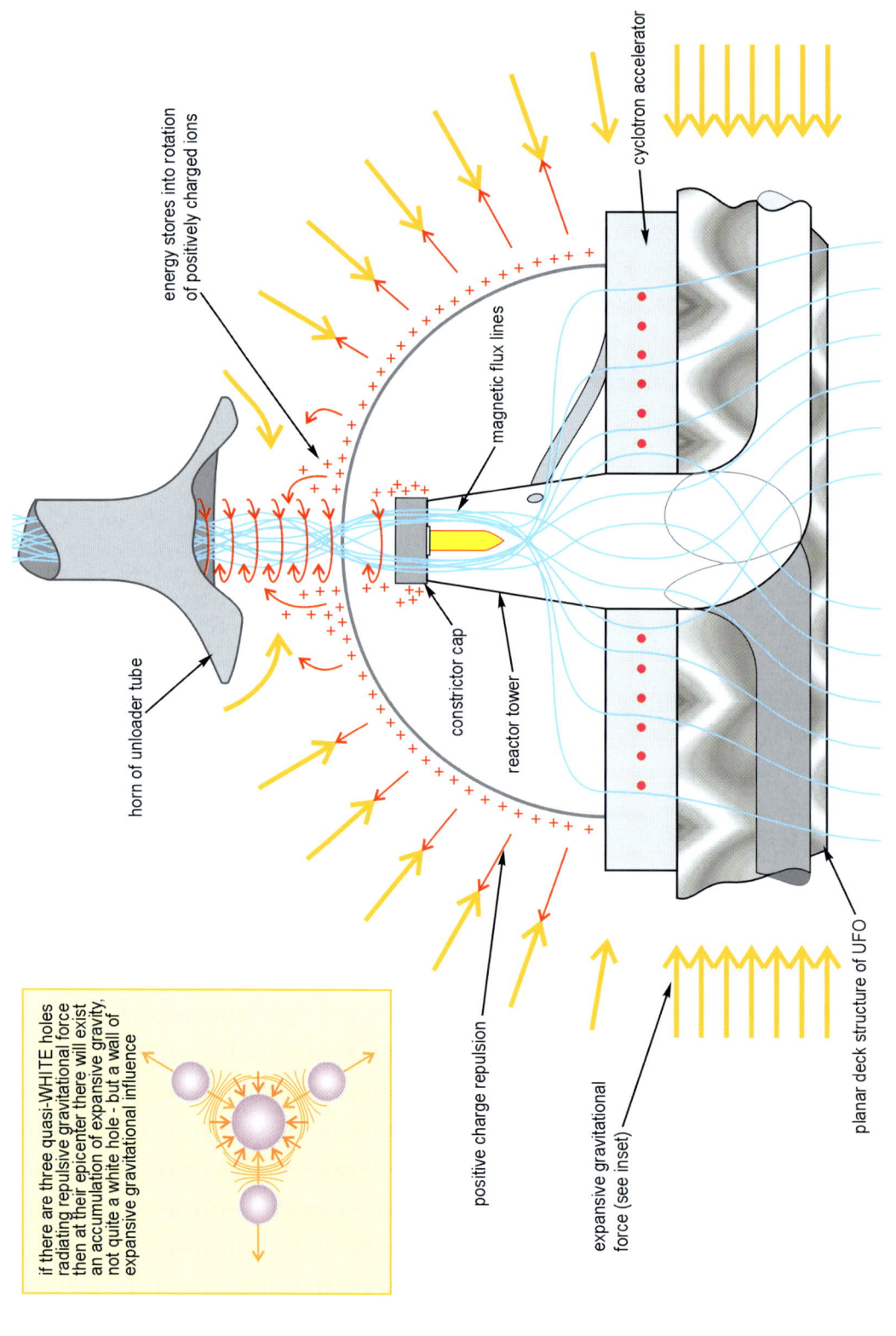

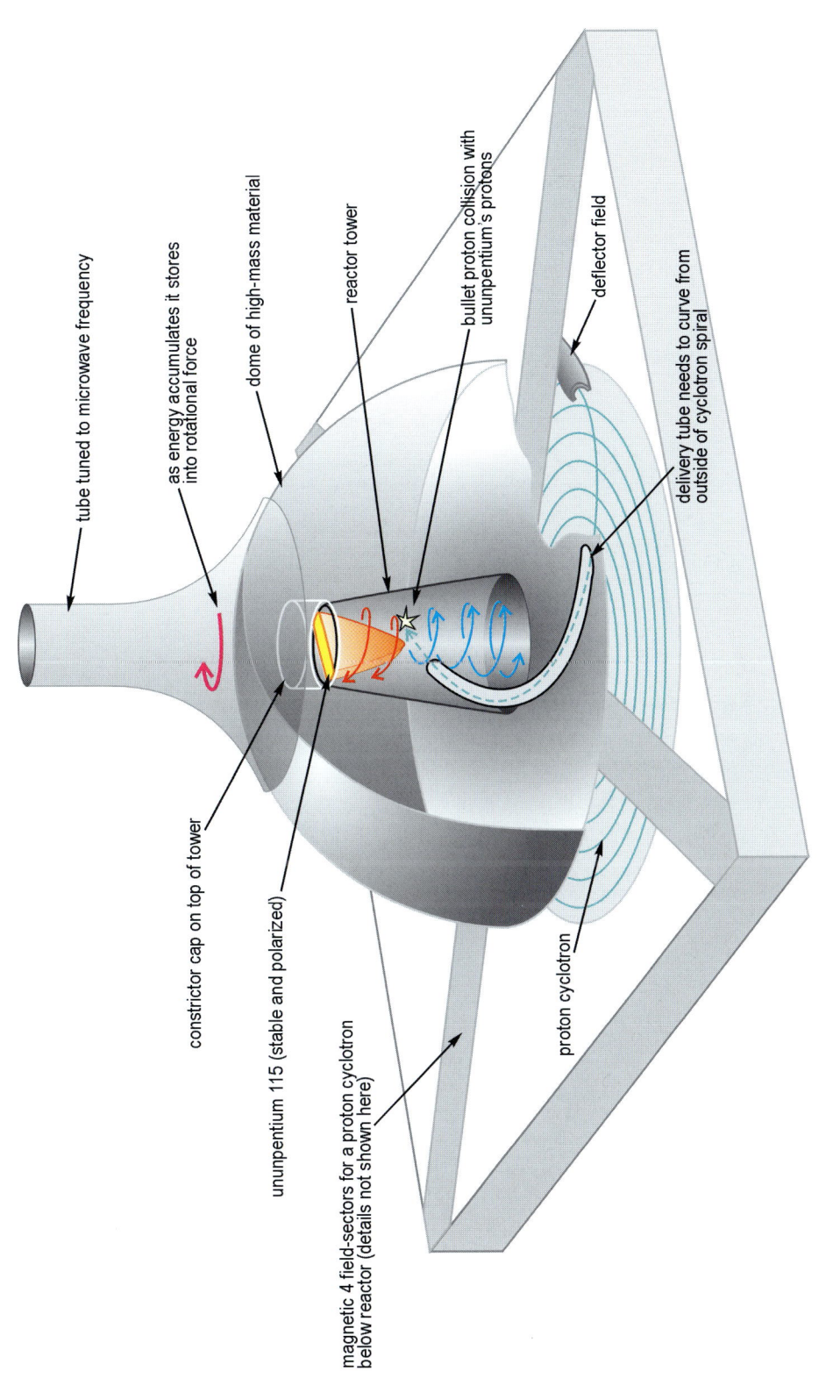

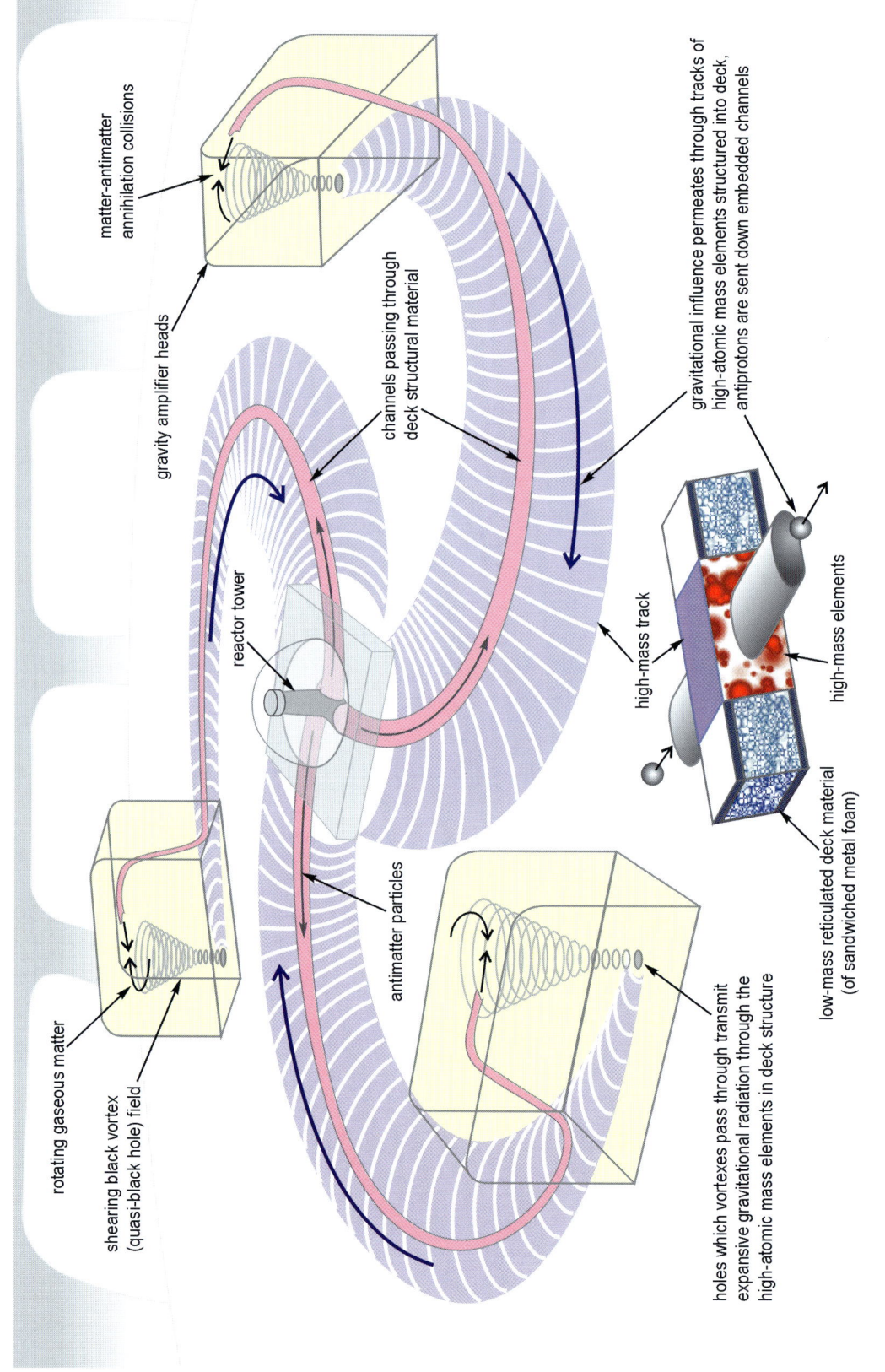

This very detailed episode was published in Ray Fowler's book "The Watchers" pp53-85, and I have given two very short accounts of it already (in my note 21 and note 30 for chapter 6), and in this continuation of that episode of what followed next after that UFO blew out its electrodynamic circuit, Betty Luca went on to mention that one of the Greys communicated to her that in order to get that UFO up and running again he, and his fellow-engineers, were going to have to re-line the "cyclonetic trowel" inside of it... Well, this obviously is not a phrase known to this scientific world, so on first hearing it doesn't say a lot about what the Greys were going to do that can be easily recognized, and apart from this one phrase and maybe three or four other technical words given by that Grey to Betty, which Betty didn't really understand at the time, that's basically all that was remembered about the Grey's explanation in Betty's recollection. Those who have studied this field will already know that Greys don't communicate that readily to their abductees anyway usually, and so I suppose we should be thankful for the few words that were recalled through Betty's hypnotherapy. However, there is sufficient data in the rest of that recollection of what Betty perceived of that scene to figure out what the Greys were doing to their UFO and why.

The first tell-tale sign of what was going on was when Betty had to be smeared all over the whole of her body by one of the Greys with electro-insulating gel (as did another woman and the man who had also been abducted that night by the same Greys of this same UFO). This was so that these three humans could be moved outside the stricken UFO along with the other Greys that made up the crew of that UFO. The crew of Greys wouldn't need the gel smeared over them because they had on their electro-protective skin-tight suits and insulating boots. Next, the Greys set up around their UFO about twenty electric-charge-grounding jacks, setting these contraptions into the ground (these each utilized a glass-like sphere in them as if to direct electric charges in only one direction, ie, into the ground, so they would appear to have worked like lightning-bolt grounding rods, but they would have enhanced the potential of the ground through their spheres somehow, so that they amplified the ground's potential in a way that would have provided the highest-potential target, so they'd readily attract and drain away any rogue electric charges coming off the UFOs). All of the crew and the three humans were then made to stand together outside that placed ring of stakes to watch the unfolding proceedings (see figure 63).

What was going to happen to the stricken UFO would entail the lower deck area being pummeled by a high-powered rotating field of charged particles. This would be to clean out all the residual pockets of electrostatic charges which would have spread inside that UFO—and that would have dispersed into all the wrong places—as a result of the short-circuit flash-over it had just gone through in its so-called accident. This pummeling and cleansing process would also cause any water in that UFO's toroid to mobilize and to heat up (so obviously the crew and the abductees needed to be brought outside this UFO to a safe distance before the Greys could perform this neutralizing process).

In fact, judging by this episode it would seem to make perfect sense that what those Greys were calling the craft's *cyclonetic trowel* was its central vortex of rotating force. Because, that the Greys were re-lining that UFO's cyclonetic trowel it would then explain the next strange scene which took place, and which was observed by Betty (and drawn under hypnosis in figure 63), of where it shows that another UFO (same type as the aaUFO) was summoned into the area by the Greys, and was depicted by Betty as coming in and positioning itself above the stricken craft.

Returning to Betty's narration, Betty then described seeing the second UFO hovering itself into place over the grounded UFO before offering down a number of grab-arms, with which it attached itself firmly to the lower UFO. She then reported that the second UFO proceeded to spin-up two gyrating energy fields all around the lower craft. The first field's flowing energies were highly visible because they were radiating flashes of light (somewhat like synchrotron radiation), and after they were spun in one direction they were then followed by another field which was spun in the opposite direction, Betty says. This lit up the immediate area and caused a lot of heat to be generated and radiated into the air around the UFOs (and to be felt by everyone watching). Puffs of steam were seen to come out of the lower craft, and at the same time the air all around the two craft began to breakdown electrically, leading to sparks flying all over the place and micro-lightning discharges to flash between the two UFOs and into the night sky. Here, obviously was the reasoning behind the alien's ring of grounding jacks and the insulating gel smeared over the two women's and the man's bodies. Then from these mini-tornadoes of light-flashes something happened to the air's pressure or air density around everyone standing by the lake watching, because moisture began to collect in the surrounding air and a thin mist formed which turned almost immediately into droplets of warm rain (this rain must have been the result of the steam blowing out of the lower UFO, rising into the air and cooling). Spectacularly, as the misty-rain mingled into the light flashes being produced by the upper UFO a whole array of colorful rainbow effects spread all around the whole area for several minutes, until the energy forces from the driving UFO slowly subsided. When everything calmed down, and when the lower UFO appeared to return to good working order the upper UFO unattached itself from it and lifted up, but then, as if to station itself in readiness just in case anything else went wrong, it remained close-by in the air directly above the rejuvenated craft.

To continue my suggestion of what was going on there at South Ashburnham lake, I would say that the upper UFO was extending down its rotating field of electro-kinetic energy, so as to induce rotational force into the ionised fluids inside and around the stricken UFO's ergosphere A and B. Firstly to make sure that all the residual electric charges that had dispersed into the wrong places were all unified into one and mobilized to move in the same direction.

And then that upper UFO would have finally rotated its energy field in such a way that would have rotated the fluid in the lower UFO's toroid counter-clockwise.

This final action would be what the Greys had referred to earlier as the re-lining of that UFO's cyclonetic trowel, which would have meant re-flowing the toroid fluid to energize its magnetic field and routing that magnetic field through the craft's base disc to establish the craft's magnetic funnel (to shape its rotating energy into a vortex). After that, with all of its energies flowing again and flowing in the right direction, the lower UFO would have been safe to run in its tick-over mode and it would have been ready to power up again and fly off back into the night sky on its own.

Although, in fact, that episode didn't quite finish there, and the rejuvenated UFO didn't quite fly off as expected, and what happened next caused a great deal of astonishment to Betty. Indeed, what she recalled next brings in a wholly new and hitherto uninvestigated avenue of science, and this unexpected development will warrant a fresh look at some known, but largely under-acknowledged, principles of astrophysics (and this new avenue of astro-science will be gone into in greater detail, and with reference to other UFOs—such as the British Rendlesham and the Russian Dalnegorsk UFOs—that have also been observed to undergo this same effect, in a later chapter of this study, of course).

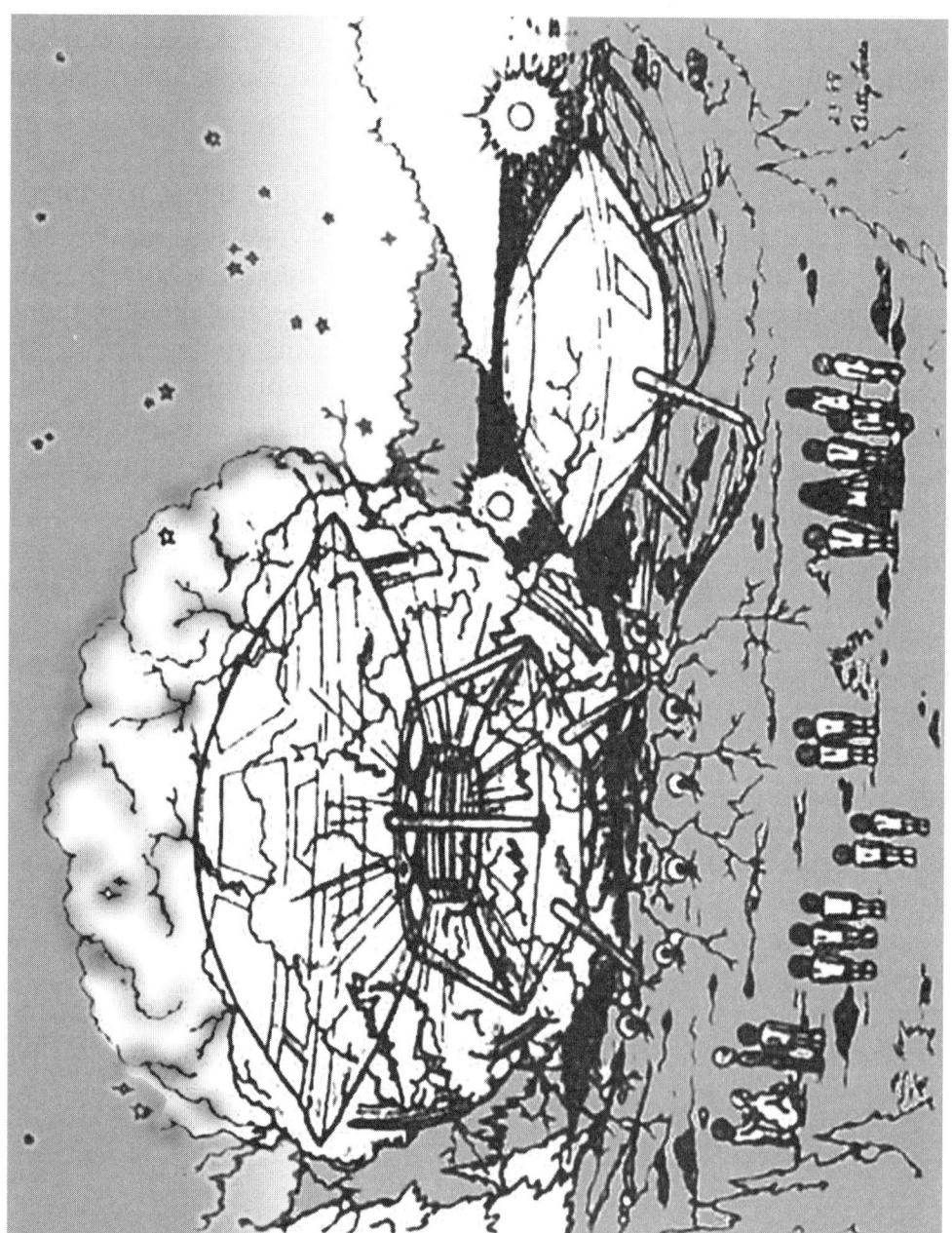

Figure 63. The stricken aaUFO is revitalized by a sister ship after the lower UFO's electrodynamic power system and central vortex short-circuited (from "Watchers" fig.15)

Returning to the current subject of black hole miniaturization, it is quite obvious, from the amount of effort that had to go into re-creating this cyclonetic trowel vortex inside the craft which abducted Betty Luca, that this procedure wasn't something that needed to happen every day on every trip. It would seem a tremendous amount of work to switch the vortex off and then have to get it up and running again just to make one trip at a time. So without doubt, this episode recalled by Betty through her hypnotherapy sessions confirms irrevocably that the aaUFO's vortex must be running all the time, and this in itself indicates that this UFO's power output is variable [note: 14]. So therefore, the bonus to this realization is that a much more dynamic power-drive system can now be envisioned, for it can subsequently be assumed that if the vortex is always there, in place, then likewise the toroid's magnetic field will also be there in existence on a permanent basis. And with that being the case there will need to be found the mechanism somewhere in this UFO which variably powers-down the black vortex's dynamics and which will need to be under the control of the aircraft's crew, possibly as a semi-automatically controlled switching devise. One such switching mechanism, of course, was evident in the variable permittivity caps that cover the top spheres (as noted in section 4.12) [note: 15]. But will there be found any other power-controlling devises on this UFO, I wonder?

Well, there are THREE areas where the aaUFO's power can be controlled; one of those is, as I've just mentioned, through the caps which cover the top spheres and these caps control how the magnetic flux converges through the stem coils. Another is where the magnetic fields pass over and around the inner ring-tube, which controls how fast the telemeter wheels are allowed to rotate (the telemeter wheels will get their rotational torque from the Lense-Thirring force inside the central ergosphere but the inner ring-tube will be used as a BRAKING mechanism on that force—as explained in section 6.12, and shall be further explained in section 9.1). The third method, of controlling the craft's acceleration and deceleration, will be alluded to in a later chapter (sections 10.3 and 10.9 below), but I should say here that for this method it will be the rate at which the telemeter wheels pull on and break the toroid's magnetic field lines, and how fast they fire their pulses through the toroid wall, that determines the UFO's power output. So while the caps covering the top spheres could be seen as the craft's on-off switches the most basic governing mechanism for the control of the craft's black vortex, and ultimately for the control of the UFO's propulsive power, will be the rotational speed of the telemeter wheels.

7.5 Lining up the Sphere-sets to the Central Vortex

It was interesting then, to visualize what might have happened to these all-important telemeter wheels in that UFO when it short-circuited.

The arms of each of the four sphere-sets, and their telemeter wheels automatically align to the toroid's inner wall. However, there appears to be nothing holding them in that alignment. The lower spheres have to revolve otherwise they will develop hot-spots from the gravitational force of the central vortex (as was mentioned in section 3.17, in reference to the so-called tidal force of tension upon them). But if the stems (with their arms holding the telemeter wheels) were fixed to the spheres then they would rotate about the axis of each sphere-set, which would clearly be unworkable and nonsensical.

In other words, the stems or the clamps will have to be manufactured in such a way which would allow the lower spheres to revolve. But they will have to be manufactured so that they will always point the telemeter wheels outward, away from the central vortex (so that the wheels will always face the toroid wall and turn on the inner ring-tube). Mechanically this could be achieved if four rods were made to inter-connect between each of the four stems (and this might be what is being indicated in figure 37).

Otherwise, the clamps might be manufactured to ensure that one of the clamp's four legs was made of more massive materials than the other three, so that by having a greater degree of mass one leg would always be attracted to the central vortex more than the other three. But then, as another of Betty's drawings clearly shows, from the way she was shown these clamps, they appear to have a round boss on top of them which looks very much like it would form an electrical connection to the stem coils, and that the stem coils would sit over that round boss (and they would work similar to how a commutator rotates inside an electric dynamo so that it remains in constant contact electrically while turning).

These images (see figure 64 below) would indicate though that there is a path of conductivity between the lower spheres and the stem coils—and that it runs through these clamps, and through these commutator bosses specifically so that one can rotate independently of the other.

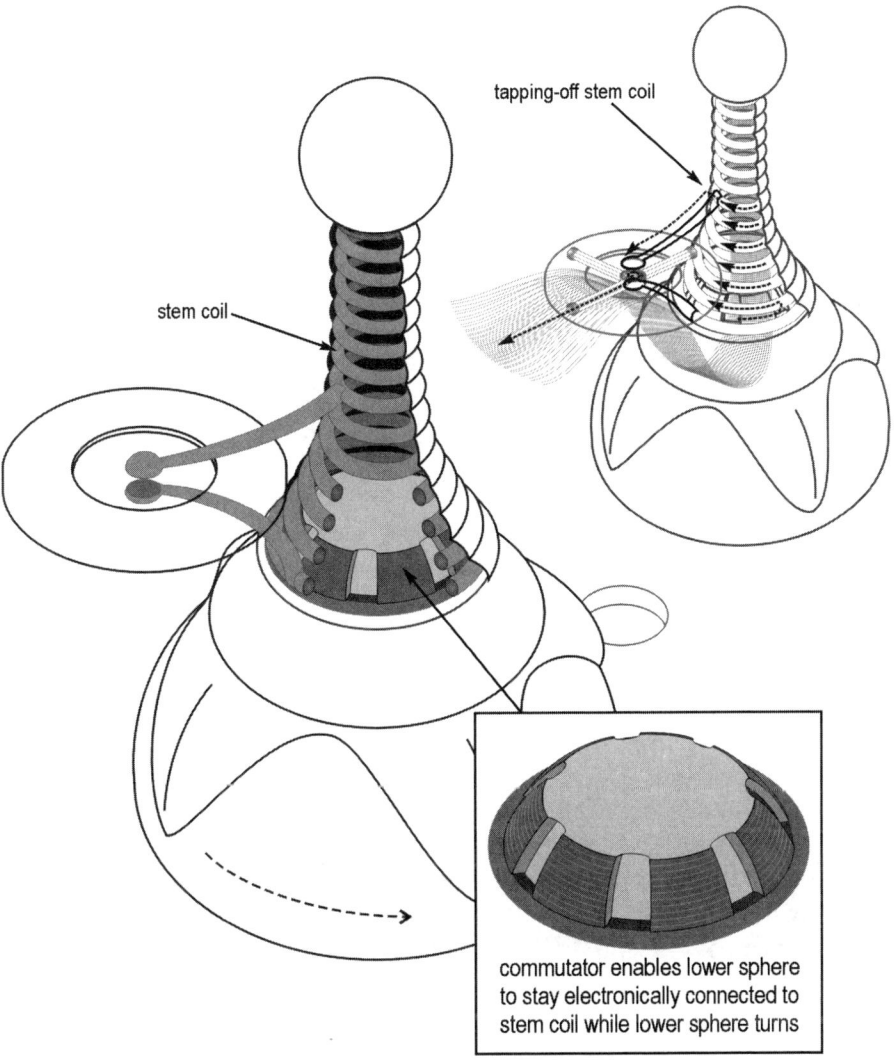

Figure 64. Showing the commutator-type arrangement which enables the clamp and lower sphere to revolve independently of the stem coil, but to be in constant electrical contact

Ok, so the next available solution will be to have the stem coil imbued with extra mass on the axis-facing side (opposite to the side the arms are on), and this way the stem (and therefore the arms holding the wheels) will always face toward the toroid wall, so that the telemeter wheels of that sphere-set will always roll upon the inner ring-tube.

But heck, maybe there are simply four rods running between each sphere-set stem keeping all four sets of wheel-arms pointing in the right direction, keeping the wheels aligned to the toroid wall, and that's probably all that will be needed—with no need to make things over-complicated!

The term telemeter wheel was, of course, one of the terms used by the extraterrestrials in the lake incident, and as was delineated above, these would deliver pulses of microwave oscillations into the toroid's water to help separate that water's positive and negative charges.

7.6 Seeing a Complete Range of Frequencies Inside the UFO Craft

The telemeter wheel pulses would contain microwaves in the 3-4 GHz range, or at least, the same frequency of microwaves as the lower spheres. But aside from the plasmonic frequencies for the lower spheres which I have arbitrarily set between 3 to 4 GHz (as alluded to in chapter 5) there would be several other frequencies being established inside the aaUFO.

For instance, there will be the pulsing system of rhythmic power generations emanating from around the center of the craft which would repeat at about 38 Hz (38 times per second), as a direct result of the beams being fired from the telemeter wheels through the toroid wall [note: 16].

Another range of oscillating frequencies to consider will be the ones resulting from the gyrations of the electric storage field circulating through the planar waveguide, which may be approximated between 3.75 GHz and 2 GHz (from my arbitrary rendition of Betty's drawings and from my scaled technical drawings, which indicate a 8 cm to 15 cm depth for the aaUFO's planar waveguide).

The mechanism to oscillate those fields of charged particles up to RF frequencies would come from the Comptonizing oscillations which will in turn come from particle collisions happening at the central part of the UFO's engine (as mentioned in an earlier section, from particles being ejected out of the black vortex by the Penrose-Williams collisions at the photon orbit ring and bouncing through that planar waveguide).

Interestingly, as a result of their microwave oscillations when those energies gyrate toward the circumferential duct, on their way to exiting through that duct where that planar waveguide converges down to a depth of only a few centimeters, there would then be affected a reduction in their oscillating wavelength, and an increase in their frequency (ie the 3cm exit depth at the circumferential duct would equate to a RF microwave frequency of 10 GHz).

I have an idea though that at the circumferential duct these energy waves will be grouped into wave-packets, and so these wave-packets could undergo a stretching process at that duct as part of a final amplification before they deliver the craft's propulsion force [note: 17].

7.7 Gyrating Storage Field Feeds Rim Field (Current Ring)

The latter point will be further expanded upon in later chapter because there is so much going on in this area of the craft, and basically I have just been bringing a few of those details to the reader's attention here, in this bridging chapter, in the form of a savory *hors d' oeuvre* to prepare the way for the more hearty meals that awaits the reader throughout the remaining chapters.

Also before this chapter ends I should do a brief recap on the craft's ergospheres: There are three internal ergospheres, and there will be at least one ergosphere of rotating energies outside the craft which will have useful forces to consider. Of the three inside the craft the inner-most one (ergosphere A) will be that which immediately encloses and surrounds the central sphere-set assembly. The planar wave guide area will be designated ergosphere B which is where the storage field gyrates. The toroid and its fluid will be ergosphere C. Whatever mention is made of the ergosphere outside and below the craft will be called the outside field ergosphere (or ergosphere D). And basically, the whole area below and inward of the outer rim (comprising ergospheres A, B, and C) will be what constitutes the 'engine' of this aaUFO (as shown in figure 54 above).

While it has been seen that a powerful generating system exists at the center of this craft's lower hull it should also be recognized that a vast amount of the craft's energy will be stored inside the craft, by it being gyrated inside this craft's planar waveguide (this is why I have called it the gyrating storage field).

This waveguide area, termed ergosphere B, is the ergosphere which most resembles the accretion disk surrounding the astrophysical black hole, although I would also have to say that because its energies comprise of two flows, one outward flowing while the other is inward flowing, then it might be better to regard it as being more like a toroidal field which surrounds an astrophysical black hole, especially as within it there will also be substantial electron-positron pair-production activity going on, from charged particles bouncing off the solid confining walls of that waveguide and creating secondary Comptonization collisions. This planar waveguide area will be live with energy, not chaotic, but rhythmically and constantly excited.

Ergosphere B actually divides into two parts towards the outside of the craft before it reaches the circumferential duct, for, while the main energy storing field gyrates in the gap between the toroid shell and the upper deck, there is an auxiliary mini field of energy which branches off from that main storage field and circulates into another circular channel just above the planar waveguide, between the planar waveguide and the shell of the upper dome. I call this auxiliary field the rim field current ring and it is facilitated by extending a moveable section of the upper deck's floor over the craft's gully:

The bottom of this channel is always there as a permanent fixture, and it looks just like a gully which runs all the way around the perimeter of the craft, while the top part of this channel is facilitated by extending the deck floor up and over that gully (so when the deck floor is extended it encloses itself around that channel, so the current ring of plasma can flow inside it fully contained). This is because the current ring cannot be a permanent or constantly operating field, it would be needed for when the UFO is flying, but it would have to be switched off when the craft had landed, and for when the entrance or the waiting chamber was being used, and it would have to be entirely controllable by the pilots of that UFO.

As can be seen in these figures (figure 66 and 71), by forming this channel around the full circumference of the craft, a field of electromagnetic and gravitational energy can be made to circulate around the craft, and which can be composed of charged particles that will be supplied from the main gyrating storage field on a continual basis (but that energy field will not be welcome when people are standing inside that craft's waiting chamber).

The significance of the gully with the floor extending over it came up as a result of studying a number of UFO interiors drawn by Betty Luca (see figures 62 above and 65 below) which show a temporary raising of the floor. Originally, Betty had not known what this gully was for, and obviously, she wouldn't have seen it when it was fully operational and containing a flowing plasma inside it. But what was so memorable about it was the fact that when Betty encountered the floor extension covering over that gully she was puzzled about how she was going to climb over it to get out of the UFO's exit hatch, when she was being taken back to her home [note: 18].

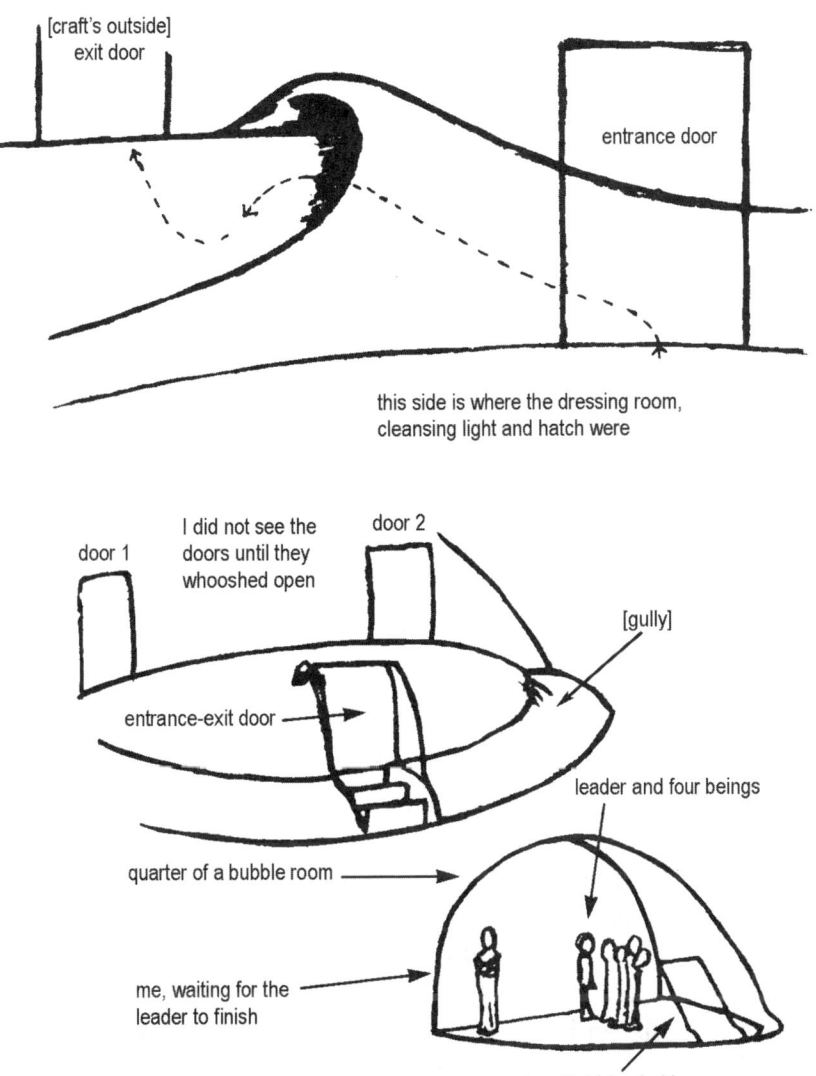

Figure 65. Betty's drawings of the quarter-section of the aaUFO which facilitated the entrance and the waiting chamber, note the gully that curves around the outside of that chamber (from "Andreasson Affair" fig.10 & fig.39)

Indeed, this raised floor was only noticed on Betty's return journey, and it was definitely not there when she was first taken into that same UFO through that same entrance on her outward journey. It would suggest that the part of the decking around the waiting chamber would be made of a flexi-metal or a polymer material which could be controlled in some way, possibly electromagnetically, to raise up and down when the UFO was preparing for flight [note: 19].

From the way Betty described the UFO shooting off after taking her home on that particular occasion, the idea was presumably to just land momentarily, to offload Betty, and to quickly take off again, and so presumably that's why the crew didn't want to straighten up the floor, because they wanted the craft to remain in flight mode.

This is why I would presume that at the outer perimeter of the upper dome shell within its hollow rim there will be directed this gyrating field of electrical plasma from the main storage field, and I would say that its angular momentum will be induced directly from the central vortex through their co-mutual magnetic field. Consequently, while this current ring of plasma rotates around the inside of the rim (and inside the gully) it will generate a localized magnetic field around the whole 360 degrees of the craft's outer perimeter—specifically as a pull-down field to pull upon any charged particles that are directed out of the circumferential duct in order to curve them up over the rim towards the dome's apex.

This may be one of the reasons why other designs of UFOs have a more acute angle at the rim (giving the rim a much flatter profile), specifically this would be to ensure that the exiting energy, just before it exits the rim, that it induces extra magnetic force into the magnetic field of the rim, which in turn would raise the intensity of the rim's Lorentz force upon the exiting propulsion energies, so that those energies would curve around the rim (rather than flow straight out of the rim to spill into the surrounding air). Once the propulsion energy curves onto the upper dome then its magnetic field will take over, and by the frustum effect those propulsion energies will then simply curl around the dome to the craft's apex [note: 20]. On the aaUFO craft, with its rather obtuse-angled profile, this pull-down process will be supplemented by the magnetic pads embedded into the curved surface of that upper dome shell, which would help control and shape the propulsion plasma into a helical trajectory. Likewise, the direction and the acceleration of these propulsion fields can also be adjusted by how these magnetic pads are configured [note: 21].

7—Black Vortex of Immense (Silent) Power

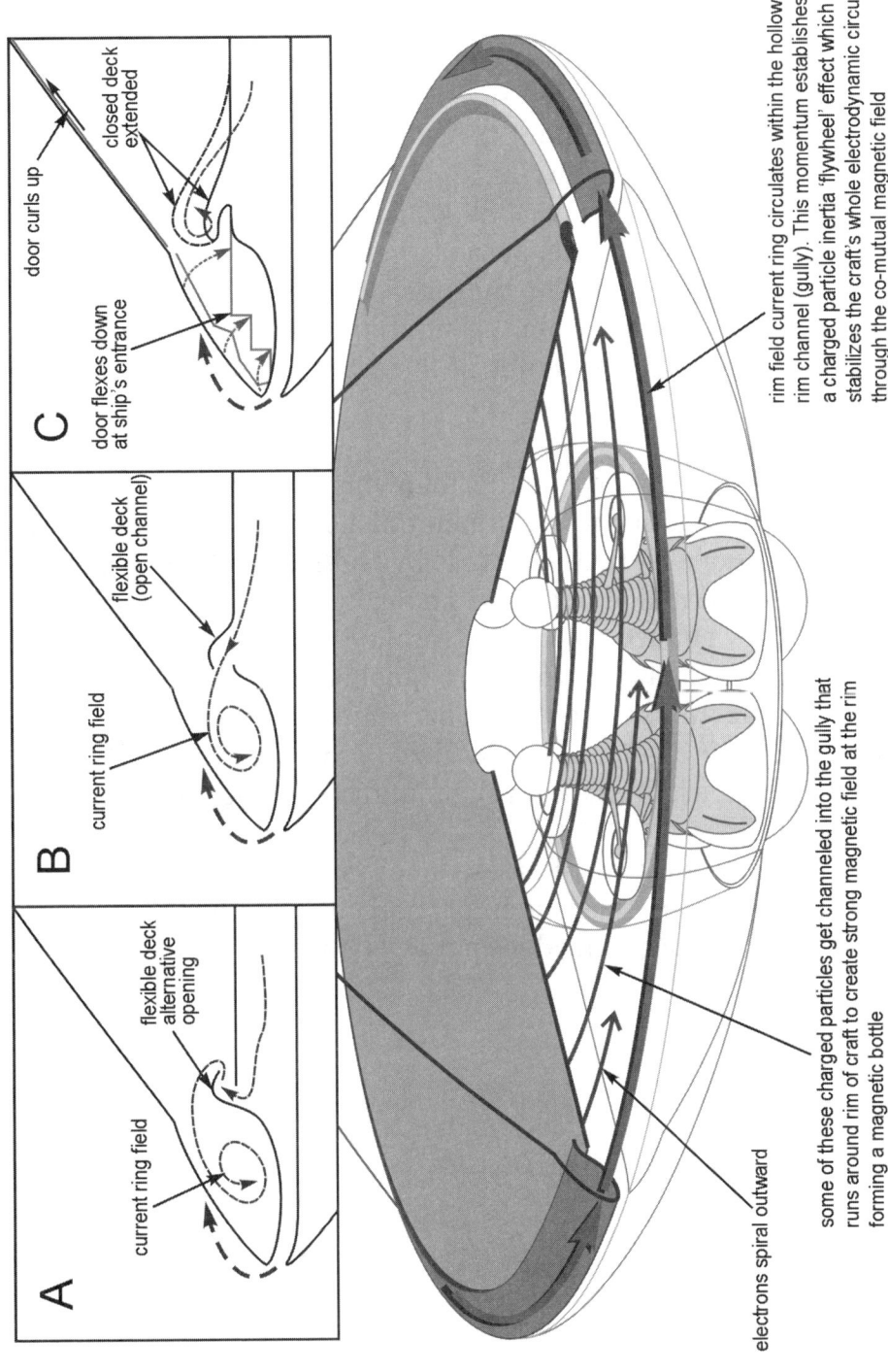

Figure 66. Using a flexible material (such as a magnetically shaped polymer) the deck can be raised to allow energy to rotate out of the planar waveguide into the gully to create a current ring around outside of craft

7.8 The Gully Paradox of a Small Gap Leading to Bigger Things

While the above section explains the gully well enough if the reader looks carefully at these memory sketches (in figure 65) it might be further noticed that there are a few rather unusual features about them. Because, these apparent unusualities may well point to a special refinement (as I mentioned in section 7.3) in the way the internal environment of that UFO is controlled. There would for instance be two questions that would spring to mind here, because in these gully drawings why are the inner sections of the UFO sealed off from the entrance area? And why have a gully all around the UFO which is fed energy from the gyrating storage field (which in turn is fed by the black vortex) when at the center of the craft that same black vortex is open to the upper decks?

These questions will be answered in the following chapters, as will another interesting question, this time relating to the gravitational forces inside a UFO: Why is it that after all the thousands of observations made by abductees that there has never been mentioned anything abnormal about those abductee's body-weights when they were taken inside a UFO? No abductee has ever recalled floating inside a UFO craft, yet it is well known that cosmonaughts and astronaughts are obliged to experience zero-gravity inside their spaceships once those spaceships leave earth's atmosphere [note: 22]. This says that the UFO generates a localized gravitational field inside of it, and easily keeps it at a fairly constant level throughout the UFO's journeys.

I like the extraterrestrial physicist,
to him or her space is not empty and not symmetric,
and everything does not equal zero...

8: Gravitational Pull and Push Forces

Contents:
8.1 Acoustic Black Vortex Gives UFOs the Stealth of Silence	259
8.2 Black Vortex Establishes Horizontal Gravitational Force	260
8.3 Difference in Aeronautics With Gravitational Buoyancy	267
8.4 Base Disc Divides Gravitational Field Dynamics	271
8.5 Negative (Repulsive) Gravity Below the Base Disc	272
8.6 Beyond the Buoyancy of Radial and Negative Gravitational Forces	276
8.7 On My Way Home After a Hard Night's Abduction...	278

8.1 Acoustic Black Vortex Gives UFOs the Stealth of Silence

Any sounds created around a vortex where the rotating fluid is axial-pointing and the fluid's velocity is greater than the local speed of sound then those sound waves will be swept inward into that vortex flow and will be trapped between the core of that rotating flow and the enclosing surface surrounding it: Which, on the *inside* of the aaUFO will occur between its black vortex and the enclosing surface of its central ergosphere A, and that of its planar waveguide ergosphere (of ergosphere B). While on the *outside* sound trapping will occur between the smallest bore of the projected helical filament-tube and the rest of the UFO's energy envelope.

This sound trapping phenomenon would result from the Lense-Thirring frame-dragging going on inside the aaUFO's central ergosphere, from the continual influx of energy being pulled into the black vortex and being fed into it from the top spheres (and from ergosphere B), which will force that energy field to flow faster and faster (as shown in section 5.1). Indeed, wherever the fluid velocity of that UFO's electrodynamic flows exceed the local velocity of sound then none of the sound waves in their vicinity will escape out of those localized supra-sonic flows. For the aaUFO this phenomenon may actually be taken as a foregone conclusion because, as mentioned above (in section 7.1), there has been the over-riding observation that light trapping has also been evident at the core of the aaUFO's vortex (as indicated in figures 37 and 62), which straightaway suggests that the rotating fields inside that UFO were not only supersonic but that they had velocities greater than that of the localized speed of light too [note: 1].

But certainly, while this engine in the aaUFO has already taken on numerous space-time properties of a black hole here will be yet another of those properties, for while its surrounding ergospheric fields will all contain rotating energies to the degree that those energies can be accelerated to localized-supra-sonic speeds, most importantly throughout ergospheres A and B, then so will that UFO craft's motive power drive be surrounded by an acoustic horizon. This is the beauty of an aircraft having a circular shape with internal cylindrical containment walls. Quintessentially, the reason why the vast majority of UFOs are reported to run SILENT will be because they are cylindrical and they utilize rotating energy fields [note: 2].

8.2 Black Vortex Establishes Horizontal Gravitational Force

Another advantage of an aircraft having a cylindrically contained internal structure will be that its gravitational forces will be radial toward the center of that craft. This follows on from the gravitational polarization effect mentioned in sections 5.1 and 6.10 above, which alludes to the fact that the horizontal gravitational force of the black vortex can be tapped-off either at the base disc and transferred into the toroid fluid (and into the craft's hull), or it can be taken advantage of as it radiates away from the black vortex through the polarization effect in the air and through other fluids, so that overall the whole lower section of the UFO would be permeated by a radial gravitational force directed inward toward the black vortex.

Furthermore, because all matter will be gravitationally attracted horizontally toward the central vortex, meaning that everything in ergospheres A to C will be gravitationally aligned toward the central vortex, then through this alignment the vortex's influence will be transmitted concentrically all the way out to the rim of that craft. Therefore this horizontal gravitational force will predominate inside this craft over any exterior gravitational field (ie the Earth's) for as long as the craft's black vortex is operational.

This in-flowing force of gravitational attraction though will not be what makes a UFO rise up or move! All it will do is cancel-out the earth's gravitational pull and make that UFO buoyant. But in making the UFO buoyant, and making it, for all intents and purposes, *weightless* (because its carrying its own source of gravity around with it) it would render that craft extremely easy to maneuver, and most importantly maneuverable by the slightest of supplementary force fields.

8.3 Difference in Aeronautics with Gravitational Buoyancy

UFO propulsion can be very technical as will be seen in later chapters, but there is no reason why a UFO cannot once it acquires its buoyancy move by simple mechanisms as well, using for instance, the same simple mechanisms that winged aircraft use. Some of these have very little energy expenditure if only slow momentum is required.

Importantly, the buoyancy factor would allow the UFO to make movements simply by creating thermodynamic imbalances, or air pressure imbalances, or air density imbalances, between it and a near-distant location. These essentially would be engineered as artificially induced pressure-differentials or density-differentials, where if the pressure-density of the UFO's own envelope can be made to differ with that of the pressure-density of the air above or to the side of that craft then that craft would rise or move sideways.

Indubitably, a weightless UFO with a domed shell would be perfectly shaped to project such differentials even further into the air ahead of itself. The UFO would only need to project into the distance a moving column of rotating air to set up a frame-dragging motion in that air, so that the air-layers can then form into a rotating cylinder which attracts toward it layers-upon-layers of more surrounding air. This could be done in a few seconds, and while that rotating column is busy pulling in more layers of air it will also be establishing inside itself a low-pressure zone, so then by the adiabatic principle the UFO's own higher pressure envelope immediately surrounding it would be drawn upward into it. Providing this sort of mechanism was used as a pulsing mechanism then as soon as the craft moved through one air-cylinder it would dump it and then establish another air-cylinder and traverse that, and so on, and as it will be the UFO that generates that air-cylinder so the next cylinder will always be moved progressively ahead of that UFO to prolong that craft's momentum. The UFO will not travel at any great speed in this mode, just by using pressure-density differentials and by riding on simple convection currents, but it will ride these columns with very little energy-output for as long as these pressure-density imbalances with the distant air can be maintained [note: 4].

Otherwise, another mechanism of reduced energy expenditure is the mechanism of a cyclone where an airflow column extends high into the earth's atmosphere, where at the top of such a column the air is more rarified, and the channeled air after rising through hundreds of meters of the atmosphere will undergo a divergence to spread radially outward into the upper levels of the atmosphere, and this radial divergence at the top of the column will be what will 'anchor' that whole air-flow system and create a pressure-gravity current, which would move that UFO into the lesser pressurized plateaus of the upper atmosphere very quickly. Indeed, at the top of that column the more it spreads and diverges into the upper atmosphere the greater will be the anchoring effect that the craft can use to develop a pressure-differential between the craft's surrounding envelope, and that upper anchored area (particularly if that upper area is stratified and fairly calm). And all the craft will have to do is maintain the differential through the column's narrow bore, to ensure that the column remains established [note: 5].

With such buoyancy a simple method such as the application of heat to the air surrounding the craft would also cause it to move, as in the case of the UFO enclosing itself inside an air parcel and applying heat to the leading edge of that air parcel (so that the air parcel lowers its density), because then the UFO would move toward any cold rarefied air ahead of it with a velocity proportional to the rarefaction and coldness of that remote air. Even by heating one side of the UFO's upper dome shell (around the rim area) it would make it rise upward or move sideways. This method of movement is quite evident in the photograph of the Grangemouth UFO where the heat trail from its electrodynamic energy suggests that the craft was moving very slowly above the guys who photographed it. This latter case brings up an interesting point to be made, because if some of the energies generated by a UFO's high-powered propulsion drive can be harmful to living organisms positioned too close to them when they power up (ie some propulsion drives will involve UV, X-ray, and gamma-ray photon emissions when they are fully operational), then out of consideration to the living beings of this planet this type of auxiliary 'safe' propulsion would be beneficial to have as an alternative to their main drive force. Obviously, not all UFOs are so considerate, but I would think it has been a point of consideration for most UFO designers, that some simple ways to create momentum would generally be made available for UFOs because these would be preferred for UFOs flying close to populated areas, to allow those craft to move a safe distance away from any humans that might be immediately below them before those craft engaged their main drive forces [note: 6].

This might also have been the case with the UFOs hovering over Mexico City in the 1990's where several of the UFO craft filmed by local people look as if they have a heat trail just above them to accompany their colored fields of synchrotron radiation seen around their perimeter rims. Obviously, once the UFO achieves buoyancy, like I say, all it would then need to do to initiate these small in-and-out-of-the-clouds type movements would be to encase itself inside an envelope of spinning charged particles and then radiate a microwave field into small sections of that outer envelope, to heat the air to create imbalances in the surrounding air density. So the technique to project heat radiation into their surrounding energy fields may be something the extraterrestrials use as an alternative to their main propulsion drive. Certainly, a whole list of simple ideas could be drawn up and this might help to explain the diversity of movements indicated in UFO reports; and from the standpoint of someone trying to fully understand how these craft work it would have to be considered a nonsense thinking of them as having only one form of propulsion.

Indeed, it might be a point of consideration here that perhaps the diversity that comes with UFO sightings is the very reason why scientists have been obliged to ignore UFO research in the past. A scientist will always look for the perfect correlation between observed phenomena and known phenomena (known through that scientist's education and experience), and so if two different reports for two different UFOs differs one from the other by the signatures of how those craft move then those scientists will dismiss both those reports as contradictory. Scientists, after-all don't have any database on UFOs with which to work from so it would be ridiculous to expect them to know anything about such craft. Certainly, if they were looking for a solution to UFO physics and they didn't know about the buoyancy mechanism which takes away their weight factor, then those scientists would never be able to understand how a UFO can maneuver in the air with such agility.

One case similar to the Mexico UFOs, of what I would regard as a UFO 'showing off,' was in an incident relayed by a witness in an interview broadcast by my local radio station in 2004, where this person from South Africa had observed a UFO above his village, and he described how he and several of his friends watched this UFO fly in low and come to a halt in mid-air above their village, and after the UFO hovered motionless in the sky for several minutes, almost as if its crew were waiting for the gathering audience below it to swell, this UFO then proceeded to zigzag back and forth in the air above its audience at breakneck speeds, and then, as if it was satisfied it had performed its show, it then flew off back into the sky at colossal speed and was never seen again [note: 7].

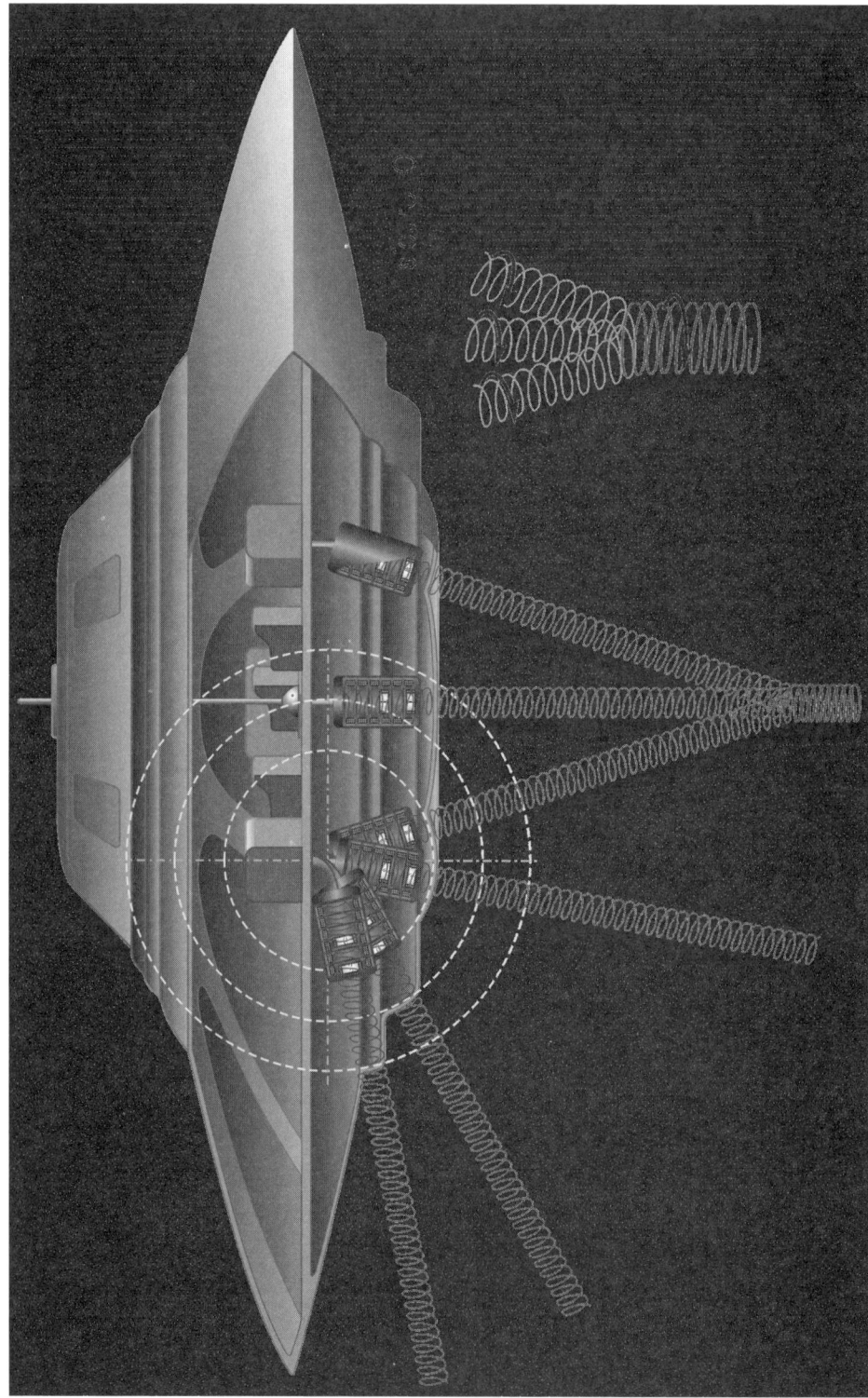

Figure 67. The UFO observed and worked on by Bob Lazar at Area 51 would propagate a rotating field of repulsive (or negative) gravitational force

This sort of maneuver might entail using more than just buoyancy and heat though, this could be similar to how the Bob Lazar UFO works by using a process quite different from the aaUFO to hover in our atmosphere by. The Lazar UFO does generate a horizontal gravitational field for itself but it then also develops spinning fields to establish gravitational epicenters which are distant to it and below the craft so that it can use those epicenters as its centers-of-gravity to pivot on when it hovers or moves [note: 8].

The Lazar system would effectively be re-locating its center of gravity to below its hull and this allows it to 'punt' the air at those momentarily established epicenters. Such a UFO by extending a beam of force into the air, can then move basically as fast as it can send out those beams, and if it takes two seconds to create a gravity epicenter then that's how long it need wait in one location before it moves to another. G-forces, so disastrous for the pilots of earth's aircraft would, inside these UFO craft never be anything other than unity because the crew's relationship with their source of gravity never changes even if that craft undergoes these wild zigzag moves! Earth's aircraft are superb but their top speeds will ALWAYS be hampered by G-forces on the pilot.

Incidentally, what the Lazar UFO uses in its rotating fields are negative gravitational forces—forces of repulsion not attraction (more on this UFO in a later chapter).

8.4 Base Disc Divides Gravitational Field Dynamics

The ability of creating a force of gravitational repulsion will be quite a unique phenomenon to earth's scientists (as would that of creating an attractive gravitational force, it should be noted) but with reference again to where the effect of gravitational polarization was noted to permeate throughout the craft's central ergosphere (in sections 5.1 and 6.10), this gravitational polarization effect can be associated with another of the curious observations made by Betty Luca, which was that long after the ETs originally showed her the workings of their UFO (in a quite separate incident several years later) the ET Greys showed her how one of these sphere-set assemblies came apart and how it could even hover in the air and maneuver in the air on its own, quite independent of any UFO craft's power drive system [note: 9].

This tantalizing observation might offer the intriguing possibility that these crystal spheres can become gravitationally charged as a consequence of them orbiting so closely around a craft's black vortex, either through the vortex's spatial radiation, or through the lower spheres' contact with the base disc which locates them, perhaps in a way similar to how the base disc transmits gravitational radiation from the black vortex through to the toroid through its high-mass layers of bismuth-lead (alternating between layers of magnesium). Both of these two mechanisms would then appear to be the product of gravitational polarization on a molecular scale, through which process these glass spheres might be expected to temporarily hold within them a residual amount of that vortex's gravitational force.

Although, might it be wrong to expect this influence to be just a short-term and transient effect, judging by the spheres that were shown to Betty it might be that the gravitational forces imbued into them do not purely depend upon whether the spheres are close to the central vortex when it is powered up so that they are actively being polarized by that vortex. There may be a way in which gravitational forces can be made permanent, or at least for a useful period of time, on a molecular scale inside these glass spheres—perhaps in conjunction with their plasmonic oscillations whereby an absorption effect would allow those molecules to *retain inside* them some of their gravitational re-alignments (much in the same way that their ferroelectric molecules can retain their electrical polarizations—as shown in chapter 5 above).

Whichever of these possibilities is applicable, for imbuing gravitational force into those spheres, if it was conventional gravitational force (of the positive-attractive variety) being imbued into them then it would mean that those spheres would act like mass-magnets—which they don't—they're the opposite, they hover in the air! So presumably they would have been polarized with the negative-repulsive form of gravity (in a particular configuration that directly opposes earth's gravitational field).

8.5 Negative (Repulsive) Gravity Below the Base Disc

The most likely solution to these spheres, is that the repulsive gravitational force being generated by the aaUFO below its base disc is made to permeate into them through the agent of the craft's magnetic field. Around this underbelly of the aaUFO it can be safely pointed out that whatever has been established for this UFO above its base disc is certainly not what happens below that base disc. Firstly, there is a substantial difference in the dynamics of the upper and lower magnetic fields of the toroid, and this is even more pronounced above and below the base disc area. For below the base disc the field lines are not being dragged nearly so much as the upper ones are (although they are still being stretched, broken, and reconnected as the upper magnetic quadrant fields are). The lower toroid field lines are oscillating into and away from the central hole in a fairly robust manner, as converging-diverging non-uniform flux lines, which is similar to how those same field lines will be creating the magnetic funnel inside the aaUFO to produce its black vortex, but there is no containment wall surrounding the ergosphere of energy below the craft, merely does it conform to a poloidal magnetosphere. So the rotating charged particles flowing into that magnetosphere will flow fully around the lower perimeter of the craft.

But then, this poloidal field is unique in that there will be no equivalent anywhere else around the UFO, it will be a magnetic field spreading outward from the center of the base disc and after it spreads outward it will curl back on itself like a big bubble, and then those magnetic flux lines will loop back into the toroid about half way up the toroid's outer wall.

So the point to emphasize here is this field's inhomogeneity, because, as is well known in electronic engineering the actual attractive force of magnetism and electromagnetism comes from where the lines of a magnetic field are convergent-divergent, and so the question to ask here might be, would the electromagnetic-gravitational force of the black vortex as it passes through the hole in the base disc be more powerful here, because of this inhomogeneity, than at any other location around that UFO? In gravito-magnetic theory, of course, it would be conjectured that this would indeed be the case.

If so then this displays a whole new grove of tree-stumps to overturn and lay bare. Particularly as up until now the aaUFO has been favorably compared with a black hole system as an aircraft with its own gravitational field and its own means of curving the spacetime geometry directly below it. For gravitationally speaking both the astrophysical black hole and the aaUFO's black vortex are very similar except for only one thing, and that (as alluded to in chapter 7), will be that while the black hole is postulated to have a point of singularity the aaUFO does not have to follow that particular line of speculation, it can have a much more powerful dynamic than that, because what the aaUFO will have is a rotating converging field which rotates faster and faster into a core of high mass-density energy above the base disc, around which would gather a very intense accumulation of magnetic field lines. This gathering forms the black vortex's magnetic bottle-neck above the base disc—but it also forms a magnetic bottle-neck below the base disc of a completely different configuration. And to go back to a point that was made in section 5.1 above, inside the vortex at its core the atoms will succumb to the core's shearing forces and gravitational pull and will break down through atomic fission reactions, and from this fissioning there will be released huge amounts of energy, much in the same way atomic binding forces are released to power the black holes of space. This makes the forces around that core, and around that accumulation of magnetic flux lines, very powerful indeed.

This power will also pull through that core any electromagnetic waves that get caught and trapped by those rotating forces (as will be shown in chapters 9, 10, and 12). But in the black vortex dynamics where there is no point of singularity, this means that whatever particles and waves that do exit through this hole, through this magnetic bottle-neck, those particles will be forced to exit the craft through the agent of the poloidal magnetic field lines spreading outward below the craft. So the charged particles, which will carry with them substantial angular momentum and gravitational force, and electromagnetic energy, will be obliged to follow the influence of that poloidal magnetic field which will be anchored at, and by, the base disc [note: 10].

But then here is the big difference between what happens above and below the base disc, while inside the craft the fields move radially inward, below the base disc the gravitationally charged particles will follow through an expansive curvature [note: 11]. So, it would follow that by the craft having a source of electromagnetic-gravitational force feeding into an expanding magnetic field of intense power, when any of its particles (and the raw energy from the vortex's core, which will be charged with gravitational and electromagnetic forces) become affected by that powerful magnetic field then they will be induced, by Lorentz forces, to follow a curvature which is expansive in nature. So in other words, the collective gravitational force flowing out from the core of the black vortex below the craft will therefore also be outward-moving and REPULSIVE. Although, it should be said that this is certainly not the end of the story, as shall be seen below (in chapters 9, 10, and 12).

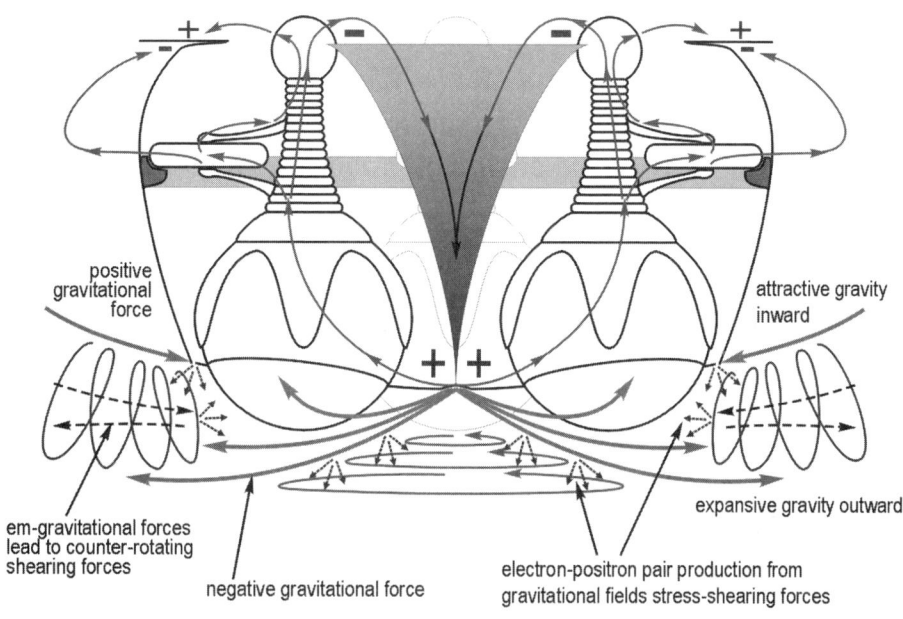
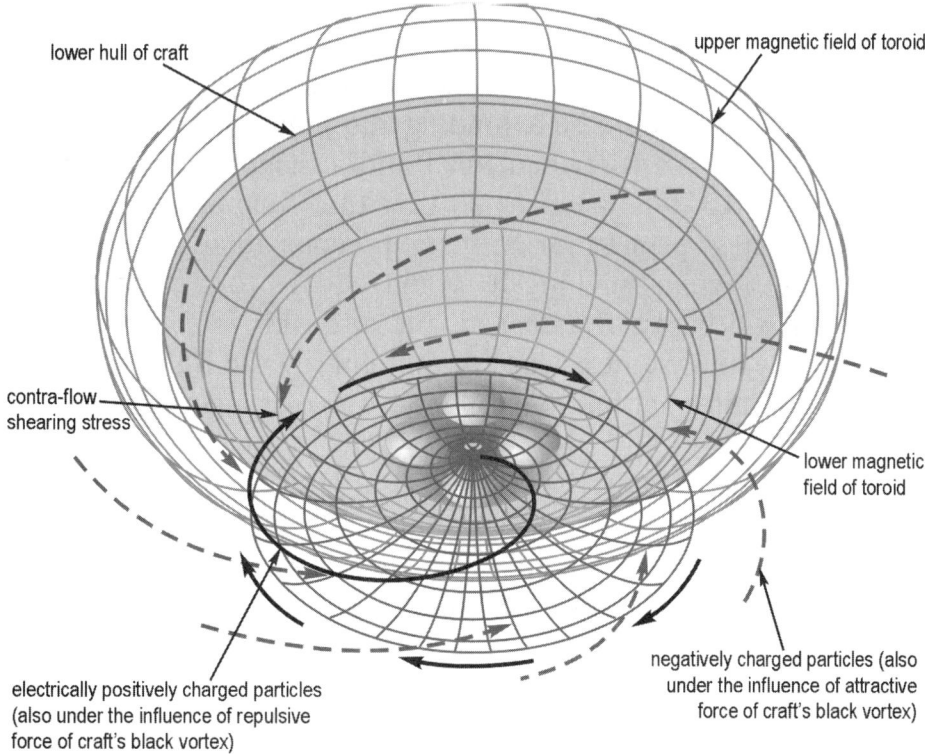

Figure 68. Some of the negative gravitational force can be induced into the lower spheres, and a shearing force will be established beneath the craft to produce electron-positron pairs

8.6 Beyond the Buoyancy of Radial and Negative Gravitational Forces

With the assumption that the UFO generates both a horizontal gravitational field which is a gravitational force acting radially inward centering on the craft's central black vortex, and an out-flowing expansion of gravitational forces below its base disc, and the possibility that both flavors of that force might be permanently (or semi-permanently) held in the lower spheres, then, I repeat once again, here in UFO technology is a new rock to establish concrete footings upon... from which gravitational research on this planet can make significant advances and can gain much benefit.

Gravitationally speaking, if the craft can hold its position in earth's atmosphere impervious to earth's gravitational-pull with its own vortex gravity, and in what can easily be engineered as an intrinsic mechanism (where the energy development chain takes care of itself with no input needed from the crew), then this frees up both the crew and the UFO's auxiliary propulsion drive mechanisms to prepare other energy fields for a range of different purposes, either for flight or for the transportation of objects.

For instance, while the UFO is in the air with no concern for our planet's gravitational force the crew could be extending and pumping up a filament-tube into a metastable state of tension (between itself and a distant location of the sky—as alluded to in section 2.10, for instance) so that that filament-tube's stored energy could be held in readiness primed for an instantaneous adiabatic jump by that UFO. With the filament-tube being so long it would equate to a very large volume of accumulated energy waiting to be released, and with this factor in mind it could be speculated that during an instant release of that energy the UFO might accelerate to speeds of hundreds of miles per hour—while its power-drive is switched off!

But there would be so many benefits from a craft being held up in the sky in gravitational buoyancy not requiring to continually propagate any propulsion forces, such as through its main helical filament-tube. Most intriguing of all though for a UFO possessing and carrying with it its own horizontal axially-pointing force of gravity would be that, even though it might be made to barrel-role through 360 degrees in the air, or made to fly through the air at Mach 9 and be suddenly stopped, or made to turn at a right-angle through 90 degrees, or even made to fly upside-down—still the environment inside that UFO would remain just the same gravitationally to its occupants with no additional G-forces whatsoever!

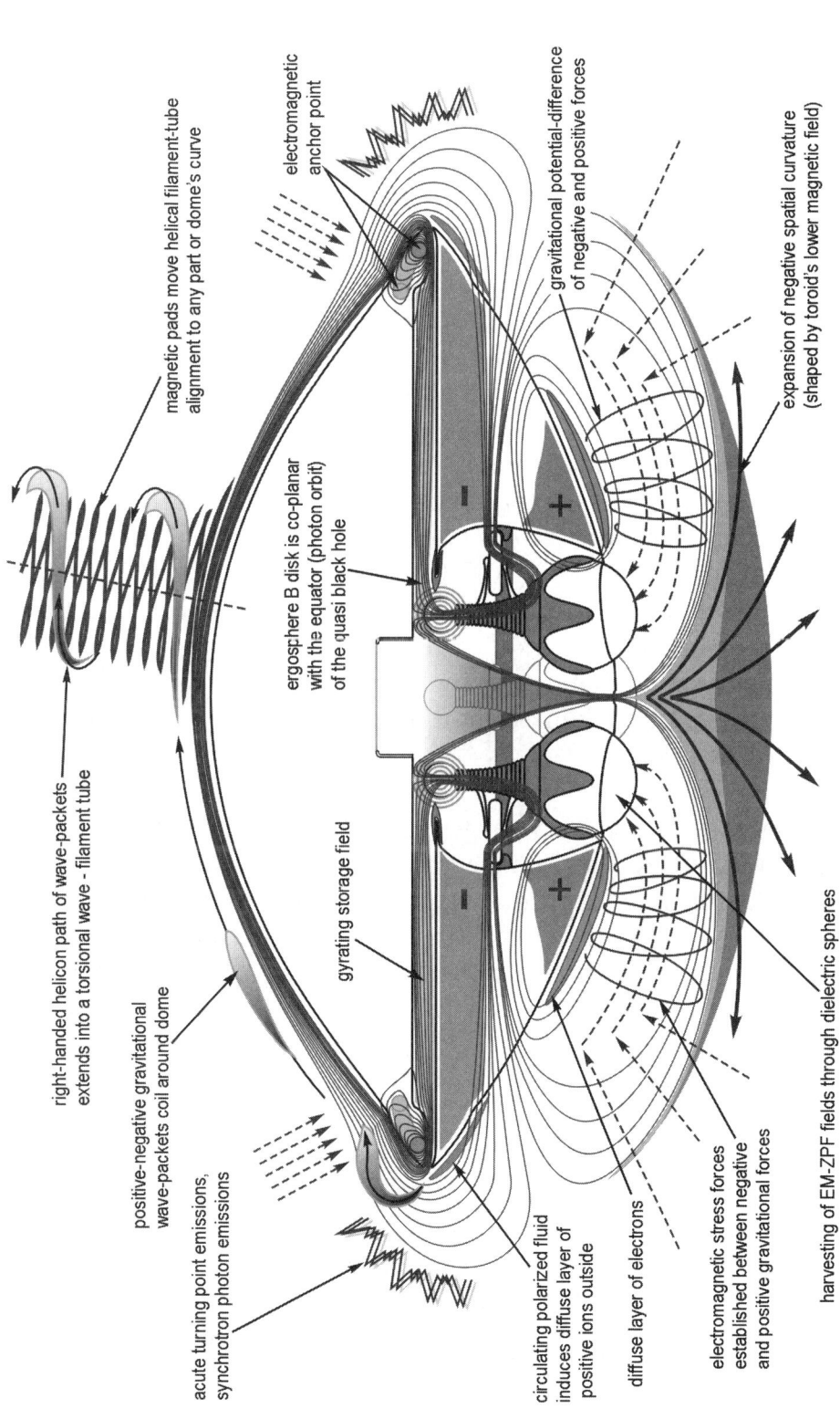

Figure 69. This shows the negative gravitational force propagated below the aaUFO craft and how the helical filament-tube can be established at any part of the upper dome's curve

So in this respect, while comparing these UFO mechanisms with that of, say, the 'old-slugger' rocket destined for Mars, while that rocket fights the elements and blasts against its ambient surroundings all the way, and in doing so expends an inordinate amount of energy (necessitating that it takes along with it a huge payload of fuel), the UFO will wrap the ambient surroundings into a filament-tube and then fall into that tube as if it were a bed with silken sheets…

8.7 On My Way Home After a Hard Night's Abduction…

I began this chapter with the observation that a vortical flow can develop sufficient frame-dragging as to trap any sound waves in the vicinity of its power drive system. And likewise I will end this chapter with an interesting corollary about trapping sound waves, found in the experiences of the intrepid Betty Luca.

After Betty was returned to her bed following one of her many abductions the Greys duly departed from the back of her house in their UFO. But Betty on this occasion was still awake, and although drowsy and tired she listened to the noise of the alien's craft just before she drifted off to sleep. And as Betty Luca later recalled…"I'm in bed and I hear whirring and whirring and—um, starting up something like a big motor or roaring. Like a whirring, roaring noise. I don't know. And its coming over from the right-hand side, by Becky's little bedroom…And its not roaring anymore, just the—like a 'dink-dink-dink'…I don't know." ("Andreasson Affair" p130).

This short comment by Betty Luca in actual fact speaks volumes about the driving fields being generated by that UFO, because judging by this description of Betty's, of the drive system beginning with a 'whirring' it sounds very much like the UFO was at first generating an energy field which was gyrating around inside it. This it would do after powering up its central vortex and gyrating its storage fields, and most importantly this craft's gyrating field was propagating outward an audible field of sound waves… which Betty heard as a whirring roaring noise… I'm emphasising this factor because it is quite unusual to hear any sound coming from a UFO, other than perhaps a slight hum. But then, almost mid-sentence, Betty says she recalls hearing the whirring sound cease and just before the UFO sped away change to a 'dink-dink-dink' sound [note: 12]. This quite clearly indicates, by the whirring or roaring dying away, that the sound waves coming from that gyrating field were gradually being absorbed by the UFO's black vortex field. In other words, the gyrating plasma once it was established into a vortical flow (inside the central ergosphere) would then have speeded up so much that its vortical flow velocity became greater than the velocity of sound… thus in a situation that mimics the event horizon of an acoustic black hole the UFO's gyrating field, by flowing faster than the speed of sound, developed an acoustic event horizon around itself… and so all its sound waves were then after unable to escape outward from the UFO.

But of further intrigue is the dink-dink-dink sound developing out of the UFO as it sped away from Betty's home and off into the night sky, and this is the most telling of the sounds because it would represent an entirely different field of energy propulsion pulses newly emanating out of that UFO, suggesting that a transformation has taken place in the aaUFO's energy, and that it had developed into a series of fast-repeating energy-compression wavepackets, wavepackets that were being braked and slowed somewhere inside the UFO before being re-accelerated again out of it, or rather that perhaps these wavepackets were compressing against something before they were being re-accelerated out of the UFO's rim. Because the dink-dink-dink sound would only be heard if the UFO's energy fields were suddenly slowed down to subsonic speeds...

9: Superradiance Amplification inside UFOs

Contents:
9.1 Rim's Gyrating Current Ring Inductive-coupling to Vortex	275
9.2 White Wall Repulsion Force Inside Rim	280
9.3 Fourth Amplification: Superradiance	283
9.4 Modelling an Ergosphere for a Slower Light Speed	286
9.5 Moving Fluid Slows Light Speed Even More	289
9.6 UFOs Establish Electromagnetic Anchor Within Their Ergosphere	292
9.7 Superradiance Plus: Another Plasma Pump (Reconnection Pt III)	293
9.8 Stabilising the Back EMF on the Black Vortex	297
9.9 Did the Greys Mention Superradiance During 1967 Abduction?	298

9.1 Rim's Gyrating Current Ring Inductive-coupling to Vortex

Closely coupled to the current ring mentioned at the end of section 7.6 is the effect that the gyrating storage field will have at the perimeter rim of the UFO craft.

Just by the fact that the gyrating field does not immediately flow out of the craft's circumferential duct, but that it circulates as a storage field just inside the rim is a master-stoke of engineering design. This marks a substantial improvement over the astrophysical model on which the overall power-drive system of this UFO is based. Because whenever the circulating energy moves around the rim area it will intensify the magnetic field at this part of the craft. And with BOTH the gyrating storage field and the current ring field circulating at the rim their combined magnetic force fields will affect a greater influence over the upper dome's magnetic field (and this will be extremely useful when that dome field is later used to amplify the craft's gravitational forces, as will be shown below and in the next chapter).

It has already been mentioned that the current ring is fed (and continually replenished) by the gyrating storage field (see figures 66 and 71). But the most intriguing factor here will be that these two outer rim fields will be inductively coupled to the inner ring through the toroid's magnetic fields—as too will be the vortex's electrodynamic spin force. In such an arrangement the co-mutual magnetic field will impart potential energy from one part of that circuit into another part of it (as happens in space between the accretion disk and the black hole, through their co-mutual magnetic field).

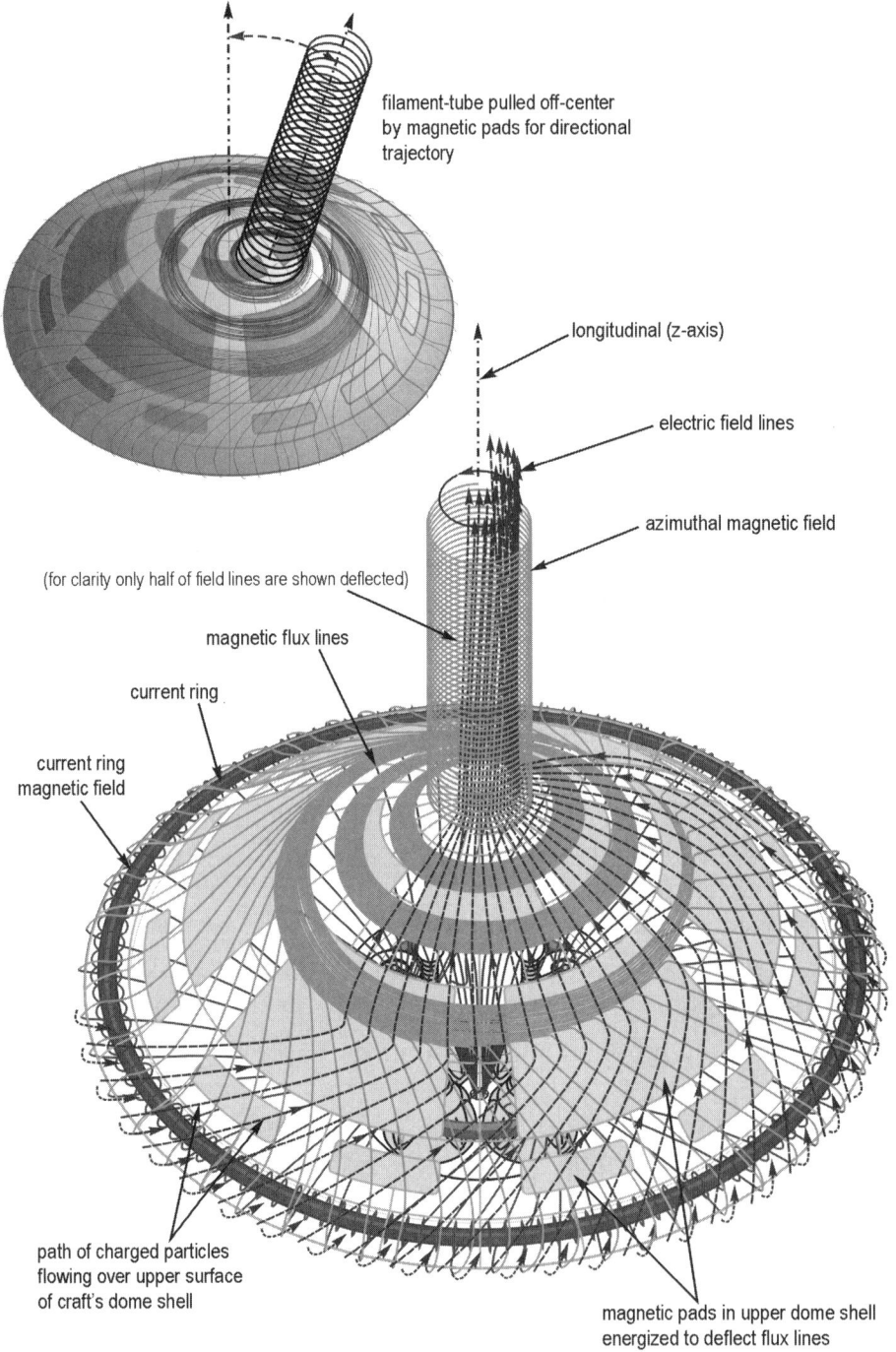

Figure 70. The electric field will extend straight up if the magnetic field lines are spiraled around the upper dome

But because of their difference in radii the effective result will be that whatever electrodynamic force is generated at the outer ring this will be greatly amplified and fed back into the inner ring, and into the black vortex, with the greatest ratio of amplification at the vortex's photon orbit ring because of its smaller radius of curvature than that of the inner ring-tube.

This refers back to something I have already mentioned in a previous chapter (section 3.14), where I point out the three areas of the aaUFO which have the highest intensities of magnetic field lines: and in that same previous section to where it is also pointed out that two of these magnetic intensities are formed at the only exits where energy can flow into or out of the craft. The greatest bottle-neck of converging flux lines will be at the bottom of the central vortex, but another congestion of flux lines will be at the rim, through the whole 360 degrees of the perimeter rim where the toroid's magnetic field meets the current ring's magnetic field; for here all around the rim there will exist a magnetic bottle-neck which will prevent charged particles from exiting through the rim. So what the gyrating storage field and the current ring are doing is strengthening the magnetic field intensity of that bottle-neck at the rim—preventing any energy from being dissipated away from the UFO's engine.

Also, as long as this outer rim field (the current ring in particular) gyrates and has enhanced mass then its momentum will want to keep carrying it in the same direction (ie counter-clockwise) as it flows just beneath the upper dome shell, or in other words, it won't want to reverse and go clockwise, and so it won't allow the magnetic field to suddenly reverse polarity either. The intriguing thing about this circulating current ring will be that because it stores so much electrical energy and kinetic energy (and will have mass-inertia) so it will act like a sort of forward-biased diode-capacitor while its plasma circulates around the gully, and so it will dampen any sudden gains or discharges occurring within the UFO's power circuit. Indeed, the faster the storage field and the current ring circulate around the craft, the more they will help steady all the electrical fields in the craft's electrodynamic generating processes. This factor will be particularly beneficial when the vortex emissions do begin to occur, when propulsion energies are ejected, so as to lessen any back-emf effects that might work back toward the black vortex to try and 'pull out' that central vortex. That there are so many of these damping mechanisms dispersed all around the aaUFO's electrodynamic circuit it is obvious that they have been specifically designed to counteract an electrical ejection mechanism which regularly discharges packets or pulses of energy.

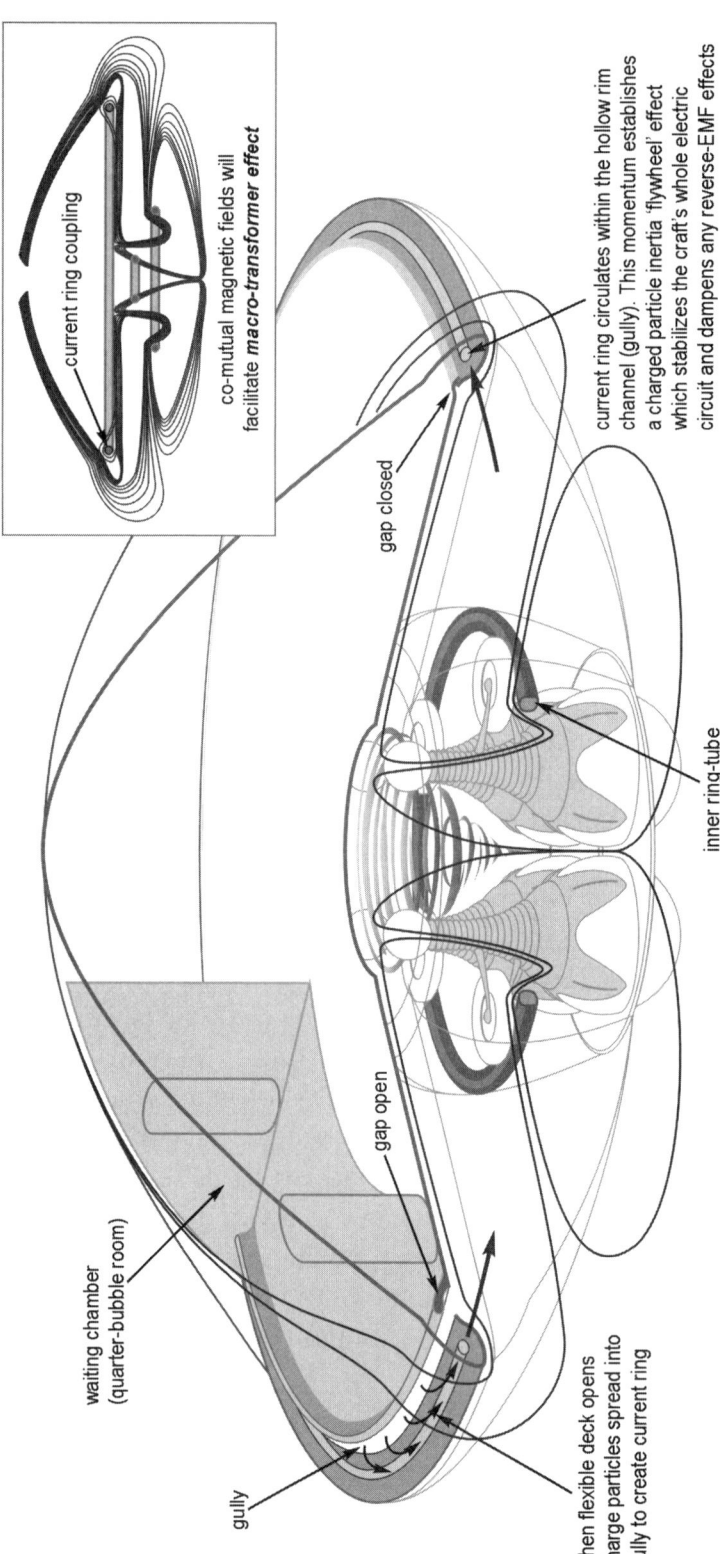

Figure 71. The central vortex, the inner ring-tube, and the outer rim's current ring will all be inductively-coupled together—this setup will dampen the reverse-electromotive forces generated when the wave-packets leave the craft's rim

9—Superradiance Amplification inside UFOs

So the extraterrestrials are putting to good use here a sort of 'flywheel effect' into their electronic circuit, and this is a trick our electronic designers (on earth) haven't cottoned onto yet, because our designers only work with linear current flows and can only use capacitors to dampen their back-emf flows, but the extraterrestrials by using rotating energies are using that energy's momentum, or inertia, to produce their damping effect [note: 1].

Another benefit to be gained from the gyrating motion of the stored energy in the planar waveguide will be that the faster it spins inside the craft, and the more it's magnetic field will induce energy into the current ring circulating just above it, the more this current ring's magnetic field will intensify through the upper dome and grab at and 'pull-down' on the flux lines that span over the upper dome, and control the helical-tube producing magnetic field which extends up to the apex of that UFO (see figure 70). Meaning that when these dome flux lines are needed for the next stage of the propulsion drive process they will be pre-intensified around the rim (and made more ready to be formed into a filament-tube around the dome's apex) so that whenever any charged particles do manage to exit from the rim, they will be forced around the rim, and by this same magnetic field will be forced to curl up toward the dome's apex.

Furthermore, with the inductive-coupling established between these two outer rim fields and the inner ring-tube field it is feasible that one ring could influence the other electrodynamically. This would mean that control of the magnetic fluid flowing through the inner ring-tube would control the power in the outer rim fields, which again, furthers the notion that the inner ring-tube will be the primary controlling mechanism for the craft's power amplification, and consequently would form the primary mechanism which controls the UFO's acceleration. Again, if this assumption can be maintained throughout all of this study then the whole craft will STILL be under the control of ONE solitary transmission line—and I see that as being very impressive!

As an aside, I did wonder if a commercial form of electronic capacitor could be devised and manufactured to take advantage of this hitherto unexploited 'charge inertia flywheel' principle... perhaps even as an electricity generator... using a gravitationally and magnetically charged core orbited by a field of free-electrons. It might then be interesting to conjecture that this anisotropic energy storage principle might even be extended all the way down to Planckian rotons (and to work on Higgs bosons, perhaps, too).

[Author's note: As you ask—so you shall receive! Just as we were going to press I came across an article about a crystal-rectangle energy-device (CRED) that has been documented as having been retrieved from the crashed Roswell UFO craft back in 1947, and which somehow generates electricity on-demand! According to the Los Alamos input in these DIA documents, this extraterrestrial device was taken into space for testing several times and was finally taken onboard the doomed space-shuttle STS-107 Columbia in 2003, for the purpose of carrying out on it further electronic current tests (it has already had 165 tests done on it since it came out of storage in 1957, apparently). From the details of the device given in the DIA documentation this CRED would appear to be a micro-matrix of rotating rotons assembled inside a chamber permeated by H4 gas, so that the heavy hydrogen protons are being separated from those atom's electrons by the localized gravitational forces of the rotons. If this is correct then this extraterrestrial CRED could indeed be seen as a microscopic version of the above 'charge inertia flywheel' principle. The supply-on-demand part of the CRED would come from the electrical load mobilising the electrons and thereby contorting the matrix to a variable degree, and producing small amounts of displacement current inside a variable number of affected atoms, so that when all those individual displacement currents all accumulate together they produce an electrical potential which exactly matches the applied load. Sure beats what Nikola Tesla came up with!] [note: 1a].

9.2 White Wall Repulsion Force Inside Rim

It has been well established through the previous sections of this study that the standard model of the astrophysics ergosphere is readily discernible in the aaUFO's design, in both its inner (ergosphere A) and its outer ergoregion (ergosphere B). But further, in accordance with the Schutzhold-Unruh 'gravity wave basin' delineation of a black hole's rotating amplification mechanisms, the aaUFO's magnetized rim fields of charged particles gyrating just inside the rim, by them preventing any energy fields from passing through the rim duct, those rim fields will act very much in the same way as that Schutzhold-Unruh model's white hole reflecting horizon which also prevents energy from passing beyond it.

For in astrophysics there is a white hole phenomenon which is the exact opposite to the black hole, and whereas in the black hole nothing can escape after passing over its event horizon, in the white hole nothing can penetrate beyond its event horizon and progress any further. Meaning that around the event horizon of a white hole all the incident waves (of gravitational and electromagnetic energy) pile up against it, or, as would be the case inside this UFO where those waves will collide against the magnetic bottle-neck just inside the craft's rim, they will bounce away from that white hole [note: 2].

Furthermore, in the case where the white hole is clearly a circular feature outside of a quasi-black hole, and being used in accordance with a magnetic bottle-neck, which is what is being established inside this aaUFO, then the white hole is more akin to a circulating reflecting wall and the gyrating charged particles constantly being driven against it would effectively be empowering that circulating wall—and would forcibly prevent any energy from escaping beyond it out of that

UFO's planar waveguide. So, there is an apparent contradiction here, in that the more the rim fields are accelerated the harder it will be for any of the energies from the gyrating storage field to pass through that white wall and exit through the circumferential duct!

This rather unexpected mechanism (the seeds of which, the reader might recall, were referred to previously in figure 36 of this study) is further emphasized by the strange characteristics of this planar waveguide, because while the gyrating storage field would be made to accelerate through it near the center of the craft, that storage field is clearly made to drastically slow down at the outside perimeter of it and to curl into a circular orbit inside the craft, at the outer edge of that planar waveguide—and this would appear to be counter-intuitive. But, then again, with a more appropriate consideration this apparent contradiction will prove to be another master-stroke in extraterrestrial engineering design.

The most telling clue to what the ETs are doing is in the shape of the planar waveguide where it meets up with the upper dome. The reader will notice the odd-looking profile of the upper deck where its shape curves down to form a sudden contraction at the outer end of the planar waveguide (just inside the circumferential rim), this is known in fluid dynamics as a vena contracta. Such a contraction is used to decelerate the flow of a fluid, as will also be the case inside this UFO with its gyrating plasma [note: 3]. In this case the contraction coupled with the magnetic bottle formed around the rim collectively creates a sort of 'one-way pressure valve' and establishes a substantial power-pressure-threshold which will have to be surmounted by any waves of energy before they can eventually blast their way out of that UFO's rim duct.

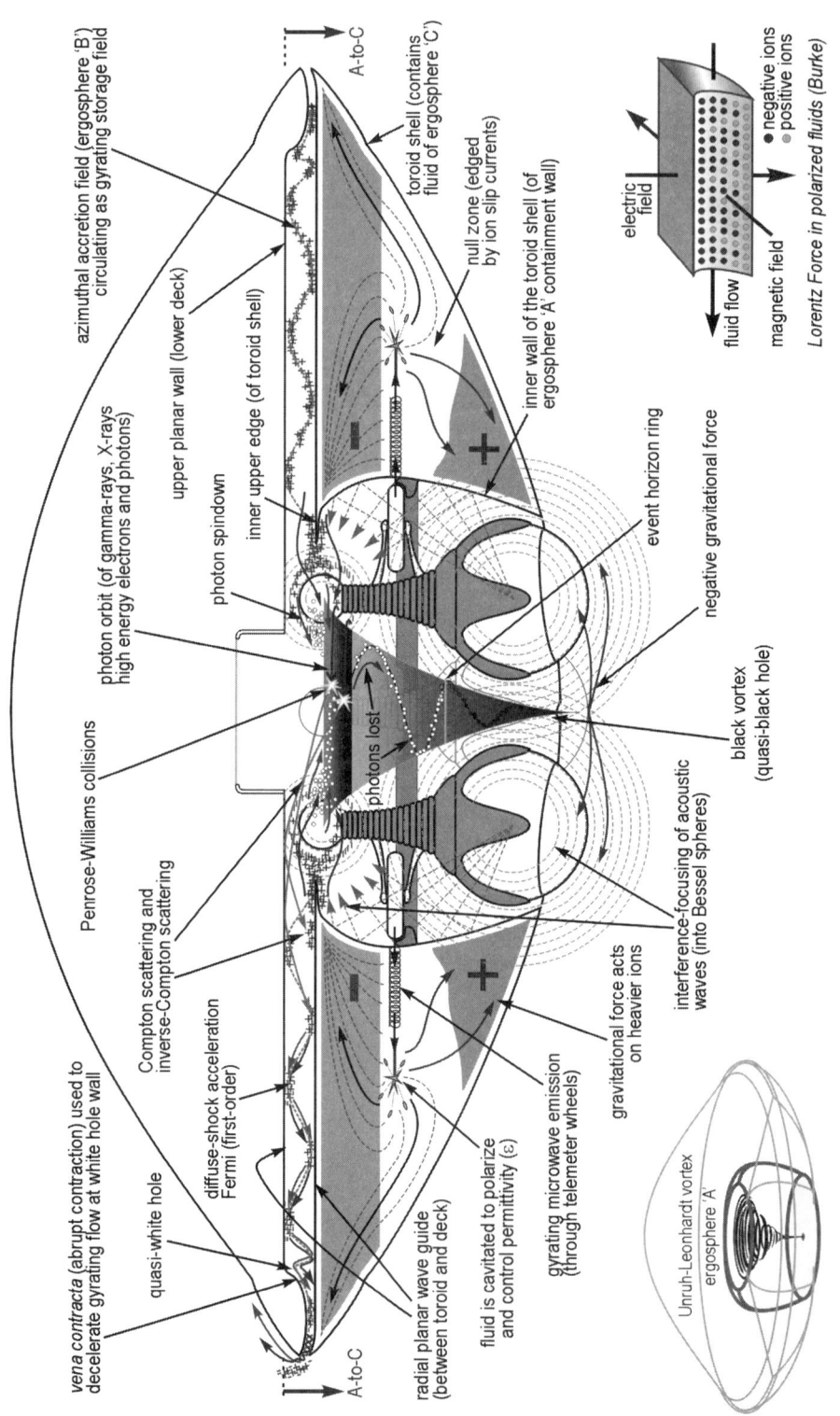

Figure 72. Many mechanisms will generate spontaneous emissions especially in the planar waveguide

This power-pressure mechanism will tend to bunch up the forces of those waves of energy as they gyrate into it, and this will bunch them as a group of force-waves, and as will be seen throughout the next chapter, this will be one of the ways in which groups of waves (or wave-packets) of energy can be amplified at this reflecting white wall. Because, very cleverly, by the current ring empowering the very white wall which those wave-packets of energy will have to burst through, so the more powerfully those wave-packets will have to be amplified before they can be ejected through that white wall and out of the craft (see figures 25, 69 and 70). This won't happen out in space for an astrophysical black hole system because it won't have any energy containment walls, and so, again, this will greatly improve the overall efficiency of the UFO's system over that of the astrophysical black hole.

9.3 Fourth Amplification: Superradiance

When those energy-packets of force don't pass through the white hole wall, from not having acquired a high enough power level (amplitude), they will simply be reflected as waves of energy away from the white wall and back into the central black vortex, this will be for both electromagnetic energy and gravitational waves which will strike against the ergosphere's white wall and bounce back again into the vortex. But then, immediately those waves shrink back into the center of the UFO the photon orbit ring spinning around the outside of the black vortex's event horizon will in return smash those waves of energy back outward again, significantly amplifying them with extra rotational force and extra mass in the process (through the Penrose-Williams amplification mechanism) [note: 4].

So then the gyrating storage field, which was initially described as a field of charged particles circulating around the planar waveguide, consequently will become much more energetic from these rings of waves rippling through it, as both from when they expand out of the vortex, and after they too hit the rim's white wall and are imploded back inward again. Because while these wave-packets bounce backward and forward through the planar waveguide these energy waves will take up more and more angular momentum from the black vortex and they will pass that rotating energy into the gyrating storage field, to increase its angular velocity; and so the charged particles in the storage field will expand and gyrate more and more energetically away from the vortex too (and the harder they collide against the white wall the more they will energise its reflecting force, and each time add a greater amount of impenetrability to that white wall's power-pressure-threshold). Similarly, because the returning waves of energy in-flowing into the vortex will be curling down into a smaller radius of curvature every time, and through a converging magnetic field, then the vortex too will continually absorb more and more energy from this ringing action and will continually be amplified. Essentially, this amplification process will ensure that the black vortex will actually GAIN energy immediately after energy is radiated away from it [note: 5].

In fact, this can be directly linked to a known amplification process in astrophysics, one that was seen to happen in the acoustic equivalent of the Hawking black hole radiation mechanism, where emissions were discovered being produced around the supersonic flow of a black hole, and the salient factor of this type of mechanism—which would be transferable to a vortex that constitutes a quasi-black hole placed inside the middle of an ergosphere which surrounds it as a reflecting wall—is that this mechanism will dictate that any energy waves entering into that vortex will be reflected outward from it at an increased amplitude, and that as those energy waves will be continually reflected back-and-forth into that vortex they will be amplified again and again [note: 6].

The basic phenomena of bouncing waves of energy in and out of a rotating black hole and having them bounced back from a curved mirror-wall of particles which surrounds that black hole, was first proposed by [Press-Teukolsky] in 1972 to explain why vast amounts of energy surround a black hole out in space. Importantly, this was the first study to tackle how an energy wall surrounding a black hole could be visualised as one that could reflect energy back into that black hole so that, instead of being purely seen as an absorber of energy the black hole could then also be seen as an amplifier of energy [note: 7]. This black hole amplification mechanism was then-after known in astrophysics as superradiance. But I would say here though that it would be the Schutzhold-Unruh model of a black hole's superradiance mechanism that gives a more appropriate description of the inner structure of the aaUFO with respect to it possessing both a quasi-black hole and a quasi-white hole [note: 8]. Obviously the UFO's mechanism is a miniaturisation of the astrophysical, and while miniaturised versions of black holes are readily accepted in astrophysics, and laboratory-sized versions are presently being developed, I would think the aaUFO model will rival them all in every department because of the purposeful ways in which its superradiance ringing is channelled through the UFO's other energy fields.

Indeed, within the aaUFO, looking at what the to-and-fro action of superradiance does overall to its collection of disparate energies and oscillating wave radiations, the aaUFO's superradiance amplification is a process which drags with it all those disordered radiation and energy emissions emanating from its various energy mechanisms, and coalesces them all into a more unified and rhythmic coherence of electromagnetic (and gravitational) energies, that work together like a single field to regulate and harmonise all those different sources of energy. With this sort of energy coherence it then becomes much easier to extract energy—from almost anywhere in that energy system—and take advantage of its coalescence of powers. With such a superradiant power-drive for instance any amount of rotational energy would be on hand to be extracted and used in the craft's propulsion drive, especially if it were released in short pulses [note: 9].

One particular mechanism that the extraterrestrials have engineered to take advantage of this superradiance ringing in and out of the aaUFO's black vortex, which offers one of the attributes of this craft's rhythmic coalescing mechanism, would be where each bouncing ring of waves will have to pass through the focused acoustic orbs of excited energy emissions surrounding each of the top spheres: For each ringing of waves will collide into those four orbs and pick up additional charged particle energy from every one of the collisions they make with the particles flying around those energy orbs. Indeed, from the fact that these orbs are being constantly supplied with energy, which comes ultimately from the ambient energies outside the craft (the ZPF fields) through the lower spheres, then this, again, is another one of those master-strokes of design-genius. Because this ensures that the UFO's black vortex will take in energy from a 'fuel-tank' which is wholly inexhaustible—that of the sky, or the open space, which surrounds that UFO [note: 10].

Another point to make is that the emphasis with the UFO's superradiant energy amplification mechanism will be that it will keep on pumping up the UFO's energy until it reaches the white wall's pass-not threshold, and obviously, once that threshold is surpassed those energy waves will have increased their power enough to burst through the bottle-neck areas at the craft's perimeter rim, to work the UFO's propulsion. This will form the basic amplification mechanism and ejection mechanism for this aaUFO's propulsive forces, but I wonder though, might it be advantageous to engineer into that UFO's rim a series of special weak points so that these ejections of propulsive forces might leave the craft in short *controlled* bursts? If it could be precisely controlled the rim would provide a way of determining exactly how these forces leave the UFO, so then, perfect control could be afforded upon accelerating and directing that craft. And this will be looked into through the next chapter.

The only other equivalent in the studies regarding astrophysical black holes, to this swing-back-and-forth energy field amplification, would be the twisting of magnetic field lines; twisting them up so much that when they are released they snap back (like a back-emf force) and release a powerful shock-wave of force which, in the case of astrophysical jets coming away from black holes (such as in quasars and pulsars), would transport energy through that jet at superluminal speeds [note: 11].

The technology behind superradiance though dictates that certain parameters will need to be established within the environment of the ergosphere(s) which surround the black vortex. The craft's energy reservoir will grow exponentially and the wave's amplitude will be amplified to the tune of superradiance inside a UFO—but only if that UFO's reflecting wall's radius from its vortical core is a certain critical distance from that core or is of a certain dispersive composition. If that distance inside the UFO is smaller than what is required for this effect to occur (and it most certainly will be) then the medium of that UFO's reflecting wall will have to be modified in such a way so that it will SLOW down the velocity of the electromagnetic and gravitational waves which need to be reflected back through it into the craft's black vortex [note: 12].

For instance, the most suitable volume of material which can be used to slow down these waves would be the fluid flowing inside the aaUFO's toroid shell because it completely surrounds the black vortex, but that fluid will need to be modified in a way which will make it dispersive before it can slow down these waves as they pass through it. Only then will it produce the correct mirroring effect and will reflect those waves back into the central vortex at a low enough velocity. This, by anyone's standards will be a difficult challenge, but then these extraterrestrial physicists do have a penchant for advancing science!

9.4 Modelling an Ergosphere for a Slower Light Speed

The optimum radius for this ergoregion's reflecting wall (with an appropriate modification to the electrical permittivity of its toroid fluid to make it more dispersive) can be calculated to determine both the frequency-range of electromagnetic waves (and the gravitational wavelengths) that it will effect most, and the power amplification factor it will produce. This gain coefficient will be relatively small for each pass of the waves through the reflector but because the amplification cycle will be repeated over and over again it will compound that gain coefficient very quickly. This superradiance amplification process has been formulated and calculated in several studies for an astrophysical ergosphere, in particular where the intention is that a miniaturised model of a black hole undergoing this process can be reproduced in a laboratory environment. In one study of this exponential amplification process [Cardoso] for a large-scale astrophysical black hole it has been shown that an incident wave field's amplitude can double every 0.6 seconds, and its energy can grow to 10^{14} (one-hundred-million-million) times its initial content—in just 13 seconds! Although, a black vortex undergoing the same process with a smaller degree of mass than an astrophysical black hole (such as that of the aaUFO's) would respond to its exponential amplification in a much shorter timescale than the astrophysical's (ie it would be almost instantaneous) [note: 13].

The slowing of electromagnetic (and gravitational) waves would be done by modifying the permeability ($\square$) and permittivity ($\square$) of the material surrounding the aaUFO's central wave generator, and significantly raising, in the case of the aaUFO, the permittivity ($\square$) of the outer jacket of its reflector (which would be the toroid) that those waves will have to pass through and reflect back through. This would follow the [Bekenstein-Schiffer] model where the rotating source is surrounded by a high-permittivity jacket which is then surrounded by a cylindrical reflecting cavity [note: 14].

So, with reference to the telemeter wheel beams mentioned above (in section 6.6) which are fired into the toroid's fluid to modify its permittivity, while on the one hand, the permittivity of a material (or in this case the toroid fluid) determines how it will polarize in response to an electric field, and so the higher the permittivity the more electrically orientated it will become and the more electric charge it can hold before it conducts away that charge, there will also need to be considered for that fluid what its dispersion relation will need to be in order for it to contribute favourably to the superradiance process.

In any dielectric material when an electric field is applied near its ionization threshold it excites the material's atoms in a way that increases the permittivity of that material. But also the effect of the magnetic field running through the middle of the toroid will induce additional ionizing effects on the neighbouring atoms in the fluid which will substantially increase both the permeability and the permittivity of that fluid [Kouropoulos]. In the case of the aaUFO's toroid, ideally, what would be required would be that the middle part of its fluid should have a high permittivity to act as an insulator, and at the top and bottom sections where the fluid is required to be polarized (into positive or negative polarities) it would be best if these charges don't conduct or leak away but that they simply accumulate. The latter prerequisite would fill the corner-edges of the toroid with polarized charges, negative to the top edges and positive to the bottom edge (as alluded to in section 6.10), and it will be the diffuse charges induced by them on the outside surfaces of the toroid walls that will be doing all the work for the UFO's electrodynamic circuit.

Really, the toroid should work not as a conductor of electrical charges but as a capacitor of electrical charges, and the more intensified and the more concentrated the top and bottom polarized zones are the better, to ensure that the middle neutral part of the toroid fluid should be as wide as possible so that its high permittivity factor can be utilized to the greatest advantage in the superradiance reflection process. This is why I have suggested that the fluid inside the toroid should have as its base plain water [note: 15].

Studies have shown that the permittivity of water varies with respect to temperature and with respect to the frequency of the electromagnetic field applied to it. One study for water indicates that a permittivity peak corresponding to an applied frequency of 9.6 GHz will occur between 20-40 degrees Celsius (68-104 F), while at around 0 degrees Celsius water's dielectric permittivity will peak at 3 GHz—indicating that water's permittivity peaks are very varied [note: 16]. That proton mobilities are highest in water molecules at around 25 degrees Celsius should also feature into the final equations to determine what frequency the firing beams will need to supply, to effect the most appropriate permittivity and dispersion relation for whatever fluid needs to be chosen for the toroid.

That's if they are to regulate the permittivity directly, the extraterrestrials may be firing these beams into the water just to ionize the water and to regulate the temperature of the water, and relying on the molecules to rearrange themselves through the polarizing process mentioned in section 6.10. If my assumption is correct about the frequency of the beams firing out of the telemeter wheels and that they are indeed as high as 15—30 GHz (conjectured throughout chapter 6) then such beams would certainly ionize the water immediately surrounding those beams (because the water molecules won't have the ability to stay in tune with a microwave oscillation above water's relaxation frequency and so those volumes of water will readily ionize).

Although, might there be something else happening with those telemeter beams, for, in conjunction with the above conjectures these beams may well have more than one purpose, they may also be used at a particular frequency within the above range to instigate a switch in the toroid wall's material from opaque to transparent... thereby performing two tasks through just one single action (the relevance of this conjecture will be explained in section 10.9).

It will also be particularly useful (in readiness for the next chapter's processes) to bear in mind that electrical permittivity and magnetic permeability of the aaUFO's flowing mediums will ALSO determine the wave group/phase velocity of any electromagnetic radiations that will pass through those mediums (especially when that electromagnetic radiation travels through those dispersive mediums as wave-packets), and the reader might need to back-track on that after he or she has read through the next chapter (well, this is after-all UFO science not rocket science—so its complicated).

Because now, the interesting question in the aaUFO is, just how many reflecting cavities does it have? Certainly there are at least TWO stages to the aaUFO's reflecting mirror... the toroid's inner wall (the wall of ergosphere A), and then there is the toroid's outer wall, between which would be its rotating fluid (of ergosphere C); and so all three of these walls might couple together as one reflector. But then there is the white wall of plasma circulating around ergosphere B just inside the aaUFO's perimeter rim where the rim's magnetic bottle is formed, and that will form the other major reflector [note: 17].

Although it is outside the remit of this study to suggest any supercooling mechanisms, by way of comparison and to show just how far electromagnetic waves can be slowed down it might be worth noting that light speed has been reduced to just 17 m/sec through an ultra cold gas [note: 18].

9.5 Moving Fluid Slows Light Speed Even More

Studies based on Zeldovich's original hypothesis for a fluid flow around a rotating cylinder suggest that the minimum radius for the reflecting wall which should be able to slow the velocity of light would equate to a ratio of;

$r_{crit} \sim c/m\Omega$

Where (Ω) is angular velocity of the black hole, (m) is its multipole moment (of angular momentum about an axis), (c) the speed of light).

Therefore, the faster the UFO's vortex rotates and the greater the multipole moment (relative to its inertial mass), then the lower the speed of light (c) will be, and of consequence, the lower the radius r_{crit} will need to be [note: 19].

Ulf Leonhardt has suggested that even ordinary water, if moving, will drag light, and that the light waves will curve away from their straight-line path when they pass through a moving dielectric because that dielectric will appear to those light waves as a curving of space time, and so the curved space-time will act on the light much like a gravitational force [note: 20]. So then the speed of light travelling through that moving medium will depend on the refractive index of that medium and the flow velocity of that medium. Dielectrics used to slow light are known as dispersive dielectrics and their motion will contribute to the slowing down of light's velocity.

Light rays travelling through a slowly moving dispersive media then behave somewhat like electrically charged particles (or an electron wave) passing through a magnetic field, being deflected by it and acquiring a vector potential in that magnetic field. Although the effect on the light waves will be to phase shift them and cause them to interfere (somewhat like a beat frequency). The effect is small says Leonhardt, but just like the superradiance effect, this phase shifting effect by the dispersive medium then multiplies every time the light waves pass through and are reflected by that dielectric [note: 21].

Interesting effects then come with the use of dispersive dielectrics around a rotating source of radiated waves, such as the black vortex. Because of the fact that the rotating source will only rotate one way, obviously, and so it would follow that around the UFO's black vortex, and throughout its central ergosphere, there will be some emitted waves that will be flowing with these rotating fields and some that will be flowing against them. The ones with positive angular momentum will be riding with that rotating flow, but the emissions going against that rotation will have a negative angular momentum, and be slowed down so much they will be trapped by the vortex, or in some cases REVERSED by that vortex [note: 22].

It should be noted in this 73 figure (below), particularly in the top rendition, that the two rays of (light) entering into the toroid would undergo interesting paths of refraction and reflection through that three-arced toroid and its fluid. According to this sketch those rays will undergo destructive and constructive interference, meaning that some of those rays will go through darkened patches and produce an Aharonov-Bohm effect.

Another physicist who might have liked joining in with this discussion is Fresnel who (in 1818) formulated that the speed of light (v) in a moving medium would be;

$$v = c/n + (1 - 1/n^2) u$$

Where (n) is the refractive index and (u) is the medium's velocity. Or, in yet another frame of reference, the velocity vector of light would equate to the velocity of light as it passes through that medium, plus the medium's velocity times the Fresnel dragging coefficient [Leonhardt-Piwnicki 2000c].

One additional factor in this rather unique branch of physics to consider is of when a wave propagates outward only to return. Because, if an electromagnetic wave (or gravitational wave) can be assumed analogous to an optical wave then after it propagates away from the central vortex, when it enters through the aaUFO's toroid wall it will be defracted by that transparent wall and toroid fluid in a CCW direction. And, because of the Aharonov-Bohm effect in a moving fluid the electromagnetic wave will then be further dragged CCW along with the moving flow which will also be CCW (see figure 73). Then when the wave bounces back toward the center and passes into the central ergosphere it will be caught into the inertial frame-dragging occurring inside ergosphere A, and will then be influenced by that flow because it also will be a moving dispersive medium.

In this way for instance, it can easily be seen how sound waves reflected back into the vortex will have their velocities slowed down to much lower than their atmospheric air speed, and become trapped by the rotating fields of both the vortex AND the central ergosphere.

9—Superradiance Amplification inside UFOs

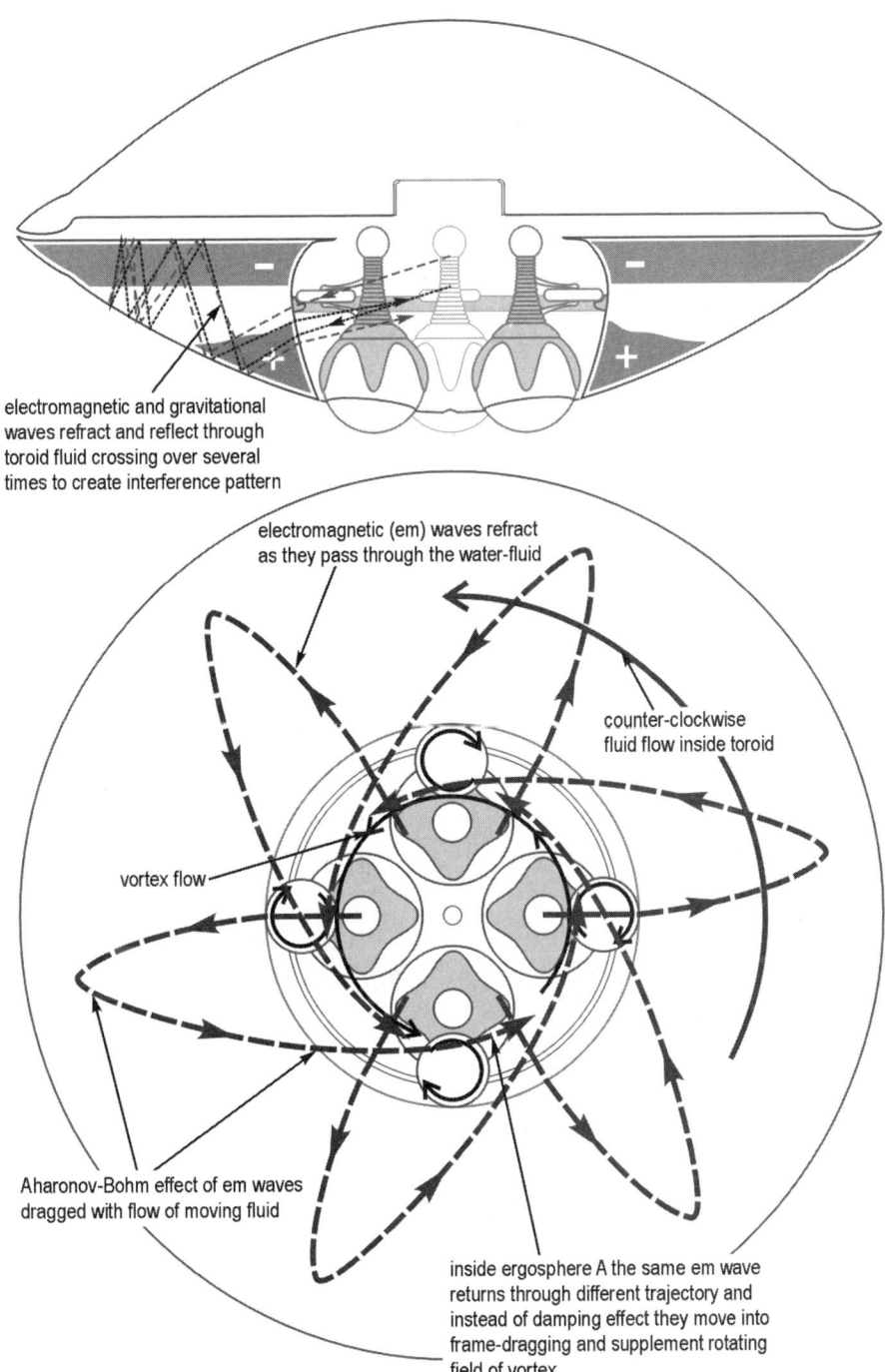

Figure 73. Electromagnetic and gravitational waves radiating away from the central vortex will pass through the toroid fluid and be slowed down, they will also be refracted along different paths by the moving fluid

But then, this craft's toroid shell may have even more to it. Looking at the side view, the electromagnetic wave will permeate into the toroid fluid and after striking against the outer hull wall will bounce UPWARD and continually reflect through the moving toroid fluid, and be slowed down substantially on its way back into the central ergosphere.

Possibly this factor of the dispersion relation of the UFO craft's internal energy fields can be illustrated in the dink-dink-dink episode (given in section 8.7) of the UFO craft when it powered up with a loud swirling noise but then flew away quite silent with just its dink-dink-dink sound. For initially the waves radiating in and out of the craft's black vortex wouldn't have entered into the superradiance mode straightaway, they would have propagated into the toroid's fluid for several seconds before altering that fluid's dielectric dispersion ratio to bring down the rebounding velocity of those waves. Only after this were done could the proper reflecting walls be established and the superradiance have come on song. The dink-dink-dinks being the wave-packets being ejected through the craft's rim.

9.6 UFOs Establish Electromagnetic Anchor Within Their Ergosphere

The main electrodynamic components on the inside of the aaUFO which its charged particles will be affected by are: the 1) magnetic field intensities; 2) the physical walls of the ergospheres; 3) the rim's white hole wall (and its magnetic bottle-neck); 4) the central black vortex; 5) the concentricity of the ergospheres relative to the central black vortex that form rebounding walls for the accelerated charged particles to bounce off; 6) the magnetic field anchor mechanism situated around the outer rim of the craft enabling the toroid's magnetic field to be stretched and twisted by the sphere-set assembly.

While numbers one through five of the above components have already been explained in the previous sections of this study the last of these components has only received scant mention up till now, because it will feature in the secondary superradiance power amplification process (delineated in the following section). That the ETs have utilized the substantial anchorage at the outer rim to such good effect also shows how deft they are at electromagnetic design. To date no study on black hole dynamics has ever established a definitive anchoring mechanism [note: 23].

As can be seen in the various diagrams for this craft (particularly in figures 25, 36, 69 and 75) the electromagnetic anchorage at the perimeter rim of the toroid allows the toroid's magnetic fields to work in a way analogous to a mechanical spring and be used to store electro-mechanical force. And this copies what happens out in space around a black hole, where an electromagnetic anchor is also found. But mechanical springs have to be held at one end so that they can be torqued up—otherwise you can turn the key to try and wind them up from now until the day one of

Searl's rotors come back down to earth and you won't get any torque out of it! In space, the astrophysical 'spring' which resides between a hole and its accretion disk relies mostly on an abundance of charged particles acting like sheets that can't quite follow the free-twisting of that system's magnetic flux lines (as alluded to above in section 2.11), although, it has to be said, this method of anchorage is not the most efficient. But here in the aaUFO craft, its magnetic anchor will be held very firmly in place by the two rim fields and by the power of the white hole's wall of plasma energising the rim's magnetic bottle-neck, and by the inductive coupling between all of the major rings of gyrating energy inside this craft, and so the more those rings are powered up so will the anchor be held fast into that coupling at the perimeter of the planar waveguide.

9.7 Superradiance Plus: Another Plasma Pump (Reconnection Part III)

Quite unexpectedly, within its confined volume the planar wave-guide will assist yet another amplification device: Because of the gyrating motion of the storage field of charged particles, those particles will form into a spiral with concentric ribs within that waveguide (which field will visually look like an uninterrupted circular sheet of vibrant energy), and those ribs will themselves accelerate other charged particles. But when the sphere-sets are allowed to rotate and they pull on the magnetic flux lines of each of the four quadrant fields, because those fields are firmly anchored at the rim, and the subsequent break and reconnection process occurs to each flux line, there will at each of these break-remake events be produced ionizing shock waves that will spread outward from those four break-reconnection locations. These will constitute MHD shock waves and will accelerate charged particles outward from each of these four epicenters to spread through the confined planar waveguide, causing these four groups of ionizing waves to ripple through the concentric ribs of the gyrating storage field. This will generate a continuous frenzy of mini-explosions at each of the four locations that will effectively produce multiple Comptonized collisions within the storage field (and within the diffuse layers of charged particles accumulating on the upper and lower walls of that waveguide—which will act like 'seed fields' to these break-reconnection accelerating forces), which will generate copious amounts of free-electrons to supplement the storage field's highly energised gyrations and will *comprehensively* amplify those diffuse layers.

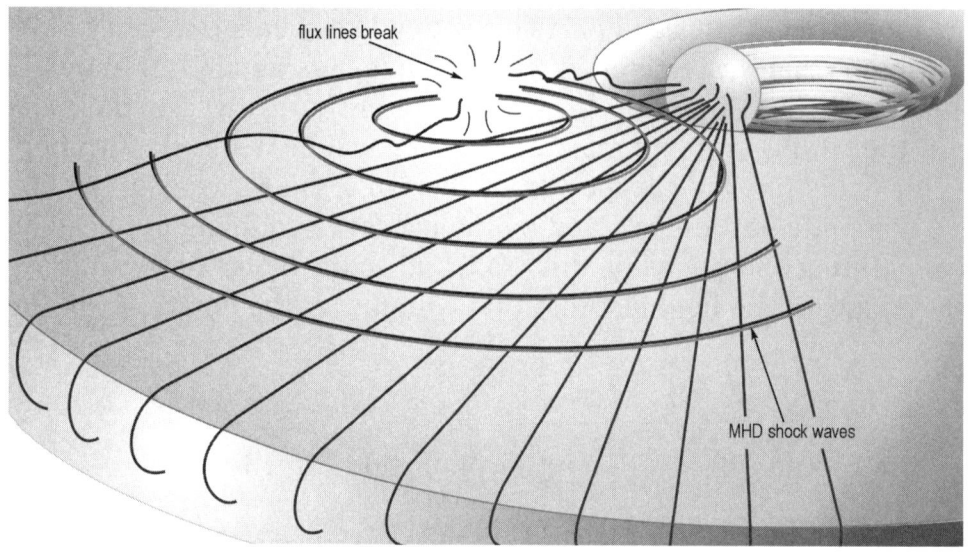

Figure 74. The breaking and reconnection of magnetic field lines will generate pulsing shock waves through each of the four quadrant fields

In the astrophysics models similar break-remake processes causes runaway electrons to be spontaneously emitted and heat to be generated. This is also similar to the phenomenon known as diffuse-shock (and first-order Fermi) acceleration, and which happens around a black hole system where charged particles are forced between shearing plasma fields and are repulsion-accelerated to increase their energies quite substantially. The ribs in this UFO's planar waveguide should be expected to do a similar thing and become particle accelerators themselves because of their inhomogeneous field strengths (from their spiralling radial electric field moving from the center outward). The aaUFO's planar wave guide mimics the very strong magnetic fields which sandwich an astrophysical accretion disk, but it also provides the added factor of confinement within its narrow circular channel [note: 24].

9—Superradiance Amplification inside UFOs

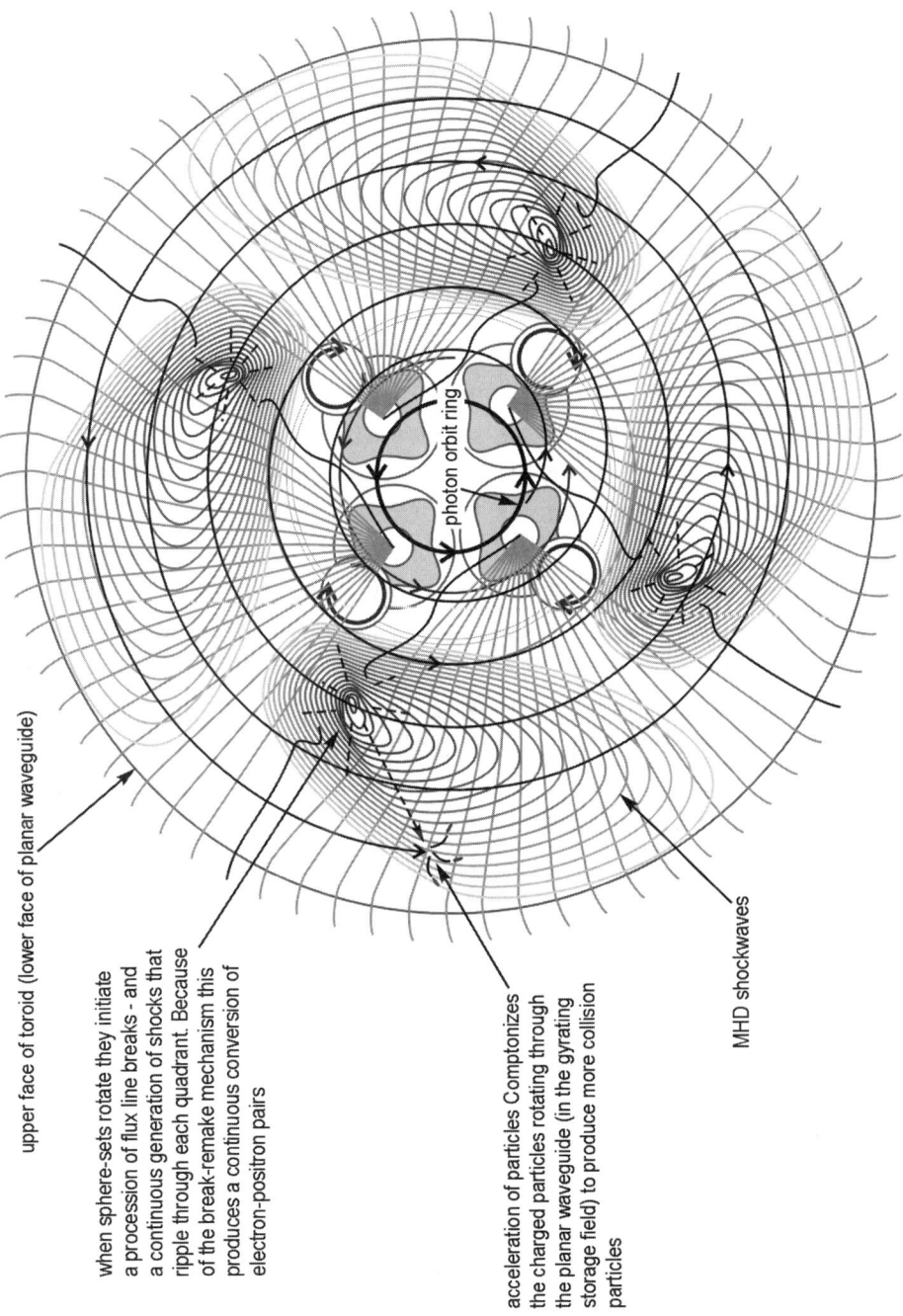

Figure 75. Four quadrant fields will undergo break-reconnection shock-explosions to accelerate the seed fields and generate further emissions simultaneously

Adding to this will be the pulsing (at about 38 Hz) which will come from the beams being fired into the toroid fluid. These beams, when they pulse into the toroid fluid and trigger their ionizing mechanisms, will subsequently induce reactions into the electric fields accumulating (as the diffuse layers) outside the upper toroid facing, which will be exactly where the quadrant fields will be cycling through their break-reconnection processes within the planar waveguide; as a result the shock waves coming from these epicenters will themselves interfere with the energies flowing in and out of the central vortex. Needless to say such a frenzy of collisions around the four epicenters of shock waves will have to be orchestrated, and this will be done through the confining factor of that waveguide, and through the influence of the storage field gyrating through that confining waveguide [note: 25].

This confinement will concentrate the Comptonization effect, for instance, that the gyrating storage field itself will be undergoing through the whole of the planar wave guide (and most especially toward the vena contracta where the waveguide contracts even more), and this will orchestrate the frequecy that the storage field will oscillate at down to the microwave range. As mentioned in sections 3.13 and 3.18, this field will be gyrating inside a waveguide of around 15 cm deep which then contracts down to around 3 cm deep (so, that frequency will be in the 2 Ghz to 10 Ghz range). That they will be doing so in the same place where there will be an intensifying of the rim's magnetic field, then by turning the central sphere-set assembly to pull on these quadrants, there will be provided a tremendous amplification factor in the craft's electrodynamic forces.

Briefly, if the reader can picture at the center of the UFO where the sphere-sets are rotating, and at the outer rim of the craft the gyrating storage field bright and busy with a fast-flowing plasma of charged particles streaking around the craft just inside the rim, and the electromagnetic waves pulsating out of the black vortex in a rhythmic resonance, it might then be possible to visualize the spiralling ribs of the storage field as their gyrations spiral in and then spiral out of the central photon orbit ring. Then too, the four groups of shock waves coming from the quadrants, turning at the same speed as the sphere-sets, rippling their waves through that spiralling sheet of bright ribs with even brighter epicenters, and that's just what it would be like inside this "Andreasson Affair" UFO's electrodynamic engine [note: 26].

When the superradiance begins ringing back and forth between the center of that engine and the outer rim, and because the overall resonance inside the planar waveguide will be greater than 3 GHz, there will automatically occur a polarizing effect between all the charged particles (ions) and the electrons, and this will be because while the electrons have greater mobility these will eagerly flash backward and forward in and out of the vortex, but the heavier (positive) ions will spiral much more slowly through the planar waveguide and sit mostly as a storage field above the toroid shell. And because the craft's gravitational force is carried by its positive ions then that force will be extended outward from the central vortex and reside throughout the UFO's planar waveguide. Furthermore, by extending the vortex's gravitational radiation in this way there can only be one result relative to any nearby mass—and that will be a greater pull on that mass. So, to a certain extent the strength of the UFO's gravitational force will be manipulatable relative to the relaxation frequency of the storage field's plasma. Although, obviously, the strength of the gravitational forces which the UFO craft can utilize for its propulsion will also be proportional to how many times those gravitational force fields are reflected backwards and forwards between the outer rim and the central vortex.

9.8 Stabilising the Back EMF on the Black Vortex

The sudden emission of the superradiant wave packets out of the circumferential duct would in most circumstances attempt to drag everything out of the center of the craft with it, possibly even suck the black vortex out of existence, if it were not for the circumferential rim effecting both a threshold limit on the energy being emitted in each wave, AND its magnetic mirror performing the task of a sort of 'one-way valve' that would act like a reverse-biased voltage diode does in an electronic circuit. Obviously, this illustrates that it would be important to not allow the black vortex to expel all of its energy with these superradiant bursts, and it should show that it would be beneficial to have the black vortex somehow pull back some of the outgoing energy allowing it to recoil back into the black vortex circuit [note: 27].

This recoil would have the effect of generating a back-EMF self-inductance which, as all electronic experimenters will know, leads to a sudden and temporary amplification of a circuit's power, as happens every-time a circuit is switched on and off and is usually regarded as a detrimental reaction because it temporarily loads the circuit components, but in this case it might be a beneficial reaction as it would be instrumental in setting up a to-and-fro oscillating rhythm. By purposely encouraging that back-EMF in this part of the electrodynamic circuit there may then be found a way to exploit this effect so as to continually supply the black vortex with an implosive collapsing force [note: 28].

Ideally both the emission and the back-reaction can be made to couple together in a resonant harmony, where one feeds the other, particularly as this toing-and-froing (or ringings which might be a better term) will be required to amplify the superradiant wave as it bounces between the black vortex and the circumferential rim. With so much emphasis on resonant frequencies, particularly in the lower spheres, the final aim would be to design this circuit toward a resonant frequency that would make all the mini-circuits sing in the same key (metaphorically speaking) even before the superresonance kicks in.

9.9 Did the Greys Mention Superradiance During 1967 Abduction?

In view of my discovery that the "Andreasson Affair" UFO generates superradiant energies, the question might be, were the Greys referring to a superradiance mechanism when the accident took place to one of their craft which involved Betty Luca back in 1967; when they referred to it as a "rotating series of semi-full swing back," mechanism which developed the energy for their UFO craft? And did the scientific investigators which attended Betty's hypnotic regression sessions in the 1980's recognize the technology she was describing?

As I mentioned at the very beginning of this study there were scant few scientific datums offered by the Greys to Betty about UFO dynamics during her abductions; there were a few extraterrestrial terms used by the Greys which were subsequently recalled during her hypnotherapy sessions, but these at the time weren't understood. But fortunately, because Betty's recall—as administered through her drawing ability—has been so good the reader can now perfectly understand what the terms that the Grey's did use relate to, in relation to our earthly sciences, and the reader should now be able to see for themself how extraterrestial engineering works. For those readers who are not familiar with the strange alien terms recalled in the hypnotherapy sessions and published in the *Andreasson Affair* series of books I shall include them here.

Most of the words and phrases that Betty remembers which were relayed to her phonetically by the Greys, were included in the following passage (from "Watchers" pp76-77):

"...and the being says they are purging and lining the cyclonetic trowel." And, "...the being says: 'balancing the oscillating telemeter wheels and leveling.' Ah, I just can't understand some of that."

Followed by:

"...rotating series of semi-full swing back... liquid line? Magnetic rings and the depolarized rim."

The first series of words is explained in section 7.4; the telemeter wheels (which should probably be telemeta wheels) have been featured throughout chapter 6; the semi-full swing back would relate to the craft's superradiance amplifications; liquid line will be the toroid's reflecting fluid; and the depolarized rim is that which is explained in section 6.10. So at last throughout this study, following one observed similarity after the other between the astrophysical models in space and the purposeful designs of the extraterrestrials, this great multiplicity of energy-mechanisms operating at the power center of this aaUFO now make perfect sense!

In this portrayal I have been able to show that this UFO's power center is excitable to the extreme, even to the level where the energy at the heart of this aaUFO will be pumped up to state of metastability, from mechanisms such as the gyrating fields, the unusually high intensity of magnetic fields, and of course the continual generation of angular momentum from the UFO's black vortex. These mechanisms would surpass those found in the astrophysical black hole simply because the black hole has no physical confining enclosures or walls around it with which to form its ergospheres, while the aaUFO has a whole series of ergoregions which are electromagnetically coupled together. Certainly, it can be seen now that the aaUFO's central power-drive area exhibits a continuity of design features that synchronizes numerous oscillating energy fields into a grand cohesion to produce a unifying superradiance of acoustic, electromagnetic, and gravitational forces, all centered around its quasi-black hole atomic engine.

What on earth will it do with those energies next?

10: Gravitational Momentum Field Projections

Contents:
10.1 Toroidal Frustum of Hoop Stress	301
10.2 Gravitational Momentum Sent into a Curved Geometry	304
10.3 Gated Rim Ejects Electromagnetic-Gravitational Wave-packets	306
10.4 Dome's Curved Geometry Compresses the Wave-packets	309
10.5 Sixth Amplification: The Dome's Apex Control of the Tube's Bore	312
10.6 Before Einstein's Cosmological Constant Caper	318
10.7 Repulsive Gravitational Force and Negative Spatial Curvature	321
10.8 Using the Repulsive-Attractive Forces	329
10.9 Chirping the Semi-full Swing-backs into Wave-packets	331
10.10 Perturbations to Vary the Optical Transparency of Spacetime	335
10.11 A Successful Mechanism to Harvest ZPF Energies By	336

Take a ball a long way into space, rotate it to generate some gravitational force to hold it in space (so that it generates some curvature in space-time). Then move a large amount of that space-time curvature a short distance away, where will that ball go?

10.1 Toroidal Frustum of Hoop Stress

Rocket science is all about hoofing huge amounts of blast-force out the back-end of a long cylinder (because our present aerospace engineers like to work only with *linear force* drives, ahem). UFOs work on a system of propulsion which is a little more refined and technical. For they operate by propelling a helical field of a combination of forces ahead of them in a beam, and then by using a natural contraction of those forces, which does all the amplification of them, the craft is pulled toward the leading part of that beam. Therefore, to a UFO the most important field is that which develops forward energy rather than that which develops backward energy.

Analogously then, that beam mechanism which a UFO might require would be something like in fluid dynamics where the plug from the drain is removed at the bottom of a spherical canister, so that when the canister's water rushes down that drain-pipe it would, by a natural process of flow dynamics tighten into a spinning flow that would lend a certain amount of organization into that flow as it followed its course of gravity. Indeed, in a drain-pipe the water will not flow straight down as a cylinder with the same cross-section as the internal dimensions of that pipe; a vortex will evolve so that the cross-section of that draining water will always be much smaller than the diameter of the pipe. By turning this analogy over, by inverting that canister and giving it a magnetic field, and a different course of gravity, it will be seen that a plasma fluid, one which in this analogy would initially hug the outside curve of that spherical canister, would eventually leave the canister's surface as a helical flow.

Fluids then have a tendency of organizing themselves with a high degree of efficiency, and it is that efficiency which needs to be recognized and amplified for the UFO.

It is relatively well known in astrophysics that one of the reasons behind hoop stress, axially-directed stress in a collimated and rotating flow, is that it will be induced from the toroidal magnetic field which surrounds a black hole system and which develops a plasma jet [note: 1]. If the reader inspects the 'donut' shape of the toroidal field its easy to see why this might be the case, the loops of magnetic flux around a toroid will form around a smaller radius of curvature on the inside of a toroid than they will at the outer perimeter, so the intensity of the magnetic force will be greater on the inside, and the bigger the toroid the greater the difference there will be in curvature. Thus, while the Lorentz (J x B) force on particles moving in that magnetic field will be stronger toward the inside so there will be a stress or force directing particles from the outside toward the longitudinal axis of that toroid. This is what the hoop stress is, an axonpetal or axis-seeking force.

A similar stress will result for a charged particle rotating around a converging (non-uniform) magnetic field, because when all those field lines converge radially toward a center then the magnetic intensity will always be much stronger toward the center than away from it. Charged particles in this field configuration will naturally drift centripetally into the stronger part of that field—toward the center.

This is the trick with most all UFOs, whether they have a saucer shape or are spherical orbs or whether they are round upright cylinders (and, indeed, even if they are designed like Searl's IGV), they will all develop some sort of toroidal energy field around their outer perimeters—so that their rotating energies can turn into a central column. Triangles if you look at the images of them are comprised of arrays of circular energy fields (usually five, six or seven), again, these are simply multiple rotating energy fields clumped together to work the ambient around them. In saucer-shaped UFOs the diameter of the disk actually becomes a significant factor in that craft's amplification process, because it will extend outward the diameter of its toroid field, and energy transferred from a large diameter will always spin faster and become more intensified after it curls down to a smaller diameter.

Such a perimeter-extended toroidal field used by a saucer-shaped domed craft I would call a *frustum* magnetic field because its basically a conical-shaped toroid. On the inside of that frustum the curvature will be smaller than on the outside so there will be a magnetic potential-difference. Charged particles will always curve inward toward the axis of a frustum to a lesser radius of curvature through magnetic drift, and indeed, will usually cause that frustum-shaped magnetic field to be stretched (drawn inward towards its central axis), which, at the same time will cause those charged particles to spin faster to facilitate the spinning ice-skater effect around its longitudinal axis of rotation. Therefore the most significant principle with a toroid field of energy, will be that whenever its charged particles become amplified, they will drift toward the inside of that toroid, and as they spin around that toroid's axis all the more faster they will undergo a relative increase in momentum. But, as in the grandiose examples of quasars and pulsars, that increase in momentum energy around a smaller radius of curvature can go nowhere other than up or down along that axis of rotation, and so the rotating particles with their increased momentum extends the toroid's magnetic field up or down the toroid's rotational axis so that the toroidal field sprouts a tube-field (see, for example, the inset diagram in figure 22).

This was the great and wonderful thing about the Schauberger turbines in the 1940's and the Searl IGV craft in the 1970's (see figure 22), what they did, essentially, was generate a toroidal-frustum electro-gravito-magnetic field around them (at their large diameter perimeters), and then they both produced copious amounts of charged particles, in their own unique ways, which they then polarized and immediately fed into those developing toroidal-frustum fields, and when those particles rotated en-masse (and they automatically drifted into the central axis by this frustum effect) they would then collimate upward into an organized helical flow above those craft, to lift those craft up when the particle rotations reached above a critical angular velocity— because as those (positive) mass-particles spun faster and faster the gravitational forces imbued into them would increase suddenly to a greater power level than that of the gravitational force flowing into earth.

The beautiful point to make about rotating fields which contract down to a smaller radius of curvature, is that whatever potential their particles were charged with before they contracted down, those potentials will be carried with them and amplified in proportion to that reduction in curvature radius. Spin velocity will be increased, electromagnetic force will be increased, and so too will those particle's gravitational force (by amplification through confinement).

The reader should note that the more efficient carriers of gravitational force will be ions not electrons, ions with a positive electrical charge—which is something not to be confused with *positive gravitational force* (which is attractive), or with *negative density*, or with *negative gravitational force* (which is expansive), or with *negative radial momentum*, all of which terms will be alluded to more fully through the following sections of this chapter.

And so the obvious point to make here will be that it will be very useful for a UFO to be able to extend such a tube-field outward from its surrounding toroidal field, because then it will have the means by which its electromagnetic and its gravitational forces would automatically become more powerful whenever they are extended through a smaller diameter bore to any distance ahead of it.

This toroidal field mechanism, it should be pointed out, is quite different to the one where a helical field can be generated through geometrical means alone (as I briefly touched upon in section 2.10). A well collimated helical trajectory can be propagated through geometrical effects alone (such as negative radial momentum), and this trajectory can be instilled into the particle's momentum through the pitch-angle of their helicity. This intrinsic type of curling-trajectory enhancement would seem to suit a UFO because it has a specifically curved dome with which to propagate those particles. This was the same theme strongly expounded by Reva Williams in her studies on black hole related quasar and pulsar jets (mentioned briefly in section 2.11), and delineated in the paper by [De Felice-Carlotto]. By using negative radial momentum, the suggestion for a UFO would be that when charged particles do escape out of its quasi-black hole system, because of the negative radial momentum imbued intrinsically into their orbits they would escape (in wave-packets) along a rotating vortical trajectory (filament-tube), and the final trajectory of that flow of particles could be determined by the geodesic-geometry they were given by the devise which propagated them, or in other words, whatever curl they were shaped with as those wave-packets curled around the UFO's dome—so that they would automatically form into a well collimated filament-tube once they left the UFO's dome—without the requirement of a toroidal magnetic field (although in section 10.5 below I will conjecture that the aaUFO may combine BOTH of these different mechanisms) [note: 2].

10.2 Gravitational Momentum Sent into a Curved Geometry

Earlier into this study (in section 3.3) I alluded to the Thorne-Sagan model of an anisotropic stress and gave some examples that could be exploited in order to provide abundant sources of energy. I also hinted that such gains can already be found in the fundamental laws of electricity—if only an appropriately designed mechanism could be made to exploit it. The same can be said of the spinning ice skater effect with regard to a field of mass-particles imbued with an electromagnetic force coupled to a gravitational force and which has angular momentum, when those particles become influenced by a non-uniform radial magnetic field by conservation of energy they will rotate with a curvature of reducing radius, causing that field of particles to rotate faster and take on more mass, and so the gravitational forces that field is carrying must also become more intensified [note: 3].

Where this factor relates to the filament-tube being derived from a combination of electromagnetic force and gravitational force the rotating electromagnetic field component is that which provides the circulating field (together with its axis-seeking angular momentum), which forms a collimated helical filament-tube ahead of the UFO. It is then this collimated helical filament-tube which focuses, or rather compresses, and propels the gravitational force component of the pulsed wave-packets (of gravitational and electromagnetic forces coming away from the craft's black vortex), into the distance when those wave-packets are pulsed out of the craft (out of the rim) and up the dome. Correlating this with the above, is also the factor, discovered through astrophysical studies, that the rotating geometry of a filament-tube will drag into it any surrounding magnetic field lines in just the same fashion as it would drag into it any charged particles—so then there would result an additional Z-pinch force, as well [note: 4].

Obviously, as the bore diameter of the filament-tube decreases the mass-gravitational force will increase in inverse proportion [Mirza-Saleem]. And, by borrowing from this latter study for a black hole system the calculations to find the gravitational force on rotating bodies of various diameters (and using those calculations on a UFO with a 30 feet diameter rim utilizing a 1 foot diameter filament-tube), the ratio of gravitational force acting between them will be in the region of 1 to 900, so with this sort of gravitational force amplification ratio the craft will be forever falling toward the leading end of its filament-tube to wherever that tube is directed (and please note I haven't written falling into its filament-tube—for this is not a wormhole that the craft traverses THROUGH such as in the Thorne thought-experiment) [note: 5].

As an aside, I think this is where the advocates of gravito-magnetic forces come unstuck, because in astrophysical studies of jets of particles imbued with gravito-magnetic force those particles are treated in those studies more-or-less in the same way as ordinary electro-magnetic particles (ie positrons-electrons); so that only their electrical and magnetic qualities (through the Lorentz force) are formulated into their equations: And so, by those advocates only working on Lorentzian principles it is always assumed that these particles get propelled by electromagnetic forces and are *pushed* down a jet's filament-tube. But as I have mentioned in chapter 2 and will explain in the following sections of this chapter this isn't the full story with gravitationally-bound particles, there can be an additional factor of a tensile gravitational entrainment force being established by the wave-packets of those particles traversing through an astrophysical filament-tube, which would automatically equate to a negative-resistance factor to these filament-tubes. This is because, particles bound with gravitational forces don't work like individual electrically charged particles, if a particle is being ejected out of the black vortex to which it is gravitationally bound (for instance, through the Penrose-Williams hypothesis) then it will be gravitationally bound to the center of that craft: So if that particle is ejected out the rim of a UFO craft it will first of all want to hug the dome as it spins around it to the apex, but when it moves to any distance from the craft it will still have established between itself and the craft a force of attraction to that craft.

Therefore, a whole new set of circumstances needs to be envisaged by anyone wanting to understand what happens with gravitational forces, particularly when considering what would happen where the filament-tube is squeezed down to a smaller bore diameter, so that the effect would be to greatly intensify the particle's attractive gravitational force: But, what would also need to be considered is that in a well-collimated cylinder of electromagnetic energy and angular momentum there would be only one direction in which a wave-packet of similar such particles could go to expel its own extra accumulation of kinetic energy—and that direction would be down the tube [Sushkov-Khriplovich] [note: 6].

So once the aaUFO's filament-tube of electromagnetic and gravitational force is collimated and stable (see figure 76 below) then a gravitational differential will be set up between the projected wave-packet of amplified gravitational force being accelerated down that tube and the mass of the craft, to which that gravitational force is bound (which craft, it should be borne in mind, is already being held in mid-air by the black vortex's own radial gravitational field which is neutralizing the earth's gravitational field and giving that craft buoyancy).

It might be tempting to point out here perhaps, that by understanding how a UFO can amplify its gravitational force through a negative-resistance, and by it developing a tensile-stress along its extended filament-tube, it may serve to explain the hitherto unfathomable mystery apparent in many astrophysical studies as to how charged particles can travel at relativistic velocities through space—seemingly without depreciating any of the energies they were originally propelled with from their black hole cores—to form the incredibly long jets that span for millions of miles!

10.3 Gated Rim Ejects Electromagnetic-Gravitational Wave-packets

In figure 76 I have arbitrarily given the rim five ejection gates, and these five gates would give 72-degree steps by which the craft's trajectory could be manipulated through different directions when airborne (the number and positioning of the gates is unlimited, but for ease of explaining how this directional system works I've plumbed for just these five gates). All that would need to be done, so as to allow a wave-packet of the gyrating forces from inside the craft to be ejected through the rim, would be for the upper rim's current ring to become slightly up-routed away from the outer edge of the rim, pulling it up just enough for the wave-packet to force its way through the magnetic bottle-neck and through the circumferential duct. The plasma will still be circulating through the ring generating its orthogonal magnetic field, but by lifting the ring's flexible channel fractionally and momentarily it would take away some of the strength in the rim's white wall of repulsive force, weakening just one section of it to provide a gap in the rim's magnetic bottle-neck. This would allow some of the power in the gyrating energy fields, being pumped up in bursts from the superradiant oscillation process inside the craft, to squeeze through the rim as a short pulse of high-powered energy so that when it rounded the rim tangentially it would then curl up into the magnetic field of the upper dome. This is why I have divided the rim into five gate sections because it would only need one gate's width of the rim's magnetic bottle-neck to be weakened to provide a big enough gap for a short pulse to pass through (while the rest of the rim would still retain its magnetic constriction and the rest of the white wall would remain intact).

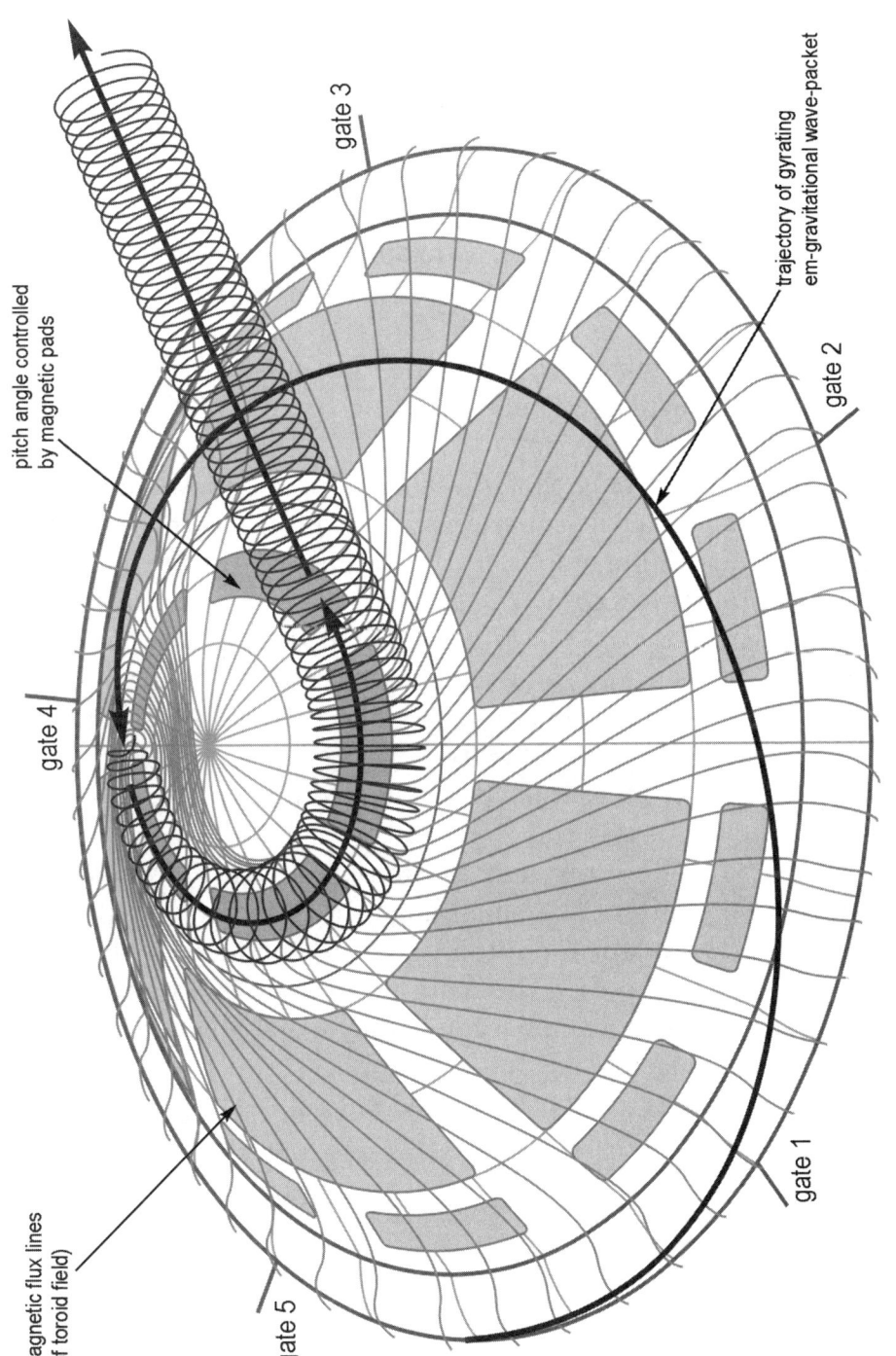

Figure 76. Wave-packets of gravitational energies and angular momentum will curl around the upper dome and into a helical filament-tube

The degree of arc employed by each gate, and the time-duration that the gate is opened for, would allow for the particular bandwidth of wave frequencies that the ejected wave-packet would be comprised of, and a quantity of electromagnetic-gravitational plasma, to be ejected out of the rim. The full 72 degrees of that gate need not be used, but obviously, the greater the proportion of that gate being opened, along with that gate's time period, the greater will be the amount of energy that passes through that gate. The craft's acceleration then, would be controlled by the length and intensity of the energy wave-packets ejected through the rim, although strictly speaking this mechanism will not be the primary mechanism which generates those wave-packets, but what it will do is work in concert with the craft's internal superradiance-pulse mechanism which will provide the main structuring for the energy wave-packets (as will be explained below).

However, during the time-periods that the rim fields are weakened at these gates the passage of those wave-packet emissions when they leave through the rim would momentarily strengthen the magnetic fields around the rim (from their own Lorentz force generating a similar force into the current ring—see section 9.2), which will cause those emitted particles (mostly protons, positrons and positive ions) to turn around the rim through a smaller radius of curvature (as a result of the extra 'pull' generated in the pull-down field, see figure 25), and this will result in a certain amount of synchrotron radiation being emitted from those particles through curvature scattering [note: 7]. And as this principle would work the same way through a wide range of power levels the synchrotron radiation emissions would range from lower energy red-yellow through to blue-white for the higher energy levels. Such radiation will be emitted at a small angle of trajectory off the curving wave-packet fields, so these synchrotron photon radiations will appear to leave the craft's rim in a more radial direction than the main propulsion energy (but because these photons will loose energy in a few milliseconds they won't go very far, and so as the main wave-packets are being pulsed out of the UFO's rim these synchrotron emissions will appear visually to be a brightly colored field of fleeting energy flashing *tangentially* from the craft's rim—as was captured in the video footage of UFOs over Mexico in 1991-1993, incidentally).

This rim-gating facility on the Andreasson UFO will, unfortunately, mean that our pilot will now have one extra transmission line added to his or her control panel and another dial to turn. That tally of transmission lines for the aaUFO is now fully up to TWO!

10.4 Dome's Curved Geometry Compresses the Wave-packets

One might think that in the various designs of UFOs what constitutes the upper dome is merely used as nothing more than a curved surface that keeps the elements away from the crew. But I learned very early into this UFO field of study, through back-engineering the Onion-drive UFO with Jeff Savage in particular that nothing in a UFO's design is perfunctory, and indeed, using the Onion-drive as an example, as well as the depictions of the UFO that abducted and returned Travis Walton (as rendered graphically by [Etreed] on his website), both of these types of UFO can be seen to develop fields of electromagnetic-gravitational and angular momentum forces in drive systems that utilize a specially curved dome of a specific radius and construction (of variable magnetic permeability material), so that their helicon energy fields generated within those craft can butt-up against, and interact with, the underside of that domed surface, to produce an induced energy field on the outside of that material, through a co-mutual magnetic field which develops on both sides of that domed material. And with this Onion-drive type of helicon energy field production it might be a good candidate for the drive-mechanisms used in other more unconventional designs of UFOs, such as the upright cylinder UFO crafts and some of the smaller sphere-orb crafts [note: 8].

What the upper dome does in the aaUFO craft is provide a curved geometry for the outside of the craft which converges the magnetic flux lines and draws them into the apex above the craft, so that there will be established a large magnetic potential-difference between the craft's rim and the apex. For, as the charged particles curl round to the apex these flux lines establish themselves into an inhomogeneous field which strengthens toward that apex, this was the point I was making in section 10.2 above, that the charged particles will be influenced by a non-uniform radial magnetic field (by magnetic drift) to always rotate into a curvature of reducing radius. This is what the magnetic flux lines are doing in figures 76 and 77 all converging toward the apex adding together greater and greater magnetic (Lorentz) force. Subsequently, when the wave-packets leave the rim as elongated pulses of energy waves they curl around the curve of the dome and by turning through an ever-smaller radius of curvature, through the converging magnetic field, they accelerate toward the apex at an ever-faster pace (which in turn induces more Lorentz force into that magnetic field ahead of them to intensify that converging magnetic field—thus the magnetic field covering the dome will become pumped up).

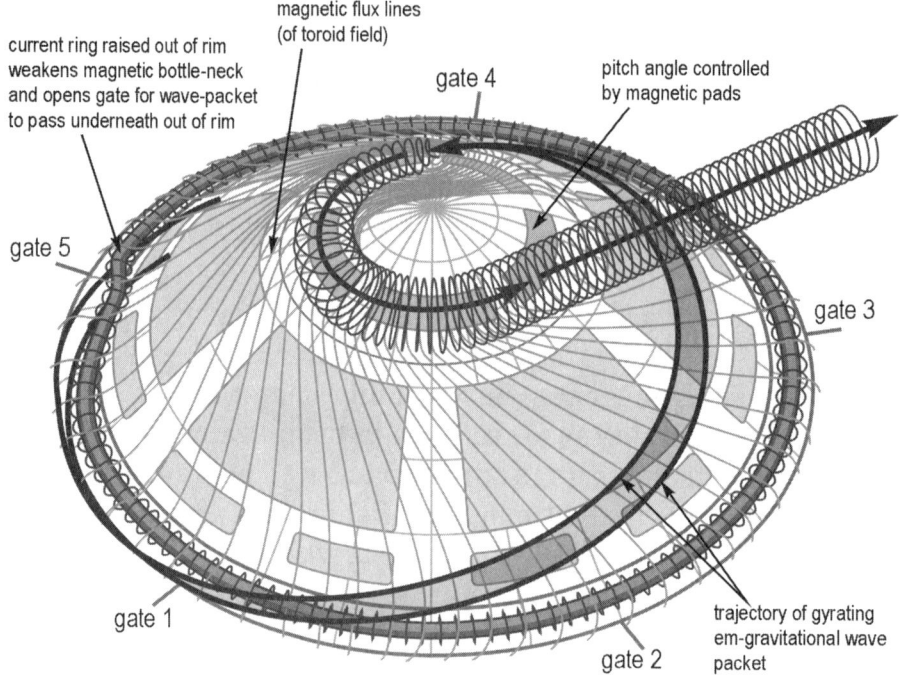

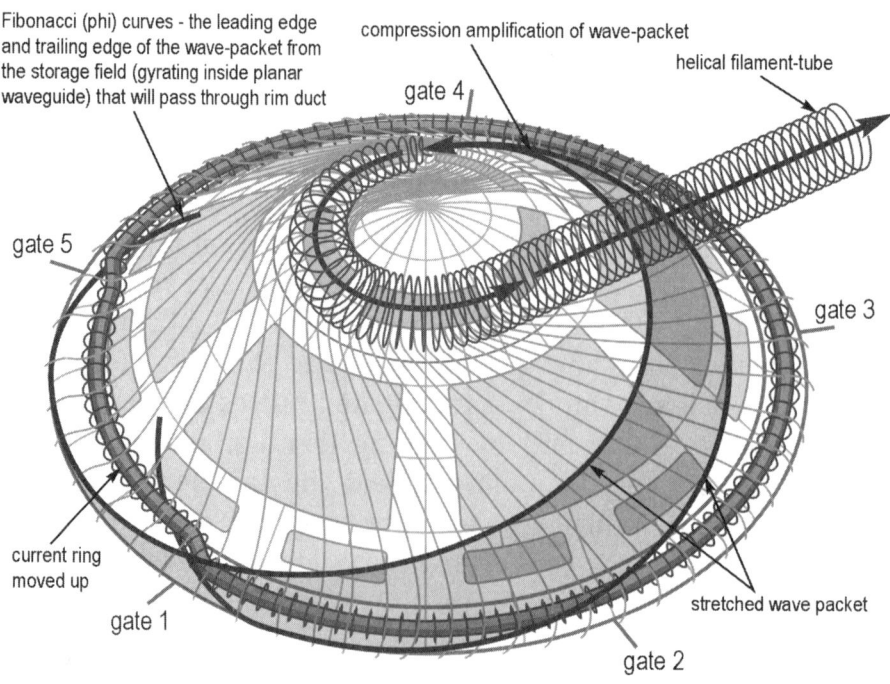

Figure 77. The position of the rim gate and its degree of arc will be determined by what direction the craft needs to go in and how much energy is needed to go into each wave-packet

And here again is shown the cleverness of the extraterrestrial designers, because, there will be a co-mutual factor between this outside section of the craft's magnetic field and the engine section of that field as being part of the electrodynamic circuit inside the craft—so the movements of these wave-packets around the craft will add to the whole circuit's resonance. And if a ring could be imagined being placed around the apex, of small diameter, through which these converging magnetic flux lines could all be concentrated, then this section of the craft's field would develop almost as much power-amplification as that occurring at the core of the craft's black vortex—in proportion to what diameter that apex ring might be.

From my understanding of this type of amplification process, and from the photographs I've seen of it the dome's curved surface will need to provide a 360 degree rotation (one complete turn) for those wave-packets to move around [note: 9]. Any more than 360 degrees would be pointless as the wave-packet will have achieved its maximum empowerment factor after it curls once around the dome, so immediately after its arrival at the craft's apex it needs to be sent on its way—with its gravitational force fully empowered—down the craft's filament-tube.

This being the case it suggests that the filament-tube will need to be initially developed in the air 180 degrees from where the rim is gated, and that the gate chosen will be on the backside of the craft, directly *opposite* to where the craft wants to go. This would be how the craft's trajectory needs to be determined, by first selecting which of the gates was used around its rim to release the wave-packets through, and secondly by directing the guide beam which guides the filament-tube's formation, so as to line its path, into the air ahead of that craft (more on the guide beam in section 10.5 below).

What lends itself best to the dome-curve form of amplification is that the electromagnetic-gravitational energy being ejected out of the craft (imbued with angular momentum) will be divided into pulses, or wave-packets, because then those 'packages of energy' would be quantized, and by quantized I mean they will have a finite quantity to them—in other words, they will have a precise beginning and a precise end to them (that they will have a precise end will cause them to generate an entrainment or tensile force behind them, for instance, which will equate to a negative-resistance being generated inside the filament-tubes). Here is a hypothetical for-instance to show what I mean: Those wave-packets could be generated inside the UFO's engine within a finite time-duration, and those packets could contain in them a fairly similar average number of waves (as an example, and, for ease of calculation; a 10 second packet containing 20 waves); these packets of waves could be created in the center of the craft repeatedly one after the other from the waves bouncing around the central ergosphere. But, when they arrive at the white wall at the rim that white wall can then deal with that wave-packet as a finite number of waves, and so the rim could (if it were rotating fast enough) shorten the time-duration of that wave-packet of 20 waves down to, say, one second, and there would still be those 20 waves in that wave-packet but now they would be compressed waves. What is more likely though, given that the white wall rim will always SLOW DOWN the gyrations of energy coming from the center of the craft (as explained in section 9.2), is that the white wall will stretch that wave-packet (lengthen the packet's time duration) and in this hypothetical case it might lengthen that packet to 100 seconds; the number of waves will still be 20 but now those waves become a group of stretched waves. Of course, in reality the above example would be more applicable if instead of seconds these wave-packet times were in milliseconds.

Taking this hypothetical case around the dome for its 900:1 amplification process, the stretched wave-packet after being ejected out of the UFO's rim will round the dome, and while going through its reducing radius-of-curvature spiralling at the apex it will also increase its rotational velocity; and while it is doing that the stretched wave-packet will behave just like the spinning ice-skater and draw into itself—and will then compress its package of waves into a much shorter time-duration. Thus the wave-packet that was stretched and which contained waves of electromagnetic-gravitational-angular momentum energies of low amplitude will then be compressed into waves of electromagnetic-gravitational-angular momentum with *much higher* amplitudes.

To summarize what has so far been discovered about this UFO's power-drive; there are four excitation-amplifications used by the aaUFO: of the toroid flow producing its non-uniform Lorentz forces; the magnetic field intensities around the black vortex and the white wall rim; the ferroelectric emissions from the upper and lower spheres; and the superradiance resonance (with its reconnection shock-waves). Then there are two geometrical amplifications for this UFO; whereby the fifth amplification is the rim-dome curl; and the sixth is the apex geometry which shapes the filament-tube to a particular bore-diameter. But this fifth amplification where the magnetic field lines converge toward the apex so that the electromagnetic field component of the wave-packet increases the intensity of the gravitational force, as the wave-packets curl toward the apex, is the most interesting for it is the most recognizable in astrophysical studies in the jets which shoot away from a quasar or a pulsar.

10.5 Sixth Amplification: The Dome's Apex Control of the Tube's Bore

How the magnetic pads around the craft's apex (the small ones at the top of the dome) are configured to control the pitch-angle, or angular frequency, will determine how fast the compressed wave-packets will leave the craft and how fast they will be accelerated down the helical filament-tube that the craft will need to extend ahead of it, and because the craft will be drawn along by the gravitational force that is amplified in each wave-packet rotating down the filament-tube, the trajectory of that filament-tube will need to be controlled by the craft. So the filament-tube, and hence the exit direction of those wave-packets will need to be directed through specific angles and this will be determined partly by how these magnetic apex pads are configured, and this would provide one of the ways by which the UFO could be made to fly at any angle of trajectory between side-ways and straight up.

Obviously, a UFO's upper dome should be regarded as a curved waveguide that has the same radius of curvature over its whole surface area, so that the dome's apex can be situated at any part of that curvature, and by adding the larger magnetic pads around the dome into the apex ones the overall magnetic field they produce can be configured to send the filament-tube off in different directions (see figures 7, 8, 23, 70, 76). From the type of field used (in figures 7 and 8), which converges almost linearly toward its apex, like the spokes of an inhomogeneous field, these magnetic flux lines could then be configured into a field that spirals around the dome to the apex (examples of this are shown in figures 23 and 70): the former configuration would send the wave-packets of electromagnetic and gravitational force into the filament-tube at a slower velocity than the latter configuration which would cause the craft to rise upward very quickly.

Whatever configuration is used the filament-tube sheath will be developed in much the same way as an astrophysical jet's tube is developed and collimated by the curling negative radial momentum in the particles of the wave-packets, and once the initial wave-packets have rotated into the ambient medium they will set up the helical field for subsequent wave-packets to traverse and strengthen. That rotating field will, on the outside of it, be developing a Lense-Thirring frame-dragging effect in the air which will attract toward it electrically charged particles from the surrounding atmosphere, and just like in the astrophysical sheaths where the electrical currents collect mostly around the walls of those sheaths so there will also occur powerful currents around the outer wall of that UFO's helical field.

In fact, this action helps to explain some of the observed phenomena around UFOs, because what the frame-dragging does is it establishes its own electric field between that rotating energy-sheath and a vast expanse of air surrounding the UFO. This action effectively polarizes the atoms of the surrounding air, causing that expanse of air to become ionized. What the torque of the filament-tube then does is wrap layers, or sheets, of those ionized air atoms around the outer wall of that rotating field—and traps them—this accumulates more and more electric charge in the sheath wall (like a Meissner sheath effect) so that a high-potential electric field becomes established between the rotating tube and the surrounding air.

This would be the reason for the air around UFOs appearing to be electrically charged, and the reason for magnetically-based instruments in nearby cars and planes going haywire, and for their combustion engines misfiring or stopping altogether—from the fact that the proper current-paths (which should be running through copper wires) in those electronic circuits are being dispersed randomly throughout the ionized air particles surrounding those electronic apparatus, because that ionized air will have just the same conductivity as the copper wires. So for the internal combustion engines, of both cars and the old prop-planes approached by UFOs, this would mean that the contact-breakers would instantly become ineffective, and ignition-coil condensers wouldn't be able to accumulate electric charges (they'd suddenly have a 100% leakage rate), and the high-tension coils would have no currents running through them to generate the magnetic fields necessary to generate high-tension voltages for their spark plugs. Without high-tension sparks no fuel would be ignited, so no power would come from those car and airplane engines!

But most importantly, as the wave-packets twist around in the filament-tube sheath a dragging-limit will be reached for those layers of particles so that they become trapped and will then form into a single shearing field and connect electronically into the surrounding atmosphere (to drag even more layers of air to be trapped into that walled sheath) [note: 10]. Correspondingly, as this happens the outer-wall sheath will then be further strengthened by particle creation, and also by the pinch-effect, which occurs between two adjacent volumes of electrical potentials (the rotating sheath-wall being one and the surrounding air the other) forcing a continuous influx of charged air to flow into it [Greyber 2005b].

In fact, with this in mind it should also be noted, that as this rotating action will produce magnetic flux lines which will be broken and reconnected (through the rotation of the filament-tube), which is essentially the same pair-production process as occurs out in space around astrophysical jets, then it would be straightforward for a UFO to travel through our atmosphere and then to continue all the way into space by using this same filament-tube propulsion.

But to return to earth's atmosphere, once the tube's sheath begins increasing its particle density it will additionally generate its own gravity forces, which will set up a converging stress-field all around and all along the whole length of that filament-tube, and this gravity field will also haul in toward it the surrounding air's neutral particles (because of the converging gravito-electromagnetic field it will create with the surrounding air). So, the rotating filament-tube will become fairly solidly 'hooked' in the air and all throughout the stratified density gradients of the upper atmosphere— but only as long as its rotation is actively accelerated by that UFO (as alluded to in chapters 2, 5, and 8) [see note: 10].

The higher-density and warmer air coming from the UFO craft will move through the tube to the higher altitudes of the tube where the air is colder and more rarified, and this air flow will serve to keep the tube open and create a pressure gradient between the craft and the leading end of the filament-tube. At the tube's leading end it will loose its rotational power and shape and diverge outward, but this will help anchor that leading end into the surrounding air in much the same way a thundercloud develops an up-draft by channeling air from lower altitudes into higher altitudes to cause cyclones and twisters.

The tube-sheath won't have any weight (and nor will the craft because of its buoyancy—as section 8.2) so it won't be under any enormous mechanical stresses, and it will only be needed to endure for two to three seconds at most, and as was shown in the video footage of the Mexico hyper-jump sky-tubes, once the UFO has traversed that tube it will eventually stop turning and dissolve itself back into the air.

The fact that the sky-tube doesn't collapse immediately after the UFO has traversed it (something I made note of in section 2.5 above) confirms the rotating nature of that tube's sheath, for as a result of the craft passing up through that tube the drag force behind that craft will momentarily increase the spin of the particles in the sheath (from the adiabatic pressure-gradient inside of the tube pulling air up behind the craft), but then once the craft has passed through and the pressure flow dissipates inside the tube those particles will slow down and the sheath will loose its cohesion and fall apart. So even though the UFO has passed through it, and obviously has stopped empowering that tube, it will take a few seconds for the particles rotating in the sheath wall to slow down and to disperse back into the air.

But it will be the precise direction of WHERE that filament-tube is set to follow in the sky that will be crucial to the craft's trajectory. That will need to have some sort of initial guidance so that the UFO's filament-tube extends away from the craft at the prescribed angle. One way this might be achieved is through a very simple microwave beam (circularly polarized) or a laser beam, lining an electromagnetic path through the air which would then act as a guide for the helical formation of the main filament-tube, and once formed that alignment-beam could be switched off and the tube would then take care of itself. This may be possible in the aaUFO through beam tubes set into the craft's magnetic pads (for instance in the five apex pads shown in figure 77). Interestingly, the whereabouts of these beam tubes may have been shown in one or two of the scenes remembered and drawn by Betty Luca of UFO craft very similar to the aaUFO (and I shall include one of those drawings below (figure 79), as well as one of my own—(figure 78).

Looking at this rotating canopy at the top of the aaUFO in some of Betty's drawings, it shows that A) the top canopy rotates independently of the main dome, and B) the top canopy has provision to direct beams of light (or microwave energy) ahead of it into any direction [note: 11].

As I say, these beams, probably of microwave oscillations, won't create the craft's filament-tube, all they will do is line a path for it in the air when it forms (and then they will switch off). When the filament-tube is established the larger its diameter the weaker its potential force will be. As the tube diameter is decreased there will automatically be an increase in the angular frequency, and when it passes a critical threshold there will be generated supplementary gravitational forces, so the beam itself becomes an amplifier of gravitational force (as in section 10.2 above).

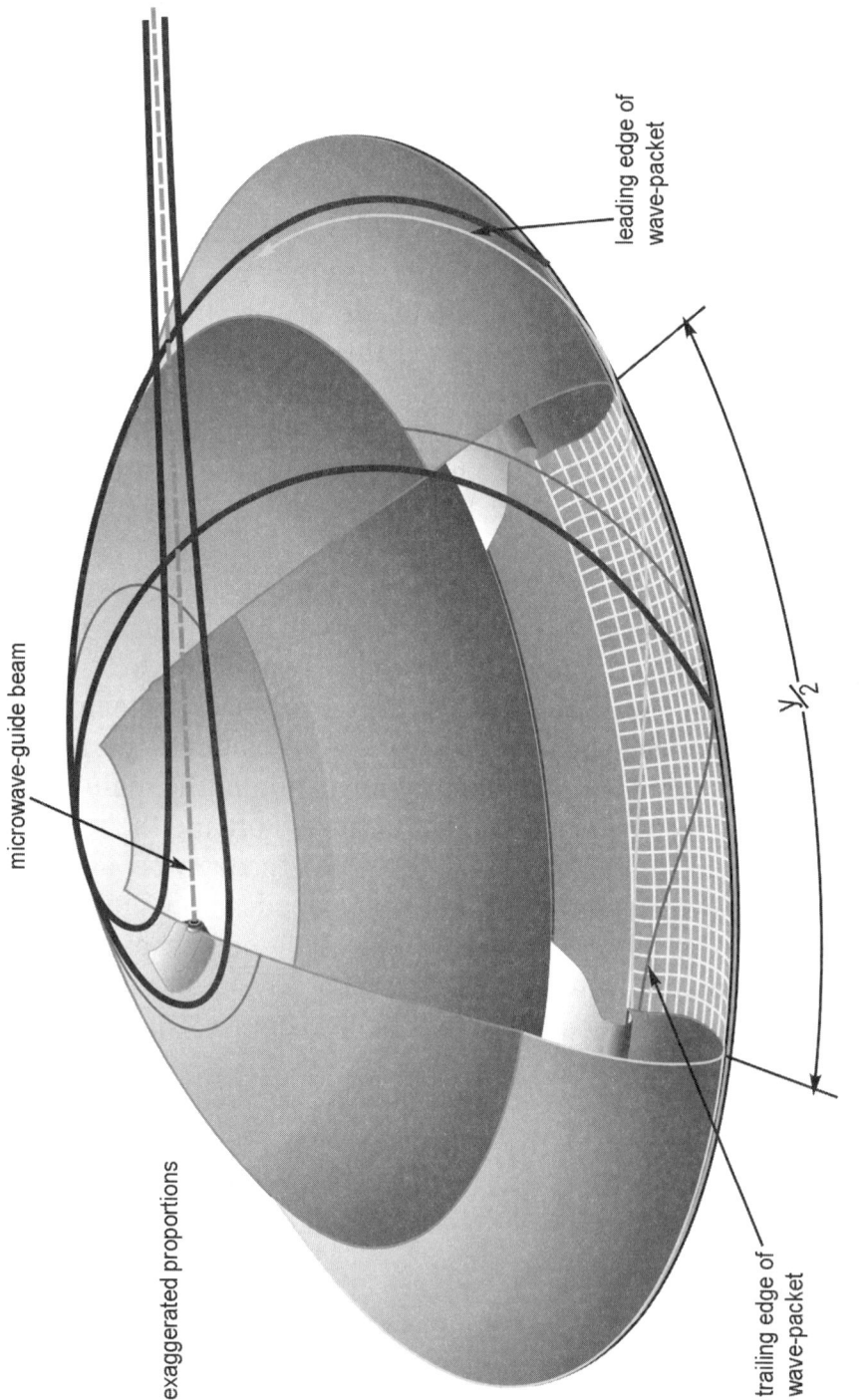

Figure 78. The physical contours of the rim can determine the bandwidth of wavelengths which will make up the wave-packet. At the apex a channel can be established in a precise direction by a simple electromagnetic beam

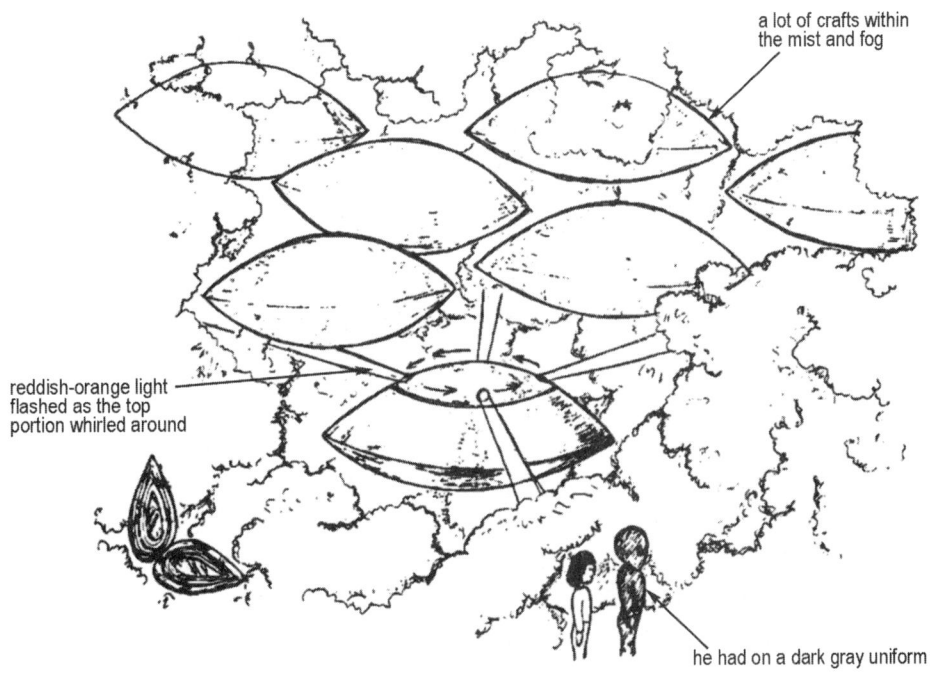

Figure 79. These light beams (of red or infra red photons) were noticed in one of Betty's abduction experiences (from "Andreasson Affair—Phase Two" fig.38)

In view of the above tube-creating techniques, an interesting point to also make here is that the rotating electromagnetic-gravitational waves as they enter into the filament-tube cylinder should have greater velocity than the incident waves rebounding away from the black vortex inside the craft (bearing in mind the slowing-down effect on those waves occurring at the aaUFO's white wall before they squeeze through the rim). This progression of velocity-increases from the outer rim onwards will ensure that there is a progressive amplification factor to those rotating energy packages which will serve to lead the craft [note: 12].

Obviously, the beauty of the UFO is its circular shape which it uses to develop such progressive amplifications through the simplest of procedures, and electrodynamically it shows that the UFO on every count is the PERFECT SHAPE to amplify rotating energy fields... but why this principle isn't used more on earth God only knows!

10.6 Before Einstein's Cosmological Constant Caper

To have two very powerful electromagnetic amplification mechanisms outside the UFO especially to amplify the rotational and gravitational forces in the wave-packets, which the craft produces and ejects from its rim, certainly is highly efficient designing. But, in order to do justice to such efficiency what would be needed for the UFO to send through those marvelous momentum-multiplying mechanisms is a type of energy that when amplified would both shoot down the filament-tube AND would pull the craft with it. That is a pretty tall order... However, the very requirement of that energy will help to define what gravitational force manipulation is all about and capable of.

In order then to see the full potential of this novel form of gravitational manipulation a few backtracks might be required—but they won't be long and arduous. When Morris and Thorne came up with their wormhole project in 1988 they realized that they would have to find a way of explaining what a negative energy density was to other physicists. After some deft mathematics they came up with the formulae for a very unpalatable situation where they found that negative energy density could only be described as being when the force of tension (t) exceeded that of the density (q) of mass-energy so that ($t > q\, c^2$). This unpalatably high factor of tension was difficult to stomach, so Morris and Thorne made the mouth of their wormhole the most important part of their paradigm because it would only be at the mouth, with its extreme out-flaring curvature of space-time, where enough radially out-flowing force could be found to equate to a high enough value of radial tension they needed to satisfy their sums. Density it seems is much more understandable to earth's physicists than tensile stress is [note: 13].

This corresponds to what Nature does on a cosmological scale and can be seen in the lore of cosmological expansion, which is indicated through Einstein's lambda (Λ) constant and in inflation cosmology. In Einstein's view of the universe gravitational force plays a big part in keeping it to an orderly shape. When Einstein calculated the forces of gravity, pressure, and density, against the mass of the universe he at first came up with the formula $G = 8\pi G\, t$, where (G) is Newton's gravitational constant and (t) is a stress-energy tensor. But then he quickly realized this must be wrong and reasoned that something must be happening in the universe which prevents all of that matter and gravitational force collapsing down into a big ball of massive matter. That something, Albert concluded, must be a very powerful expansive force pervading through the universe which counter-balances matter's tendency to fall in on itself, and which must exist to keep the universe in perfect balance. So Einstein came up with his cosmological constant (Λ) lambda, the expansive force which keeps the universe inflated against gravity. And in 1915 he formulated it into his calculations for the universe's energy as $G + \Lambda g = 8\pi G\, t$ so that finally physics could go forward with a static wellbalanced universe, or so they thought. However, when

Alexander Friedmann in 1922 came out with his modification to Einstein's GTR which suggested that the universe might be expanding, and this was confirmed by Edwin Hubble in 1929 when his powerful telescope made deep space scans and discovered red-shifting stars, then it was realized that the universe was not held static in that state of perfect balance that Einstein had predicted. This meant that the mathematics in GTR for the lambda force AND Einstein's gravitational force were all wrong!

After this astonishing revelation had sunk in (it took several years) Einstein finally admitted he'd made a mistake and had to abandon his General Theory of Relativity assumption that the universe was unmoving and acutely balanced. Physicists and astronomers have now grown accustomed to the fact that the universe is expanding in proportion to the curvature of space-time, and indeed, that it is expanding at an accelerating rate [Ostriker-Steinhardt].

That the universe is expanding through curvature correlates with the fact that this part of the universe has an apparent vacuum. This is not the same sort of vacuum though that can be obtained inside a bottle in a laboratory experiment, it is a displacement field of relative tensile force or negative pressure, caused by the expansive nature of space-time (like that lab bottle getting bigger, if you like). Its a space-time curvature effect, much like the Morris-Thorne paradigm of the throat of a wormhole, where its radial tension is created by the out-flaring curvature of space [note: 14].

While gravitational force subtends to space-time curvature if the spiraling trajectory of that curvature is expansive and out-stretching then that gravitational force must be REPULSIVE.

This means that the force of curvature (which was there long before Einstein came up with his Lambda constant) for this part of the cosmos is a repulsive force... a negative gravitational force.

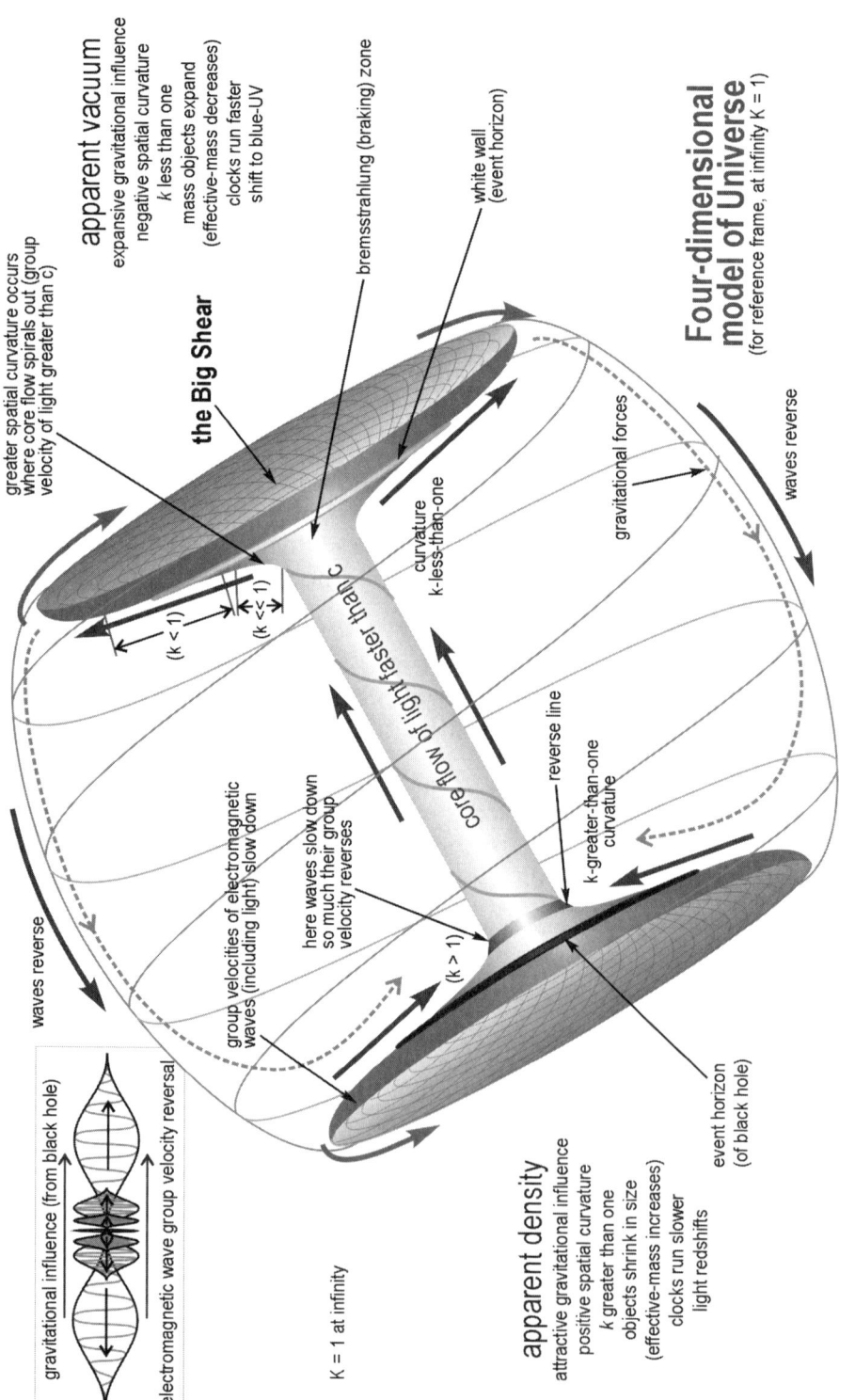

Figure 80. This model of the universe is both a solution to the flat space problem and it can be seen as a diagram to show how UFO's manufacture both positive and negative gravitational forces above and below them

10.7 Repulsive Gravitational Force and Negative Spatial Curvature

This diagram (figure 80), which I will refer to as a paradigm (model diagram), has been designed mostly to show the reader how positive and negative gravitational forces work, although it has not been drawn entirely without cosmological correlation [note: 15]. So in it might be seen our section of the universe (or rather, the only part of the universe so far observed by our scientists), and in it might also be seen how our area of space, and our particular galaxy, can have what is termed negative spatial geometry, so that an apparent vacuum will result from this repulsion-gravitational force which is expanding through our part of the universe. Looking into this model at the opposite end there can be seen what could be termed a quasi-black hole (which, strictly speaking, is not a true black hole because it has no point of singularity)—it will be though a matter-cruncher and flowing from its core there will be a superfluid motion, and so at this diagram's quasi-black hole there will be a positive shearing pressure, a compacting gradient of density, and a slow implosion of positive-attractive gravitational force.

Light waves on that density end would be pulled upon and slowed (relative to the increasing density and curvature), while at the core of our vacuum end where the earth has been residing light waves would be pushed faster (relative to a decreasing gradient of density and curvature). And, it might be interesting to conjecture, that if ever our immediate location of space-time ever changed to a different curvature (ie if we moved to an area with a greater or lesser degree of curvature so that the effect of that curvature became more or less pronounced than it is now), then the change to the wavelength of light we would then experience would likewise be greater, or lesser, than that which we are experiencing right now.

In fact this subject has lately been broached by physicists Joao Magueijo and Andy Albrecht who initially proposed in 1998 the Varying Speed of Light (VSL) theory, in that rather than having always traveled at the same fixed speed of (c), in the distant past just after the so-called Big Bang—which, for the purpose of this study will be referred to as the Big Shear—light traveled faster than it does now (although, as this VSL hypothesis does say, it *will be either c that changes*, or it will be one of the *other components* of the fine structure constant that will change—see following chapter). However, for the purpose of this study I will continue to assume that electromagnetic wave velocity (front velocity) will be constant throughout all areas of the universe (and that it will be solely the *group-velocity* of electromagnetic waves that will be manipulatable).

For instance, from an earth-person point of view, in earlier times where k is much smaller than 1 ($k \ll 1$), so that the effective-mass of objects was/is a fraction of what it is now, then it would mean that a mass object would have been, would be, greatly expanded at the Big Shear core relative to its dimensions now. Consequently, as space moves out to earth's location where k is only just smaller than 1 (k < 1), it means that our planets, our solar systems, and our physical bodies have slowly become more dense—and smaller.

Therefore, in that primordial k-much-less-than-1 (k << 1) and more extreme state of curvature, clocks would appear to have been running faster nearer the proximity of the Big Shear core than they are right now (as figure 80 shows). The wavelength we attribute to Light would have been different too, from the fact that timed events would have taken more of our scale of time to complete (clocks run faster), but because the distances traveled by that wavelength would have been far greater—then light would have been blue-UV-shifted [note: 16].

To see how a further aspect of this curvature effect might work the reader should take a look at the above diagram and then visualize what happens to a group of electromagnetic waves while they are being pulled upon in the positive end of this universe model. The group velocity will be slowed-down continually as those waves are being pulled down toward the rotating core of the quasi-black hole, and then, between that quasi-black hole's event horizon and what I call the reverse-line, the wave's group velocity will REVERSE so that those electromagnetic waves will appear to go backwards (they will get slower and slower and then, rather than coming to a stand-still and stop, the group velocity of those waves will simply go through a reversal, as shown in the inset). Note that this would be because in this system there is no point of singularity where time, it is alleged, would otherwise need to be seen to come to a complete stop.

So in this model of the universe, where this model's axial bridging tube might be likened somewhat to an Einstein-Rosen bridge, the big difference will be that where the 'worm' physicists have worried that in the instance of someone passing through such a bridging tube to the 'white hole' which exists on the far side of it, that that person's Time would then-after appear to run backwards, according to this model that is absolutely NOT THE CASE. Indubitably, this universe model shows that, as far as the concept of 'traveling back in time' is concerned, we humans are living proof that time does no such thing as go backwards when our light waves pass through a white hole!

In other words, for our electromagnetic (light) waves a series of changes will be affected upon them which are wholly unfamiliar to the mental representation, as was presented by Einstein, of how light is supposed to travel through space. For those waves will proportion themselves according to the funneling curvature of that hole's spatial geometry, which in the case above, they will be forced to propagate within that 'dark matter' (higher-density matter) end at a lower group velocity than (c)—and to redshift. So, firstly, this discrepancy shows that white light is not the be-all-and-end-all yardstick throughout the universe that it was thought to be by Einstein. Light speed would have been the yardstick and a universal constant had it existed in a 'flat' universe, which, of course, is what Einstein originally assumed was the case. But we know that assumption is now wrong and was corrected with Friedmann's mathematics and Hubble's telescope observations. As an aside, the reader might well ask whether is it because Einstein formulated his General Theory of Relativity AND his calculations for gravitational forces—on that naive presumption that the universe was flat—that present day physicists have failed so gloriously to detect these forces and to discover how to produce gravitational force?

Secondly, this suggests that while electromagnetic wave velocity is universal it is PULSED (ie light waves are modulated by another frequency). More to the point for this immediate study though, would be that the same principle, of group velocity and modulation, goes for gravitational forces (as I have shown in this and the previous chapter of this study).

However, back to the model of the universe; after these 'real-time' back-flowing groups of electromagnetic waves have vanished down into the black hole (and have gone beyond its event horizon), and have reversed through the black hole's core they will immediately be pulled into the bridging tube and will quickly pick up more and more speed and become accelerated by the curvature of that tube, as photon wave-particles, to a velocity that will be much in excess of (c). But can this really happen? Certainly, it would be impossible according to contemporary laws of physics relating to the wave front velocity of light, but these are electromagnetic waves heading IN REVERSE, and so these velocities can be infinite—as has been proved by the experiments carried out by Günter Nimtz in Germany, which produced electromagnetic wave transmissions at a grand velocity of 4.7 times (c) [Nimtz].

Exiting out of the superluminal tube, at the Big Shear core of this model where the curvature opens out, the group velocity of the electromagnetic wave-particles will drastically decrease in velocity as the effectiveness of the space-time curvature on them begins to change very dramatically at the superluminal tube's end, and this will cause them to brake appreciably. So obviously, here at the core's opening, which would seem to suggest where the supposed *moment of The Beginning* occurs, where the electromagnetic waves and lower effective-density matter particles emerge from their faster-than-c-travels, there would result an enormous amount of energy dissipation. Because, those high-energy supra-fast matter-wave-particles would have to shed a HUGE PROPORTION of their higher energies—almost instantaneously—in that curvature changing process, and remember, this is where clocks are running faster and event-times are compressed, so huge amounts of heat and light in explosive emissions of energetic radiations would be given off at that core during that slow-down process. Indeed, the consequence of this *ongoing* bremsstrahlung-like braking process would be that it would give the appearance from afar that there was a huge explosion-event of gigantic force occurring. So, this might offer a new look at The Big Bang which has been assumed to be a one-time event! Also, this may suggest where cosmic rays get their extremely high energies from, for these could actually be the core's primary projections of supra-energy particles which have given off little of their original energies, but which have merely had their trajectories re-directed as they exit the core tube!

But because of this braking effect upon those faster-than-light energy particles and waves, these vast amounts of energy dissipations going on would continually feed energy back (by the agent of the universe's inter-stellar magnetic fields) to the positive end of the universe to power the huge black hole at that end. Indeed, it would feed the whole system, so that the whole system would be self-sustaining.

Intriguingly, this self-sustaining energy explanation would finally offer a solution to the age-old problem in Science as to why the universe is still expanding after being in existence for so long after its supposed 'moment' of violent creation, a point which has never been justifiably explained by any advocate of the Big Bang model, whereupon the theorists for that model have always pleaded that the whole of the driving energy, to both form the universe, and to keep it expanding was generated, in its entirety, just from the one and only enormous explosion it was theorized as having in the beginning! Certainly, this new-look positive-negative universe model would mean that all the electrodynamic energies needed to drive the universe, to keep the galactic spirals moving, are being continually generated, and the beginning is still being generated. And if this model I am presenting here shows more feasibility than the presently authorized one-off explosion then the Big Bang hypothesis was, and always will be, an undignified nonsense.

What might have fueled the assumption that there was only one Big Bang (and this was a factor that was touched upon in the Magueijo series of papers), is the WHITE HOLE WALL effect, whereby its ring of influence around the Big Shear's portal prevents us looking back through time to anything before that ring, to any event which has taken place between it and the bridging tube's portal; for just as a black hole allows nothing to escape once it has passed over its event horizon, so that white hole's event horizon allows nothing to penetrate it, and therefore it is this time ring which prevents any event information beyond it being inspected by us. So this is why I say physicists have wrongly assumed that all the expansive gravity which shapes our part of the universe came from just one event—because that expansive event is ongoing and its taking place right now!

Returning once again to the case in point, our reversed electromagnetic (light) wave-particles, and their grand entry into this part of the universe, will also be accompanied by matter that has been dissociated into subatomic particles (by the matter-crunching black hole) much like the primary cosmic rays, and these will be accompanied by gravitational waves. The gravitational waves, just like the electromagnetic waves, will also emerge from the universe's core running in the reversed direction. This will mark the essential attribute of the white hole and it will establish its greatest antithesis to the black hole—for this part of the spiraling universe will have NEGATIVE spatial curvature, and a gravitational force as that of a white hole and EXPANSIVE. This also means that we are provided with a vacuum in our part of space. But, to reiterate the point made earlier, our vacuum will only be comparable to the vacuum found in the lab test-tube because our location of space is being continually expanded into an ever-increasing volume of space-time.

But judging by this universe paradigm (as shown in figure 80), as the curvature gradient spreads out toward the (k < 1) state, toward where our location in space is, it can then be seen that light wave velocity would become more and more attributable to the (c) speed constant that we have been accustomed to and which has been our constant for many millions of years. So that, yes, for our (k < 1) section of space and for our finite span of the universe's curvature, Einstein was perfectly correct in assuming that electromagnetic (light) waves will predominantly travel at (c) velocity (but only from the fact that our 'seconds' and our 'minutes' and 'hours' were devised for us while we existed in this same (k < 1) state—which is something that our Albert took for granted because of his presumption that space was flat). Although, having said that, this sort of diagram does suggests that the curvature immediately around us may have been altering VERY SLOWLY over time, so that our time and our effective mass-density very slowly and very subtly during those millions of years may also have been changing, albeit imperceptibly by us, of course. But, it would be so interesting to conjecture that had our Albert as a boy been taken by a long-range UFO to a distant part of the universe's curvature, and was then encouraged to measure the speed of light at that spot where he arrived, that he would certainly have been obliged to raise his eyebrow at the differential he would have discovered on his little slide rule.

Another interesting question, subsequent to the aforementioned conjecture that light waves have not changed their group phase direction since they left the other end of the cosmos, will be, are we all existing at this end of the cosmos with our primordial light waves still traveling BACKWARDS? Because that is what the gravitational and the matter waves will be doing...

But, going back to the original point I was making at the beginning of this section, and to the point I was making even before the beginning... at the end of the previous section, if the reader cares to glance yet again to the universe diagram (figure 80) then there will be noticed a certain similarity between the electrodynamic structure of the aaUFO and this paradigm model of the universe (also see figure 69); for the UFO has inside it a black vortex corresponding to the positive end of the above universe model, and below that UFO there will be developing in the craft's lower magnetic field an expanding spatial curvature—which will be generating a repulsive gravitational force.

And so, by flipping this universe diagram around it will be seen how at the bottom of the aaUFO craft there would occur a localized manipulation of gravitational forces and, just as gravitational physicist John A. Wheeler once said, while matter will tell space-time how to curve, space-time-curvature will tell matter how to move... [note: 17]. And out of this negative spatial curvature generated below the aaUFO will come negative energy density—which in UFO technology is manifested through a gravitational repulsion force.

In fact, UFOs are not the only electrodynamic mechanisms on a scale much smaller than the cosmos that can generate negative energy density, according to William Unruh and Ted Jacobson this repulsing force is similarly generated through the in-flowing force of an astrophysical black hole—even though it has a positive spatial curvature. For just outside a black hole the waves of electromagnetic and gravitational force passing too close to that black hole's event horizon are having their group-velocities reversed all the time. For the rotating gravitational force will be so powerful on those waves that as they try and propagate away from the black hole the black hole's pull will slow their group velocities down so much that they too will pass through the zero velocity line—and as they can't stop at zero they will reverse and travel backwards in the opposite direction! So immediately outside a black hole some of its electromagnetic-gravitational waves go backwards and expand away from the event horizon just as fast as they approached it [Jacobson] [Schutzhold-Unruh] [note: 18].

This is nowhere more perfectly illustrated than in what happens inside the aaUFO, as explained above (in section 9.3), when the superradiant ringings take place, it will be these negative gravitational forces that will empower the outward spiraling energy fields, which, when they gyrate outward those fields will collide with the rim's gyrating white wall, slow down and reverse (again), and then bounce back (again) into the central black vortex, into the rotating force of the photon-orbit ring—only to be reversed yet again and to gyrate outward. And this will happen again and again, over and over again. The proviso, as mentioned in [Schutzhold-Unruh], is that for the white wall to slow down and to then reverse those waves there will have to be (for the reflecting quality of that white wall) a particular refractive index, and that it will be undergoing dynamic motion, which, in the aaUFO, will be determined by how the energies of its white wall are gyrated just inside the craft's outer rim.

Not all wavelengths of these energies around a black hole will be affected by this action and so in such a small model of a quasi-black hole, as in the aaUFO, it would be interesting to see how the different ergospheric compartments will contribute to the negative waves and which to the positive ones [see note: 18].

Another interesting feature to consider with negative waves (negative refractive index waves) would be that as the phase-front of the waves is negative and for all intents and purposes going backwards, there will be no limiting constraint upon their phase velocity, other than that determined by whatever refractive index is engineered into whatever media it is traveling through—which for a UFO will always be a variable because it can manufacture its own media (ie inside its filament-tube). To reiterate Nimtz's example of 4.7 times (c) then both the gyrating plasma and the fluid moving through the UFO's toroid can have their permittivities and hence their refractive index adjusted by that UFO.

Lastly, this universe model of positive and negative spatial curvature will provide a satisfactory explanation to the gravitational dynamics of yet another of the well known UFOs; that of the captured UFO at Area 51, allegedly seen and worked on by Bob Lazar back in the late-1980s. In this UFO craft there would be encased inside each of its three gravity amplifier heads circulating (shearing) flows of massive particles setting up in each of those encasements a miniaturized quasi-black hole. These quasi-black holes would be configured as dual amplifiers of both positive gravitational forces and negative (expansive) gravitational forces so that they could transform these forces into a beam which could then be propagated away from the craft to create instant movement, because, wherever that rotating field was directed it would locally affect a curvature of space-time beyond that craft, and because that craft was gravitationally bound to whatever forces were in that beam, then by wrapping that space-time it would locally generate a repulsion gravitational force which the craft would either ride upon or propel itself from, just as if it were creating an epicenter of gravitational force below or beyond that craft. This new hypothesis for the Lazar UFO, it should be noted, is a comprehensive departure from the one Bob Lazar himself originally expounded (and in particular this localized epicenter of gravitational force created by the craft's beam(s), strictly speaking, wouldn't be a force that would be intended to oppose the earth's gravitational field, as was hypothesized originally by Lazar) (see figure 67), as I explain in my supplement to this study's chapter 12 [note: 19].

10.8 Using the Repulsive-Attractive Forces

However, going back to the aaUFO and to its two powerful amplification mechanisms operating on whatever special sort of energy can be sent down its filament-tubes (from the beginning of section 10.6)... the thing about the craft ejecting a wave-packet of purely attractive gravitational force (from its upper dome) is that it will be gravitationally attached to the craft, and if that wave-packet can be projected away from the craft then at some point, while the attractive force is amplified by it rotating through the tube's smaller diameter, that amplified force will pull on the craft to move it. It would be like the bungee-jumper stretching a rubber bungee cord which develops a tensile force between the jumper and wherever the cord is anchored, then at some stage all that tensile force stored into that cord will result in a contraction of that cord to pull the jumper backward. Only, in the case of the aaUFO its wave-packets will be projected ahead of that craft and so their anchor will need to be developed at the forward end of the craft's stretching 'bungee cord,' so that when its threshold of forces activates between the craft's anchor and its attracting gravitational force and it contracts, it will be the craft that gets pulled forward by its wave-packet's contracting forces. Note that the craft will NOT be going down the tube after its wave-packets, the amplified gravitational attraction force will be doing all the negotiations between the mass of the craft and the forward end of that gravitational attraction. All the craft will need to do is supply a center of gravity for that movement to take place.

For the above notion of being like a bungee cord's contraction I was initially going to describe its action as being somewhat similar to a backscatter chirp. Whereby, with a chirp system a backscatter pulse is made to return into the forward propagating wave-packet acting like a pump pulse—so as to compress the forward pulse to greatly amplify it and in the process bunch together the whole bandwidth of frequencies of waves in the wave-packet, which in this case would be composed of the UFO's electromagnetic, gravitational and rotating kinetic forces. But because there are negative gravitational forces involved this whole notion warrants a slightly different approach [note: 20].

For unfortunately, negative energies are not quite as wonderful as they seem and certainly not as straightforward as would be liked, and dealing with them needs that extra degree of thoughtfulness. To this end it has previously been formulated in quantum field theory that an observable restriction which must be acknowledged and ascribed to negative energy density, is that for every pulse of negative energy density there must be following immediately after it a compensating pulse of positive energy density (or otherwise if there is a delay then the time delay between the two densities will always work out to be inversely proportional to the amplitudes of those pulses). This forms the basis of the quantum interest conjecture, whereby the outcome is usually that the positive energy density which trails the negative will always be the more amplified of the two and will always overcompensate for whatever amount of energy there is in the negative pulse [note: 21].

Because of this seemingly self-defeating compensating factor the recent research into negative energy density has faltered somewhat, mainly because, I think, the scientists working on it have yet again tried to compartmentalize that negative energy into a separate component. Remember how I was saying in chapter 1 of this book, that the extraterrestrials work on multi-functioning and never on separate mechanisms... these scientists working on negative energy density have assumed that before they can work on that energy they somehow have to separate it the from the positive energy density, and which of course, has proved impossible. As it says in the [Ford-Roman 2000] article all attempts to split the negative energy away from the positive have failed, and so they will always fail because of the fundamental laws of Nature [note: 22].

Looking at this problem from an entirely different perspective, through extraterrestrial UFO technology, the greatest advantage can be gained by *specifically* using the composite wave-packet of both negative energy AND positive energy! The above agreed-upon sticking point with negative energy density—that it will always be accompanied by a packet of positive energy which will dominate over it—is precisely what this aaUFO is taking advantage of to fly through the air with!

This will be the case with a wave-packet of gravitational force and angular momentum which is propagated down a helical field. As I say above, such a mechanism is allied to astrophysical jets powered by black holes inside quasars which extend such long distances through space—and these thrive on negative energy! A huge factor built into the dynamics of a quasar is the integrity of the rotating walls of those jets, in that those walls don't bulge to the particle ejections passing through them and will direct all their accelerating forces up to the forward end of those filament-tubes. Indeed, the helical field can be the most stable energy field there is—it can deal with unimaginable forces (as delineated throughout chapter 2 of this study). In a helical filament-tube if negative energy, expanding gravity, is sent down a helical filament-tube first it will accelerate ahead through its expansive momentum and then drag after it the positive attractive gravity.

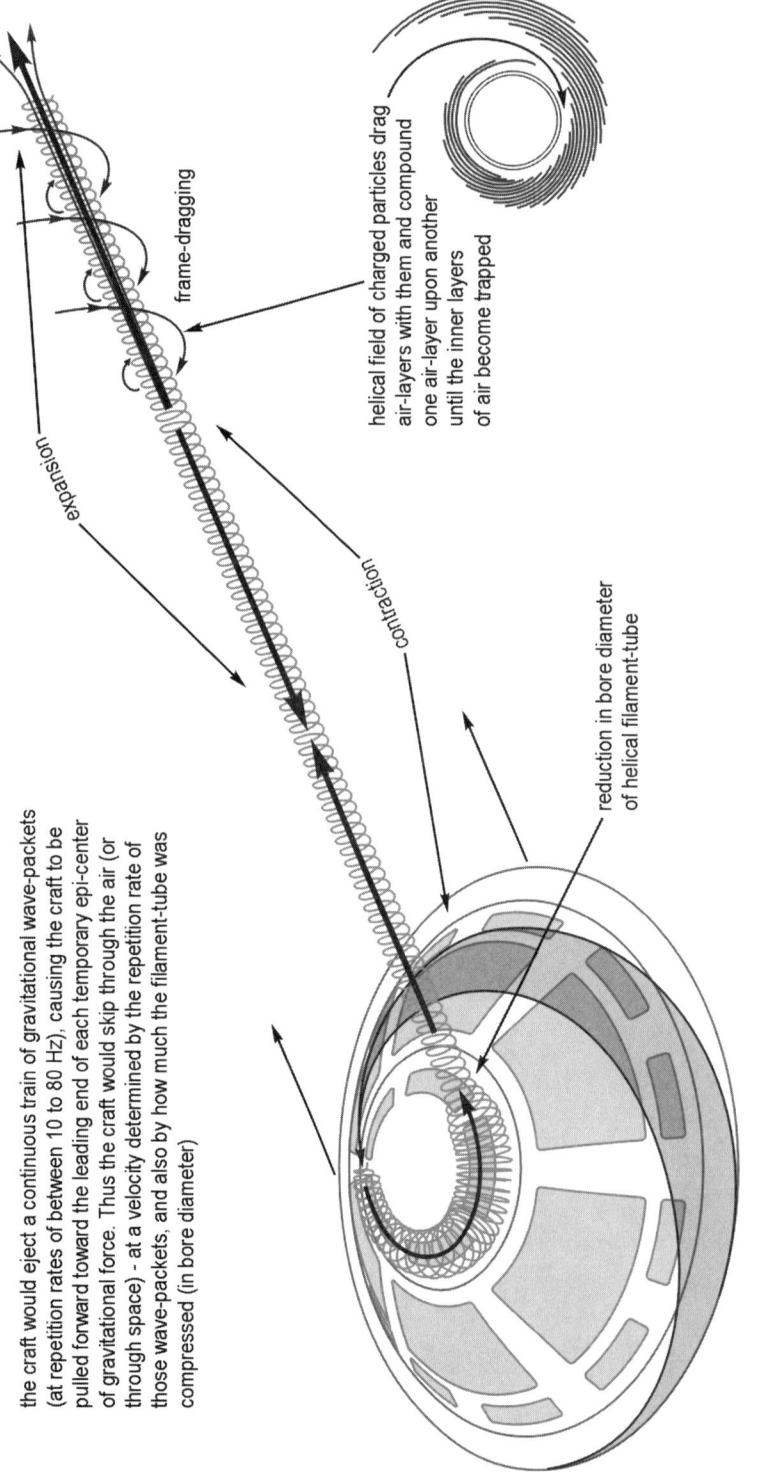

Figure 81. The wave-packets ejected from the UFO will have both negative gravitational force and positive gravitational force, the former will be an expansive force and the latter will be the trailing attractive force

In the case of the domed aaUFO that craft's filament-tube will be strongly collimated, and with its small bore will provide a powerful amplification factor to the forward waves in the wave-packets coming out of the UFO's rim and curling around the dome to the apex. The first bunch of waves will be the negative gravitational force, and as they spin into the tube and are amplified, and become more expansive, their force will accelerate down as a spinning force toward the forward end of the tube. This expansive force will work as a forward-moving anchor being confined inside the cylindrical wall of the filament-tube (from it being a rotating repulsive force continually being intensified by the tube's tight bore). And when the positive part of the wave-packet follows and is pulled through the filament-tube by the acceleration of its expanding partner, its force then becomes amplified inside the tight bore and its attractive gravitational force will then begin to contract.

This contracting (attractive) gravitational force though will be bound to the gravitational center of the UFO craft (to the black vortex) and so as those waves contract they will pull on the craft. And judging by the rules of quantum interest stated above where the positive attractive force will be greater than the leading expansive force, the final acceleration of the craft will be even greater than the speed with which the forward part of the wave-packets have sped down the filament-tube into the distance ahead of the craft—so consequently, the UFO's speed will be governed not merely by the forward propulsion forces but also by the contraction velocity of those gravitational wave-packets.

At the head of the filament-tube where the tube's integrity falters the forward part of the wave-packet will momentarily expand and cause the air there to rarify, to lower in density, and to heat up, and just as will happen all along the tube as it rotates in the air (as above 10.5) this forward end will draw in from around it in-flowing currents of air that will serve to hold the end of the filament-tube to that location until the craft reaches it, or until the craft sends out its next wave-packet to continue the process further onward [note: 23].

This will give forward momentum to the gravitationally buoyant UFO craft and when consecutively repeated, at say, 38 times a second (38 Hz), those wave-packets will accelerate the craft through substantial distances [note: 24].

10.9 Chirping the Semi-full Swing-backs into Wave-packets

Although, through chapters 6 and 9, I have shown how the telemeter wheels work and why they should be firing pulses into the toroid's fluid, I'm going to add some more refinements here to those initial suppositions, because I don't think the telemeter beams are used only to adjust the water in the toroid, to heat it and to change its permittivity (and to speed its flow); these beams can do this anywhere upwards of 2.45 GHz, and the water will go on being affected by microwave oscillations up to 200 GHz, so the beam's frequency for it to produce an effect on the water is not that critical, it just has to be 2.45 GHz or above. So I think those beams are ALSO being used to change the character of the toroid's inner wall—to render that wall transparent to electromagnetic and gravitational waves. Because, if the toroid's wall is made transparent to these waves coming away from the black vortex then it provides them with the perfect switching device with which to transform the toroid and its fluid into the high-permittivity reflecting mirror alluded to above (in sections 9.4 and 9.5), which will slow down those waves so that they can develop into the craft's superradiant ringing pulses.

This induced transparency would be determined by the toroid's inner wall (or at lease the middle horizontal band of it) being made of a switch-able dielectric material, and again, this would suit a metamaterial that had a permittivity transition frequency in the microwave region, for instance around 6 GHz. So that when that material was radiated by a microwave oscillation at that 6 GHz frequency it would then switch from opaque to transparent to allow the electromagnetic and gravitational waves coming out of the craft's center (and rebounding out of the center) to pass through that central band in the toroid wall, to permeate into the toroid water, and to reflect back through the wall again, and to then return into the central ergosphere (as in figure 73 above). Manipulating the toroid wall in such a way would considerably SLOW DOWN the waves passing in and out of the central ergosphere which, as the reader will recall (from section 9.4), is the pre-requirement for the establishment of the superradiance amplification mechanism—because slowing down these waves equates to reducing the distance needed between the craft's black vortex and its superradiant reflector.

Because, and here is my main point of consideration, rather than having the rim gates divide the craft's gyrating energies into wave-packets at the rim, while they are being ejected out of the craft and coiling their way around the dome to be compressed at the apex to produce their propulsion momentum, it would be better to have those gyrating energies already formed into wave-packets BEFORE they arrive at the rim. This way the UFO would be provided with a more efficient way of stretching those wave-packets, because it would use its rim's white wall to slow down the waves in those packets and to spread them into a more elongated wave-group before ejecting them.

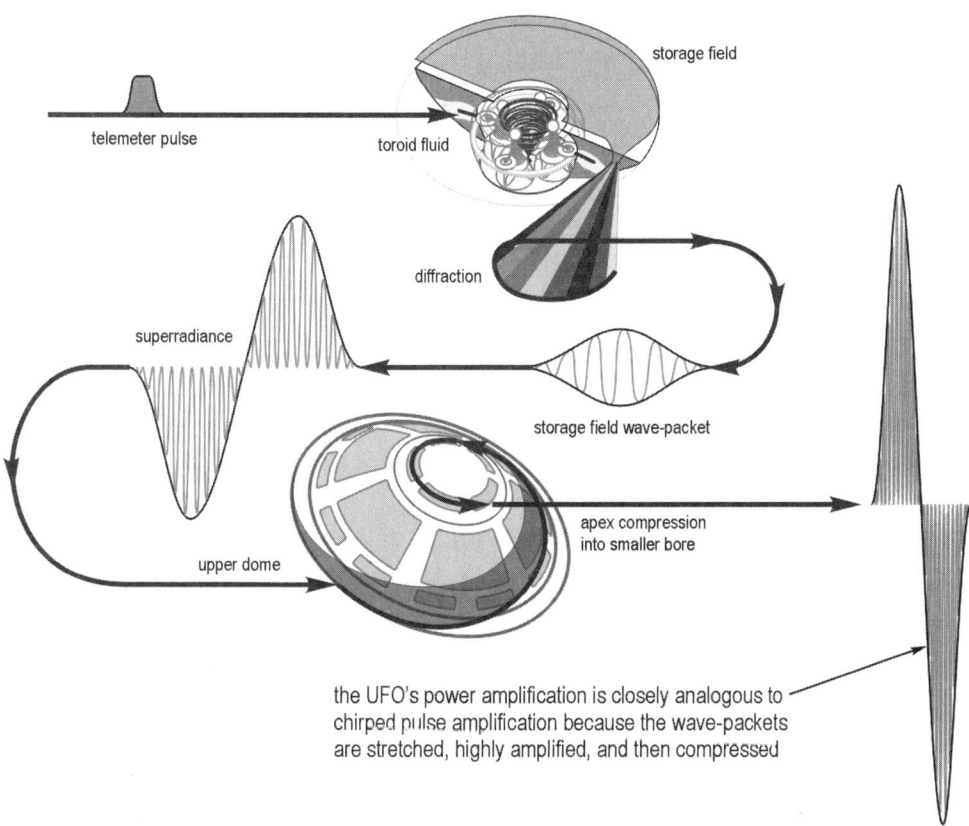

Figure 82. The electrodynamic arrangement of energy inside the aaUFO is analogous to the mechanism of chirped pulse amplification

Also, if the wave-packets were created inside the center of the craft, they would be taking advantage of this very precise switching mechanism performed by the telemeter wheels. The waves coming away from the black vortex either by radiating away or rebounding away from its photon orbit ring will be continuously oscillating waves, and left to their own resources they will gradually pump up by bouncing (by ringing) to-and-fro between the rim's white wall and the central black vortex (as has been explained all through chapter 9), and the same thing will happen with the waves inside the central ergosphere, of course. But if those waves can be placed into envelopes by the telemeter wheel pulses activating the toroid wall, thereby switching the toroid wall—through which those waves have to pass—from opaque to transparent, then, as I say in 10.4 above, then they become quantized, and those groups of waves will have a finite quantity to them that can be manipulated in a way which will eventually amplify them enormously—firstly, by stretching them as they gyrate out of the planar wave-guide and pass through the rim, and then by compressing them after they have passed out of the rim and have curled up around to the apex of the UFO's dome (see the schematic in figure 82).

By using the telemeter wheels to propagate into the toroid wall material their microwave beams (which up till now have been conjectured to possibly have an operational range reaching as high as 30 GHz—as section 9.4); they could be made instead to deliver microwave beams oscillating at just above the transition threshold (the tentative 6 GHz) to make that material transparent, and then that wall, and the toroid fluid reflector, and subsequently the very triggering of the superradiance mechanism itself, will be switching on and off under the control of the telemeter wheel beams, and these switchings will be governed by how fast those telemeter wheels rotate. In other words the pilot, by controlling the inner ring-tube (see chapter 6) which allows the sphere-set assembly to rotate at a pre-determinable speed, will control the firing cycle of the telemeter wheels, which will make the toroid wall intermittently transparent, which in turn will determine that the superradiance amplification ringings will be governed by a specific repetition rate—meaning that if the pilot controls the inner ring-tube to allow the telemeter beams to fire at 38 pulses per second then the superradiant amplifications will also be set precisely at 38 ringings per second.

Essentially, to control the toroid's inner wall the telemeter beams would be operating a technique which induces transparency into that material by changing its refractive index (which effects different materials at different frequencies, mostly between 5 to 10 GHz), where the wall material which would normally be opaque beneath a critical threshold frequency would become transparent when these beams were applied to it at a frequency just above that threshold (the tentative 6 GHz). This technique, known as Electromagnetically Induced Transparency works on semi-conducting dielectric materials and is achieved by propagating an oscillating field orthogonally (at right-angles) into that material—which coincidentally is exactly what the telemeter wheel beams are doing in relation to the toroid's inner wall [note: 25].

The required effect would be for that material to allow waves of a fairly wide bandwidth of frequencies of electromagnetic and gravitational waves coming out of the black vortex, to pass through it during the time-periods when the telemeter beams of microwaves are passed through that wall material. So the toroid wall would provide a switching effect.

The repetition rate of these switching pulses of microwave beams, which I have arbitrarily set throughout this study at a nominal 38 pulses per second, can be varied to be more or less than 38/sec by speeding up or slowing down the rotation rate of the sphere-set assembly by the controlling pilot (by him or her using the single transmission line running to the inner ring-tube which controls the sphere-set assembly—as in section 6.2). Of equal importance, is that the duty cycle of those pulses (each of which delivers a microwave beam of, as in this tentative example, 6 GHz microwaves), could also be adjusted by the pilot, and this latter switching mechanism would mean that the waves of electromagnetic-gravitational forces could be placed into long or short wave-packets, but with the same repetition rates—so that the amount of waves collected into them could be varied [note: 26].

By introducing this pulse interruption into the superradiance amplification it won't mean that the superradiance will start-stop, start-stop (and so on) at the start and finish of each and every telemeter wheel pulse, all the pulses will do is gate the superradiant oscillations—so that they peak with metronomic regularity and synchronize to the telemeter wheel pulses. In fact, this would really be the best way that the superradiance could be controlled to administer acceleration and deceleration and to initiate the UFO's power-down sequence [note: 27].

Thereafter it will be how the UFO sends those superradiant wave-packets of energy out of the craft and into its two powerful compression-amplification mechanisms around the dome (and where they come off the dome) that will determine how fast the UFO moves. The rim and dome's mechanisms of stretching and then compressing those wave-packets will lend themselves most readily to an amplification method called chirped pulse amplification, which is very similar to the system I have mentioned above, of stretching the packets of waves and after ejecting those stretched wave-packets outside the craft, amplifying them by its dome compressing them so that a huge amplification occurs in their forces [note: 28].

10.10 Perturbations to Vary the Optical Transparency of Spacetime

On the need for a wormhole idea for FTL communications and travelling, if a new approach is to be addressed with regard to space-travel it should be addressed as a whole package, not only of a new form of craft but also, and perhaps most importantly, as a new form of manufactured media, through which the craft would need to travel, be it out in space or while flying through the earth's atmosphere. Quite specifically also, that media would require to be bridled by a new form of channel, as if it were a containment field, which can be engineered immediately ahead of that new form of craft, to give that media a variable degree of negative spatial curvature so the craft can be attracted toward a ($k > 1$) potential-difference. For only then could the light speed, or the electromagnetic and gravitational wave package which enveloped that craft, have any hope of being accelerated to superluminal speeds [note: 29]. This, in fact, is not such an 'unheard of' as one might imagine, considering that airforce servicemen actually reported seeing this very process with their very own eyes during World War II—where a rotating field was engineered ahead of a craft which would foreshorten into it with colossal acceleration and disappear into the distance in an instant [note: 30].

Certainly, the fact that light is bent when it passes through a media with higher refraction index than air or a vacuum is well known, and it has been accepted that in such circumstances light waves will slow down because the higher refraction index equates to a higher density and less transparency than air or in a vacuum. What has also been accepted, rather rigidly at the behest of Einstein and Newton, is that light travels through a vacuum at the highest velocity there is in this universe (apart from Cherenkov radiation which is faster than light but only by a small amount), and so we all see that if light travels in a straight line then that straight line will be the most direct path between two points. But is light speed, while light travels in a straight line through a vacuum the true speed? Not according to the negative spatial curvature model of the universe (as shown above in figure 80).

What if the light waves were to pass through a medium with a transparency factor GREATER than that of the vacuum, the path of light wouldn't appear to get any straighter so how would the light waves travel through that supra-transparent medium? One thing that might happen is that the distance covered by the light waves will be greater than that covered if the medium were air or the vacuum. Ludicrous perhaps for the fans of GTR, but in a channel engineered as one that creates a greater curvature than the curvature which presently regulates our section of the universe, then that channeled environment will offer a greater speed than light has through our present location in the vacuum.

This indeed, is shown in this universe model, in between the positive spatial curvature and the negative spatial curvature there would, in proportion to the differential potentials of both, be a much GREATER transparency factor in that channel spanning between those two opposite space-time geometries than through normal space-time. If a UFO craft engineered around itself a positive spatial curvature and aimed an energy envelope at, say, a distant star with a negative spatial curvature to lead at the head of that energy envelope, once it was set up that channel between the two would operate at faster-than-light speed, and when the craft would be hauled into that channel that craft would travel faster-than-light with almost no energy-expenditure toward its propulsion...

10.11 A Successful Mechanism to Harvest ZPF Energies By

The next big question will be... How does a UFO harvest energy while it is out in space? The exact same way a black hole harvests the virtual field and produces abundant energy for its jets... by gravitationally and electromagnetically stressing the space-time fabric. And too in our atmosphere by stressing the surrounding ambient to produce electron-positron pairs [note: 31].

I have already alluded (in section 6.10) to one mechanism that will allow outside energies to be produced and brought into a UFO craft's central ergosphere, where the rotating base disc will drag on the toroid's magnetic flux lines, and will stretch and break them to stress the ambient below that UFO, to produce abundant supplies of electrons-positrons.

But there is an even greater supplier below the aaUFO, for just as the aaUFO's upper dome is used as an amplification device so too will the lower dome be used as an amplifier. But because the lower dome and the lower magnetic field are configured differently to the upper dome, that amplification will manifest itself primarily as an in-spiraling of electrically charged energy, in a way that will produce an increase in potential toward the central axis of the craft. Obviously, this will result in a shearing force below the UFO, but also, in consideration of the differentials in potential between the in-spiraling positive and the negative gravitational forces radiating away from the craft's black vortex, AND by considering the magnetic field spreading out below the center of the UFO then this shearing force will be the product of a whole combination of gravitational, electrical, magnetic, density, and kinetic forces.

Indubitably, looking at this area below the UFO's base disc it can be seen that it will be absolutely loaded with extremely powerful electromagnetic stress forces. So much so, that it can be assumed that the result of all these stresses, on the exterior ambient or space-time fabric will be the copious production of electron-positron pairs. Subsequent to this pair production and kinetic momentum there will be induced additional magnetic forces, which in turn will haul inward toward the lower spheres, for instance, electrical charges from that exterior ambient, or space-time fabric (see figures 68 and 69). As is also shown through these two diagrams, while the charges flowing out of the anti-vortex below the craft will be electrically positive, then because of their electromagnetic force-generation it would follow that the spiraling flow inward, from the surrounding medium (atmospheric air, for instance), would be electrically negative.

What these diagrams show also is that the in-spiraling energy will not interfere with the outgoing energy flow of the vortex, so it will circulate around the axis of the UFO very close to the craft's hull [note: 32].

To take advantage of that energy density gradient and that inhomogeneous graduation converging toward the center of the craft just below the black vortex, ideally, there should be established a 'current path' between the exterior ambient and the interior of the UFO, and most advantageously that current path into the UFO's central ergosphere should be directed THROUGH THE LOWER SPHERES. In other words, the lower spheres would act as the most porous material (electrically speaking) below the center of the craft—which they will be anyway, as explained in section 4.9, if they are fabricated with sufficient electro-positive 'vacancy' materials around their outer surfaces.

With this potential-differential phenomenon the UFO would then be able to take advantage of the dielectric absorption mechanism within the lower spheres, to separate out the negative and positive charges harvested from outside. And again, the acoustic fields buzzing around the lower spheres will be instrumental in shaping the incoming energies into oscillating charges.

So yes, the lower spheres will convert the virtual energies of the ZPF field of space, into real-time electrical charges. Then, as mentioned in chapter 5 of this study, the plasmonic oscillation phenomenon will transfer those charges through the dielectric bulk of those glass spheres and on into the craft's electrical circuit.

But then as these tentative renditions of those circulating fields below the craft indicate, there could also result a substantial amount of electron-positron recombinations, and therefore large amounts of annihilation photons produced below the UFO craft. Then, if the two shearing fields below the craft can be spread over the whole of the lower half of the craft (such as within a double-layered magnetosphere, for instance) it would be possible for that UFO to be sheathed inside a wall of photons—like a light ball.

11: Mass Diminution and Time Dilation in UFOs

Contents:
11.1 A UFO is a Sealed Environment Around a Quasi-black Hole   339
11.2 Dimension Anomalies Experienced by Abductees   346
11.3 Deopulating Within a Mass-density Gradient   347
11.4 Matter Transportation Field Projections   360
11.5 de Broglie Wave Mechanics and the Planck Scale   363

11.1 A UFO is a Sealed Environment Around a Quasi-black Hole

Its interesting how a picture can be intensely studied or viewed for a thousand times and then one day it shows something quite new. Very late into this research project I spotted something very interesting in several of Betty Luca's drawings of the scenes she observed when she was escorted by the Greys through the entrance door and into the operational part of their UFO. Judging by what she saw and what she drew of those scenes there appears to be one or two intriguing factors that should be taken to one side and considered further. These would concern the UFO's below-deck area, of where the engine part of the craft was and where the ergospheres of rotating energy were.

The main drawing of how the lower ergospheres looked from inside the craft is the one in "Andreasson Affair" (p38) (see above, fig. 62) which shows that the engine area has several levels to it, three in fact, with the most evident level being that of the black vortex. Considering the quasi-black hole hypothesis delineated throughout the previous chapters it makes perfect sense that she saw three levels; as surrounding the black vortex in this aaUFO (but looking to Betty like the deepest level), would have been the base disc of the central ergosphere above which the frame-dragging energies would be circulating around the outside the black vortex. Spreading outward from these two rotating fields, and slightly higher (and looking a little more solid), would have been the top face of the toroid (which is also the lower wall of the planar waveguide). And although it isn't mentioned in Betty's descriptions, above that, there would have been yet another level that would have been the level of vibrant energy shimmering just above that top face of the toroid, which would have been what I have been referring to as the gyrating storage field, the rotating field of charged particles analogous to an astrophysical black hole's accretion disk (this level would only have been apparent when the engine was fully operational during flight though) [note: 1]. Finally, the fourth level in Betty's drawing would have been the level she would have walked over after she'd passed through the door from the UFO's waiting chamber, it would be the lowest of the upper-deck levels which, while it establishes the horizontal demarcation between the engine's ergospheres and the crew's operational area, it also forms the upper wall of the engine area's planar waveguide.

But the most intriguing factor is that Betty could actually see through that deck level to the vortex below it!

For this to happed it would infer one of two things, either the Greys had made that deck-flooring transparent, which is plausible because as was shown in the previous chapter metamaterials in scientific laboratories can now be changed from being (RF) opaque to (RF) transparent by the flick of a switch (so the next stage onward for scientists would be to develop a material that can be switched to being optically transparent—possibly through terahertz wave oscillation), and this could be what the extraterrestrials were utilizing inside their UFO. Or, it could be that the UFO's deck floor had an opening directly above the black vortex and ergosphere A, where all the subatomic energy was being rotated [note: 2].

I believe it would be highly beneficial, from a scientific standpoint, to speculate that the deck was OPEN even though just below that deck is a fully operational black vortex. After-all, this arrangement would not be a huge danger for the crew because, as has already been noted in an earlier chapter (in section 3.17), there will be no direct vacuous force immediately above the UFO's quasi-black hole, it won't have a longitudinal gravitational force that draws matter straight down into it—its only in-flowing force will be rotational, acting at a tangent, horizontally. Therefore, if it can be quite feasible to suppose that this craft's black vortex was indeed fully open to the upper decks, then consequently the whole of the upper deck areas would be fully involved in all the bulk-effects that could be generated by that black vortex. And with the ability of that influence to permeate all through the various rooms and compartments of those upper decks then those upper deck areas would be fully subjected to whatever adjustments that vortex's density gradient would make upon the parameters of the dimensional constant inside that UFO.

The UFO is the quintessential sealed environment, it has a material shell with powerful electromagnetic fields controlling the rim duct and the base disc vent hole and the small gap between the base disc and the toroid shell, so whatever the dimensional constant is in the world outside there could be engineered inside that UFO a dimensional constant with quite a different set of values. The perfect dynamism to control pressure, mass-density, environmental dielectric constant, electromagnetic wave velocity, and gravitational force within that sealed environment is the UFO's black vortex—and these are all components of the fine structure constant. As a result of this control, it would be quite possible that inside that craft there would result a proportioning effect upon whatever material objects existed inside that craft. And indeed, because the upper decks would form an environment completely sealed off from the outside world, then the same proportioning effect would be experienced by whatever living bodies were present on those decks, be they aliens comprising the ship's crew... or humans being conveyed as passengers.

That the deck was open (and perhaps having only a safety wall around it—see figures 85 and 87 below) this would then explain why Betty said the air she breathed inside the UFO had a strong smell of ozone to it and felt as if it were electrostatic in nature. And this might explain why the Greys have been observed as wearing a one-piece skin-tight shiny suit and why they have to wear such thick boots, which, as Corso discovered from the US engineers who back-engineered the Roswell craft (which I believe was a craft very similar to the Andreasson UFO), because those engineers told Corso that they'd found out that the Grey's suits were made from a material specially woven directionally (in such a way as to be electrically conductive), so that any air-borne electric charges within their UFO could be directed to the deck safely on the outside of those special suits, while at the same time causing no harm to the body inside of them [note: 3]. That this protective wear was needed to take care of any long-term effects upon the Greys from exposure to an electrically charged atmosphere, would also suggest that the adjustability of the electrical permittivity of the air inside their UFO was a necessary part of that UFO.

This will go a long way to explain exactly why so many abductees have reported that the environment inside the UFO they were moved in appeared to be proportionally much larger than when those craft were observed from the outside (which means that it explains the commonly reported effects associable to UFOs and links them to the once-fictitious concept known through the British "Dr. Who" TV series as Time-And-Relative-Dimensions-In-Space—TARDIS, or as it is known of in science, topological diminution, and transposes that concept into a bona-fide reality).

Surprisingly, its easy to see why this effect should happen, for it comes down to the unique properties of a black hole's topological gradient regarding its mass-density and gravitational in-pull, and because these effects are most pronounced *just outside* the rotating black hole's event horizon—between the event horizon and what's known as the stationary limit (where, uniquely, there occurs the HIGHEST EFFICIENCY for any process which produces a mass-to-energy conversion).

And furthermore, for the UFO craft which engineers a quasi-black hole inside of it, and develops for itself a miniaturized version of those special mass-to-energy dynamics found just outside its very own event horizon, and which it wholly contains inside its purposely-built structure, which it uses to operate its own rotating-shearing energy field, then this effect can be especially powerful and effective. Yes, astrophysical black holes are gargantuan by comparison, but with the astrophysical black hole there is no physical barrier sealing the energy environment surrounding it, and there are no containment walls guiding and constricting the rotating fields within its ergospheres out in space... at most there might be a torus energy field surrounding a black hole in space which might slow down electromagnetic and gravitational waves propagating through it, but the astrophysical topology will essentially be an open one, so consequently the TARDIS effect will not be quite so apparent around a black hole out in unbridled space.

A UFO however, will have solid containment walls around its energy ergospheres—so that the more that energy is pumped into them the faster those energies will rotate in the center of those craft! And, of course, the UFO itself is a sealed environment...

Herein is the beauty of the UFO, if a topological environment can be created and perfectly sealed within the confines of the featured domed aaUFO craft then a brand new hallway of doors suddenly swishes open and leads to exciting new scientific fields—into quantum mechanics, into variable dielectric-constant fields, into de Broglie wave mechanics, and even into the proportional Planck matter constant. With this manufactured mass-density-gradient featuring throughout the whole craft just as much as it would be inside the UFO's ergosphere A, then, as per Einstein's $e = mc^2$ divided by the spatial dielectric constant (as noted in section 10.7), the electrical permittivity of the air-atmosphere within the UFO's upper decks would also increase significantly and lower the phase-velocity of electromagnetic waves across the whole frequency spectrum [Rabinowitz].

This would have a knock-on effect inside the UFO because then the black vortex would become even more massive, from the continual drawing-in of increased-mass-energy particles coming through the central ergosphere from these upper deck volumes. So not only would the UFO's black vortex establish an increased mass-density gradient between itself and the rest of the sealed environment surrounding it—but it would keep on recycling that air-energy so that it would continually build on those increments and environmental changes. On the outside this increase in intake would be viewed as a speeding up of that vortex's azimuthal velocity, but from inside that altered environment some things would appear to slow down (as will be explained below). So, by conservation of mass-density the black vortex would therefore have changed the whole mass-density gradient inside that sealed environment—and it would have affected physical dimension alterations to all of its mass objects, because, as I say above, all the factors which make up the dimensional constant (∂) will have been influenced by the UFO's black vortex.

The subject of the dimensional constant has been discussed from many different angles by physicists, particularly in the [Magueijo-Albrecht] paper with regard to their varying speed of light theory, whereby changes have occurred to one or to all of the component values in the fine structure constant ($\partial = e^2/4\pi\hbar c$). From a universal point of view the dimensional constant we experience in our everyday lives, presently, is different now to what it was in our distant past (as was noted recently in the red-shift signatures from deep space energy fields around quasars) [note: 4]. The most important cross-reference to emphasize here from these universal points of coordination (because these effects will also work on the local gravitational potentials that are generated within rotating fields), is that the most pronounced variations in the dimensional constant (fine structure constant □) are known to exist near massive objects, such as around black holes, where those variations are proportional to the local gravitational potentials surrounding those massive objects [Magueijo 2003].

In fact, it is highly significant that there is such a broad array of recognizable data to be found in many UFO reports, but also such a diversity of unexpected alterations in this dimensional constant phenomenon that scientists have clearly not ventured to think about. The reason for its significance is this, theoretical physicists have long conjectured that if there were changes to this fine structure constant, those changes wouldn't be noticed by anyone because everything related would all change in the exact same proportion; as noted in the words of Poincaré when he asked, "What if you went to bed one night, and when you awoke the next day everything in the world was a thousand times bigger? Would you notice anything?" To this question physicists in the past have said in one voice, NO, because they have opined, that all the measuring rulers and other objects of matter would all be effected in exactly the same way, so thanks to our physicists and their introspection little has been learned about this sliding gradient [Krogh].

But here we have the data from those people who have gone through these dimensional changing experiences with the extraterrestrials and who have the empirical knowledge to show how this effect works. Therefore, the reports by abductees have provided an invaluable contribution to the wider understanding of this TARDIS effect. Indubitably, it can now be verified that Poincaré was quite wrong in his assumption about not noticing any difference when changes are made to the fine structure constant.

Because, as a result of piecing together the evidence of this phenomenon taking place inside UFOs from abductee reports, it appears that not EVERY ONE of the mass-space-time components change through exactly the same proportional gradient, as will be further alluded to below (and in following sections).

For instance, although at the time of her abductions Betty Luca didn't notice any changes in her own mass density, she did notice certain anomalies and she did record anomalous differences when she drew her drawings under hypnosis of the crafts she was in. We are fortunate to have in Betty a perfect artist, and one who 'drew it like it was' (even though some of the things she experienced didn't make much sense to her at the time). But because so many drawings were produced, and because her abductions were numerous, many scenes in her collection can be compared, and used to provide stand-alone evidence or used to confirm accounts reported by other abductees. With regard to the present subject whenever she was INSIDE the UFO craft she was always correctly proportioned, just the same as if she were standing inside the room of a house (ie she would be standing up straight and there would be space between her head and the UFO room's ceiling). Yet whenever she was outside next to that UFO craft she viewed herself as being proportionally larger than she should have been to fit comfortably inside it. This factor has been noted by numerous other abductees who have been aboard UFO craft. Frame of reference is always different inside to that of outside. There is usually evidence of time dilation too but that subject will not be gone into in any detail in this study.

Figure 83. How did the Greys, and Betty Andreasson, AND the tall Elder all fit comfortably into this UFO? This drawing shows Betty being brought before the Great Door beyond which she saw Beings of Light (from "Watchers II" fig.37)

This sizing theme was even one that Bob Lazar mentioned, in that he was puzzled too at the extremely small amount of space available inside the UFO he was asked to back-engineer at Area 51, speculating that only small aliens would be able to travel in such confined compartments.

According to the variable metric theories (expounded by physicists such as H.A. Wilson, Dr. Hal Puthoff, and J.G Depp, for instance) this is perfectly feasible. Moreover, subsequent to the understanding of how this UFO craft's ejected wave-packets might be comprised of slowed group-velocity waves, it might be worthwhile tracking back a notch here to take a closer look at the energies bouncing around inside this craft. The electromagnetic waves radiating away from the central black vortex that pass through the toroid (ergosphere C) are being slowed by their multiple-reflections through the moving toroid fluid (see figure 73), due to its increased refractive index (sections 9.4, 9.5 and 10.7). And too, the whole volume of ergosphere A around the black vortex is a frame-dragging high-permittivity environment. So too is ergosphere B (the storage energies gyrating through the planar-wave guide). And while these moving field dynamics in all three of the craft's ergospheres are slowing down all electromagnetic waves passing through them—EACH TIME they pass through them (so this will be a compounding effect)—then these actions will all be contributing to the establishment of a ($k > 1$) environment throughout the UFO's upper decks (also see figure 80) [note: 5].

11.2 Dimension Anomalies Experienced by Abductees

Would Betty Luca have noticed she was getting smaller, shrinking down in size when she was entering the UFO? Not necessarily if the changes happened slowly. Certainly, I know she wasn't consciously aware of the process happening, although she did record its effects. And this account will explain another oddity I've noticed in the design of the aaUFO (see above, figure 65) where a whole quarter-section of that UFO is given up to serve as the entrance area. Its just a waiting chamber yet it is sealed-off from the inner part of the upper decks (and the control area), and its also sealed off from the central ergospheres (except for the gully in the deck that runs all the way around the inside of the craft's rim). This when every other area of that craft is jam-packed-full with ancillaries in purposefully compartmented areas busy with some action or another. Yet, this entrance section is nothing more than a featureless volume which is sealed off from the inner control areas by the swish doors that blend unnoticeably into its walls when closed. My conjecture is that this will be a holding chamber that will be used by the ETs in much the same manner to how a decompression chamber is used by deep-sea divers when they come back to the water's surface, in which divers have to spend time adjusting themselves to a change in pressure half-way to the water's surface [note: 6]. While the environment inside the UFO will be manufactured to a different density to that of the world outside then whoever enters the UFO through the dome shell entrance hatch, and intends to pass between the two different density environments, will always have to spend time in that UFO's holding chamber so as to have their bodies adjust to the different mass-density being maintained within the sealed part of that UFO.

What supports this conjecture are some hitherto unrecognized details that can be found in Betty's statements describing, under hypnosis, the time when she was brought into the aaUFO craft by a leader Grey accompanied by three other Greys ("Andreasson Affair" p37) where Betty says, "We're in like a half-bubble, or quarter of a bubble, room. And he [the leading Grey] has withdrawn himself with the others, and they are standing over there talking….[see figure 65 above and figure 85 below] I'm just looking at this room. Something goes down on the sides of the bubble…Where the steps come up, it goes down [the gully]…I feel very weightless and icky. My hands and my legs feel like they are asleep or something. And they are still talking over there, and they glance over at me once in a while…Oh, hurry up! And I'm crossing my arms now. I'm tired of just standing there…I feel weightless. Oh, my feet are pins and needles or something—even my arms and my hands. He's still talking, and…about two or three of them are leaving. That door whooshed open, and they are going in—and its closing." And in ASBT-II p28 Betty says, "The whole craft I saw in 1967 was about thirty feet in diameter, and once inside and past the waiting chamber it appeared much larger than first expected."

From these recollections of when Betty thought she was just waiting for the leading Grey to finally stop communicating to his associates, although she couldn't then understand what was going on and grew evermore tiresome of the waiting, its quite obvious the leading Grey was accounting for her body to acclimatize itself to the craft's inner mass-density environment, just the same as the aliens would have been doing for their own bodies during that period of waiting. Hence her feeling of weightlessness and her pins-and-needles (or paresthesia, which in itself denotes her environment changed to that of a different pressure). Once the waiting chamber had been hyper-densified the inner doors could have been opened, and only then would Betty have been taken into the heart of that UFO.

Actually, in Phase-One of the Andreasson abduction experiences (as published in "The Andreasson Affair") a tentative reconstruction was made by technical artist Fred Youngren of one of the UFOs which Betty went inside, these were the diagrams I originally based my 2000-2001 website diagrams upon, and which, for some strange reason, are still being portrayed on the websites of ufologists (which does suggest to me that while ufologists are disposed toward reporting about UFOs they are not necessarily disposed toward thinking about UFO technology). But as the reader will notice Fred Youngren's technical drawing shows that he too was having problems trying to fit everything Betty described as seeing into that UFO which, to an outside observer would appear to have been only thirty-feet in diameter [note: 7].

11.3 Deopulating Within a Mass-density Gradient

Again, here is another variation on the theme of this mysterious thing about the ET's using mechanisms to alter the size of material objects; referring to this image (of figure 84 on the next page). Betty Luca said, continuing on from the episode of the UFO's rejuvination at South Ashburnham lake (section 7.4), that after the second UFO re-powered the stricken UFO it then hovered in the air over that craft, and while Betty was standing on the ground she watched the upper UFO extend down a cylindrical energy field all around that UFO (which had brought Betty to the lake), and as a result of that energized field the lower craft shrank down to the size of a car—even though the man and the woman (see figure 63), and a number of Greys had gone inside it!

In contemporary investigative science, in the field of UFO and extraterrestrial scientific research (for instance, at Vallee-NIDS, the National Institute for Discovery Science), this mechanism is called *topological diminution*, or space-time diminution, and again, the key factors involved relate to how material objects become subordinated to whatever mass-density gradient is manufactured around them. This is because when changes have been made to the component factors of the dimensional constant (fine structure constant ∂), which have to operate through energy equivalence principles, so the effective mass-density of objects also changes, and that change equates to the scaling of objects [note: 8].

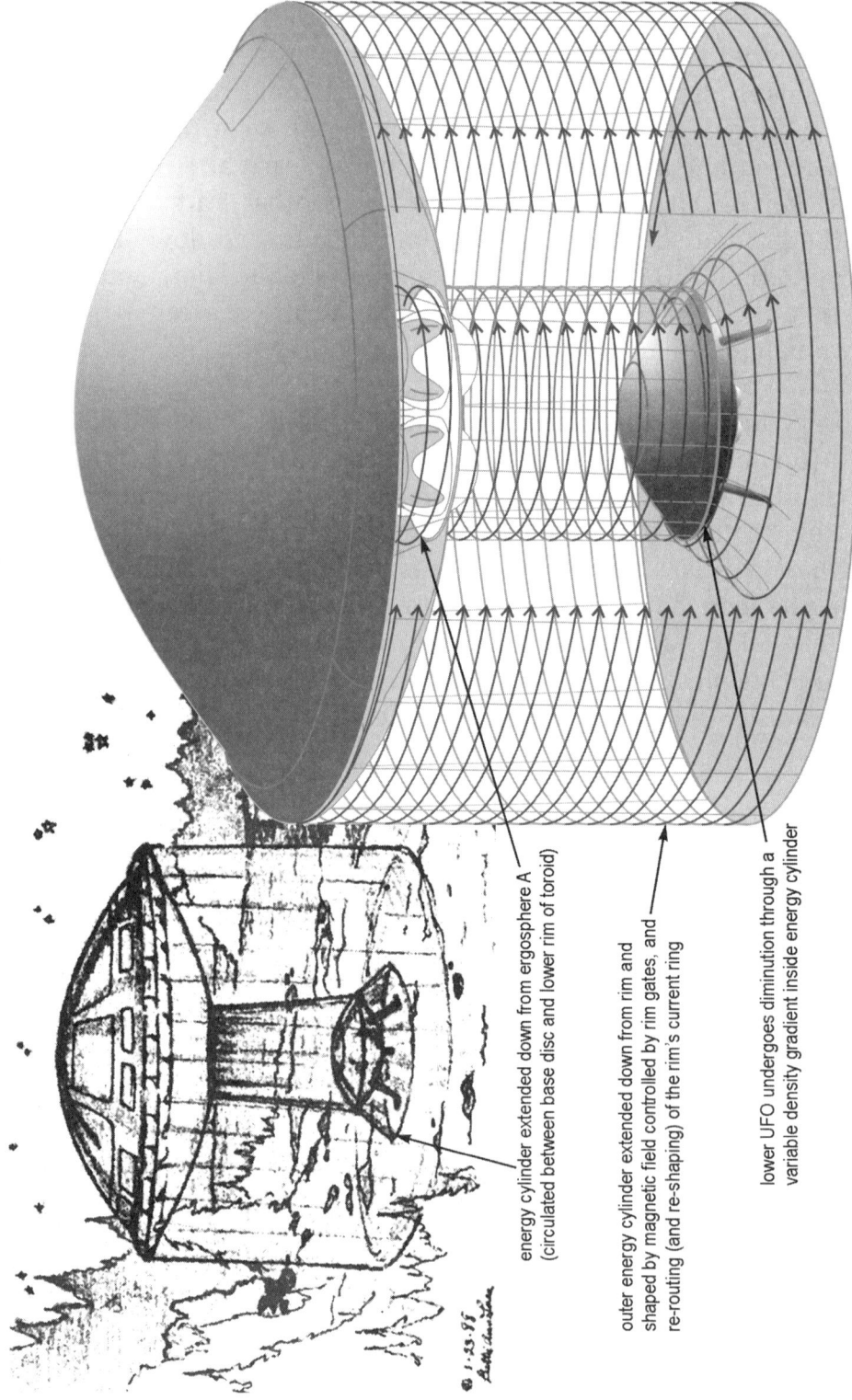

Figure 84. A UFO establishes a skirt between itself and the ground and then engineers a mass-density differential between itself and the lower craft. Inside the lower craft when it shrank were two humans and a crew of Greys! (from "Watchers" fig.16 and "ASBT-II" drawing 52)

The Greys call it "opulating"... and in this instance this was done by them transferring the mass-density altering influence of the craft's black vortex inside to the outside of that UFO. Through, what Betty perceived at the time, the establishment of some sort of energy-skirt that appears (as shown in figure 84) to have been extended down from the upper craft's circumferential rim to the ground.

At first I thought this skirt may have been required to hold the upper craft steady in the air, but while that craft has the gravitational buoyancy mechanism in operation via its central black vortex then that steadying action would be unnecessary. So by extending this skirt down it obviously marks the boundary of an energy containment field, such that while this skirt is in place that UFO will be generating a perfectly sealed cylinder around itself—with the cylinder bounded by the UFO's rim and under-belly, the skirt's wall, and the ground. This would exclude the ambient influences of the outside world while another skirt is sent down inside it from the base disc of the craft.

They would be doing this by taking energy from the rotating storage field and from the rotating fields around the central vortex, and they would be projecting them both down from the craft, sending ergosphere B as a rotating energy field downward from the circumferential rim, and they would be projecting ergosphere A down through the gap around the base disc (through the gap between the base disc and the toroid's inner rim), and they'd either be establishing one outer and one inner cylinder, or they'd be having both of these rotating energy fields meet at ground level to seal together as a double-cylinder.

Either way, these cylindrical walls coming down from the inner and outer rims of the toroid, would follow the basic principle of the Meissner effect where in a helical energy field the magnetic field and the charged particles it accelerates are predominantly grouped as a thin outer wall or sheath around that helical field. To keep the walls of these (two) cylinders straight the extraterrestrials would simply be copying the collimation process of when helical sheaths are formed out of black holes (as explained in section 2.10), with hoop stress of inward magnetic tension balanced by an outward centrifugal force (of the particles rotating in the walls), and with this balance they will keep both cylinder-walls collimated. Once the inner wall of that double-cylinder was in place the UFO would be able to confine the pressure-density effects that resulted from whatever permittivity-influencing plasma was being sent down by the ETs into that inner cylinder (as a continuation of the black vortex's density gradient), in the same way that gradient can be developed and allowed to permeate into the UFO's upper decks, so that the ETs can quite simply re-organize that volume's dielectric constant and consequently the dimensional constant of any physical object inside that inner cylinder's perimeter wall [note: 9].

The clues to what the extraterrestrials were doing in that scene, can be found in another instance recorded by Betty, of when she was being taken on board one of the small UFO crafts, where she did actually register that the 'opulating' and 'deopulating' processes may have been carried out on her, and where she records,

"The exterior of this craft did not seem as large as the interior, however the Watchers mentioned one of their words and its meaning to me. They said they could "opulate and deopulate," which they revealed meant, they could make more or less of, increase or decrease size, appear and disappear!" [ASBT-II p27-8] [note: 10].

Also, this will explain some other comments made by the Greys, recalled by Betty when under hypnosis, which inferred that these extraterrestrials have come from an environment which has a greater mass-density than ours, one which is much heavier than ours ("Andreasson Affair" p144). Well, if it's a heavier environment than ours then no wonder they are comfortable existing in an environment smaller than we humans normally exist in!

In fact, there are many scenes drawn by Betty which have humans, Elders, or Greys, standing close to the UFOs they had just been traveling in, and they all suggest that the UFO is far too small to accommodate them (as is notable in figure 30 above).

In UFOlogy this down-scaling factor has led many observers to believe that the majority of UFOs that have entered our atmosphere are only scout craft, too small to possess inside them any living crew members—so quite clearly that scout-craft theory can be thrown right out the window! What this down-scaling factor would help to explain also, blowing away the Roswell cobwebs again, would be the strange observations made about the four alien crew members in the civilian reports about that Roswell craft; from those observations that the craft those extraterrestrials came down in was disproportionately smaller than it should have been relative to the size of its crew.

11—Mass Diminution and Time Dilation in UFOs 351

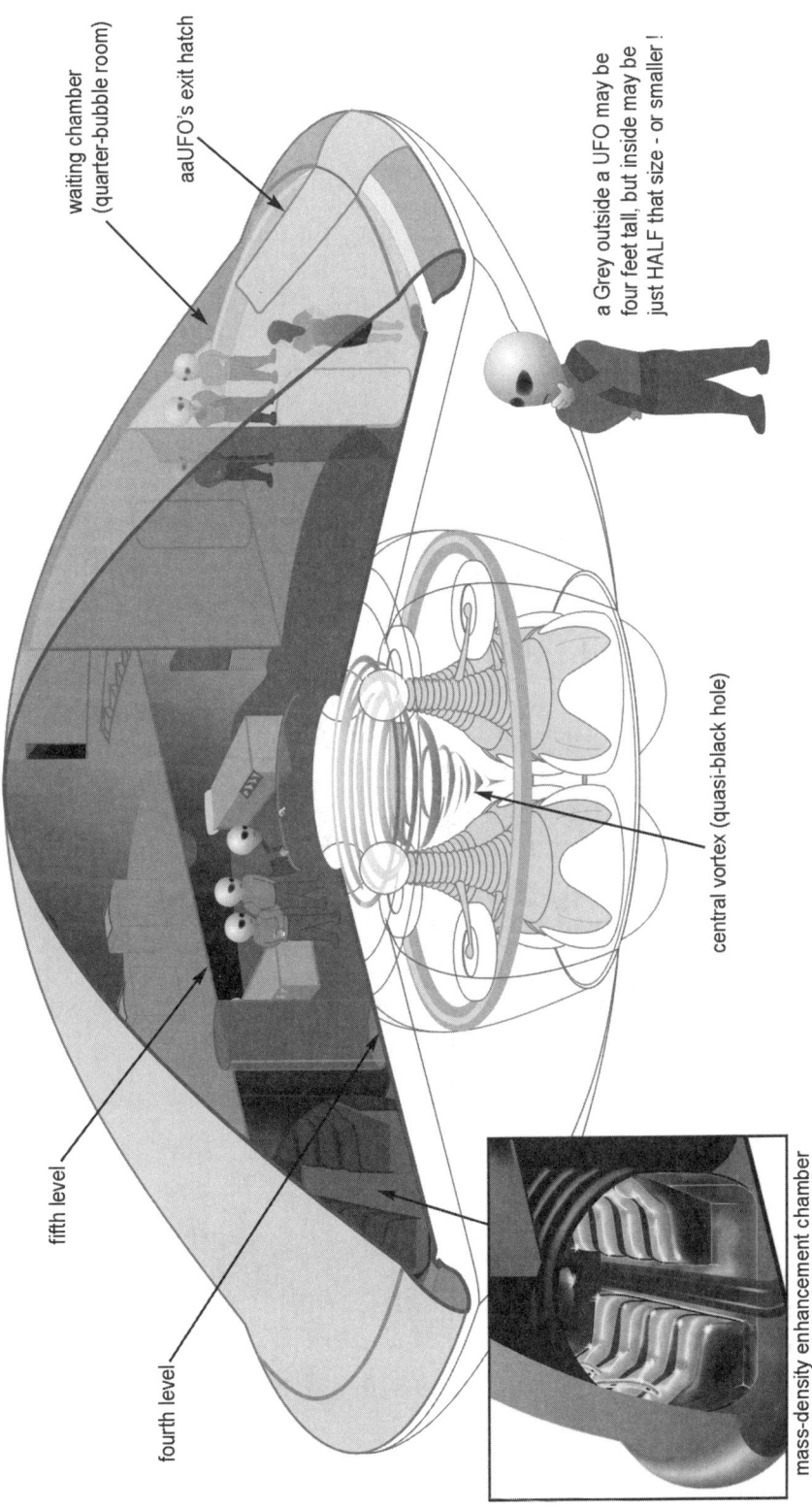

Figure 85. The secret to a 'TARDIS Effect' inside a UFO has to do with its sealed environment (see also figure 54 of this study)

Which suggests that while outside their craft, in the earth's atmosphere and within the constraints of our dimensional constant, those Greys are about four feet tall, but INSIDE their craft with its interior density gradient established and sealed off from the outside world then those Greys might only be half that size—or possibly even smaller than that!

One slightly different account of a UFO being opulated, or having its effective-mass-density increased (so that it is obliged to reduce its physical dimensions)—and this refinement will give the researcher even more information with which to better understand this effect—will be that of the Dalnegorsk UFO. The main details of this UFO seen in 1986 have been stated in section 5.10 above, so I won't go into all of its details here: This UFO was observed by many witnesses, and even photographed, to have landed on a hilltop close to the town of Dalnegorsk, in Russia, where it stayed for two hours while its crew affected a temporary repair to one of its power units. From these observations, and also from subsequent magnetic field readings taken from where the UFO landed, it was calculated that the UFO was only 1.5 to 2 meters in diameter all the time it was on top of the hill, and that it had an energy orb around it which seemed to inter-phase with the UFO so that a vacuous environment could be established between the UFO and the outer wall of that energy orb. This vacuous environment was conjectured because of the numerous samples of carbon-glass, wood, and ferroelectric metals that were later found at the site which while they showed obvious signs of being surrounded by intense heat (of above 3500 degrees Celsius) they had not ignited and some of the wood hadn't even blackened, which gives the suggestion that all the oxygen had been taken out of the environment between the UFO and the outer wall of its energy orb (there are a whole list of chemical effects which were observed by the scientists who analyzed those UFO artifacts too, which I'm not going to list here because they will be the subject of a more comprehensive publication at a later date, but suffice it to say that where oxygen should have been in many of the samples it was missing).

And so, in view of this hitherto unrecognized facility of UFOs to manufacture their own mass-density gradient then I would postulate that the vacuous environment noted around the Dalnegorsk UFO was the result of a density-differential established between the higher mass-density of the UFO itself and that of the outer wall of the energy orb which surrounded the UFO; and that the outer wall of that UFO's energy orb was used as a confinement-barrier to differentiate between the UFO and the mass-density environment we exist in on earth.

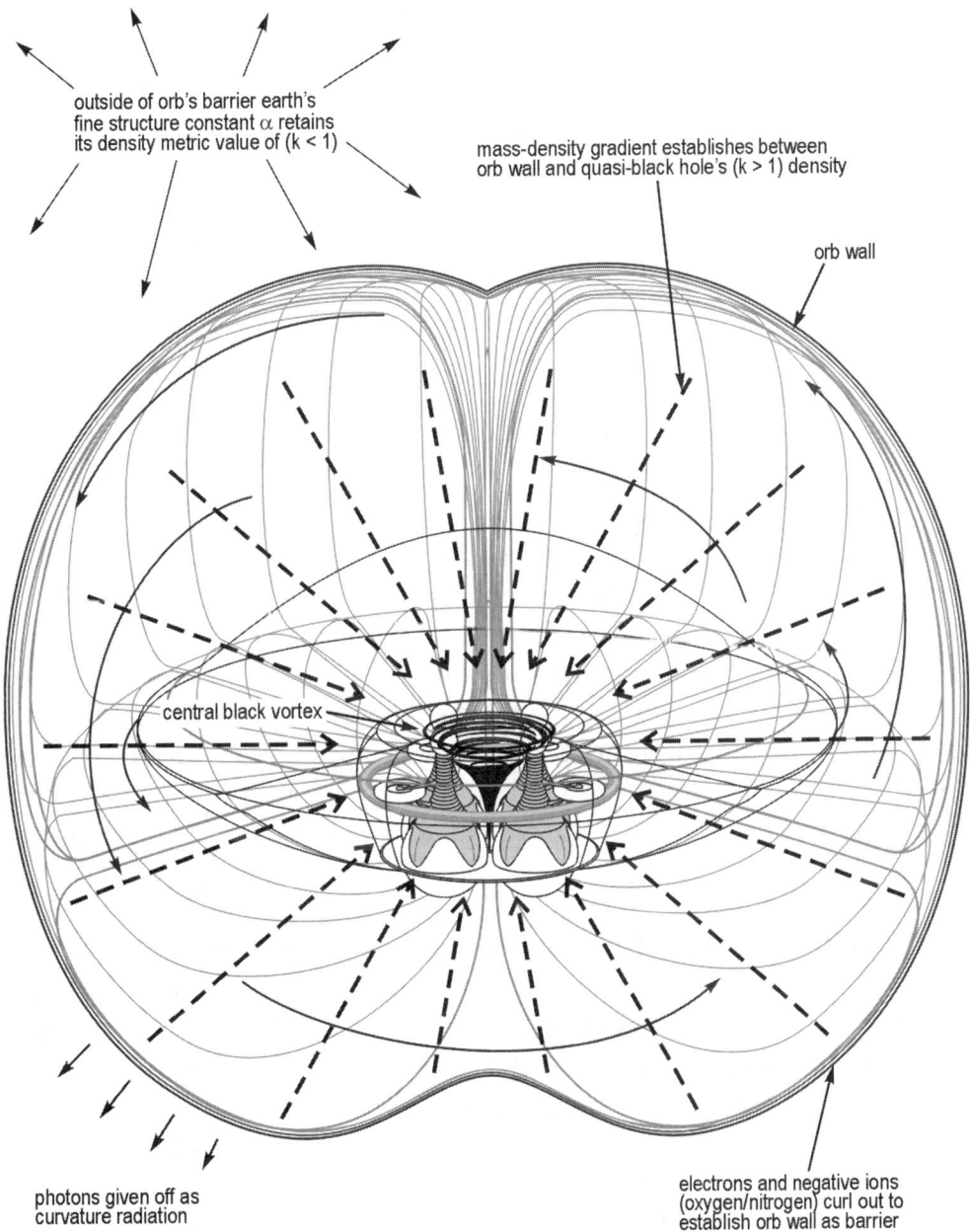

Figure 86. The energy orb surrounding the Dalnegorsk UFO would have provided a barrier, or shield, between the UFO's higher (k>1) density and the earth's (k<1) mass density

Interestingly, when the effects of this opulation mechanism were better understood it was then realized that these effects were also there to be found in the British Rendlesham UFO incidents, where it was reported by two US security patrolmen, Jim Penniston, and John Burroughs, that the first UFO they encountered was the size of a small car (of around 2 meters in diameter). Furthermore, when patrolman Larry Warren was describing the second Rendlesham UFO he said he actually watched it reduce its size down from 30 feet to 20 feet [note: 11]. Most significant of all though, is the fact that in their separate reports Penniston and Burroughs mention that the first UFO they encountered was surrounded by a bright misty plasma and that as they tried to approach it the air surrounding the UFO densified more and more as they got closer to it, until, they said, it was like trying to walk through treacle!

A corresponding feature of this different density environment will be that timed-events (when inside a UFO) will take less of our time to complete (ie their clocks will run slower) so that 10 hours spent inside a UFO would equate to just ONE hour back on earth, for example. But does this apparent time-dilation inside a UFO craft apply to all timed events? Incredibly, it would seem not to, for in our correspondence Betty told me that she remembered once, when they were transporting her back in their craft, that she happened to be watching a group of Greys who were leaning over an open cylinder, and she could see they were fascinated by something going on down inside that cylinder, but from where she was positioned she couldn't see for herself what was in the cylinder, so she asked one of the Elders standing near her what those Greys were doing, and the Elder informed her that they were watching molecules moving around..!

Now, the only way those Greys could have been watching the movements of molecules was if those molecules had become comparatively larger, or otherwise, this was an indication of timed movements slowing down. It is more likely though, that those molecules must have been moving slower than how we would expect to find them, and if so then the old Heisenberg Uncertainty Principle flies right out the same window as the ufologist's scout-craft theory! If they were watching molecules they were most probably watching molecular collisions, and observing with their own eyes—in real-time as they happened—events our scientists have only managed to record through still photography by using photographic emulsion to produce snapshots of!

Figure 87. Betty arrives into a room where Greys are watching molecules! Does this mean that electromagnetic energies were slowed down because that environment's permittivity had been changed? From "Watchers II" fig.53

This observation, of course, will have interesting ramifications for the Magueijo-Albrecht conjectures about the variable speed of light!

Furthermore, and to confirm that this time-dilation effect can be associated to the field of higher mass-density around a UFO, in the reports of the Rendlesham UFOs, especially for the second UFO, security patrolman Larry Warren reported that after the second UFO energized itself—and after it reduced down in size—the energy field surrounding that UFO completely distorted his perception of motion and time—so that everything he saw was in slow-motion, at half speed. And so additionally, this comment would indicate that Warren had entered into the UFO's higher-density field gradient and he himself had become part of it, and that he was observing whatever was going on around that UFO from inside that UFO's energy orb [note: 12].

But just to clarify this thing about clocks slowing down when the effective mass increases: Is this molecular slow-down and object slow-motion to be expected if clocks are said to run slower? Of course not, the latter observation is counter-intuitive, if clocks are running slower and our timed events are taking less time then subatomic particle movements and particle collisions should be moving all the more faster—but apparently the opposite is happening and particle movements are happening in slow motion. So this would indicate that object diminution does take place and clocks will run slower inside a UFO's higher mass-density environment—but that electromagnetic wave velocity REMAINS THE SAME as it is in this lower mass-density environment.

So this is why I say that here is empirical evidence which suggests that not ALL the factors of an altered fine structure constant seem to be changing through the exact same proportion inside the UFO's altered environment. And, almost as importantly, it has to be said that there are always two ways of perceiving these differences—as a person who looks into that altered environment from the outside, or as a person who looks outward from that altered environment. What the scientists need to do is actually experience for themselves these altered environments before they compare the two and do their mathematics on this new field of science. And going back to Poincaré regarding his thoughts on this effect, the answer to his question would have to be... yes, we would notice the difference [note: 13].

But what are the full ramifications of these alterations to perception and environmental integration? What can earth's physicists expect to discover with such altered perception when those physicists undergo environmental integration into another mass-density? Those physicists would do well to delve into the astrophysics of black holes in order to get a taste of what they might find. Miniaturization of mass objects, slowing down of electromagnetic waves, and wholesale changes to the Planck constant which would equate to different inter-atomic spacing for a start... and here is another savory morsel to tempt those physicists... the latter changes would equate to less disruptive binding forces within atomic nuclei.

These changes, of course, are environmental factors that the extraterrestrial black hole physicist would inherit from the elasticity of space-time. But further, consider what happens when that physicist can negotiate between the different environments of altered mass-dimension-constants? Well, there are numerous references that suggest, nay confirm, that many of the extraterrestrials enjoy an operational basis that has some degree of control over the space-time matrix that we think is a rigidity. This rigidity's hidden secret though is all about bandwidth, and the ETs will be using this enhancement of their perception to discern energy dynamics through a much wider range of the energy spectrum than is currently available to our earthly scientists. Here is an example, the reader will see its complete relevance in due course through the supplementary chapters of this study, while here it will show a possible way in which the forces inside an atom can be exploited when materials can be manufactured in one dimension and utilized in another.

Because, through the availability of two different mass-density environments, two different modes of the Planck constant (h) as it were, there will exist an energy excitation gradient between the two, and so it may be possible to exploit certain materials that might be more easily manufactured at a greater density than can be manufactured here due to them having lower inter-atomic reactions. I'm thinking here of superheavy elements (from elements 110 to 118). From the possibility that the atomic nuclei of those elements can be made more 'cohabit-able' than they are here when synthesized through a completely different set of parameters, and in particular through a comparative lessening of their atomic binding energies, these materials so formed could then be brought into this density realm with inherent stabilities hitherto unknown on this planet. This might be a way in which these so called 'exotic' new elements might come into this realm and be used on this planet to great advantage. I give as the perfect example the synthesis of the superheavy element ununpentium 115. The synthesis of this element on earth has so far resulted in difficulties, from the fact that existing R/D while it can provide this new element with a nucleon count of 115 protons that same research can only combine with them 173 neutrons, and this results in a sub-standard element with a half-life of just 87 ms. Yet the island of stability for this element has been calculated to be around Z-299 (115 protons and 184 neutrons) and while several attempts have been made no stable synthesis of ununpentium 115 has yet been achieved [Geng 2003]. However, with the availability of another mass-density laboratory environment that superheavy atomic material might exist in a 'calmer' state, less aggravated by the inter-atomic forces such as the strong nuclear force and the Coulomb repulsion force. In other words, it might be more conducive to the stable synthesis of these superheavies if they could be manufactured within a greater mass-density than ours, and then transferred into this density environment for our use in the fission and fusion industry.

Also, from the fact that electromagnetic waves operate at a relatively SLOWER scale (slower group velocity) for the extraterrestrials in those higher density environments, it means that it will be much easier for them to discover energy mechanisms occurring at higher frequencies. These sorts of frequencies will be a lot higher than the frequencies that our scientists are comfortable with given our present database on chemicals and materials, and so our scientists may only have a fleeting knowledge of these far-end-of-the-spectrum mechanisms.

What might this mean? Well, for one thing where our scientists have been preoccupied for so long with GHz electromagnetic waves, ie the microwave frequencies, the ETs will be working with THz wave frequencies and above, and so their scientists will have discovered and developed materials which work at infrared wavelengths and UV wavelengths for instance. And this would mean they will be developing and engineering machines which will be working through plasmonic and phonon energy interactions to affect their power transmissions (rather than how we just use electron-charge transfer through copper wiring).

We do have an awareness of these advanced sciences, even though for us they are still at their infant stage, so on the one hand, although the ET's world will be entirely different to the one our scientists work with, on the other hand it will be fairly easy to see what areas they have been working in by the way they construct their craft.

What correlates with this are the sightings of UFOs shrouded in infrared radiation, for example, and what also correlates with this are all the observations in the present study where the ETs are using plasmonic and phonon interactions within the actual structure of their UFOs in order to transmit energy (and this by the way, will be why the ETs don't use separate copper wiring arrays the way we do in their electronic transmission lines). For the astute designer this is something to bear in mind when designing advanced aircraft for near-future travel too, so that our space travel vehicles can start to come out of the stone-age (or is it wire-age..?). Rocket science development may be regarded on this planet as the bee's knees but it still uses wiring looms by the truck-load, and I would think that for the outside engineer who uses metamaterial transmission lines in the terahertz range, knows precisely how gravitational force is generated through massive rotating fields, uses rotating field power generators, and works every day through different fine structure environments then rocket science to them is really only at the beginning of a long journey (although, I'm sure the extraterrestrials wouldn't be so reformational as to say that for themselves).

The ability of the extraterrestrials to control their topological environment, at will, may also now explain the red and green environments that Betty Luca was passed through on her way to the so-called 'crystal forest,' as alluded to in chapter 1, because it now becomes quite clear that the Greys were using this density facility to 'transpose' an abductee to what was supposedly a different environment in an entirely different location. Betty drew several views of this separate sealed chamber which had been constructed inside the aaUFO she was conveyed in (as published in The "Andreasson Affair" figures 24—28, and see the inset in my figure 85 above). This was the chamber fitted with rows of seats and which had cylindrical walls especially corrugated for extra strength. It was from this chamber that Betty described being taken, in the dark, along a track, firstly through a red environment and then through a green environment (as described in ch1's note 13). Within that sealed chamber inside a UFO energized by that craft's black vortex the environmental metric would have become more dense ($k>1$), meaning that the electromagnetic waves carried through would cover a greater distance, and they would redshift. For the greenshifted section there might be two ways this might occur: If the already redshifted electromagnetic waves are then carried through to a less-dense ($k<1$) environment then these red waves will speed up to green (and, correspondingly, any visible white-light waves will be shifted up to UV). This, if nothing else, would suggest that the topological diminution effect can be performed through several separate stages on a physical body. But then, another possibility for the greenshift might be that if the new environment Betty was being taken to already existed at an electromagnetic wave velocity greater than (c), or if the EM wave velocity of that new environment was modulated at a higher pulsing-rate than the one that exists here in our presently-experienced material world... and so by saying that I'd be saying that the new environment existed as another dimension entirely... then, when those redshifted waves were carried through into that 'faster' environment, and Betty along with them, they would have covered less distance—and so they would have changed color and would have been greenshifted. I would like to think what happened with Betty was the latter, and that's why in section 1.3 I have proposed that the crystal forest holographic vista was operated at a UV frequency (and this would have been why Betty and the Grey that accompanied her were made to wear the glass bridging shoes). This latter mode of dimension-shifting would then also inter-relate to the alien movements described by abductee-experiencer Jim Sparks where he observed the Greys performing this dimension-shifting as they passed from our outside dimensions into our earth-realm physical world, whereby, as happened with the Greys that literally walked through the closed-door of Betty Andreasson's house ("Andreasson Affair" fig.3), here is obviously a branch of physics already used by the aliens which is ready and waiting for earth's researchers to discover more about, especially where, as Jim Sparks says, "The E.T.s can be 5% to 10% in this dimension on this physical plane while the rest is in another dimension. There are dimensions out there like layers of an onion... The E. T.s molecularly change the vibration of something so they can be in another dimension and solid, but be projected here in only 5%" [Sparks].

As an aside, within such an altered density sealed environment, where events are completed on a different time-scale to that experienced normally upon earth, this would explain why abductee-experiencers have reported time-dilation and time-diminution events as if it were the norm when being taken away from this world. Importantly, it would explain how some people who have been injured in these experiences, or who have undergone complicated surgery procedures on UFO craft, have experienced 'miraculous' healing before they have been returned to earth by the ETs (I cite as one example the Travis Walton case where his chest area had to undergo intensive healing for several days before he was brought back. But most especially this would relate to the gestation-related operations carried out on female abductees). And it would be interesting, although quite straightforward, to hypothesize how physical healing times would be beneficially advantaged subsequent to surgical operations carried out on this earth, if such altered environments were constructed and made available on this planet [note: 14].

11.4 Matter Transportation Field Projections

Coupled to the above-mentioned mechanism of producing helical flows out of the bottom of the craft, in reference to the image of this being used to bring living beings into the UFO (as shown in figure 88 below and as depicted on the cover of "The Andreasson Affair"), the same helical mechanism would equate to something that could loosely be described as the 'body transporter' mechanism. For, it would seem that the extraterrestrials are using a cylindrical energy field to transport objects up into their UFO through short distances of 20 meters or so.

This mechanism may well involve the generation of a helical tube of around two meters in diameter which extends down from the bottom of the craft and which shields whatever matter is inside it from the effects of ambient gravitational forces, and which imbues weightlessness into any physical body affected by its influence. Otherwise, and it would obviously need a lot more research to find out more about the parameters of this, but it possibly directs the black hole's angular momentum and gravitational force into the walls of that cylinder (again, using the Meissner effect on filament-tubes as mentioned above), which would shift the terrestrial gravitational fields to the outside of that wall, and after it created the necessary sealed environment around whatever object, or person, needed to be transported, the craft would then establish another gravity-density cylinder inside it, most likely as an inverted conical-field of the black vortex's positive gravitational influence, and this field would then be directed around the body or bodies to be transported, so that they would simply rise upward (or fall upward) into the craft through gravitational attraction [note: 15].

But furthermore, because with this mechanism gravitational influence is so interwoven with the dimensional scaling coefficient of physical mass objects, then there is good reason to suppose that when the beam is activated it could also alter the size of the bodies in that transportation process. Certainly, with regard to figure 88 the scale of the hole which Betty and the Greys are poised to pass through (and the reader will note that this hole is inside the protruding glass spheres in this drawing of how Betty remembered that particular experience), would suggest that the physical size of those bodies would not be the same for both the person(s) actually involved in this mechanism and the distant observer of this transporting mechanism taking place!

Figure 88. Either the UFO's vent hole opened up by a huge amount to allow this to happen, or Betty's and the Grey's physical forms underwent diminution from an alteration to the mass-density environment surrounding them. Note how this beam is on the INSIDE of the spheres. This image is from "Watchers" fig.9

The actual technicalities of how these fields can be established, or generated, by a UFO may be understandable from something I've already made note of about the circulating energies inside the aaUFO's central ergosphere. Particularly, for instance in chapter 6, where I have shown that the central ergosphere will have Lense-Thirring frame-dragging, but also Berry-type counter-flowing currents inside the toroid wall whereby these epicyclic rotations would equate to anti-votices (which also serve as kinetic energy reservoirs for the main vortex). These anti-vortices may yet prove to have other uses and there may be a way of controlling them independently of the main vortex, by amplifying them and extending their influence below the craft.

For instance, if these oppositely-rotating eddy currents can be amplified while they are inside the central ergosphere and above the base disc, they could then be used to induce a single circulating field that would surround the central vortex (somewhat like a flat donut outside of the vortex just above the base disc). Then, because there is already substantial hydrodynamic momentum operating around the lower part of the central vortex, just above the base disc where the vortex's angular momentum is at its greatest, this induced rotating 'donut' field could be pumped up to a substantial force (see figure 55). And then by another induction mechanism the force of this flat rotating donut field above the base disc could be mirrored on the outside of the base disc, underneath the base disc. Then, that secondary induced rotation could be extended down below the craft to provide the above-mentioned sealed environment (and again, this would probably involve a curl-developing electro-gravitational rotating field). Alternatively, this induced donut rotation above the base disc could be used as a holding field, to keep the UFO's electrodynamic power circuit operating within the central ergosphere, while the central vortex filament itself was widened and propagated downward below the craft [note: 16].

As mentioned above, such a rotating beam of angular momentum and gravitational force would develop a substantial shielding effect at its perimeter wall where most of its rotational energy would be concentrated. The secondary (density altering) rotating field could easily be sent down the center of that tube and depending upon it being configured with positive or negative momentum it might then be used as a transportation force, to pick up or to set down, any material objects found within that beam. For the person(s) being transported it would be like standing under the center of the 'eye of a tornado,' where the physical body would not experience rotational torque but would just experience that field's bulk-effect, of a lessening of earth's gravitational force, and then an increase in a 'gravity' which caused them to slowly fall upwards.

11.5 de Broglie Wave Mechanics and the Planck Scale

In the field of quantum mechanics the frequency of the de Broglie wave field and the de Broglie wave structuring determines the size of atoms, which, in other words, means they effect the size-scaling of anything that constitutes a material object. With regard to the interior environments of UFOs this is very interesting considering that de Broglie waves, in turn, are influenced by gravitational potentials [Krogh].

Similarly, the Planck constant for mass (as also previously mentioned in section 11.1), because it forms one of the components in the spatial dielectric constant (the fine structure constant ∂) hypothesis, any changes made to that Planck constant will consequently alter the way electron charges, quark exchanges, and inter-atomic forces, will be proportioned inside all the atoms and molecules that make up material objects.

But also, as Roger Penrose says, the Planck constant (∂) is proportional to particle momentum for a particular wavelength (relative to its frequency), so if the momentum decreases, as would happed when electromagnetic waves are slowed through a (k > 1) refractive medium, then the Planck constant increases; and through a k-less-than-one (k < 1) medium the Planck constant decreases. Which means that by changing the Planck constant, or by changing the relationship between a particle's frequency and its wavelength, then changes can be made to the effectiveness of any given quantity of energy [note: 17].

The refractive index of any medium is determined by the permittivity and the permeability of that medium, and changes made to these will effect how an electromagnetic wave's wavelength will correspond to a particular frequency. For instance light travels slower through earth's atmosphere than it does through this part of space because of the differences between those medium's refractive index (and light travels even slower through glass or through water for the same reason, so that light's frequency downshifts to that of colored light).

Therefore, the frequency parameters by which this present (k < 1) environment is established all around us will depend upon the Planck constant (h), which in turn depends upon the fine structure constant (∂), which in turn will depend upon the permittivity and the permeability in operation for this part of the universe. We have to ride along with that Planck constant and that fine structure constant, er... or do we have to? Planck time, for instance, is regulated by whatever values the gravitational constant (G) and the Planck constant are, so time becomes a variable just as does the speed of light when either the gravitational constant or the Planck constant or the refractive index are changed—as would happen inside an enclosed UFO. This would be why the extraterrestrials don't see any rigidity when they look at time, and its why many abductees have learned that the extraterrestrials just don't register time in the same way that we look so subserviently at our time-scales on this earth. Time to the extraterrestrials inside their sealed UFOs is a completely different commodity to the stuff our wise teachers have been informing us about and showing us how we should run our lives by.

Does this mean there can be time-travel excursions back into the past? I don't think so, although, I can see that the ETs could in theory slow time down to a complete standstill. As our Watchers at this particular time in our evolution the extraterrestrials would be more concerned about our futures than our past, and I would see this as their overriding concern as they interphase with us humans. So I'm sorry to say that Dr. Who is probably the only wise old owl that should be asked that question about traveling back in time! There is a way, however, that the events we recall of the past can be changed by these extraterrestrials (and this partially was what the alien abduction program was all about—but more on that another time, perhaps).

A more profitable avenue to go down, regarding topological environment control, might be to change the operational frequency of the material universe that we are currently subservient to, this may then offer us an entrance into a quite different dimension—if all these parameters can be understood and used to our advantage to affect some sort of shift. Because, if a person relocated to a different frequency and that person had a different Planck constant (for instance, the extraterrestrial's home environment would have a different value of (h) for its heavier density), then that would change the values of the gravitational and mass-density constants too. Hence they'd be accessing a mechanism that can either opulate or deopulate (as the ETs refer to it), and they could even blink in and blink out of existence—which presumably shakes down to simply shifting out of one frequency of material existence and shifting into another frequency of material existence!

This then, might mark one of the fundamental differences between extraterrestrials and homo-sapiens, we're imprisoned in this rendition of material existence and we are constrained to abide by 'our' designated rules of space and time (because these have been set down in stone for us by the likes of Einstein and Newton, and of course through the wagging fingers of lesser physicists). But the extraterrestrials have free-will, and they aren't constrained because they've long ago woken up to these variables [note: 18]. Or are the homo-sapiens about to wake up too? Is there being shown to us here a doorway that we might open for ourselves, one that perhaps might lead to a more ethereal existence?

12: Area 51's S-4 UFO Reinterpreted

Contents:
12.1 Lazar's Superheavy Gravitational Field	365
12.2 The Bob Lazar S-4 UFO Through the Looking Glass	369
12.3 The Central Reactor's White Hole	372
12.4 Fishing for the Gravity A Force	376
12.5 Bob Lazar's Technology Reconfigured to Make More Sense	377
12.6 Quasi-black Hole's Delicate Stomach	378
12.7 Bob Lazar's Element 115 is Here to Stay... its Official!	385
12.8 Constant Energy Delivered from Polarized Ununpentium 115	388
12.9 Proton Cyclotron Beneath the Dome	394
12.10 Pitching in From Three Angles to Load the Dome	398
12.11 Holy Greyl Found Hidden Inside Lazar's Frequency Parameters	405

12.1 Lazar's Superheavy Gravitational Field

From Bob Lazar's description of the S-4 UFO's element 115, where he says that he was shown an energy field extending around the wedge-shaped piece of 115 that the alien's used for fuel, Lazar might have been slightly mistaken in his understanding that this was the field of gravitational force which that UFO had to amplify and to propagate into a beam, in order to provide that craft with a propulsion force. Recent researches into the synthesis of hyper-nuclei elements have found that there does indeed exist around those elements a halo or thin mist of loosely bound nucleons (of either protons or neutrons), and this will be because the binding energies of those nucleons in their outer orbits have been found to be disproportionately weaker than the binding energies of those atom's inner orbits [note: 1].

Physicists have postulated that this factor of reduced binding energy might be very beneficial for lower energy initiations of atomic fission and fusion reactions, and, of course, for the proton-antiproton annihilation process; and some have wondered if this same halo effect can be found to be even more prominent in the new superheavy elements—such as in elements 110 through to 118—now being synthesized by physicists all around the world. However, while the original expectation had been very high to find this same exotic phenomenon, to the dismay of these researchers it seemed this new superheavy group of elements was throwing some curve-balls from the very unique way these atoms are structured. Highly intrigued, this has led physicists to go beyond conventional wisdom about atomic nuclei, and to discover a great deal more about how the heaviest of atoms can be synthesized through new and unconventional means. And it is only now, now that more has been discovered about these strange elements, that physicists are beginning to believe that these superheavy elements too might yield an enhanced halo effect around them, just as had been found around the hyper-nuclei of lighter elements.

Indeed, because the rewards are so high the exotic phenomenon of superheavy nucleon skin is now actively being pursued by physicists who are looking at ways in which it can be created and enhanced around the elements such as the ununpentium 115.

Was the strange light-bending skin effect around the alien's element 115 that was shown to Bob Lazar back in the late 1980's at Area 51, the same as this new nucleon skin effect that is being so rigorously sought after now by twenty-first century scientists?

More to the point, does this mean, even though it has taken all of twenty years for the world of physics to catch up to this one small factor propounded in UFO energy-generation-dynamics, that we can now assume that the world of physics can offer fuller explanations when it comes to the question of what is going on inside UFOs?

Well, yes it should be so, surprisingly, quite a lot of UFO technology can be found in the physics of today in 2008, but I would have to say that the overall understanding of UFOs, particularly the propulsion-drive-systems of UFOs, such as in the UFO that Bob Lazar worked on and experienced at Area 51, will be more meat-and-two-veg to the average astrophysicist than to a physicist. This is because the array of technologies used by the extraterrestrials in UFOs is very 'universal', much more expansive than the sort of mundane sciences our physicists encounter in their everyday researches here on earth. So perhaps I should qualify the above mentioned observation by saying that the understanding of UFOs will be better assimilated by those who have made efforts to study into the greater array of sciences encompassed by astrophysics, while this understanding, alas, will not necessarily be so plain-sailing for the physicists or for the electronics engineers who have been educated to understand only terrestrial sciences.

Now, although he did design rockets at one stage Bob Lazar was not an astrophysicist, and so what I am asking here is this, did Bob Lazar completely understand for himself how the UFO which he worked on at S-4 in the 1980s actually work? Because, after a lot of research into this particular UFO (which I will refer to as the sfourUFO), and from my knowledge of UFO astrophysical propulsion dynamics, I would have to say, judging by the presentation he gave of it, that I doubt whether he did fully know how it worked. In light of this somewhat startling discovery the intention of this chapter will be to offer a great deal of new information about Lazar's UFO, from both the fields of physics and of astrophysics, which will allow this sfourUFO to be better understood by the dedicated researchers in this field.

First off, if you read through the George Knapp interviews of Bob Lazar (c. 1989) you might be forgiven for quickly coming to the conclusion that Lazar's knowledge of some parts of the sfourUFO's energy conversion system, when he was asked questions about it by Knapp, was almost that of a layperson. Sure, his knowledge of the ununpentium (atomic element 115) was twenty years ahead of its time, and his background knowledge of the nuclear physics of the 1980s was quite good (although, as will be seen, not exceptional), and Lazar's descriptions of how that craft's gravitational forces could be manipulated, to engineer the necessary form of space-time curvature in space for travel, was also fairly good. But, it has to be said, Lazar's knowledge of what went on in between those two main functions of the UFO he worked on was vague at best, and at worst—misleading! But this is not totally unexpected, very few physicists have across-the-board wisdom throughout all subjects. Even they will agree that physics is split into a whole multitude of disciplines and anyone who works in physics tends to specialize in one of (or in only a small number of) such disciplines. So physicists often don't cross-pollinate that well into other areas of science. For instance, a nuclear physicist will know little about biology, and an electrochemical physicist will know almost nothing about aerodynamics. And, more poignantly, most all physicists will know very little about astrophysics and astrophysical energy dynamics—in fact, between these two disciplines in particular there is often found, what might be regarded as, dare I say it, a certain amount of elitist rivalry.

This compartmentalization tendency amongst physicists might then help to explain why throughout much of the Bob Lazar information 'package,' and whenever he was in interviews describing mechanisms operating within that Area 51 UFO, that he seemed to be relying on things he was either told about that UFO by his co-workers at S-4, or which he had recently discovered from the information packs on extraterrestrials he was given access to in his facility's briefing room situated beside that UFO's hangar. Like he said on his website, he was given specific tasks to do on that UFO craft, and as security was tight around him those tasks were all he was allowed to do. So, for the rest of the sfourUFO's workings, whenever he spoke about them after he left the Area 51 facility he was presumably using a certain amount of guesstimation in those published explanations of that craft. This, like I say, is quite apparent in the transcripts of his various interviews (for George Knapp and others). And to be fair Bob Lazar does say on numerous occasions that he really didn't believe anyone, not even the guys at Area 51, knew then precisely how that UFO worked. So presumably, the data he presented on his website and video was best-fit guesstimation he had pieced together for himself from what he could recall of the snippets of information about the craft's gravito-electric dynamics, as spoken about by his co-workers at various times during his employment at that much denied secret facility at S-4.

Obviously then, because some of the sfourUFO's technological presentation is based, in one or two places, on best-fit suppositions made by Lazar it has been quite difficult to finally understand how that UFO does actually work. This has led to great difficulties when unraveling a proper and correlated working-hypothesis for this particular craft even after many other types of UFO were finally shedding their secrets, particularly, as so many crucial components were simply left out of Lazar's presentation, for whatever reason. Perhaps Lazar knew nothing about those technical details, or it may have been that it wasn't in his remit to say anything about them then. But whole mechanisms were missing; how for instance, did the craft's central reactor tower do its job? And how did the gravity-A wave, as Lazar refers to it, get amplified and configured into a faster-than-light propulsion mechanism? And how did the energies from the antimatter-annihilation events get transferred into the gravity amplifier heads? For anyone viewing that presentation the graphics were absolutely wonderful and superbly detailed, but as far as the scientific theory was concerned there were a lot of blanks and so obviously a lot of the sfourUFO's workings were left to the imagination.

Don't get me wrong, I'm not saying that Bob Lazar had no UFOs to work on at Area 51, and nor am I saying that whatever he said was bunkum, I think Bob Lazar did a sterling job with the knowledge he possessed and with the intricate way he portrayed the strange things he saw and experienced. What I am saying is that the various parts of the sfourUFO, as they were presented by him, like the reactor technology and the thermionic conversion process, were somewhat misinterpreted by Lazar (or by Lazar's production team) as to how they really fitted into that craft and how all those processes worked together as a manipulator of gravitational force. I'm convinced that UFO worked, and I've traced a very large proportion of the sfourUFO's attributes to recognized sciences, and I'm very pleased to say that most all of the elements in Lazar's description do make sense, only, its just that those elements seem to be configured wrongly into the wrong sorts of energy transmutation mechanisms. This is to say that Bob Lazar's theoretical breakdown for that UFO, as it stood when his website was up (c. 2000), is a bit mixed up and comes out like several jigsaw puzzles all amalgamated into one but with the separate pieces fitted together with a big hammer! And, as a result of my researches to check out everything that Lazar spoke of I would have to say that there have been discovered one or two red-herrings that seemed to have been implanted into Lazar's descriptions of this sfourUFO, and I can only think that these were, perhaps, put there to throw budding researchers off the scent from finding out just how this craft does work.

But that's good, that's a very good point to make here, red-herrings, blind-alleys, unanswerable questions; misinformation... it must be said, does have tremendous power. In the field of UFO acknowledgement the best way to deny the very existence of UFOs is to make sure no one can understand how they work. For as long as UFO dynamics cannot be understood by the general public, or by scientists who administer to the general public, or indeed by media producers that administer to the general public through film, television and the internet, then UFOs will always be regarded as 'mysterious' things that are claimed to fly in our air but only as if by some sort of 'magic'—and everyone knows that magic is an illusion which can be dismissed as unreal whenever necessary by any protagonist who exercises the prejudiced voice of ignorance.

Elsewhere I have presented explanations of how many different groups of ETs have modeled their UFOs on different facets of astrophysical dynamics, within which explanations (given in previous chapters of this book, for instance) I have shown how the Andreasson UFO and the Onion drive craft, and the Searl IGV vehicle, can be comprehensively explained through the same dynamic principles as those found in space. Similarly, through the following sections of this examination of the Lazar UFO, I will also show that while within the sfourUFO there can be found some of the basic principles commonly found in earth's sciences there will, more importantly, be found the very specialized principles that will only be recognizable in the electrodynamic and hydrodynamic principles of black hole systems. And furthermore, I will show that there are to be discovered in the sfourUFO, just as there are in the Andreasson UFOs, very obvious modifications to those basic astrophysical features, which the designers of these craft have obviously incorporated to give those UFOs a much more enhanced form of gravitational force manipulation—and so as to provide substantially more propulsive potential than can be found in, say, a straightforward miniaturization of the Kerr black holes of space.

To do this properly it will be necessary to present a whole new array of diagrams for the sfourUFO in a special presentation, but before those details are aired, I think it would be best to give a brief recap in this chapter on Bob Lazar's working hypothesis for this UFO, which is as follows.

12.2 The Bob Lazar S-4 UFO Through the Looking Glass

Bob Lazar's assumption that this UFO has a reactor at its heart which produces thermoelectric emissions is a good place to start.

Lazar says that built into the base of the reactor was a miniature but highly sophisticated particle accelerator. This accelerator would fire protons (with tremendous energy) into the tip of the wedge of element 115. The premise of these Area 51 physicists being that by colliding a volume of protons into the 115 an extra proton would be fused into the nucleus of a number of ununpentium atoms, so that those atoms would then transmute up to element 116. But with 116 being unstable, or rather, having a very short decay lifetime, these newly formed 116 atoms would almost immediately decay down to element 114 and in doing so release, as Lazar put it, small amounts of antimatter. Intriguingly, Lazar says this downgrading to 114 will also produce another element which will decay all the way down to lead (element 82), and even more intriguingly he says that some of the 116 that downgrades to 114 will then transmute back up again to element 115. The reason for these unusual reactions being, says Lazar, that the alien's nuclear transmutation process has a much higher level of efficiency than that known to earth's physicists. Indeed, elsewhere, specifically where he is explaining the antimatter annihilation process, he opines that such an overall process is neither fission nor fusion, but that it results in a unique process of "total annihilation reaction." Perhaps not so strangely, Lazar often calls this annihilation event an 'explosion.'

Further, whenever the elements transmute up to 116 and then decay down again they release two antiprotons, says Lazar, although in some places the wording he uses is anti-hydrogen (which in conventional physics denotes the release of an antiproton and a positron). However, no mention is ever made in Lazar's presentation of any protons being released, nor any neutrons being released, and nor any binding energy being released in these various reactions. And nor is any mention ever made of ANY type of photon, alpha particle, or electron-positron emissions resulting from these subatomic reactions—which, of course, are known to accompany such reactions in conventional explanations.

After the pseudo-fissioning of the ununpentium the resulting flux of antimatter antiprotons (or anti-hydrogen) would then pass down a tuned evacuated tube (vis, the tower) through to the lower chamber of the reactor to where a small quantity of gas was housed (where presumably it is not evacuated). He doesn't say whether this chamber was above or below the particle accelerator, and the location Lazar assumes for it is somewhat obscured in the (reactor.mpeg) animation of this procedure. But as soon as the antimatter reaches the gas (or as Lazar calls it, gaseous matter) this antimatter-matter collision event initiates and results in a total annihilation reaction, thus the matter-gas and the antihydrogen reaction event would produce huge amounts of heat, which would then be converted through a 100% efficient thermionic conversion process, claims Lazar, into electrical power. Unfortunately, no further details are given about the craft's electronic circuit, nor of how this electrical energy can be tapped-off from the total annihilation chamber and fed into those electrical circuits.

Most importantly, in Lazar's presentation, the craft's gravitational force is claimed to come about as a result of the antimatter antiprotons making their way DOWNWARD through the reactor's tower and then the sfourUFO's gravity A wave, as Lazar refers to it, is extracted when the 115's antiprotons collide with an amount of gaseous matter, so that it can be amplified and propagated OUTWARD and UPWARD from the craft's reactor in such a way that the A wave's gravitational effect initially becomes apparent at the hemispherical dome which surrounds the reactor (ie, it reacts with the atoms of the material from which the dome is made as it accumulates around the dome). This being so that when the central tube is lowered down onto the hemispherical dome and it makes contact with the dome this tube would, says Lazar, 'load the system,' so that this gravity A wave would then be forced to travel up that central microwave tube, because the tube would be tuned to the particular frequency that the wave would be harmonious to. When these gravity A waves then spill out from the top of the craft, upon exiting the upper end of the central tube (where this tube protrudes about a foot above the craft's shell) those gravity A waves would then expand outwards and at the same time curl around to encircle the whole of the craft in a somewhat poloidal (apple-core) shaped field. And after curling below the UFO this poloidal field would be punctured by one, two or three pointed beam fields that the craft would additionally propagate downward from its three cylindrical gravity amps located inside the craft's lower hull.

As I have already started to say in the previous section, after looking quite objectively at everything Lazar has said about this sfourUFO, there will be a few problems to highlight and to iron out first before such an aircraft can be successfully manufactured and operated.

For instance, for the above mentioned hypothesis to work there would need to be a constant supply of ignitable matter-gas which would need to be stored inside the UFO somewhere (or drawn in from whatever atmosphere the craft was flying through). The craft would also have to be built to resist, or to dissipate, the tremendous amounts of heat generated in this antimatter-annihilation process (Lazar conveniently gets over this problem by saying that the extraterrestrials had devised a near 100% heat-to-electrical-energy conversion process). And then, the fact that the ununpentium would need to be precisely honed into wedge-shaped blocks is certainly going to be problematical, when one considers that this material's elements have to undergo continual transmutation, and in some cases, as Lazar puts it, transmutation down to lead! Also, and most annoyingly, the only hint of the grand amplification process directed upon the gravity A wave seems to occur in this hypothesis somewhere vaguely located around the central reactor area—through some wholly unspecified process. But the single-most inexplicable factor in Lazar's delineation is that as soon as it does get amplified and the gravity A wave shoots up the central tube, spreads out of the top of the craft, and after it curls around to the underside of the UFO on the OUTSIDE of the UFO, it somehow manages to pass THROUGH the cylindrical gravity amps which are on the INSIDE of the UFO... Which doesn't make sense, and call me old-fashioned, but this part of the Lazar hypothesis, to borrow a Vulcan phrase, is truly illogical! And so the reader is left scratching their head while trying to figure out how this equates to a force which can be amplified, and how that gravity A wave force can be amplified through the three gravity amps in the lower deck of the craft in order for it to be used by that aircraft for propulsion!

I won't bore the reader here by going through all of the sfourUFO's propulsion technology because most of it will prove to be quite sound and this will be gone into in greater detail toward the end of this chapter. But certainly, I feel the above details of the reactor and the 115 material are presented sufficiently illogically to suggest that not all of what Lazar has published about this UFO is the full tin of biscuits. And while the reader might be forgiven for dismissing large chunks of the above delineation as complete nonsense, let me say, please be patient because there is some really precious platinum hidden inside this ore just waiting to be extracted and polished up.

Also let me remind the reader that what I am endeavoring to do here is not a grand denial or a dumb skeptic's invalidation of Bob Lazar, merely am I pointing out that certain parts of his theory have been mis-aligned or mis-packaged, and that these parts will benefit greatly from a small degree of re-packaging. So that, hopefully, what will be explained below will provide the reader with a brand new box of biscuits that will be worth its weight in gold-pressed platinum.

12.3 The Central Reactor's White Hole

The first misconception to deal with from Lazar's delineations is that the reactor he was back-engineering wasn't propagating away from itself a positive gravitational force—it was accumulating a negative gravitational force—it was massing around itself a white hole not a black hole. The reactor was storing and giving off a REPULSIVE-EXPANSIVE gravitational force not an attractive gravitational one: And that's why whenever he tried to move closer to the reactor he felt a force from it pushing against him. That's also why, when he was toying around with the reactor which his associate at S-4 had built (named as Barry Castillio on Lazar tape #003), and he tried to put his hand through its gravitational shield, which in that instance Lazar could see visibly as a dark field that extended out from that reactor so that it wholly surrounded it, he again couldn't push his hand past that wall of energy, describing it as being like with magnetism when you try to buck two same-pole magnets against eachother (and you just can't get them to touch). So what Lazar was pushing his hand against was a gravitational REPULSION FORCE, and this is a force known to astrophysical researchers as negative gravitational force. It is also one of the components necessary for superradiance amplification, which is a very powerful amplification mechanism known to exist around black holes (and, the reader will be pleased to know, it is an amplification mechanism put to good use by the extraterrestrial designers inside the Andreasson UFO as explained in my chapter 9 above) [note: 2].

Further, from the fact that there was a shield of expansive gravitational force around the reactor's dome, this would explain why whenever he and his colleague tried to force any material objects through that shield—as when they were throwing golf balls at it—the harder they were thrown against it the faster they just ricocheted off it like it was a solid impenetrable wall; because that shield (hovering a few inches outside the metal dome which houses the UFO's reactor) would have been where the event horizon was of that force.

With the accumulation of a repulsive force centered around the reactor a pass-not shield would have been formed, and this phenomenon of a white hole's gravito-dynamics is quite easy to explain and to duplicate (see my chapters 9 and 10, for instance, where, because I have discovered that repulsion forces can form part of a UFO's amplification mechanisms (ie, through superradiance ringing), I have coined this shield the WHITE WALL, rather than the white hole, because this unique phenomenon is better describable as an impenetrable reflector which is used in UFOs as an amplifier (against which electromagnetic energy, gravitational waves, and kinetic force can be reflected back into the craft's quasi-black hole). This type of gravitational shield or wall would be the exact opposite of a black hole's event horizon beyond which no mass can escape.

That gravitational forces can be both attractive or expansive is something that some physicists do find hard to accept, strangely, and perhaps this is down to the false constraints put on gravity by the spurious teachings they have had administered upon them throughout their curriculums of education, and from the never-ending supply of gravitational theories that have been published, which seem to have filled physics journals ever since Isaac Newton and Albert Einstein, and through unquestioned obedience to textbook explanations of gravity that seem to be quite oblivious to the fact that expansive gravity was borne out by Einstein's Lambda force (his cosmological constant), and even though it has been acknowledged that our solar system presently exists in a part of the galaxy which has an expansive gravitational space-time [see above note: 2].

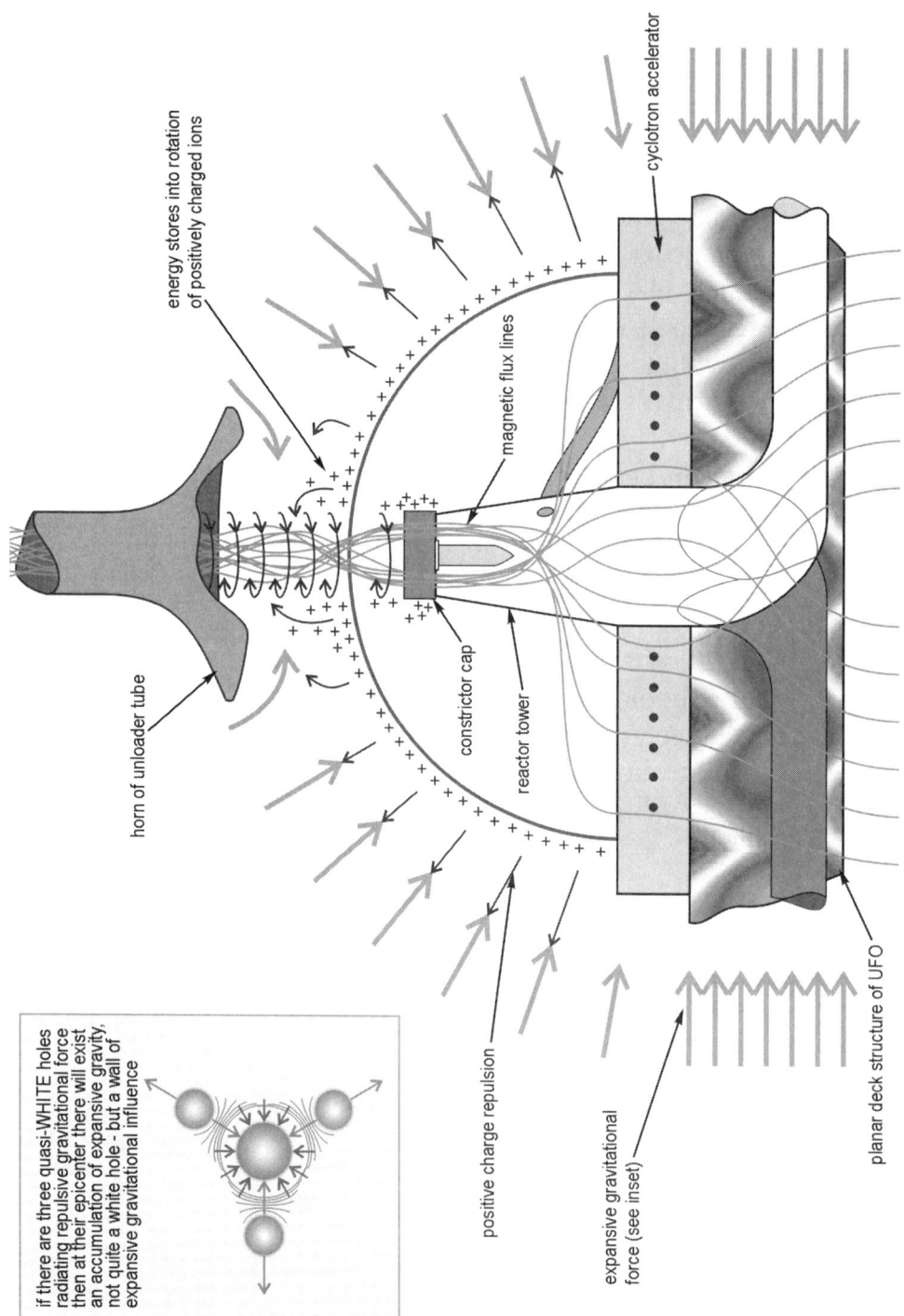

Figure 89. The pass-not shield around the reactor dome was not actually generated by the reactor—it was only where those gravitational forces met

What is most intriguing though, for this technical examination's point of view, is that Lazar made note of the fact that the above-mentioned pass-not sphere of repulsive force surrounding the sfourUFO's reactor immediately disappeared when the long central tube was lowered down toward the reactor's metal dome. For, just as Lazar has suggested, this would mean that by the central tube coming down and approaching the reactor's dome, whatever expansive gravity that was being stored around the central reactor was, after that tube got to a certain distance from the dome, being immediately propagated up into that central tube (and thenafter passed out the top of the craft). So this observation by Lazar, in fact, takes the 'magical mystery' out of at least some of this UFO's description, because this is an engineerable phenomenon which can be ascribed to a known reaction of when an oscillating field suddenly matches up to a waveguide of the same frequency. Precisely why this could happen will be explained in due course below. And so too will be answered the question which also comes to mind, of how that long central tube can be lowered down close enough to that metal dome, when all around that dome is exhibited the effects of the white hole's mass-inhibiting event horizon..!

Something else that can be better understood about this sfourUFO, through there being both expansive and attractive gravitational forces, and this will borrow from other facets of astrophysical studies, will be how and why the force exiting the craft's central tube, the gravity A wave as Lazar calls it, after it does propagate up that tube, will immediately spread outwards and follow an expansive trajectory from the center to beyond the outer rim of that craft. Indeed, what will also be understood will be how and why all of Lazar's gravitational models work through his craft sending out beams which PUSH AGAINST the ambient field below them—which, again, denotes the expansive mode of gravitational force.

Actually, having said all of the above about the craft's white hole, I have to emphasize that the reactor will not be the source of the white hole's gravitational force, the reactor will only be the location of where all that expansive gravitational force will be meeting and accumulating. So, looking into the reactor for any possible clues of where to find the gravitational energy that Lazar was on about was actually one of the red herrings in his hypothesis. The sources of generation for the craft's gravitational force were at completely different sections of the sfourUFO, and although that force does derive from the matter-antimatter reactions it doesn't get produced at the ununpentium the way Lazar says it does, as will be explained below.

12.4 Fishing for the Gravity A Force

One final point to make about Bob Lazar's hypothesis before its secrets can be divulged, and this goes back to the thing that was most central to it, the gravity A wave; this gravity A wave, according to Lazar, was described as a cumulation of the strong nuclear force which in superheavy elements (like element 115) is so abundant that it extends beyond the perimeter of that element's atoms, and because of this factor this gravity A wave could be readily accessed and simply amplified, claimed Lazar. Elsewhere he likens this gravity A field to the force which holds together the 'stuff' of protons and neutrons, and he suggests it is that force which holds together the upquarks and downquarks of elementary particles. These factors do, in fact, relate to conventional wisdom about superheavy atoms, although, beyond these basic factors Lazar's theory does get a bit muddy.

In conventional scientific theory there is a gravitational force inside the atom but the consensus opinion among scientists is that it has only miniscule strength. If the gravity A force that Lazar is referring to is similar to the strong nuclear force and to that of the atomic gluon then it can be an attractive force or a repulsive force [note: 3]. This gluon modifies forces between the quarks and antiquarks, the combinations of which go on to make elementary particles like protons, and it is known that the farther the quarks are apart the stronger will be this gluon force. If, however it is the very powerful Coulomb force that has been found recently to be very prominent in stable superheavy elements (especially when their shell structure is deformed and it exemplifies this), then once again, this Coulomb force can be both an attractive force or a repulsive force. However, what might be of greater interest here, is that just recently it has been proposed (and with it a call for Newton's old scamps on gravitational force to be overhauled) that essentially no difference will be found between strong nuclear force and gravitational force. This new proposal, even more interestingly, also sees that strong nuclear force will create dents and twists in the curvature of space-time—just the same, the reader will appreciate, as how that effect is presently ascribed to the mass-derived form of gravitational force. And so this new scientific proposal postulates that there can exist a connection between the atom's strong force and the manipulation of space-time's curvature—and it indicates that what Bob Lazar has been saying about atomic gravity could be quite correct [see note: 3].

Then also, going back to something mentioned at the head of this chapter (and the reader will see below why I am laboring this point here) Lazar did say that when he saw one of the element 115 wedges after it had been prepared and milled into shape, that it appeared to have an energy field surrounding it which deflected the light around it, and indeed, it was from this observation that he assumed that this was because some sort of gravitational field was surrounding that wedge. And although he didn't say what part of the wedge bent the light most, I would assume that it would have been around the point of the wedge (where there would have been the most pronounced field affecting a change to its refractive index) where this sort of deflection would have been most noticeable.

Certainly, it was quite obvious from Lazar's webpages that he was quite chuffed that he was only one who noticed this effect and was the first person to hypothesize that this light-bending might denote a gravitational force which might be extending beyond the radius of the 115's atoms. And consequently, to discover more about this anomaly Lazar suggested doing tests on the ununpentium to find out whether or not a laser beam of light could be bent by whatever this energy field was surrounding the wedge. As it turned out it was from those laser tests, carried out by Lazar, that he declared that this light bending phenomenon was proof that there was a gravitation force emanating from that superheavy element.

However, interesting though this whole gravity A wave conjecture is, I'm not going to go any further with it here, because I do strongly believe that this line of inquiry, of how this gravity force can be isolated and extracted from around the ununpentium 115 wedge and then amplified, is a pointless quest. This, as I will show below, was just fanciful speculation and the reader will discover that it was just another of the red herrings in the Lazar presentation.

12.5 Bob Lazar's Strange Technology Reconfigured to Make More Sense

I do hope Bob Lazar doesn't take it personally that I have so irreverently pulled his presentation to pieces, hopefully it will be seen that the result here justifies the means. Either way, I would now like to put the sfourUFO's technical explanation back together again in a way which, as I claim above, will hopefully appear to make more sense to the reader. So, starting from scratch, here is how I believe the sfourUFO works.

What is NEEDED in the sfourUFO is a mechanism which implements a whole series of nuclear reactions, and this all-encompassing mechanism has some pretty tall objectives: This mechanism needs to draw together the matter and the antimatter particles from the craft's central reactor in order to inaugurate the matter-antimatter annihilation events. It needs to safely contain the whole of that annihilation process once it triggers and converts into substantial amounts of heat and radiation (and in agreement with Bob Lazar it will have to be able to perform a 100% conversion on all the heat and radiation released—otherwise it will be like having red hot braziers inside these UFOs, which might not be too much fun for the crews operating them and sitting right next to them). This mechanism needs to fully convert that heat and radiation into useable electrical energy (and make provision for that electrical energy to be tapped off and physically connected into the craft's electrical circuit). This mechanism then needs to amplify (or would generate be a better word to use here) a gravitational force, and deliver that gravitational force to the three gravity amps that are located in the sfourUFO's sealed lower deck compartment, and to the central reactor dome so this gravitational force can then be propagated upward through the craft's central tube.

There will be only one kind of mechanism that can do all these things with a reliable degree of safety and the highest degree of efficiency. That mechanism will be the rotating shearing field—the miniaturized quasi-black hole—and inside Lazar's UFO there will actually be THREE of them.

12.6 Quasi-black Hole's Delicate Stomach

And this essentially concurs with what many other UFOs have as their primary energy provider and amplifier, for just as we on this planet have an almost total reliance upon the archaic and carbon-spewing internal combustion engine so extraterrestrials have a preference for the supremely efficient miniaturizations of quasi-black holes. This is because the miniature quasi-black hole, with its various forms of rotating accretion disk storage field, is far and above the most efficient converter of gas-fuel-mass-particles into useable energy (of electrical energy, gravitational force, and angular momentum). And just to reiterate the point made in an earlier chapter of this study, it has been calculated that a black hole in open space will have an energy-conversion efficiency rating of a around 32% which compares to humankind's nuclear power's mass-to-energy efficiency of around 1%.

As an aside, because of the way these UFOs are structured in order to separate and to confine a whole number of rotating energy fields, specifically to amplify them, to substantially improve upon the basic dynamics of a black hole, the above mentioned efficiency rating of 32% will obviously be improved upon, and, in this wise it would be interesting for the mathematicians to calculate just how near the gas-fuel-mass-particles-to-energy ratings will come up to 100% for these UFO engine-structures, particularly when those mathematicians perfectly envisage all the modifications have been exploited to the full by the off-world designers (that is, if they would like to follow my various UFO studies, for instance, and see for themselves just how the extraterrestrials have gone about implementing these modifications to exemplify basic black hole dynamics) [note: 4].

What the extraterrestrials are doing with these UFO-size black vortexes, and in the sfourUFO's case where its three black vortexes are triangulated equidistant from the central reactor, is drawing into them the matter-antimatter subatomic particles from the central reactor and converting them (through the matter-antimatter total annihilation process) into electrical energy, angular momentum, and gravitational force with the utmost efficiency. And what is so remarkable is the way the extraterrestrials are doing it—with the minimum amount of machinery (and with the minimum amount of risk to the ship's crew).

These black vortexes will be contained inside each of the three box-shaped units (which Lazar refers to as gravity amplifier heads) which are located directly above each of the cylindrical gravity amplifiers. The premise of these miniaturized quasi-black holes is very simple—a shearing plasma field initially rotated by a four-phase magnetic 'waist-coat' and then continually fed with more and more energy—so then that plasma field has no option other than to spin faster and faster to accommodate that relentless influx of energy, so that these fields end up rotating at relativistic speeds. This will be the reason why it was reported that no one at Area 51 was able to cut through the casings of those amp head boxes and open them up to find out what was inside them, for they would have been constructed to withstand the uttermost powerful forces. They would definitely not have been meant to be tampered with in any way shape or form (primarily because it is most likely that once started they were never intended to power down to a complete standstill).

The gravitational radiation would be siphoned off not at the top, but at the bottom of those vortexes, and there would be an inter-connection between all three black vortex units which would direct their expansive forces toward the central reactor, and this would be physically affected through the central decking material (actually through the atomic-cellular structuring of that decking material) and which would be part of the structural framework of that UFO that spans from the center of the craft to the perimeter rim. By utilizing the UFO's structure the gravitational influence while it is being transmitted toward the center will, at the same time, be transmitted from all three heads through that deck structure all the way out to the UFO's rim (more about this in section 12.11 below). The antimatter particles feeding into the vortexes (from the ununpentium) would be routed through the deck structure too, and would be sent from the central reactor to each of the three black vortex units, although because they need to be insulated from that structure those antimatter particles would flow through specially formed channels within this planar floor structure, and by some clever construction methods the actual carrier field which transmits the gravitational radiation toward the reactor, can provide the necessary micro-magnetic blanket inside the walls of those antimatter channels to prevent that antimatter from touching those walls (see figure 90 below).

One of the foremost representative features of these vortex units will be their azimuthal integrity, because it will be through their rotating forces and shearing dynamics that these miniaturized quasi-black holes will be able to develop powerful enough forces to both initiate, and just as importantly, to CONTAIN the proton-antiproton annihilation process that Lazar has tried to ascribe to this sfourUFO.

Indeed, a relativistic rotating force would be the *only* confinement force powerful enough to keep these 'explosions' of energy in check while these annihilation events are continually being initiated between the antimatter and the matter particles! And only a rotating field suitably confined would be able to store-accumulate all the resulting reactions being triggered that each of the annihilation events will go through as it progresses (ie through subatomic emissions, photon radiation, and further energy disintegrations of the pions, gluons and gamma rays into their final raw energies).

The more energy that is pushed into these rotating fields the more angular momentum they will produce, which would continually raise the mass-to-energy efficiency of those fields, of course. In fact, this would illustrate how *only* a suitably modified quasi-black hole could be expected to produce anywhere near the 100% efficient thermionic conversion process that Lazar has claimed the sfourUFO will be able to maintain.

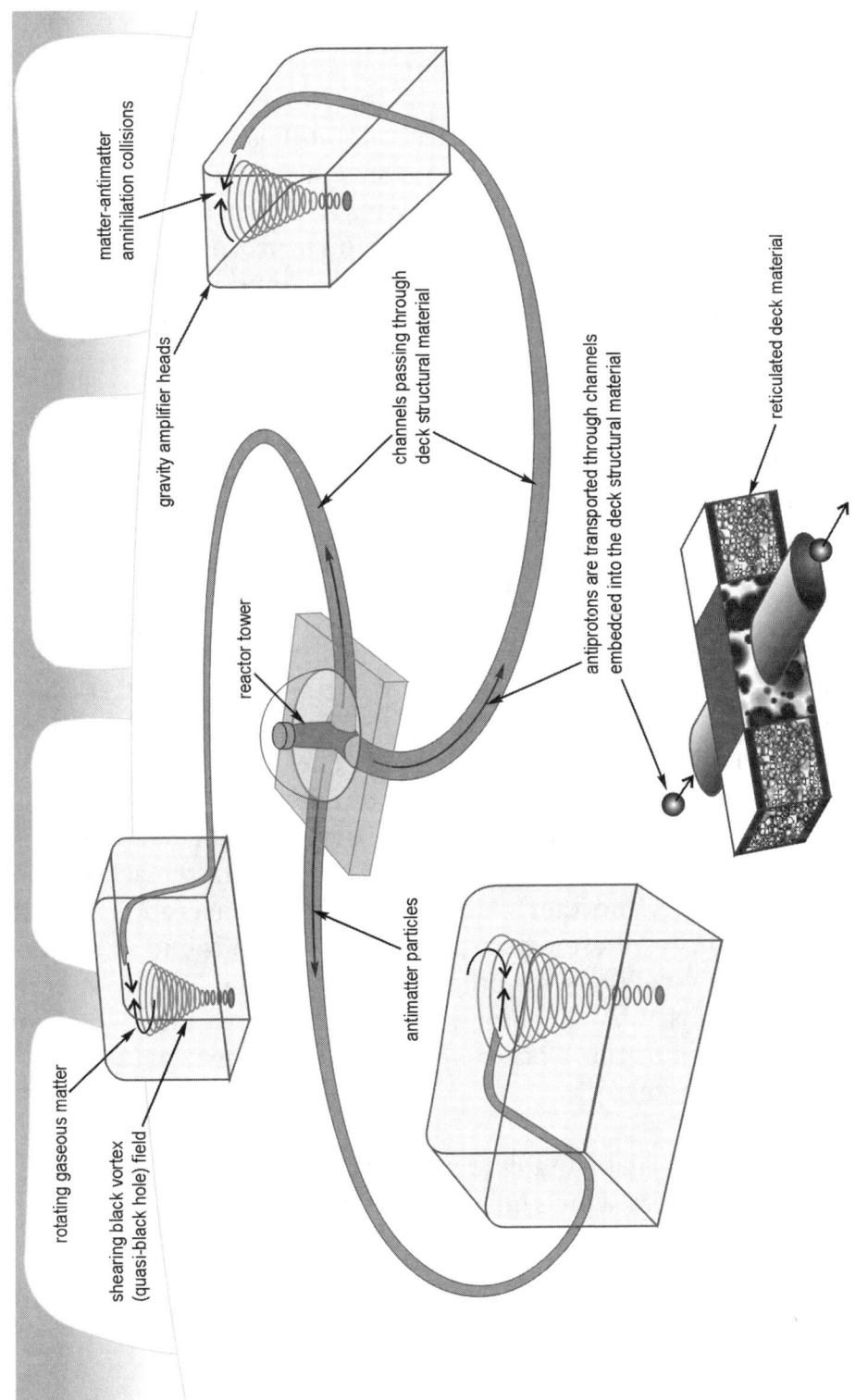

Figure 90. Anti-protons when they collide with matter are the ideal fuel for the quasi-black holes. Quasi-black holes are the ideal power system to fully contain their annihilation reactions

This energy break-down efficiency will depend purely on the cleverness of the UFO's designers and on how good they are at orchestrating precisely how the fuel-particles are prepared before being delivered into these vortexes. This, I would say, is probably what makes UFO 'engines' so unrecognizable to earth's scientists because the ETs don't use fuel the way our engineers do (or rather, our engineers don't presently know that fuel doesn't always have to be liquid-based and volatile). This cleverness of design is something that cannot be over-emphasized, and therefore it will be found that much of a UFO's internal mechanisms and structure will be seen to contribute in some way or another to the breaking down process of the incoming matter, whereby gas-fuel-mass is broken down into subatomic particles, through a whole range of oligo-dynamic dissociation mechanisms, before it passes into these vortexes (just like in the aaUFO) [note: 5].

Efficiency is all about maximizing the output of these engine-structures from how energy is put into them. So, the less energy these black vortexes expend breaking down the gas-fuel-mass-particles into raw energy the more powerful an influence they will radiate away from them and deliver into the UFO's other electro-dynamic mechanisms, which produce power and propulsion.

What also needs to be set straight here is that contrary to common speculation, and this is a major point to bear in mind, the only sorts of mass-matter-energies consumable by black holes whether out in space or inside a UFO are subatomic particles, gravitational binding energy, electromagnetic energy and kinetic energy. A huge chunk of mass floating through the cosmos cannot enter into a black hole out in space. It doesn't, mass will always be broken down first into smaller and smaller pieces, and out in space this is done through the shearing forces of the accretion disk which surrounds the black hole, and then at the center of that accretion disk there will be what is known as an ergosphere where particles are spun round at higher and higher velocities, so that they collide with other particles to produce even smaller, subatomic particles, and THEN those subatomic particles are fed into the shearing forces and gravitational forces of the black hole and broken down yet further into pure energy (as per $e = mc^2$) [note: 6].

Something which also goes against tabloid science, is the fact that black holes are not uncontrolled frenzied guzzlers which take mass-energy out of existence without return.

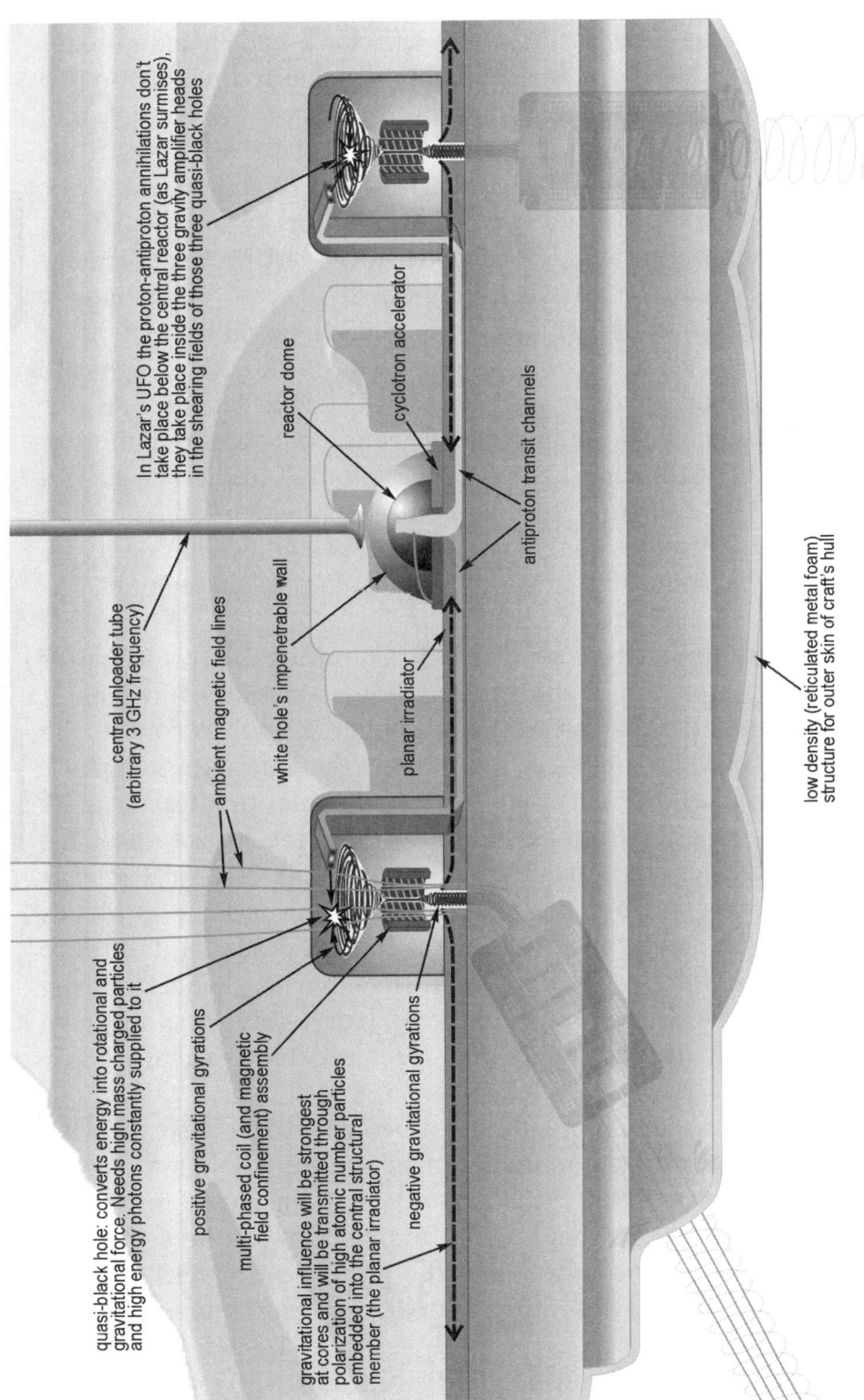

Figure 91. Very cleverly the extraterrestrials have designed their UFO structures to transmit both electrical force and gravitational perturbations

On the contrary, there is an enormous abundance of energy and high-energy particles being produced around a black hole out in space through the rotating forces which surround it, and this abundance of energy will gather around the periphery of the black hole (from the fact that there is just so much of it), and in many cases because there is so much of it produced that build-up of energy has to then be forcibly ejected away from that black hole out into space, and this is where the pulsar jets and quasar jets that accompany black holes get their abundant driving energies from. Astrophysical jets like the M87 or NGC 6251 have each been projecting relativistic charged particles millions of miles into space—all from the energy overloads of a single black hole (as mentioned in section 2.5). However, inside a UFO the efficiency of its engine-structure will depend on this breaking-down process being mostly replaced by mechanisms which already produce the smallest possible quantas of energy, and this, like I say, will be solely down to the prowess of the UFO designer to come up with the most efficient ways of doing this. So, the UFO will have mechanisms inside it which generate subatomic particles such as electrons, positrons and photons (as in the Andreasson UFO), or, as in the case of Lazar's sfourUFO, there will be a constant production of matter-antimatter-pair annihilations which as a fuel is probably the most ideal because it will probably give one of the highest mass-energy-to-raw-energy conversion ratios from rotating shearing fields [although, see note: 7].

As mentioned above, once the energy is extracted (from the matter-antimatter collisions) it immediately gets stored into the rotating bodies of those three black vortexes, and while each black vortex is continually fed with energy it is continually accelerated, and so each of these rotating forces acts like a dynamo with a gigantic torque-flywheel that gets more and more massive: So then some of that force can be tapped off as electrical energy, just like from a dynamo, at the three cores to produce electrical energy of both polarities. In astrophysical studies this electrical energy tapping is a known phenomenon attributable to the 'membrane paradigm' (it has been calculated, for instance, that there will always be the biggest potential-difference between the mouth of a black vortex and its core [Price-Thorne]); and by using electromagnetic coils that surround the vortex filaments (see figure 91) these can instantly convert the rotating particle motion into high-frequency electrical power.

Its certainly obvious that angular momentum can be drawn off these shearing units too, and this can be induced into adjoining fluid mechanisms through the agent of the co-mutual magnetic field, so that kinetic energy can be transferred into those adjoining fluids (beside the vortexes), or that angular momentum can be channeled directly downward to be developed through the craft's gravity amplifiers into the craft's propulsion forces (which will be gone into in a separate presentation).

Then, by passing their rotational powers THROUGH the UFO's structuring, through holes in a high-mass material, such black vortexes can also radiate a gravitational influence—which can be transmitted through the atomic cellular gravitational polarization effect upon all the atoms in the materials immediately surrounding those quasi-black hole cores—out to any other parts of those craft. This will be one of the sfourUFO's two gravitational forces that can be harvested and then amplified. This first one will be a collective and dynamic gravitational force though, not the individual gravity-cum-strong nuclear force that Lazar said he saw poking out of his ununpentium wedge. Also, the gravitational influence used by the sfourUFO will be the expansive type of gravitational influence, which will be harvested from where it is radiated outward from the anti-vortex (below the cores of each of these three black vortexes) and it will be projected through the craft's middle deck material (see figure 91 above, and see more on this in section 12.11 below).

This will also be how the forces from all three amplifier heads will be transmitted axially toward the center of the craft so as to meet where the central reactor is located, and from a construction point of view I would suggest that the best way to do this would be to incorporate into the deck three separate tracks of high-mass material between the amplifier heads and the reactor—and have the remainder of that deck fabricated as a low-mass sandwich of reticulated metal foam (as indicated in figure 95 below). The result would be that around the central reactor there will be a meeting of those three powerful forces of expansive gravity, so that there should exist as a result a shield of repulsive gravity around the central reactor. Because then, where these three projections of expansive gravity from the gravity amplifier heads all converge upon eachother there will be created the white wall field—the antithesis of a black hole.

This is why I say the central reactor in Lazar's UFO will not be the SOURCE of that white wall of repulsive gravitational force with its impenetrable shield. The repulsive shield is what you might call the event horizon, and in fact, what that white wall pass-not-shield is doing there is preventing the consequence of those three extreme forces of expansive gravity compressing themselves into higher and higher density.

Incidentally, by generating just outside the reactor such a huge pressure-density gradient, and then having three channels which run from the reactor's tower to the most vacuous part of each of the gravity amplifier heads, a very powerful pressure-vacuum differential will be established in those channels, through which the reactor's antimatter antiprotons would be sent at great speed.

12.7 Bob Lazar's Element 115 is Here to Stay... its Official!

So what is the fuel supply for Lazar's three gravity amplifier heads, because obviously, what these quasi-black hole rotating shearing fields need in order to keep going, is a constant supply of energy—and the Lazar UFO only uses one energy-breakdown mechanism! But astonishingly, with each of its three black vortexes encased inside its own gravity amplifier head working well within its safety parameters these would be very adequately provided for by the constant delivery of antimatter particles coming away from the ununpentium reactor at the center of the craft, and this transmutation process works more-or-less the way Lazar says it does. Lazar has claimed, or at least it has been claimed for him on Lazar-based websites, that the sfourUFO's antimatter total annihilation reactions will each produce 1862 MeV of energy. This figure compares favorably with NASA's estimate of 1876 MeV for the antiproton-proton annihilation process it wants to develop in the future. When one considers that the basic nuclear fission of uranium-235 gives 210 MeV of energy, then by anyone's standard, these figures for ununpentium 115 denote tremendously high power ratings [see note: 7].

Twenty-first century science is well tuned into the synthesis of superheavy elements now with researchers developing new ways to produce elements as high as 120, and even back toward the end of the twentieth century it had been found that a 118 element could be created through high powered cyclotron collisions (even though that 118 had a very short half-life duration). What bodes well for the Bob Lazar hypothesis is that whenever element 118 was created, after it decayed down to element 116 that element 116 would then decay down to element 114 (which is one of the things Lazar alluded to in his hypothesis). What scientists have also found is that there is an island of stability here (this again validates what Lazar said), although mostly it is centered around 114 protons with 184 neutrons (Lazar's extraterrestrial ununpentium 115 is rated as 115 protons with 184 neutrons) [note: 8].

However, right now, the earth physicists must admit that they don't fully understand what sort of process has been followed, in order for the extraterrestrials to produce their particular form of element 115, for, they don't know how it was synthesized into such a stable element. Presently (circa 2008) the research that has been done on element 115 synthesis can still only attribute to that procedure an element which ends up with 115 protons but with only 173 neutrons, so that it amounts to a low grade ununpentium with a half-life of just 87 milliseconds, which obviously is an inferior product to that of the ET's.

So what I would like to do here, is offer a few observations that haven't been aired before about this superheavy element's structure, and about the alien way the ET's 115 may have been synthesized. In the recent researches into the synthesis of superheavy elements (of, say, elements 110 to 118) it has been found that the actual structure of the superheavy element isn't necessarily that of the usual spherical shape, its a somewhat deformed shape. This deformation, either into an oblate (squashed) or a prolate (stretched) shape, of the atom's structure results from an interplay of the strong nuclear force, the electromagnetic force, the gravitational force, and the atom's Coulomb force. It has further been reported that the intense repulsion of the Coulomb force, and the strong nuclear (nucleon-nucleon) repulsion force, from there being so many nucleons jam-packed together within these superheavies, is a major contributory factor to the deformation of these superheavy atomic shells [note: 9]. At the same time though, it has been found that this deformation is no bad thing, because deformed superheavies are the most stable superheavies there are at the moment, and, the deformation can equate to higher excitation states for these atoms.

This being the case, it may be worth mentioning here that I have seen elsewhere, in other UFOs, that the extraterrestrials use gravitational forces to polarized the atoms in the materials they use. One instance that comes to mind is their use of multi-layered bismuth-magnesium-zinc (in the Andreasson and the Roswell UFOs), because what the ETs tend to do is utilize a high-mass substance, like lead or bismuth, and use it to transmit gravitational influence from their quasi-black hole more directly through those substances to other parts of those craft (more directly than it would otherwise be radiated). Whereby, in such cases, a high-mass material is used around a gravitational source specifically because the more-massive nucleon clusters in that material's atoms will be affected more by that localized gravitational source than the atom's less-massive electrons, so that as the atoms become electrically unbalanced, or in other words, electrically polarized, that electrical influence can be transmitted through to other areas of the craft, much in the same way we transmit electrical charges through plasmonic oscillation (as I explain in more detail throughout chapters 5 and 6). What also happens with gravitational polarization is that the atom's structure becomes physically deformed because the electron shells become biased over to one side of the atom (ie the atom takes on a pear-shape) [note: 10].

Would the extraterrestrial's special compound of ununpentium 115 have had its atomic structures deformed by a strong gravitational force while it was being manufactured? Furthermore, would this deformation make those superheavy atoms more stable, and at the same time cause them to develop around their outer shells a nucleon skin—comprised of nucleons which were extending beyond the atom's normal radius? Scientists of this world might bulk at such an idea, possibly from it sounding too exotic, or possibly from the fact that the scientists of this world know so little about how gravitational forces can be generated as a localized and perfectly controllable field.

But then, why shouldn't it be expected that the alien's superheavy elements can have a different atomic-structuring than that found on earth? It would surely set a strange precedent if I suggested here that the fuel for Lazar's UFO should come from this planet, and so it would be more conducive to extraterrestrial science and engineering experience, and to UFO manufacturing procedures, to take it for granted that Lazar's element 115 must have come only from another planet, and one which was expected to have a different gravitational field to that of our planet. A possible correlation toward this factor might be found through the Betty Andreasson's recollections of the Greys that abducted her in their UFO craft, because those Greys once explained to her that they normally existed in an environment which had a greater density than the one we are experiencing on earth at the moment. Otherwise, if Lazar's element 115 came from a planet with a greater mass-density than ours then it would also have been influenced by a greater gravitational force than we have here.

Or, for the astute UFO researcher, the extraterrestrial's ununpentium might simply have been fabricated inside a manufactured environment of higher density and higher gravitational influence inside a sealed metal shell in close-proximity to a quasi-black hole...

Either way, I believe the clues pointing to this gravitational form of polarization can be found in Bob Lazar's account of how the extraterrestrials gave to the American (and I presume they are American...) scientists researching at Area 51 a large amount of their ununpentium 115 to experiment with.

12.8 Constant Energy Being Delivered from Polarized Ununpentium 115

Because, in this long convoluted inspection of this alien power source, I'm beginning to see that the wedge of ununpentium while it is very important because of its super-heaviness, what it might have about it that will be found to be just as important, for unraveling exactly how the extraterrestrials have synthesized this particular form of 115, will be the unique shape it had to be given and also the unique fusing and milling processes it had to go through in order to get it into its wedge shape, before it could be used in this UFO's central reactor.

The 115 was supplied by the ETs to the scientists at the American S-4 facility as round flat discs. No less than twelve of those discs needed to be fused together by the scientists at S-4 according to Bob Lazar's notes, and after fusing them together the resulting composite cylinder was then milled into a cone shape, and after that the cone had to have two edges milled off it so that it could be given parallel sides; and so what was left was a very narrow triangular wedge with curved edges that came to a sharp point. Lazar also noted that the milling process, carried out at Los Alamos where Lazar used to work, created a tremendous amount of wastage of the original 115 material. Indeed, Lazar greatly emphasized the fact that even though the milling instructions accompanying the original twelve discs implicitly asked for all the waste material to be returned with the finished wedge to S-4, not all of it was returned complained Lazar, and you got the feeling that this was very much to the chagrin of the S-4 scientists. But then, when all things are considered, I suppose they had every reason to feel pissed-off about this apparent theft. Back in the 1980s a superheavy element such as 115 was a uniquely rare and priceless commodity, the fact that the milling procedure it required turned most of it onto the machine-shop floor was difficult enough to stomach, but then, to not be given back all the remnants of this precious commodity, well, this would have been highly unacceptable. After all, Lazar and the other guys at S-4 would have presumed that the only suppliers of that stuff were beings that—weren't exactly locals.

stack of extraterrestrial's discs

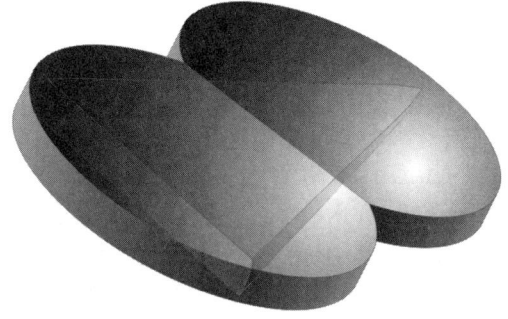

the above image shows how the ununpentium element 115 wedge could have been shaped from just two of the alien discs, while the image below shows how it was cut from the cone (which was turned from the original stack of discs after they had been fused together)

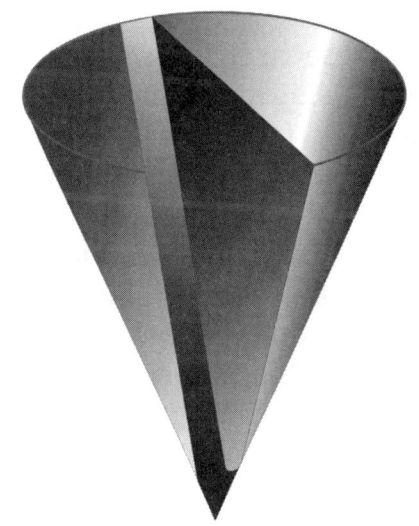

Figure 92. The alien's ununpentium 115 fuel discs had to be stacked in a special way before they were fused together and milled into a wedge

But then, this whole procedure with these twelve discs is strange, if this superheavy material was so rare a commodity why allow it to go through such a wasteful milling procedure? Because, the same wedge-shape could have come from using just two of the ET's discs, surely. If just TWO of those 115 discs were straight-edged so they could be butted together side-ways (end to end—instead of flat to flat), they could quite easily have been fused together and shaped into the same wedge as required, from simply milling that combined pair of discs side-ways, and cutting off the edges at an angle to form a triangle, and then curving round those edges and finally sharpening up the pointed end. So simple—that way there would be a substantial saving of that precious ununpentium 115, and instead of getting just ONE wedge out of those discs, crikey, as many as SIX wedges could have been produced from those twelve supplied discs! Makes perfect sense, much less wastage and six times the valuable final product. Well, it would make perfect sense, unless there was more to those twelve discs than meets the eye...

So herein resides the clue as to how the aliens synthesized that element 115. The fact that they had to be stacked one upon the other, flat side upon flat side, says volumes about how the alien's material worked, obviously, there must have been something SPECIAL in the way those original discs were manufactured by the extraterrestrials, and they must have been polarized in some way.

I believe each of these twelve discs supplied by the ETs was gravitationally polarized just as I have alluded to above and also where I've noticed a similar phenomenon in other UFOs (as has been explained in chapter 5 above). Because, where it has been suggested, in the previous section, that an atom's structure can be deformed through the four presently-considered atomic forces operating inside superheavy elements, so a gravitational transposition could be initiated upon those atoms to re-arrange those four inter-atomic forces and to re-structure those atom's shells. Obviously, the gravitational forces to do this would have to be extreme, which of course they would be in the immediate vicinity of a quasi-black hole core, for instance. In close proximity to such a gravitational polarizing force it would simply be a matter of rendering the material to be polarized into a semi-solid state, and then positioning that material near the gravitational source to exploit the enormous mass-differential between each atom's nucleon-cluster and its electron cloud (which for a heavy element such as ununpentium 115 would amount to an imbalance ratio of 4774:1). This would shift the nucleons away from the center of the electron cloud in each of that material's atoms, and this could be done either as a semi-permanent shift, or as an oscillating stress. And with a shift all that would remain to be done then would be to fix this shift in a permanent fashion (analogous, perhaps, to how an electret is given permanence through its solidification process).

In the alien discs the polarization effect must have run between the planar faces of those discs, meaning that each disc's polarization had to be orientated in the same direction when they were stacked and fused together prior to milling. This would have been in the instructions given by the extraterrestrials to the S-4 scientists (and this same instruction must presumably have been passed on to the engineers at Los Alamos too, if they did the fusing-together of those twelve discs). And so after this stack gets honed down to its required wedge-shape, that would be when the polarization effect would then become most apparent, for the wedge-shaping would be done especially to exemplify this atomic cellular imbalance. That this shaping procedure was so important would account for why the huge amount of wastage in the milling processes could not be avoided.

For the scientists of earth this is very much un-chartered territory of course, and as yet there is no precedent set for this sort of manipulation of atomic structure, no research in mainstream science has been done on it before, even though the process of gravitational polarization has been observed around black holes by astrophysicists for decades (see note 10 above). But I would presume, with regard to the sfourUFO's element 115 that the required effect would come from orientating the point of the wedge toward the furthest end of the stack with the most massive displacement, and having the wedge's blunt edge at the front end of the stack where the least massive displacement occurred.

This orientation effect of the atom's structure would stretch the atom slightly and push the electron cloud to one side of the nucleons, so on one side of the high-mass proton-neutron cluster there would be a predominance of electrons buzzing around the nucleons but on the other side just a smattering of electrons will orbit outside those nucleons (see figure 93). With all the atoms of the 115 similarly disfigured the result will be a large accumulation of each atom's electric displacement field, and a large accumulation of the Coulomb repulsive force from the fact that the nucleon's charges aren't being neutralized all the way around the cluster (ie the positive proton charges won't be fully screened by the electrons).

Anti-Gravity Propulsion Dynamics

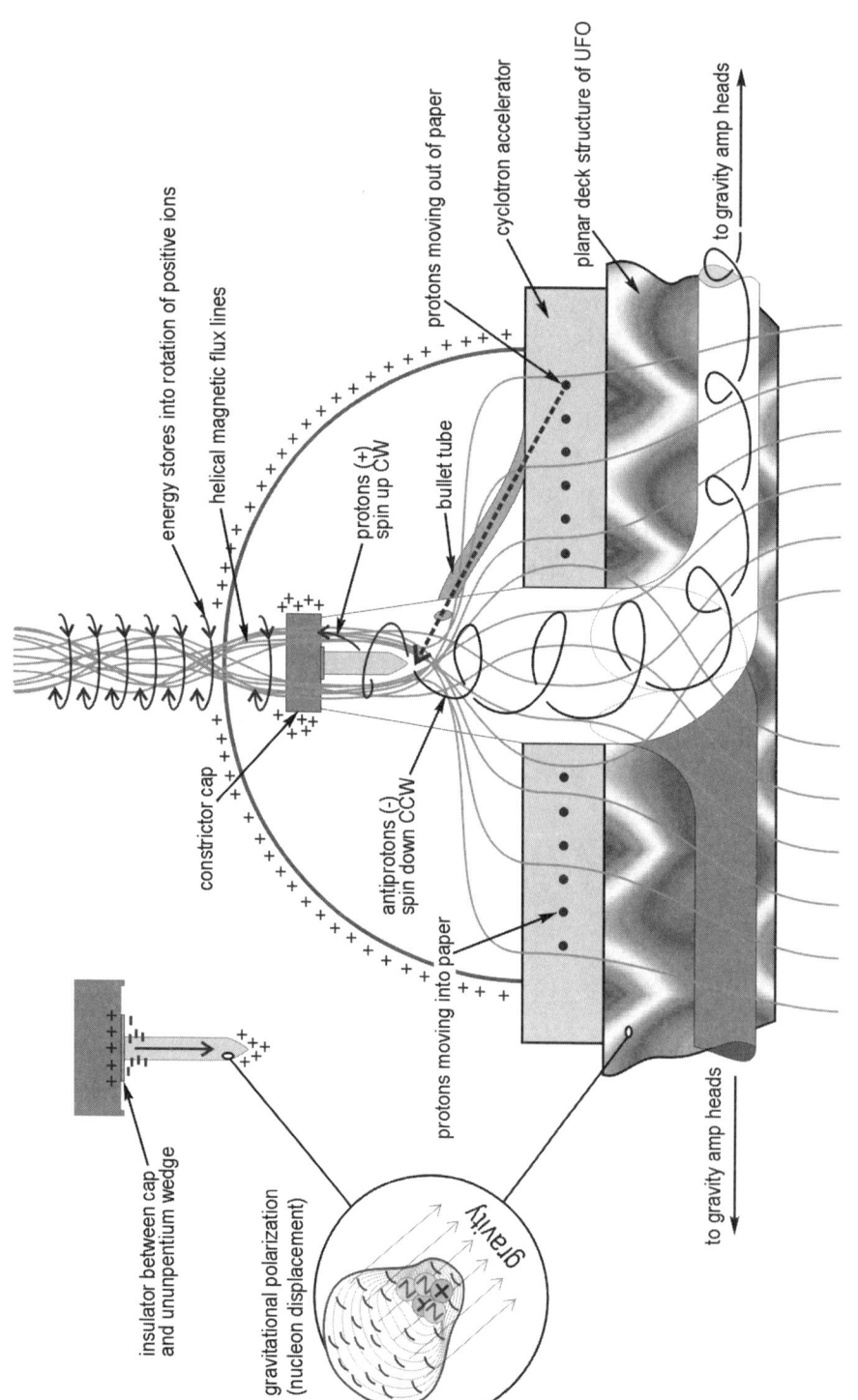

Figure 93. With the aid of the magnetic fields when the protons are fired at, and collided with, the antiprotons created will shoot downwards while the protons will shoot upwards

Given the above orientation (with the most repulsive force directed towards the wedge's point) it would obviously give the wedge a positive electrical charge at the pointed end, and a negative charge at the wide, flat end. But more importantly, this would align the atomic Coulomb and the inter-nucleon repulsion forces the same way down the length of the wedge and cause these repulsive forces to push out from the wedge's pointed end.

By having an accumulation of this out-pushing field at the point its quite easy to see how a halo effect would then be produced around the point of that wedge of ununpentium 115, and this halo effect, or mist of protons or neutrons surrounding the 115's point, would presumably be the most productive target for the accelerated beam of bullet-particles coming up from the reactor's cyclotron accelerator to strike against (rather than the nucleons within the actual atoms inside the 115 element) so as to trigger the fusion-fission chain-reactions which Lazar says will produce the sfourUFO's antiprotons [note: 11].

The reader should note this follows on from what was alluded to at the beginning of this chapter, where mention was made of the atomic nucleon skin which develops around some hyper-heavy elements, because in heavy atom reactions whereas extra protons and neutrons can be added to the atom these additional nucleons don't necessarily acquire additional binding energy, so they end up sharing the rest of the atom's binding energy and consequently become only weakly bound into the atom's outer shell (researchers have found these nucleons may have a binding force of just 0.3 MeV, for example, instead of the more normal 8 MeV) [see note: 11]. This leaves some nucleons in an almost free-floating arrangement and not rigidly confined to those outer shells of the atom, and this is possibly what Lazar was referring to, and what he was attributing to as his 'gravity A' field, when he observed that it was bending light.

One more thing, did I infer that our scientists have mastered the duplication of element 115? Well, if so then that's not quite as true as it needs to be, the element 115 that Lazar spoke of was stable—whereas the 115 that has so far been synthesized by earth's scientists has a half-life of just 87 ms (as already mentioned in section 11.3 above). So obviously there is a missing ingredient that we don't have which gives the ununpentium its long-term stability. Given that I have suggested that this ununpentium of the ETs might also be gravitationally polarized, which would put even more deformation into those atoms, then there is very obviously a synthesis procedure that our earth scientists will not be aware of, or our scientists will not have the necessary laboratory facilities with which they can carry out the same manufacturing procedure as the ETs, who can synthesize a stable version of this ununpentium.

The missing ingredient may be the laboratory environment. It may be that the laboratory that the extraterrestrials have at their disposal operates at a greater mass-density than ours—this would affect how superheavy nuclei can be clustered together, and this would affect how the four nuclear forces all balance together inside those atoms. This was a point I touched on in section 11.5 (and in section 10.7 above) in that the extraterrestrials would be privy to an environment that has a greater density than ours, therefore they will experience a different set of parameters for their dimensional constant (fine structure constant α), and they will have a different Planck constant which determines how atoms are structured, meaning that inter-atomic forces (such as the quark-gluon forces, for instance) could be manipulated by them in ways presently unknown to our physicists, so that during the manufacturing process the atomic structuring forces within those superheavy atoms would work together much more advantageously within their environmental density (the result being that when those elements are brought here, into earth's environmental density, they are stable and remain so) [note: 12].

12.9 Proton Cyclotron Beneath the Dome

Therefore, just to clarify what I believe is going on around the reactor in the sfourUFO. I would see the very exacting process of the shaping of the ununpentium 115 as an indication that the wedge itself is pushing nucleons to the very edges of the atoms around the wedge's point. And with this somewhat 'loose' skin of nucleons swirling around the pointed end of the ununpentium wedge it would be easier for the stream of bullet-protons, coming up through the particle accelerator, to collide with them and initiate the necessary reactions (up to 116 and then down to 114 and back to 115 and so forth) which will, as Lazar says, produce an abundance of antiprotons.

The particle accelerator, below the reactor, is supposed to be 'highly sophisticated,' but I would presume it is a proton cyclotron, rather than a proton synchrotron, and so it would have a fairly stationary magnetic field running through it, but it should also have an alternating electric field applied to its acceleration sectors which would probably be centered around the MHz range.

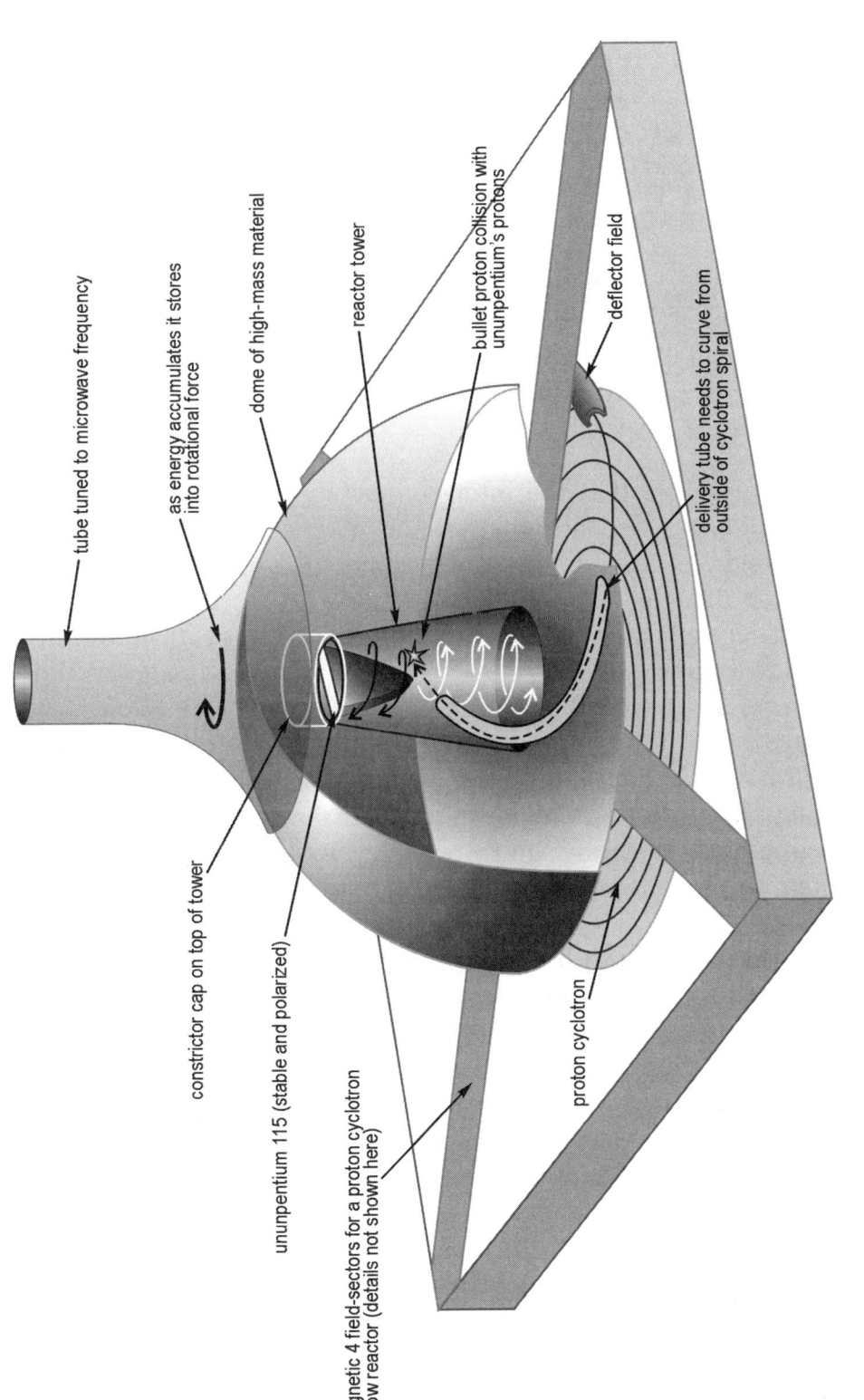

Figure 94. The box below the reactor will house a simple cyclotron accelerator

Although, to make things much less complicated overall, if this cyclotron's sectors could be designed to oscillate in the GHz range then this same frequency would then be useful to link together a whole bunch of mechanisms operating within this UFO, to 'switch them on' at the same time, if you like. For instance, if the cyclotron's alternating field could be linked into the craft's central unloading tube at that tube's resonant frequency then both could operate at the exact same frequency and one could trigger the other (and this will make more sense when it is further explained below).

The cyclotron would be encased inside the square box beneath the reactor dome, with the diagonal cross-members shown in Lazar's image denoting that the cyclotron has a four-sector bending-magnet configuration. It might be ball-parked to deliver protons at around 350 MeV (but I'm really only guessing here using conventional cyclotron figures).

My suggestion here will be that the inter-connecting 'linkage' for this central part of the sfourUFO will be its central (longitudinal) magnetic field. There will be a vertical magnetic flux field operating within the cyclotron (or synchrotron) casing to power up the protons so that they gyrate around the outside of the reactor's tower tube and become accelerated to a high kinetic energy as they drift out to the perimeter of that cyclotron box (see figure 94 above). And I would suggest that the extraterrestrials are interlacing with these cyclotron flux lines the ambient magnetic flux lines coming up through the center of the sfourUFO, which, on their way up through the craft will bunch around the craft's central tube (which I shall hereafter refer to as the central unloader tube). Having said that these flux lines will have to be directed into a helical path as they flow up from the cyclotron and pass through the reactor tower, on their way to that unloader tube.

I know from my researches into other UFOs that the extraterrestrials are huge advocates of poly-functioning, whereby different mechanisms are all tagged together so that they all operate in concert with eachother (and without separate switching circuits). And in this regard it may actually seem quite straightforward to understand how this UFO works, and how its reactor works, but to actually construct these craft the trick will be working out how all these disparate mechanisms can link harmoniously together (without half-a-ton of insulated copper wiring looms that most all of earth's electro-mechanical systems are presumed to require). On this subject I would like to say here also, that the reason why I know Lazar's hypothesis is based on genuine extraterrestrial engineering is because there are so many separate functions of this UFO which do all meld together into such a beautiful and unified workable system. This marks the sfourUFO craft as a phenomenon that could not have been designed and constructed by humans. For we humans design our machinery always as an ensemble of separate mechanisms; and where our scientists use mechanical parts (with wires) which are mechanically rotated to produce energy, the extraterrestrials use confinement and constriction of magnetic fields to rotate charged particles to produce their energies. And so all of my discoveries here for this sfourUFO confirm that a very large proportion of Bob Lazar's initial description about this UFO is sound, and that this UFO does exist and was being back-engineered.

The helical factor of the magnetic flux lines will be important because when the proton-proton collisions occur inside the reactor (at the wedge's point) and the antiprotons are generated, because the antiprotons will be negatively charged and the protons positively charged, this helical magnetic field will send the antiprotons DOWNWARD and the protons (and any other, incidental, particles flying around) UPWARD, thereby separating them apart. Obviously, the more concentrated this helical magnetic field is the more intensely this task of separation will be performed, and so, as is indicated in the accompanying diagrams, this will be why the reactor's cap will be so important too, for it will be used to constrict those magnetic field lines (so that they will mostly converge through it) and in so doing will provide the degree of magnetic (Lorentz) force necessary to separate the protons from the antiprotons. So, in other words, without the constricting cap the protons and antiprotons would simply not separate far enough apart and would cancel eachother out almost as soon as they were created.

Actually, my idea for the reactor's cap will be that it can also be manufactured to be an accumulator of positive electrical charges (ie from the ununpentium's protons going upward and striking into it). This cap is very much thicker than would be expected, which certainly seems to indicate that it has one or two specific tasks to perform where it is situated above the reactor, even though it looks fairly nondescript (ie if it was just a plate that capped off the reactor's tower-tube it would only need to be a few millimeters thick). So the major requirement will be that it needs to be constructed as a positive 'sink' which accepts positive electrical charges just as much as it needs to be made with a high magnetic permeability (so that it draws through it the bulk of the magnetic flux lines that run between the cyclotron accelerator and the unloader tube).

Then, and this is the *piece-de-resistance*, by inserting a thin insulating material between that constrictor cap and the upper-most end of the ununpentium wedge (so that when that cap is clamped down onto the reactor's tower and it butts up to the wedge the cap and the wedge will be separated by the insulator), then whatever positive electrical charges accumulate within the cap's material a corresponding accumulation of negative charge will act upon the ununpentium (and cause a negative repulsion at the top of the wedge). Or, to put it another way whatever negative charge is accumulated within the top end of the ununpentium wedge it would follow that there would be induced an equal amount of positive charge in the cap above it.

12.10 Pitching in From Three Angles to Load the Dome

Once created the antiprotons would be sped down the reactor tower, being turned by the central magnetic field and coursed through the inter-connecting curved channels, to each of the three gravity amplifier heads. Ideally, they would be fed into each amplifier head at the top of its black vortex, against the flow of that rotating vortex. And as soon as the highly-accelerated antiprotons arrive and smash into the matter particles (the gaseous matter, as Lazar worded it) rotating in those vortexes, the annihilation events would be initiated instantaneously and the full reaction process, through the intermediate production and destruction of pions, muons and gamma rays, would ensue and lead to the full conversion of those matter-antimatter particles into pure energy.

With each event producing 1862 MeV in regular succession a tremendous amount of energy would be stored into these three black vortexes, and THIS will be where the 100% thermionic transmutation of antiproton-antimatter into heat and energy, that Lazar spoke of, will occur inside this sfourUFO (although, where heat is mostly a non-coordinated radiation, while the overall influence within these black vortexes will be to quickly align that radiation into its rotation, that energy won't be given any chance to give out any heating effect, and it will convert straight into raw energy). The vortexes will also be where residual amounts of atomic gravitational binding force and electromagnetic energy will store (from the whole range of particle transformations), and where the highly accelerated forces of the vortexes will amplify and imbue gravitational potential into their rotating particles.

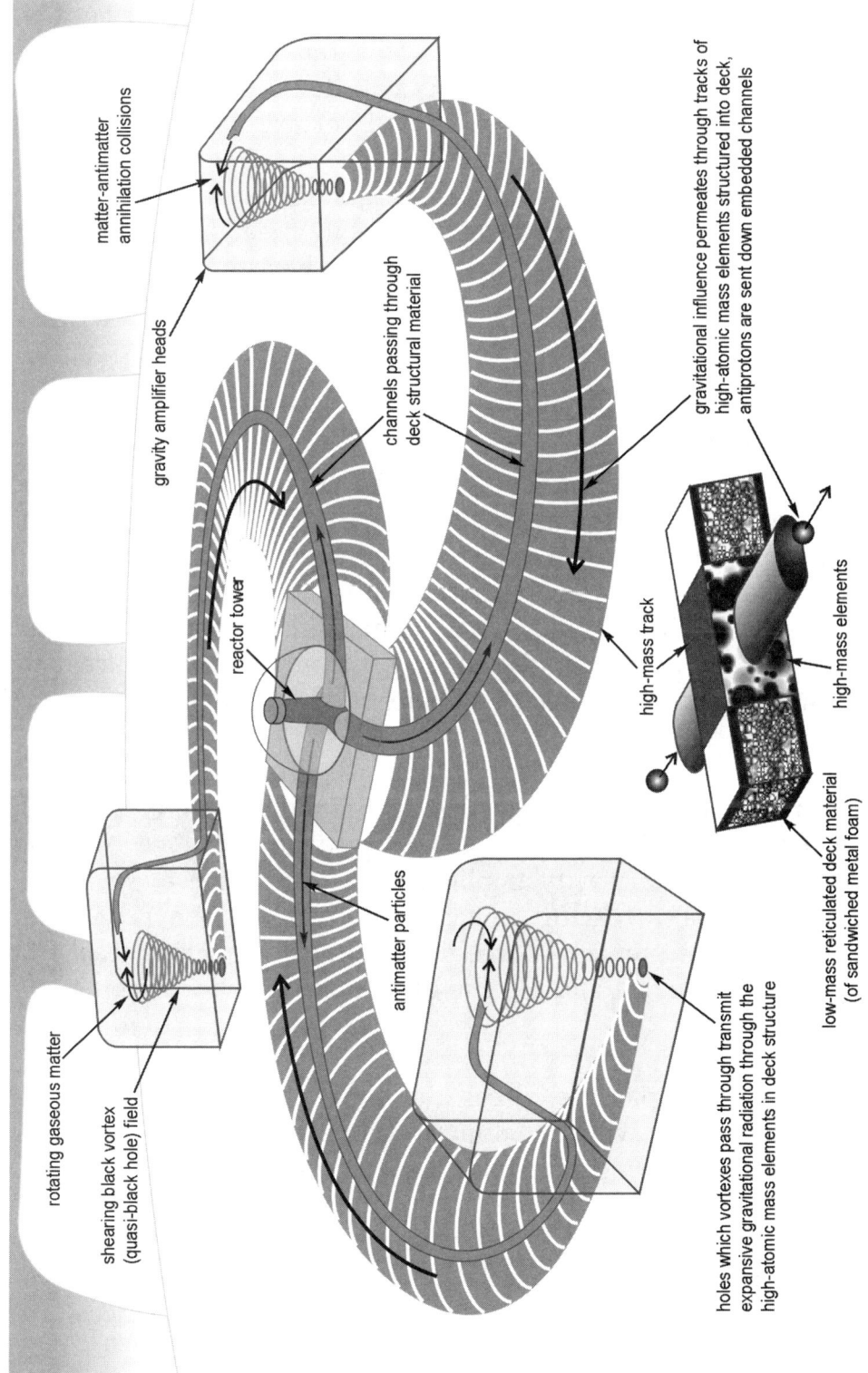

Figure 95. To keep the antiprotons away from their channel walls the gravitational polarization tracks should follow the same path as those channels, so as to give each channel wall a micro-magnetic blanket

As will be seen in the accompanying drawings, the main products of all this transmutation of mass to energy will be angular momentum and gravitational forces, and the harvesting of the expansive gravitational force will take place at the bottom of each of the quasi-black holes, below the high powered cores which are the demarcation points that determine whether the gravity force will be attractive or repulsive. Below the core the rotating fields will be anti-vortexes and the forces they carry will be expansive not attractive (and it will be this force which will be transferred into and amplified through the cylindrical gravity amplifiers in the deck below them, as will be explained in the following presentation), and whatever power is radiated at these anti-vortexes, whether downward or horizontally through the structure, will be proportional to the energy stored into the black vortexes above them.

This expansive gravitational force will be what will be transmitted, as mentioned above, through the sfourUFO's structural material, through that material's atomic cells, from the gravitational polarization effect upon those atom's massive nucleons. This is seen by the extraterrestrials as a more direct way of transmitting gravity, and it is elsewhere used around the centers of the Andreasson and the Roswell UFOs. Although, what the extraterrestrials are using in those other two UFOs is a horizontal layer of bismuth (in the form of micro-undulations of bismuth), which is a high-mass material and they're using the micro-undulations to generate a particular frequency of oscillation into these structural materials (which transports the gravitational influence).

This however, is not how the sfourUFO transmits its gravitational forces through its structure (although the principle is similar, as will be explained below). Moreover, this transmission from each of the gravity amplifier heads of expansive gravitational force will move upon the central reactor, as has been described above (in section 12.6), as a convergence of three separate force fields all pushing against eachother. Where these three wave fields meet and repulse eachother will be where the white hole event horizon will be.

However, for whatever expansive forces there are converging around the outside of the reactor dome, there will be to oppose them a proportional amount of a quite different repulsive force flowing out from the surface of that dome. This will be the Coulomb repulsion forces accumulating just outside that dome from all the incidental positively-charged mass particles oscillating around it; and compounding against them will be even more positive charges accumulating on the outside surface of that dome, in response to the positive charges at the constrictor cap (recall from science classes, that in the Faraday ice-pail experiment, a positive charge accumulating inside this dome, which will occur around the reactor's constrictor cap, will induce a positive charge on the outside surface of that dome).

So the three expansive forces from the black vortexes will be repulsing eachother, forming a shield, and just inside that hemispherical shield (the one that Lazar couldn't push his hand beyond) there will be an electrically-positively-charged force pushing out to that shield from the inside. And as the three prongs of expansive gravity will effectively be pushing any and all magnetic flux lines toward the longitudinal axis of that sfourUFO, then the electrically-positively-charged particles just outside the dome will become imbued with so much Lorentz force that they will be spinning very energetically around that dome.

And this follows on from what a black hole does with its interplay of gravitational and magnetic forces, and its why a black hole is surrounded by fields of rotating particles, because the only way extra energy can keep coming into that system is by that system rotating faster and faster. So a black hole is not the only system which uses gravitational force to turn electrical energy into rotational force, this UFO's white hole will be doing it too, albeit in a slightly different manner.

The fact that there will be a continual influx of energy to the outside of the dome, and to the constrictor cap inside the dome (from the ununpentium reactions), so that all that incoming will get converted into rotational energy both inside and outside the top of the reactor's dome, it means that the positive ions will be stored in exactly the right place to be imbued with gravitational forces when they propagate up and out the central tube (positive ions are more efficient carriers of gravitational force than electrons, by the way, a fact well known by astrophysicists). This will answer the question asked, in section 12.3 above, about how the central unloader tube could be lowered so that it enters through the white hole's pass-not shield... remember the magnetic field, which accelerates the protons in the cyclotron? Well, those magnetic flux lines will be oscillating their charged particles at a frequency of 3 GHz, and it will be because of the frequency influence carried by these magnetic flux lines, which will be flowing up through the top of the reactor and up through the top of the dome, that there will be imbued into these positive ions rotational forces which will spin them around the top of the dome. These oscillations of expansive gravity force and Coulomb repulsion and spinning particles though will still be fairly uncoordinated movements and will have no predominant frequency just yet, because their collective energies will be bouncing off eachother and will spread around the shield as random frequencies and even colliding with varied amounts of rotational force. But as the central unloader tube is lowered down and approaches closer and closer toward the reactor dome, there will come a point where this tube's horn-end, AND the white wall shield's oscillating energy, AND the helical magnetic flux lines will all suddenly lock-on to eachother and resonate to the dimensions of that central unloader tube (which I have given an arbitrary 3 GHz).

The very moment this happens and all these energies harmonize together with coherence it will be like removing the plug of a drain, and all these convergences of force and rotating positive ions will, quite instantaneously, UNLOAD themselves up the central tube.

So, this will be the precise moment when the white wall shield around the dome will instantly disappear, and as long as the unloader tube is near enough to the reactor's dome, or touching it, this shield will not accumulate its forces any more around that dome (because the innate frequency of that coherent oscillation will take it through the path of least resistance, which will be through that tube of the same frequency) and after that the expansive force (but not with its event horizon) will accumulate on the OUTSIDE of that craft.

From the top of the tube, above the UFO, the expansive gravity will expand outward to surround the top of the craft, and the positively charged particles (themselves imbued with expansive gravitational force and substantial angular momentum) will spin out the tube and curl around the expansive hemisphere surrounding the top of the craft, and with their intrinsic curl will spin round the sfourUFO's rim and back into the belly of the craft, to form a quadrupolar field, or rather, a poloidal field as Lazar has it in his images (the apple core field). What has to be emphasized though is that in order to do this, to curl down around the craft, those accelerated particles will have to be imbued with something called negative radial momentum [note: 13]. Obviously, a highly accelerated force of particle-energy directed up a narrow tube is going to fire out the end of that tube like a bullet—straight up—and that's not going to be of any use to this UFO.

So what will be needed for all that gravitational energy, gravitational waves if you like, imbued into the positive ions shooting up the unloader tube, will firstly be to rotate as it accelerates up that tube, and this will be where the central helical magnetic flux field will come into play, for its Lorentz force will help control the way these energies rotate.

Then, secondly, because it will be so important how the charged particles exit the tube through a particular helical trajectory (helical pitch angle) it will be important exactly how that tube is terminated (and how it is shaped internally). As I have alluded to elsewhere (in chapters 2 and 10 of this study) regarding how a rotating helical field can be controlled and made to do what its supposed to do is a science in itself (so I won't be delving into that here, and the reader should refer to my previous chapters to discover more about the astrophysics of this).

It will only be if these particles and forces can be given sufficient curl that they can be expected to produce the necessary poloidal field around that UFO, but when it does become established this surrounding field of expansive gravitational force will give the sfourUFO gravitational buoyancy against the earth's gravitational field, although having said that it won't give any propulsive movement to that craft. That propulsion will come solely from the gravity amplifiers in the craft's lower deck.

So, inside the craft the whole central deck area will act like a tuned waveguide once the unloader tube is lowered, with all the components and converging forces resonating in coherence to the same frequency as that unloader tube.

One further note on the central reactor, when the tube fully touches down on the reactor dome and all the convergences of force automatically re-routes up the tube and the white hole shield dissipates (again, just as Lazar said it would), once that shield around the central dome does collapse, because it will effectively leave the dome as a highly-positively-charged terminal oscillating at 3 GHz (or thereabouts), it will then-after establish between itself and the rest of that sfourUFO's whole deck area an electron cascade effect, because an electric field will be established between the dome and the air inside that deck area, and the result will be to draw electrons toward that dome from the surrounding air at high speed (and this will ionize the air). So here, the crew would need to get used to the ubiquitous smell and taste of electricity inside that craft when it was in flight.

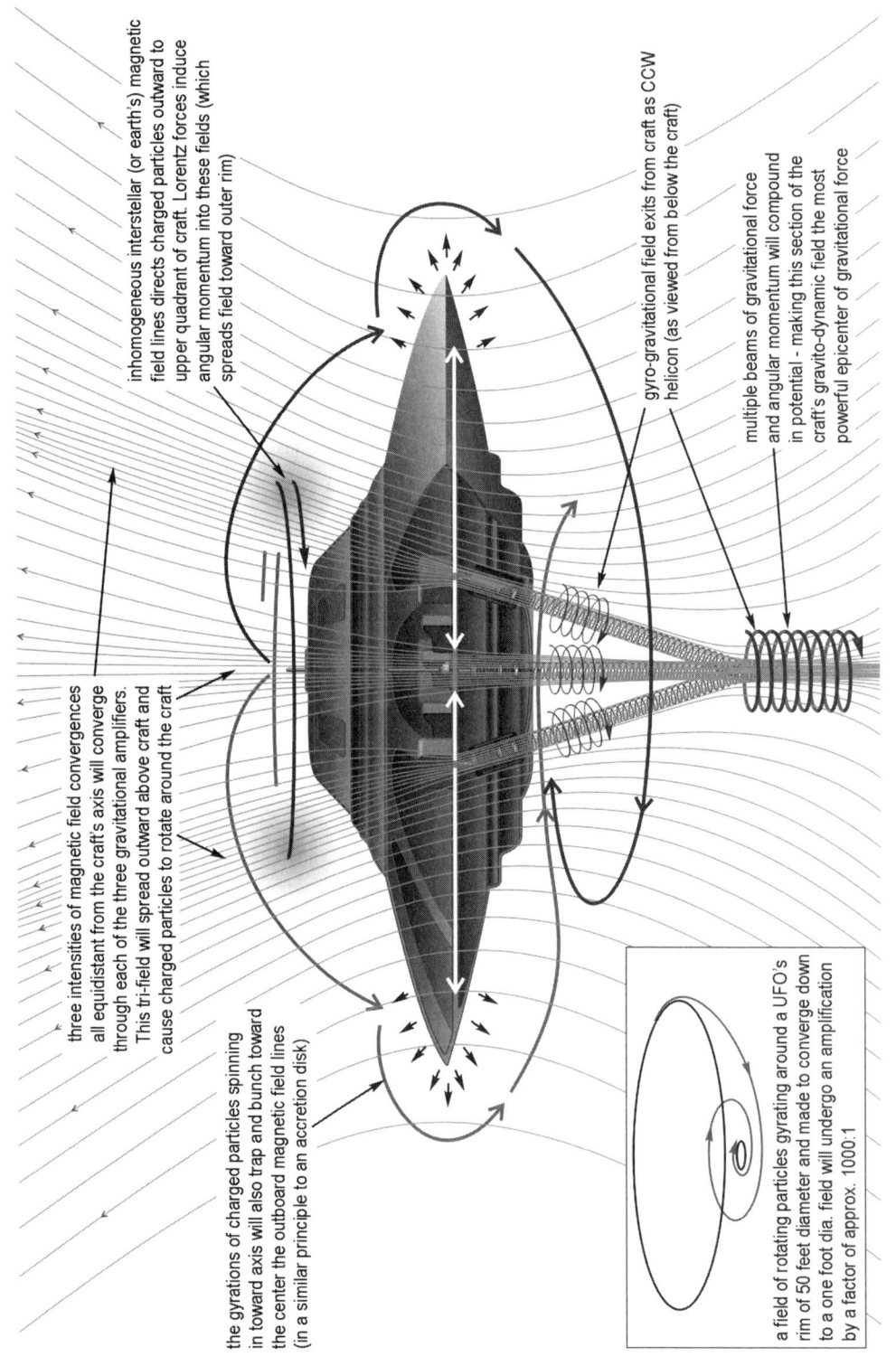

Figure 96. Stressing the magnetic field lines around this UFO will produce electron-positron pairs which will then be induced into the gyrating beams below the craft (the beams do not pass directly through the hull)

12.11 Holy Greyl Found Hidden Inside Lazar's Frequency Parameters

On the Gravity Propulsion page of his website presentation Lazar gives a very strange frequency rating for the sfourUFO reactor's element. He says it had a carrier wave frequency of 7.46 Hz with a one micron bandwidth... which, as every good UFO researcher knows, or should know, this doesn't make any sense. After-all, a 7.46 Hz frequency has a wavelength of 40 thousand kilometers, so carrying it inside a bandwidth of one micron would be analogous to putting a river into a straw! Thinking it might be some strange and secret number in disguise, I did wonder if this rating might be to do with the cyclotron, which would need to be regularly pulsed and it would also involve a one micron thickness; because the exit window of a cyclotron (for where the accelerated particles are fired out of it) usually has a one-micron thickness; and by firing the cyclotron's exit field every 7.46 Hz (7.46 pulses per second) then this configuration would provide a steady flow of bullet-protons into the reactor, so that the reactor would then provide a steady flow of antiprotons to the amplifier heads.

But then, on Boh's Omicron Configuration page a slightly different attribute is given to these pulsings (for the gravity wave amplifiers this time), where it says that the three amplifiers are individually pulsed at 7.46 Hz; and correlating this to where it says, on the Microwave Analogy page, that this seven-point-four-six times a second pulsing relates to universal vector fields, then this figure begins to make more sense. And now I also see why Lazar gave this oscillation as a wavelength not as a frequency.

What these denominations are saying is that an oscillation with a one-micron wavelength (which relates to an open-air frequency of 300 THz) is being pulsed at 7.46 Hz. If so then this is worthy of a lot more research, for a 300 THz frequency can be found in the infra-red band of the spectrum, which is very interesting when one considers that UFOs have been observed and filmed over Mexico shrouded inside infra-red energy fields [note: 14]. But then, what would be a more intriguing path to follow here would be if that one-micron wavelength was a reference to the propagation wavelength of the gravitational force itself, as it is moved THROUGH that UFO's material structure. In other words, it might be more productive looking at the gravitational polarization mechanism, the transmission of gravitational influence through the UFO's atomic structuring, in the same way that phonons are passed through a material's atomic structure at infra-red frequencies but at a much reduced velocity. And then considering, that this might also be the way the force of gravity propagates via the core holes around this UFO's three black vortexes, through its structure, at speeds which might be found to be closer to sound speed than to light speed!

And this factor, I'm sure, will have ramifications all through the industry of UFO back-engineering—because not only do the other types of UFO (the Dalnegorsk UFO, the Andreasson UFO and the Roswell UFO) portray the use of this gravitational polarizing effect through their preference for high-mass materials (as I have shown in sections 5.1 and 8.5 above)—but THIS could also show a path which leads mainstream Science to the Holy Greyl of gravitational force generation and detection.

To recap this important feature used by UFOs, it will be the action of the quasi-black hole's gravitational polarizations upon the atoms inside the materials which surround the UFO's quasi-black hole cores, which will cause the nucleons and electrons in those materials to separate momentarily, and these separations will generate terahertz frequency oscillations [Kiefer-Weber] [Oh 2004]. To take advantage of this oscillating field what the UFOs are doing is establishing ways in which these perturbations can be transmitted efficiently through the material structure of their craft—at the same time carrying with them their gravitational force. And because the greatest combination of electric field oscillations and gravitational force pulsations will be occurring around the black vortex cores, that's *exactly* where the extraterrestrial designers are tapping-off these terahertz oscillations. And by using magnetic field constriction the rotating filaments of those vortexes are simply squeezed through holes in the UFO's structure, and immediately around those holes (and where the magnetic field density is greatest) the material's cellular composition will be such that the gravitational forces can be coupled to the magnetic and electric perturbation fields also running through those materials—so that the gravitational forces can be spread out radially from those holes as a molecular gravitational polarization effect [note: 15].

This is why I have indicated (in figures 95 and 96) that the sfourUFO's main deck would have high-mass tracks running from each of its three quasi-black holes, and these tracks would direct those gravitational perturbations toward the central reactor dome.

Up until now, the consensus opinion among physicists is that gravitational waves travel at the same velocity as light waves but then this is pure guesswork by those physicists [note: 16]. But this radical and new assumption about gravitationl influence makes a great deal of sense because obviously, if a force is transmitted like a plasmon or a phonon from atomic cell to atomic cell, then it won't be transmitted at the speed of light. Gravitational influence has to travel the path of most resistance, it has to move at the speed which is governed by the cellular structure of that material.

So then, looking at this gravitational influence in the same way that electric charges can be transferred through (and, in some cases, over a material by using surface phonons to pass them over the outside of) that material, might be very rewarding. Particularly, as in the latter case, where the surface phonon might be considered as a coupler of gravitational force, and this is possibly one of the ways that this gravitational influence might be found to propagate inside a UFO, because of the fact that the extraterrestrials are using the UFO's structure to transmit their gravity somehow. This, in a way, would be similar to how surface acoustic waves move over a material (and they might be using a similar principle to how surface acoustic waves (SAWs) move excitons and electrons over that material). This being the case these wave trains of gravity force moving through or over the material structure could then be chirp-amplified (which would compress those waves to increase their amplitude). They could be chirped with the repetition rate which Lazar gives, of 7.46 Hz, and this would greatly amplify the gravitational force as it moves onto the central reactor through the UFO's structure, and the whole chirping process could be achieved automatically (without another separate amplifier) simply by constructing the UFO's material structure in a way that it would act just like a chirping waveguide does to an electrical field. This would mean taking the basic principle of acoustic wave chirping a step further and establishing wave-bunching structures (or materials) within that UFO's structure, specifically around the reactor. Then, and this is the beauty of the reactor being in the center of the craft, any concentric fields of force (coming from the three gravity amplifier heads) will automatically be amplified just by moving through a smaller and smaller radius toward the center.

However, it would be better if there was a particular type of concentrically moving field which aided this transportation of gravitational force, and as I say above, this transporting perturbation may be engineerable actually inside the atoms of the UFO's structure. Inside a solid material the atomic nucleons and electrons while they are bonded to atoms there will be sufficient movement within those atoms to allow those charged particles to contribute to collective oscillations within that material, and basically this is how the phonon and the plasmon transmits electrical charges (rather than electrons themselves) from one end of a material to the other, and so just as with an electrical field moving through that material so those perturbations could also carry gravitational force. Instead of electrical charge waves, gravitational 'waves' of influence could be moved from one part of the craft to the other. This is to say though that the gravity forces inside the UFO will not propagate of their own accord like an electromagnetic wave or even like a longitudinal wave, the gravitational influence will be carried by a moving perturbation field (be it electrical or magnetic) as it moves through that material. This then would be very similar to how the Roswell UFO directs some of its gravitational force through its structure at a particular frequency [note: 17].

The greatest difference would be to the waves' propagation velocity, as mentioned above, it wouldn't be at (c) the same speed that light travels at through the air (which would be the normal propagation velocity of electromagnetic waves), and strictly speaking, it won't be quite the same speed as the acoustic phonon or plasmons either, because these are carriers of electron charges. A gravitational force is more likely to be moved through a material's cells via the displacement of that material's heavier nucleons (which are much less mobile than electron charges), and so the movement is possibly more like that of a strain wave. Either way, in this case the gravitational wave/influence velocity would depend on exactly what metal, or what dielectric, the transmitting structure being used was made of (or, alternatively, by determining exactly what transmission velocity is required and using that parameter to dictate exactly what material that transmitting structure will need to be made of).

Obviously, this is a whole new and unique form of mechanics where transmission times through a material will depend on the elasticity of that material (and for nucleons not for electron charges). This is why, returning to the point made about the one-micron wavelength above, an oscillation can have a high frequency in air but it will have a much lower frequency in a solid. And because all materials have different compositions with different elasticity then they will have different acoustic, and different cellular-perturbation velocities even though the same oscillation is passing through them. For instance, bismuth will have a phonon-electron acoustic velocity of 1790 m/s while magnesium will have a 4602 m/s velocity, and so on. Different materials, different velocities of perturbation, equating to different resulting-frequencies being transmitted through them.

With SAWs, their wave velocity can be around 3000-meters-per-second, and so by having a wave-period of one-micron this would mean that the gravitational influence they carried would arrive at the central reactor as a 3 GHz oscillating wave [note: 18]. And as it would be quite feasible to presume that there would be a particular material that could be used in this sfourUFO's construction which would transmit this gravitational influence from its quasi-black holes through its structure's atomic cell nucleons, also at a velocity which would result in a carried frequency of 3 GHz, then doesn't this resonate well with the central magnetic field's operational frequency and to the frequency attributed to this UFO's central unloader tube..!

So, therefore, I am now wondering if the target parameter for the transmission of gravitational force for this UFO should be around 3 GHz, but at a velocity 5 orders-of-magnitude slower than (c), and that the other requirement would be that these oscillations should be pulsed (or chirped) at 7.46 Hz. If this is the case then it would, just as importantly, mean that the material the sfourUFO would need to be constructed from would have to give a cellular polarization/perturbation velocity through it of around 3000 meters-per-second. These figures, going back a few chapters, would also bode extremely well for the Andreasson UFO, because it has already been established for that aaUFO craft that it operates at a frequency of between 3 to 4 GHz with wave-packets that have a nominal pulsing rate of 9.5 Hz (as I have alluded to in more detail in chapters 7, 9 and 10 above).

With these techno-nuances working together so well (with the helical rotating field mechanics shown in the previous chapters of this book) maybe now the astrophysical gravito-dynamics of UFO propulsion technology (rather than the patchwork-quilt that has been hacked together of late using Einstein's and Newton's physics) will be afforded a little more decency by scientists who, after all, have all but drowned in the mire of existing gravitational theory...

But wait, hold the page, is there something more going on here..? Referring back to this 'one micron' denomination that Bob Lazar used, this has been puzzling me for ages and I just couldn't see why it looked so odd. But now I see what is going on, here is the solution, this was no idle mistake made by an absent-minded physicist. No-one, but no-one, gives an oscillation and bandwidth configuration as a frequency and a wavelength! Well, the reader might also say, no one gets them mixed up and placed in the wrong order either, but that's beside the point. Always these two are given together as a pair of frequencies. But, in this particular case giving the one micron as a frequency would have been incorrect, it wouldn't of explained to the researcher what Lazar had seen and was trying to tell others.

So, it means that Lazar knew precisely what I have just explained above, about that oscillation downgrading to acoustic-wave-velocity while it passes through the UFO's material structure; and he knew that the normal association of 300 THz to one-micron was not what he was trying to convey. Only if he knew to give this distinction would he have considered it more appropriate to give that oscillation as a one-micron wavelength...

I rest my case, but with one last question, because yet again I am puzzled as to why Lazar appeared to know so little when he was interviewed by George Knapp. If for the sake of argument, Bob Lazar really DIDN'T KNOW why he should provide this information in this precise manner... did someone at Area 51 indoctrinate this information about the sfourUFO into him in this very specific way? In other words, was Bob Lazar 'telling his story' about this sfourUFO as he saw it, or were these the de-classified details about this UFO that were specifically written by the Area 51 physicists, and given to him to disseminate as if to tell the rest of the world, hey—look what we guys are working on!

Afterword... on the Transformation Phenomenon

Although this book is mostly about UFOs it is also about energy dynamics and energy generation. Several new methods by which electrical energy can be produced will be developed as a result of this study, and these methods of electrical production will be found to be considerably more efficient than any existing nuclear, gas, or coal-fired turbine generators.

But, I would like to bring to the reader's attention something even more important, more important even than the survival of the human race and even the preservation of this planet of earth...

Because, if through the pages and diagrams of this study these new methods of energy production and the featured UFO gravitational force engines can prove for themselves to have workability, then perhaps some credence should also be given to the supernova warning messages which have been inter-dispersed throughout the experiences of abductees, and that have been included in the only available communications that have been delivered from the extraterrestrials—through the abductees.

Then, with that credence, not only will the extraterrestrials prove themselves to be engineers of the kind of advanced sciences that we have not looked at yet, but we might even afford to consider that these warnings portray the main reason as to why the extraterrestrials have come to earth.

For it may not be that the extraterrestrials are here to offer their technology to us just because our scientists have been considered, by that greater scientific confederation, to be dragging their heals when it comes to spacecraft utilization of available power and force dynamics—there may be extreme environmental factors of the utmost urgency that these outside scientists are trying to inform us about—which our leaders are either ignorant of, or if, while they *do know about them* they have a vested interest in keeping those factors *unknown* to the general public. So, perhaps from hereon-in one essential question to ask whenever governments enforce their stringent denials about UFOs is... *Have our government leaders known for decades that a mass extinction event for humankind is immanent—and the preparation for it is staring us all in the face right now?*

This mass extinction sequence of events, that the extraterrestrials have foreseen, of a supernova blast heading on a collision course toward the immediate location of our whole solar system, has been described to scores of abductees by different factions of the extraterrestrials as a wholesale gamma-ray bombardment of earth and incineration of all living things that presently exist above our planet's surface.

The extraterrestrials have foreseen that the human race is facing a mass extinction event, and that nothing will remain above the surface of this planet after this transforming bombardment from space other than dry scorched earth and stone boulders!

The reason, say the extraterrestrials, is because a blast force that has long ago been initiated from a distant star system that has broiled over and gone supernova many thousands of years ago, is presently hurtling through space at high speed toward our position in the cosmos, and unlike the threat of a rogue asteroid THIS blast force cannot be diverted!

Thus it has been indicated in many of the ET's messages published in Betty Luca's "A Step Beyond Tomorrow" (ASBT) books, such as in message 1.33.1 (12.29.77:8.15pm) where it says,

"Take heed, think over, the withdrawal from the harmful incision's swirling spread of swelling waves rolling from the sun-star when it's umbra aligns it's negative energy megastrike of swelling waves. Please know, we beckon to you to flee hastily within the cylindrical Ka-nuggar ship. Before the incision's spout of swirling plasma and grinding-polluting wall of anti-magnetic desolation aligns the destructive transformation of the star's swelling waves and it's nova megastrike. Take heed to organize a withdrawal from that nova's swirling umbra of alpha energy, until you identify the first indications that it has slackened off and calmed sufficiently."

How can such an alarming warning of catastrophe have any credibility without the authority of our government leaders? Are there any correlations known to our scientists that would by-pass those leaders and validate these incredible warnings? Would this be why our government scientists have been so reticent about validating any faction of the extraterrestrials as to their existence, and have so vehemently denied that their UFO craft have been seen flying over our cities, energy installations, and military bases? Is this why the testimonies of some abductees have undergone ridicule and invalidation by those who work in scientific and psychoanalysis professions?

Here I should remind the reader that it has, without doubt, been shown through the present study, that there are benign groups of ETs that are capable of engineering greater density environments, both inside and immediately outside and forward of their ships; meaning that, as in the latter case, their ships are capable of traveling faster than the speed of light (or rather, faster than the speed we consider is always attributable to light); and from this conjecture it would certainly be reasonable to suppose then that the extraterrestrial physicists have for many millennia been traveling into the far-flung depths of space to see for themselves a whole catalogue of supernova eruptions occurring in many distant parts of the cosmos.

It would also be feasible to suppose that after they had seen the very initiation of such supernovas that they could monitor the trajectory of both the light wave and the physical blast wave as it expands through space, and determine whether it posed a threat to any civilization caught in its path—much in the same way the previous three mass extinction events might have been perceived for earth's inhabitants at 11,500, 33,000 and 60,000 years ago, when the supernova eruptions from the Vela and Geminga supernovas occurred [Sonett-Morfill-Jokipii] [Frisch 1997] [Zank-Frisch].

The present mass extinction warnings spoken of by the extraterrestrials reported in the Betty (Andreasson) Luca abduction experiences describe the full process of how a star goes out of balance and broils over to initiate its supernova blast. These ETs then warn that the blast force of that supernova is currently hurtling through space and heading TOWARDS US RIGHT NOW! These messages then go through the effects that the energy of such a blast wave will have upon our sun's heliosphere, saying that it will cause our sun's protective sphere which encapsulates all the planets of our solar system to collapse to a perimeter of only one earth-to-sun distance—saying this will cause the sun's chromosphere to expand out to the very surface of this earth!

Strangely enough, the possibility of this impending mass extinction catastrophe has already been calculated, our scientists know full well what the effects of such a supernova blast when it arrives will have upon our world, upon humankind. The gamma-ray bombardment has been acknowledged and explained in the scientific papers of Dar and De Rujula, who warn of *environmental pollution by radioactive nuclei, depletion of stratospheric ozone, extensive damage to the food chain by radioactive pollution*, and gamma-ray bombardment from the jetted cannonballs (of highly relativistic jets produced in supernova explosions) which will produce *highly beamed cosmic rays by ionizing, sweeping up, and accelerating the particles of the interstellar medium [Dar-De Rujula 2002]*, which is the medium our sun's heliosphere is made up of—and all this is precisely what has been described in many of the extraterrestrial's messages! For instance, the ETs are saying in the ASBT message 1.29.3 (10.21.78) that,

"Before the star's positive electric centimeter swelling waves connect and traverse over to strike against obdurate mankind he must first pay the price and seal within. Before deliverance from the star's heaped-up swelling gossamer of waves of energy. For salvation away from this fire the squadron is circling your sphere. Employed task force awaits return..."

And in ASBT message 1.31.1a (6.15.78:12.35am) where it says that,

"A legation exists to emancipate, to transplant away from the annihilating umbra of the mega-electron dusts of the star's nova encrustation strike, of mega swirling plasma into the elastic magnetic matrix. Sew together the star's swelling solar streams of precipitation. Or obdurate Seven-mankind must flee the conduit of electric negative-polarity rays and beta radiation from the sun's swelling waves—which will be the portent of carcinogenic sickness and disease."

At first inspection two possible candidates for the impending supernova blast that the ETs have seemingly come here to warn us about are the supernova eruptions of WR 104 and Eta Carinae. These supernovas were triggered 8000 and 7500 light years ago respectively, and in the case of the Eta Carinae supernova the brightening of its distant burst was first perceived here 150 years ago. That means that while the initial light waves of those eruptions have been detected on earth only recently, the actual physical blast waves which will be traveling out from the star at a speed slower than light (so that they will take a little longer to strike us), are still traveling through space ON A COLLISION COURSE WITH US!

But then, there is even more being said by the extraterrestrials in the Betty Luca messages to consider, for they have presented *both* the offer of a squadron of motherships, with an open invitation to the people of this world for them to *physically* escape to safety to another part of this universe, *and* the ETs are offering information about the much anticipated *Transformation of spiritual beings* to the next stage of evolution... and because they are doing this it suggests that a grandiose itinerary has been formulated by a very proficient and knowledgeable body of spiritual-scientific minds especially for the beings of this whole earth realm—perhaps *only* for the more spiritual beings of this earth realm.

Indeed, if I can offer my own personal experiences from my own abductions, which took place back in the late 1950s, from when I was taken to a transparent platform out in space, and where I was shown huge electronic charts, moving star maps, an illuminated book of designated steps for this event, and was shown different landscapes on another planet, and was made aware of many different forms of "extraterrestrials" that had congregated upon this location of space, seemingly, for the sole purpose of bringing about a removal of, or rejuvenation of, as many beings as possible from our world, then the impression I got of this *Transformation* phenomenon was that it will be the BIGGEST EVER project to be undertaken, and that every part of this operation has been planned down to the n^{-th} degree, by both the extraterrestrials and the highest of spiritual entities.

Further though, that the ASBT messages do offer information about the anticipated transformation from our present 'human' stage of existence to the next evolutionary form that those souls must reside in it suggests, interestingly, that the benign extraterrestrial *Elders* (through their human-grey Hybrids and the Greys, no doubt), which many of the abductees have communicated with, will be functioning as mediators (as they have as genetic engineers in the extraterrestrial's abduction programs) between us and a particular instruction from the Source realm our religious leaders know only of as the Supreme Being.

More details on this Transformation will presumably be made available at a later date, of course, although the reader can find out more in the [Redfield] and the [Soskin] books (noted in the bibliography) which were written especially to broach this very phenomenon.

Notes for Chapters

Notes for Chapter 1

Note 1—While the cause of death for three of the Greys, and injury to the fourth survivor, probably resulted from the impact of the crash, another point of consideration has to be that those three Greys had died because of a very sudden change to the density environment within that UFO craft, as a result of the explosion ripping open its hull. As will be explained throughout this study inside the UFO is created a very unique environment, similar in nature to a pressurized fuselage of an airliner, but one which changes the mass-density of all that is contained within that UFO's shell. When its interior environment would have suddenly de-densified (as a result of splitting the craft's hull), there would have been a truly traumatic effect on their bodies. That this density environment can be generated inside a sealed craft will introduce a hitherto never-before considered branch of physics. That new branch of physics awaits the reader in the following chapters.

Note 2—That the world's first nuclear bomb, presided over by Oppenheimer, was detonated at the Trinity Test Site in New Mexico, close to the White Sands military base, on July 16 1945, may give one of the reasons as to why the ETs had begun sniffing around this area of space ever since the mid-1940's. The other intriguing question about this much denied event, was why did that UFO try and make for White Sands? Well, it's crew probably just hoped (or they had observed) that the personnel there had sufficient protocol procedure to help the UFO's injured crew members (this same protocol was to be noted at the Rendlesham Bentwaters-Woodbridge airbases [Howe] when a UFO crew had to make repairs to their craft in 1980). Unfortunately, because their ship never quite made it to that US base and it came down hard in deserted wasteland the crew didn't stand much chance of survival. Evidence collected by the residents of Roswell suggests that three ETs had died after the second crash-landing but that one of the Greys remained alive. And the inference is that the three dead bodies laid six days beside the stricken craft at the mercy of the elements, before being discovered on July 8th by a civilian surveyor by the name of Grady Barnett. Just after Barnett discovered the scene a college party of student archaeologists who stumbled into the immediate area by chance found Barnett putting covers over the dead bodies. Soon after this a US military search team arrived (after being guided to this location by Mac Brazel who had been flown over the area in a military plane to look for these survivors). The US military team then secured the whole area and removed all civilians from it. The fact that the alien dead bodies had been exposed to the elements for six days can be corroborated from the fact that Glen Dennis the civilian mortician at Roswell was later asked by the military hospital for Roswell specific questions about what signs of deterioration a dead body would normally exhibit when exposed to the elements for precisely that amount of time, and what the normal procedure would be to preserve body fluids blood and tissue *in order that they may be re-examinable at a later date*.

The 509th bomb group flew the B-29 planes which took off from the Tinian airbase in the West Pacific carrying the first nuclear bomb which was dropped on Hiroshima in 1945. After the second nuclear bomb was dropped over Nagasaki an estimated 300,000 human beings died from these two blasts or from subsequent radiation poisoning. In Nagasaki the heat generated by its nuclear blast reached an estimated 7000 degrees F which sucked in the surrounding air at over 600 mph...

Note 3—Col. Philip J. Corso worked in the US Army in a special research-and-development section dubbed *Foreign Technology* during the 1960s and 1970s under the direct command of General

Nathan Twining of MJ-12. Corso's involvement with members of Roswell's 509th in the retrieval and inspection of some of those alien artifacts (and the subsequent farming out of different sections of that UFO's technology to US research laboratories for analysis) is detailed in the book called *The Day After Roswell* by Col. Philip J. Corso (1997) pp98-116. While I have no need to claim that the type of UFOs featured in the Andreasson experiences were precisely the same as the craft that was hit by lightning and crashed at Roswell in 1947, I will though claim that there is an uncanny resemblance between the power drive shown to Betty Luca and the power drive attributes of the Roswell craft as delineated by Corso in his book. That being the case I may be able to conjecture precisely how the Roswell craft was disabled that night in 1947 (more on this later on in the book).

Note 4—In the mid-1980s an American named Jim Sparks was abducted and what is so intriguing about his abduction account is that he has managed to remember almost *everything* that happened to him (the astute reader will recall some of Sparks' recollections featured in Linda Moulton Howe's *Glimpses of Other Realities vol II*). Sparks was a first-hand witness to the way the Greys had breached a seemingly ordinary wall and had entered into his room at night [Sparks]. His origination about how he perceived the Greys were doing this is as follows, *"The E.T.s can be 5% to 10% in this dimension on this physical plane while the rest is in another dimension. There are dimensions out there like layers of an onion... The E. T.s molecularly change the vibration of something so they can be in another dimension and solid, but be projected here in only 5%."*

E.W. Davis on behalf of the Aerospace Physics & Astrophysics Div. National Institute for Discovery Science has published (in 2001) a quite comprehensive list of the most unusual of UFO Phenomenology, and also the UFO-Extraterrestrial Intelligence Hypothesis (ETH) as according to the National Institute for Discovery Science (NIDS) see especially pp44-52. This document can be obtained (2008) from the website: http://www.nidsci.org/pdf/davis_mufon2001slides.pdf

Note 5—For more information about acoustic shaping see the *Out of Thin Air* article by Bennett Daviss in "New Scientist" 1 Sep 2001 p32-5 about standing waves acting on resin particles inside a microgravity chamber; and *Radio Waves Could Construct Buildings in Space*, also by Bennett Daviss from "New Scientist" 12 Oct 2002 p15.

Note 6—You can waste a lot of time and money by obtaining books on the subject of UFOs and ETs that are written, quite frankly, by people who are not intellectual enough to work out the full landscape picture of this field of wisdom, merely do they offer their readers the pictures of a few tubs of paint. One of the better books filled with comprehensive and thought-provoking content, both from ex-military personnel and abductees, is *Glimpses of Other Realities—Vol 2—High Strangeness* by Linda Moulton Howe (1998).

Note 7—The Royal Tour yes, but she would have to have gone through all the pain inducing physical examinations and procedures just the same as anyone else who has ever been abducted will testify. Numerous drawings rendered by Betty have been published which show some of the physical procedures she has had done to her, such as the Greys pulling her eyes out of their sockets and forcing a needle probe through the eye socket into her head (*Andreasson Affair—Phase Two* fig. 45). But I have to say from what I have read of Betty's encounters and from what Betty has confided to me, that this is a woman who has the highest level of courage, integrity and valor, keeping her attention always more on what information she could recall to benefit others rather than any concern for the pain she herself had to re-experience through the recall process. In hypnotherapy there is little difference between administered pain and recalled pain, and Betty has shown remarkable endurance to what, by anyone's standard, must have been an excruciatingly painful procedure. A case in point was when Betty recalled how the ETs embedded her into an electronic-wire-ribbed circular disk filled with a gel-like substance, which when activated spun around violently before entering—with Betty—into another zone of swirling water: Those who witnessed Betty in that hypnotherapy session at the time expressed shock and the utmost concern for Betty while she recalled this incident, for as she relived this frightful experience they saw her facial features contort to the very motion that she physically experienced at the time. What those investigators saw and described were the effects of extreme gravitational and centrifugal forces pulling on the skin of her face—even though she was unconscious and sitting in a chair in front of the hypnotist (see *Andreasson Affair—Phase Two* pp103-108). Dustin Hoffman's *Marathon Man* was an utter wimp compared to this lady... and therein lays a clue to her reward for these experience-abductions.

Note 8—Not to be confused with the US government's ill-produced and pointless Blue Book inquiry, or rather non-inquiry, into UFO activity. The Andreasson ET's blue book was unusual by the fact that it was indeed a book that had pages, usually when an abductee is allowed to handle one of the ET's 'books' it is more akin to a tablet computer with a big screen which glows (ie in *Glimpses of Other Realities—Vol 2—High Strangeness* by Linda Moulton Howe, and in Australian Patricia Thomas' abduction account). Sometimes these tablet books produce holographic type presentations of actual historical scenes with sound. Sony will be wanting to patent that idea for sure!

Note 9—And before the skeptical reader calls for the truth serum, don't worry Betty has gone through all that stuff years ago, with lie-detector tests, scientific and psychotherapy analysis, she's been through the works, all proved positive.
The complete series of books giving the account of Betty (Andreasson) Luca's encounters with Greys and extraterrestrial Elders and of course UFO craft are: R.E. Fowler *The Andreasson Affair* (1979); R.E. Fowler *The Andreasson Affair—Phase Two* (1982); R.E. Fowler *The Watchers* (1990); R.E. Fowler *The Watchers II* (1995); R.E. Fowler *The Andreasson Legacy* (2000); Betty Luca has published a series of booklets called *Extraterrestrial Communications—A Step Beyond Tomorrow (ASBT)* (Pt I & II—1999).

Note 10—See S.J. Putterman *Sonoluminescence: Sound into Light* Sci. Am. (Feb 1995) 32-37. Other than the above the best place to start in this vast subject is to look at Taleyarkhan's model of sonofusion, and the best two documents I've seen on that are; Taleyarkhan, R.P. *Nanoscale Explosive-Implosive Burst Generators Using Nuclear-Mechanical Triggering of Pretensioned Liquids* USPatent appl. 20030074010 (Apr 17, 2003); and Lahey, R.T. & Taleyarkhan, R.P. & Nigmatulin, R. I. *Sonofusion—Fact or Fiction?* The 11th Int. Topical Meeting on Nuclear Reactor Thermal-Hydraulics (NURETH-11) (2-6 Oct 2005).

An interesting paper on electrogyration is Kaminsky, W. *Experimental and Phenomenological Aspects of Circular Birefringence and Related Properties in Transparent Crystals* Rep. Prog. Phys. 63 (2000) 1575-1640.

Note 11—Once it has been realized how this happens then the reader of Betty Luca's experiences will notice that those experience episodes feature a number of these 'altered density' environments.

Note 12—Throughout many of Betty's encounters these *light forms* have shown themselves as spiritual light beings but which can 'clothe' themselves at will into these light forms—sprightly forms which resemble the human body, just like we would put on a coat, and after taking those light forms off they would return to their spiritual essence of light and would transform into a ball of light or streak of light energy! Many of the things Betty saw these nimble light forms do, and the *light-hearted* games (quite literally) they would play around her, were rendered into drawings by Betty and published in the *Watchers* and *Watchers II* books. Certainly, the light forms should not be confused with the Greys, although they might be associated in some way with the Elders in an out-of-body state. They actually remind me more of the *radiant bodies* spoken of by the *Master of the Key* being whom made a visit to Whitley Strieber and whom was the subject of Strieber's *The Key* book.

Note 13—And by saying 'taken to' I'm not necessarily assigning these manufactured environments to any distant location in particular, they may have been on earth somewhere. In *Andreasson Affair* (pp71-79), in *Andreasson Affair Phase Two* (pp97-107), and in *Watchers II* (pp57-66) Betty tries to explain how the Greys took her on three occasions to one particular place, which the Greys claimed to Betty was their 'home.' But considering how much trouble they went to to disguise the motions of those particular journeys it seems to me to have been either inside the earth or inside a very large mothership out in space somewhere. But bear in mind that on another occasion these Greys explained to Betty that they normally existed in a heavier density to ours (Betty was asked under hypnosis whether she thought this particular place was on earth somewhere, but as it was coming to the end of that session the questions she answered didn't go into much detail, although Betty did hint that it may have been inside this planet but not necessarily of this dimension (*Andreasson Affair* p86)). This would make sense as through most of *Andreasson Affair* Betty was taken directly from the transit chamber of the UFO she was abducted in along a track in the dark, *through* a mirror-glass obstruction into a red environment (which was of a *vibrating* red color), and then this was followed further along the same track by a green environment, which was separated from the red by a circular membrane, suggesting that on this occasion Betty seemed to have passed twice through a sort of 'stargate shift.' Apparently there were crystals hanging in the air in the green environment (indicating light refraction again), and the strange fact that she couldn't see from one colored environment to the other (ie the red would be opaque whilst she was in the green environment—see Betty's note below). These environments are perhaps not so strange though, they can probably be engineered with laser ring technology—see the [Mallett] reference in bibliography section). Its interesting too that the most powerful lasers are in the red and green bandwidths, and presumably this was to alter her perceptional frequency (or her material frequency) in two stages. Spectrum wavelengths for red are 740-620 nm, while for green they are 575-500 nm, and UV starts around 390 nm. From the description in *Andreasson Affair—Phase Two* (p123) it seems much more apparent that she was phased through a ring laser system—Betty even gives a drawing of what essentially would be created in such a multiple beam ring laser set up, and then she is passed through a further 'rotating' mechanism before she 'arrives' at what she calls the crystal forest (here the Elders are depicted in an outdoor scenes but these have restricted views of the distance). But in *Watchers II* where Betty talks of going to this place again there is no 'red-green conditioning' this time, Betty is again taken by the Greys in a UFO and again is put into a sealed room for the journey, but this time she is immersed in gel-like fluid (to counteract a fast shift in mass-density perhaps?), but then as soon as that journey ends Betty is taken straight into the crystal forest, albeit this time through a deep mist.

Betty explains further, *"The beings had four metallic barrels that came out of the mist and shot something like straight (and pulsing) lasers to appear from them. When I was taken in 1967 into the red vibrating atmosphere there was a whirling red light that two Greys and I entered through into the green atmosphere. I could not see a demarcation of where the red environment began and ended, and the green appeared. We may have been swiftly transported through a tube like wheel from one atmosphere*

to the other like stepping through a door. I often wondered how come I couldn't see one area from the other. Each environment seemed a very separate place except at the large whirling entrance which did not seem to blend into the other color either."
Betty Luca (personal correspondence Feb 2005).

Note 14—Indeed, regarding our present-day 'plight' from the big energy corporations saying they soon won't be making enough profits from natural fossil fuels (because of its shortage), one of the Greys that accompanied Betty indicated that we on earth seem to be making it very difficult for ourselves when it comes to looking for energy sources, that Grey even told Betty that, *"Energy is 'round about man that he does not know of. It is the simplest form of energy. It is within the atmosphere—this atmosphere... It has all been provided for him... many riddles will be given... Those that are wise will understand... Those that seek will find... They must remain hidden in this way because of the corruption—the corruption that is upon the earth... If they are revealed outright, man would use it."* (from *Andreasson Affair* p122). As I hope to show throughout this book, that Grey couldn't have spoken a truer word!

Note 15—Betty and Bob Luca's hypnotic sessions were conducted between 1977 and 1994 in four phases. Actually, re-reading through *Andreasson Affair* of how Betty describes her initial recollection of her alien abductions, I can now see where Steven Spielberg got some of his inspiration from to make *Close Encounters of the Third Kind*. Although, if more had been understood about ET technology then, and had the general public been less prohibited from finding out about these things, that film would have proved even more impressionable than it was, for the reality I have found in this field is a lot more startling than ever that fictional film was.

Note 16—The discovery of electron emission from ferroelectric materials (barium titanate) occurred in 1960, and while research chugged along steadily through the 1970's and early 1980s the real progress in ferroelectric spontaneous polarization emissions emerged toward the end of the 1980s (see Gundel-1989 and Riege-1998 for summaries).

One of the scientists to compile a report on this UFO's technology was the Space Technology & Applications International Forum (STAIF) scientist Paul A. Murad. The lecture paper which can be seen of his conclusions about some of the things described by Betty Luca during her abductions is AIAA paper No. 95-2894, *"An Electromagnetic Rocket Stellar Drive...Myth or Reality? Fluid Dynamic Interactions and an Engine Concept"* as presented at the July 10-12, 1995/San Diego, CA. 31st AIAA/ASME/SAE/ASEE Joint Propulsion Conference and Exhibit. However, although Murad recognized that the Andreasson UFO does establish a quasi-black hole his assumption that by spinning up the central sphere-set assembly there would be produced a downward thrust-force through the base disc of that UFO, to give it a *rocket-like* propulsion, would seem to suggest he had not yet taken up the required study of astrophysical dynamics! It was interesting though, that Murad in later years fully recognized the valuable work being carried out on cylindrical torsion fields by the Russians, noting the rather obvious link to inertial frame dragging – which is what this present GMODC study has undertaken to illustrate [Murad 2005].

Note 17—Corso mentions that scores of small plastic-like discs were found in the upper decks of the Roswell craft, these turned out later to be silicon chips, but to the investigators of the day they might just as easily have been checkers from an in-flight game board. In those days in 1947 no-one knew even what a transistor was (strangely enough they weren't 'invented,' or should that word be 'diagnosed,' until 1948). That these silicon chips were farmed out to defense contractors in 1947 and to the Bell labs in 1961 has probably been the reason behind the introduction of computers. That it took so long suggests that the commercial scientific sector really is the poor cousin to military providence.

Corso also points out that the maser, the laser, and night-vision glasses also came from the back-engineering of the Roswell craft (op. cit. pp183). If so, then no wonder UFOs have been denied so vehemently since 1947, we owe most of this worlds advances in electronics technology to those extraterrestrials. And just imagine all the misappropriated patent royalties!

Note 18—Dr. David M. Jacobs has published several studies on the abduction phenomenon and has formulated a special form of hypnotherapy aimed at reaching the most pertinent episodes that the abductee has experienced. His published books are; *Secret Life: Firsthand Accounts of UFO Abductions* (1992), and *The Threat* (1998). The body-nutrition spray account was published by Jacobs in *The Threat,* and seems to correlate with what other abductees have been told on the rare occasions of being spoken to by the Greys, in that their bodies were able to intake whatever was needed whenever they did take in nutrition, in such a way that they would have no wastage. Apparently, the most comparable nutritional system we know of on earth is that of the frog which intakes water by absorption through the pores in its skin. If their outer layers of skin do process their nutrition this might explain the objectionable smell around them which some abductees have mentioned. This was something which had been emphasized in the original autopsy carried out by pathologist Dr. Jesse Johnson by the attending nurse assisting Johnson, in that the smell coming from the alien bodies which had begun to decompose was highly objectionable. But the strangeness of their skin was noted by witnesses at Roswell too, Len Stringfield when he researched into the Roswell alien bodies and badgered some of the original doctors that took part in the autopsy, was told that the alien's skin tissue under magnification had a mesh-like structure to it. And Sgt. Melvin E. Brown who was guarding the bodies at Roswell's 509 air base later said that the alien's skin had a leathery look to it and it was beaded similar to that of a lizard's skin [Schmitt-Carey]. Col. Philip J. Corso noted that there was never found on the retrieved UFO craft at Roswell any food supplies or any signs of food preparation for its crew. He also said that when he read an autopsy report on the alien bodies it suggested that their outer skin resembled a thin layer of fatty tissue which was completely permeable, as if the body absorbed chemicals or nutrition directly through the skin into its circulatory system [Corso] (pp96-97).

That the Greys communicate with us through the part of our brains that organizes our thinking and mental images, prior to those thoughts and images being turned into our preferred modes of communication, would answer the oft asked question as to how they seem to be able to make themselves understood to us no matter what language we speak. Similarly, they won't need to carry around with them a whole library of language-dictionaries to know what any one of us is thinking. Apparently this was a facility we humans used to have, before our so-called Fall from grace. Another little trick the Greys seem to be very good at doing, is temporarily entering into a soul's life-space (usually inside that person's head) during the abduction experience, I say this because I have been told by some people (who obviously didn't know better) that if one hums a single tone in their head they will be protected from that Grey's intrusion into their mind (I think this advice was originally written in a book on how to survive an abduction experience), and I'd have to say, from experience, that the REALITY of such a situation is not quite as simple as that! There has been substantial research carried out on abductee phenomena, and one of the common occurrences was indeed of the Greys actually entering into the abductees mind for the purpose of scanning it. Dr. David Jacobs, who studied many such case histories came up with the following conclusion about how the Greys were doing this, where he noted that, *"By exciting impulses in the optic nerve, the alien is able to 'travel' along the optic neural pathway, through the optic chiasma, into the lateral geniculate body, and then into the primary visual cortex in the back of the brain. From there he can travel into the secondary visual cortex in the occipital lobes and continue into sites in the parietal and temporal lobes and the hypothalamus. Through that route, the alien can stimulate neural pathways, travel to many neural sites, and cause the 'firing' of neurons at whatever sites he wants."* (taken from p84 of *The Threat*).

Note 19—My own favorite ET hypothesis can be found in the thought-provoking paper by Beatriz Gato-Rivera entitled *Brane Worlds, the Subanthropic Principle and the Undetectability Conjecture* (2003) (preprint: http://arxiv.org/pdf/physics/0308078). For another extraterrestrial hypothesis (ETH) delineated in a more recent paper, see [Deardorff-Haisch-Maccabee-Puthoff].

But then it was also surmised that there are beings of advanced minds in the Kip Thorne paper on wormholes, see [Morris-Thorne].

Note 20—Ok, so the USAF's super-dooper spyplane Aurora does around 3,600 mph when its pulse-detonation engines are fired up—but its structural integrity will only tolerate that speed if that plane toes a very straight line. Give that flight-path anything more than the slightest curve and Aurora's hull will disintegrate into a thousand pieces. But that was on the boards years ago... what have they got up in the air now that's what I want to know...and how closely does it resemble an alien UFO!

Notes for Chapter 2

Note 1—The complete series of books giving the account of Betty (Andreasson) Luca's encounters with Greys and extraterrestrial Elders and of course UFO craft are: R.E. Fowler *The Andreasson Affair* (1979); R.E. Fowler *The Andreasson Affair—Phase Two* (1982); R.E. Fowler *The Watchers* (1990); R.E. Fowler *The Watchers II* (1995); R.E. Fowler *The Andreasson Legacy* (2000); Betty Luca *Extraterrestrial Communications—ASBT* (Pt I & II—1999).

Note 2—While the reader might consider that a *toroidal* field is shaped like a donut, then the *poloidal* field in this study is shaped like an apple with the field lines running through its core.

Note 3—The colored light emissions around a UFO's rim will be synchrotron radiation emission from electrons excited to a very high degree and then being forced through an extremely tight curvature around the craft's rim.

Three examples of this fuzzy energy field are the Canadian photo taken by R.J. Childerhose (which can be seen at http://www.brumac.8k.com/RJC/RJC.html), the Grangemouth UFO (which can be seen at http://www.ufoevidence.org/cases/case1047.htm), and the 1991-1993 Mexico videos (see note 4 below). For evidence of a UFO rendering itself invisible see James McDonald's *Air Force Observations of an Unidentified Object in the South-Central U.S., July 17, 1957* Astronautics & Aeronautics (AIAA) July 1971 66-70 (which can be seen at: http://web.archive.org/web/20030605000634/ufophysics.com/dual/aiaareport.pdf) in this report the USAF crew following a UFO watched the UFO blink out to become invisible for several seconds and then reappear on the same trajectory. The 2004 Mexican IR images filmed by Mexican military pilots showed conclusively that those craft while they flew inside an infra-red energy envelope were completely invisible to the naked eye. Perfect stealth maybe?

Note 4—The reason as to why there were so many video recordings made of UFOs over Mexico City and surrounding towns, was because of a long-standing prediction made by Mayan astronomers foretelling that soon after an eclipse around the beginning of the twenty-first century the earth would be visited by the 'Masters of the Stars', and so when the eclipse of July 11, 1991 was taking place many Mexicans assumed that this would be that great day. So thousands of Mexicans spent the whole day looking up at the sky with cameras by their sides at the ready. They weren't disappointed! Hundreds of sightings were made of not just one UFO but of a whole fleet of UFOs. Subsequently, the scores of videos of UFOs shot between 1991 and 1993 were then collected by TV journalist Jaime Maussan who endeavored to broadcast most of them on Mexican TV. These then made their way to British TV (and no doubt the rest of the world) and in Britain they were broadcast on a TV series called "The Paranormal World of Paul McKenna" during 1997. The hyper-jump clip which I presented and explained on my website www.ufophysics.com, from which the F697 frame is taken, was taken from a video filmed on July 11, 1991 (on the day of the eclipse).

Note 5—The Mexico UFO clips I have been working from can be found in the compilation DVD called "The Best Video Footage From Mexico Is On This Videotape: Jaime Maussan Mexico Videos 1991-1994."

Note 6—For the definitive studies of black hole and accretion disk systems see [Uchida-Shibata], [Blandford-Znajek], [Williams 1995], and [Punsly-Coroniti 1990b], and for one of the best explanations of these systems see [van Putten 2002].

Note 7—There are two basic types of black hole model, the first to be proposed was that of Schwarzschild's who (in 1916) advocated that black holes were un-charged massive spheres that existed with a stationary space-time geometry (no doubt to conform with Einstein's erroneous belief that the universe was static), and the second, which is now the much preferred model, was proposed by Kerr (in 1963, but which was modified by Newman in 1965) of a space-time geometry which had rotational forces (through which it formed an accretion disk that broke down mass into sub-atomic particles). But also see for instance [Meier 2005].

Note 8—Where a toroidal field is infinitely long it resembles a tube. See [Meier 2005]. The Poynting vector (S) of a circularly polarized plane wave that is propagated into free space along the longitudinal z-axis is one that is a constant vector parallel to that axis, also see [Mansuripur 2005]. The Poynting vector and the Alfvén wave will be alluded to later in this chapter.

Note 9—The Meier documents are excellent for having collected many of the most important features of astrophysical jets. For instance see [Meier 2003] and [Meier 2004].

Note 10—The grand scale of these astrophysical jets is a common factor mentioned in most books on black holes, quasars and pulsars, but is a feature especially mentioned in the [Honda-Honda] paper on jet collimation. See also the study by [Carilli].

Note 11—[Tsinganos] and [Li-Lovelace]. The term 'poloidal' has been used mainly with two different connotations in the theories of the dynamics of black holes. But see the poloidal insert in figure 12 above.

Note 12—For the shear magnitude of force and energy production that can come from space eruptions the reader should follow the progress of Eta Carinae (in the Sagittarius-Carina Arm) the supernova blast that is so large its plasma field has taken on its very own field dynamics. Also see [Blandford-Payne] and [Uchida-Shibata] papers, references to which are in the bibliography pages at the end of this book.

Note 13—The basic gist of this mechanism can be seen in [Blandford-Znajek] p435-436, although the earlier source [Ruderman-Sutherland] p57-59 will be seen to be more informative. Other studies for electron-positron production are [Blandford-Payne], [Punsly-Coroniti 1990b], and [Punsly 2001].

For the Searl Energy Generator's toroidal energy field it may be worth looking at the Russian research program, where it describes this electromagnetic field being detected as concentric fields. And the reconnection electron-positron pair production mechanism occurs as the satellite rotors of Searl's design rotate around the central stator ring. See http://alexfrolov.narod.ru/russearl.html (and note 28 below).

Note 14—For the image and background of this photograph see http://www.ufoevidence.org/cases/case1047.htm which in my opinion is the best photograph ever taken of a UFO in flight. There is a lot going on in this photographic image of the Grangemouth UFO photographed by Phil Trevis in 1991. Around the rim are the trails of energy passing over its rim, then there are two large flow fields showing bright emissions, and two narrow guide fields which show a lesser amount of emissions. But the large photonic emissions at the rear of this UFO were giving off sufficient heat to cause the air around the UFO to have a smoky heat haze. This heat haze still left in the air and caught on this photograph clearly shows that the craft was passing overhead very slowly and jinked slightly to alter its direction.

Note 15—See Punsly-Coroniti 1990b fig.4 diagram on the synchrotron radiation spin-down into a black hole (the same chart is also found in B. Punsly, (2001) *Black Hole Gravitohydromagnetics* fig.8-6). Interestingly, on their helical trajectory away from the accretion disk as they coil around the outside of the ergosphere and then around the longitudinal flux tube, before shooting off into the distance, they will pass very close to the shearing and oppositely-rotating fields where the original sub-atomic particles would have been broken up and converted into photons, where some of those photons would have been lost into the black hole... but which now passes its rotational energy into the accretion disk to set this whole process in a cyclic continually repeating motion!

Note 16—This is the basic principle of the Penrose process and Hawking evaporation, but also see [Punsly-Coroniti 1990a] [Visser 1998b].

Note 17—Moreover, in a recent continuation of Stephen Hawking's theory about black hole evaporation it has been suggested by [Traschen] that rotating black holes by this evaporative principle will as a result emit massless particles that possess angular momentum.

Note 18—See [Carilli-Barthel] on black hole efficiency. Also, following along the [Penrose] process is the assumption that if the adjoining magnetic field is strong enough then rotational energy can be extracted from the hole at its transition region (just outside its event horizon), and obviously, if this region is also where charged particle production is prominent then the two will go hand in hand to produce rotating charged particles. A study by [Li 2000] claims that up to 29% of a black hole's total energy can be extracted.

Note 19—In the torque expression J x B, it denotes J (current density of a sheet of charged particles) times B (magnetic flux density) which together equate to the Lorentz torquing force of $F\beta \sim J \times B$.

Note 20—To justify my proposal that gravitational force can be equated to a differential in angular-momentum existing in the same spatial frame, I will ask the reader to consider that Einstein-Rosen waves were long ago proposed by A. Einstein and N. Rosen because they saw gravitational waves as being cylindrical (see J. Franklin Inst. 223 (1937) 43; after Guido Beck in Z. Phys. 33 (1925) 713). So my (Potter proposal of gravitational force) is based upon adjacent cylinders of rotation rotating at different velocities and one being trapped by the other. But this 'radical' assumption most logically follows the fact that mechanical energy when emitted from a collapsing black hole is emitted as rotational energy (angular momentum), in the form of gravitational waves (see for instance [Ferrari]). This being the case then the effect of, and detection of, gravitational force is relative to how those waves are 'packaged' (ie an isotropic emission of gravitational waves would not be as effective as the same amount of force rotated into a confining filament-tube. In some respects then, gravitational force is analogous to magnetic force being the lesser or greater to the degree that its flux lines are inhomogeneous.

Note 21—As will be seen from a slightly different angle (in section 3.17) this is precisely the same as the image of the 'cannibal' theory of galaxies [Ostriker-Tremaine] [Hausman-Ostriker] where a central hole consumes from the ionized energies of its surrounding suns by the tidal effect. This mechanism is also featured in the Scientific American article of [Blandford-Begelman-Rees].

Note 22—When this cannibalism mechanism is translated down in scale to the sort of size represented in the aaUFO's quasi-black hole, and this is where Stephen Hawking's theories on 'mini-holes' comes into play, there will occur a 'tidal effect' where the near side of the mass object will be affected much more so than the far side. Through this tidal effect the quasi-black hole's gravity will act upon the center-facing side of these four glass spheres to strip away any photon energy that these spheres produce, and then the quasi-black hole's rotational force will then spin them down into its massive core. Hawking suggests the effect on such a miniature scale will be all the more ferocious than its space cousin.

For the miniaturization (all the way down to atomic-scale) of black holes see S.W. Hawking *Black Hole Explosions?* Nature 248 (1974) 30-1; S.W. Hawking *Particle Creation by Black Holes* Commun. Math. Phys. 43 (1975) 199. But one only has to look at a hole's power equation to see why, the radiated power is proportional to the hole's density (p), which is inversely-proportional to the hole's size. The smaller the hole the higher the density, the higher the density the greater the power it radiates [Rabinowitz]. For a more modern continuation of the miniaturizing of black holes see [Williams 2002b], and of course the papers by [Schutzhold-Unruh] and [Leonhardt-Piwnicki 2000a].

Note 23—I say quite different from any other, but a similar ultra-bright emission from a UFO has been noted with such an intensity that it was likened to burning magnesium—as in the Iron Bridge case: Jacques and Janine Vallee 1966 *The UFO Enigma* (Ballantine Books) New York, pp221.

For an extra note on these 'almost tangible' bars see Betty Luca's correspondence regarding this in chapter 3, note 9.

Note 24—This reconnection process will be featured throughout chapter 3 and chapter 9 of this study. In the papers of [Kirk] and [Kirk-Lyubarsky] the same magnetic reconnection energy production process is reiterated, attributing to it the large particle production processes found in active galactic nuclei and in pulsar winds.

Another standard paper in astrophysics is [Uchida-Shibata] which points out that infalling material released at the inner-most ring of the accretion disk (where it abuts to the black hole) will, through magnetic reconnections, cause particles to collide energetically and produce thermodynamic regions producing X-ray emitting reactions.

A Cal Tech website gives another example of this little exploited wonder, found in the earth's downstream field where one example of this phenomenon can be found in the interface between adjacent twisted magnetic flux-line tubes. See the Caltech (2005) webpage on *Inertial Alfven waves, mode conversion, resonance cones* at: http://ve4xm.caltech.edu/Bellan_plasma_page/inertial.htm

Note 25—Also see fig.1 of the zig-zagging effect of spiraled magnetic field [Kirk 2004]. And compare this zig-zagging with the drawing included *ASBT* bk I diag.4 (p24).

Note 26—The study of the reconnection principle is very extensively presented in [Priest-Forbes], which lists around a dozen mechanisms of magnetic field reconnection (pp461). Other primary explanations of this principle are in Parker, E.N. *Sweet's Mechanism for Merging Magnetic Fields in Conducting Fluids* J. Geophys. Res. 62 (1957) 509-520; and more importantly Petschek, H.E. *Magnetic Field Annihilation* (in *Physics of Solar Flares*, ed: Hess, W.N. [NASA SP-50] pp425-439). Otherwise see the Blandford-Znajek paper and the paper by [Ruderman-Sutherland].

Note 27—See the report on the Searl-Effect (dated 1986) by S. Gunnar Sandberg (University of Sussex) and a note from it which was posted on my webpage aphysufo.htm still accessible at: http://web.archive.org/web/20050305132319/ufophysics.com/aphysufo.htm

Note 28—That the Searl SEG in its Russian tests developed a toroid field around it to form an ergospheric system that mirrors the effect of the toroidal ergosphere created around the sources of astrophysical jets, this would bode well for the astrophysicists such as Unruh, Leonhardt, Lebedev and Dimopoulos who are attempting to develop miniature black hole systems and jets in the laboratory. What is so interesting about the Searl generator though is that while its power-developing system has no discernible accretion disk it does develop out of its toroid ergosphere a filament-tube jet (a process which Searl aided by configuring the central stator magnet ring as a composite of single magnets joined in a ring, each having their fields angled slightly inward to produce a frustum field). Indeed, considering that a filament-tube can be developed out of such a Searl SEG toroidal field this system of Searl's would certainly be the best model to work from when constructing a gravity manipulation drive. I wouldn't use Roschin & Godin's model as it is flawed by the fact that its outside rotors are fixed and electrically grounded, and that the reconnection process will be interrupted by all the bulky fixings around the stator. The Roschin & Godin model will work but not nearly as efficiently as the original design attributed to John Searl (this will be gone into throughout several of the following chapters of this study to address the neodymium and nylon ferroelectrics Searl also used).

On closer inspection of Roschin & Godin's Russian Searl experiment report the one meter diameter rotating SEG developed a series of concentric vertical walls (to about 15 meters in diameter), and so rather than developing a thin disk it seems to have developed a series of current sheets. These concentric sheets have been noted around black hole systems, and a very similar model of these current-sheets can be found in [Kirk-Lyubarsky], and also in [Lyubarsky-Kirk]. But also, these current sheets will produce variable electric fields, and potentials, between them in the same manner as that previously noted for the spark-gap ionization effect (section 2.6 and note 13 above), so then there will be a set of potential differences which will do work on the neodymium and nylon, yes the nylon, in the center of Searl's rotors (see further details in section 4.4).

That the Searl filament-tube formation transcends through several stages of development after the rotors begin to revolve, and after it eventually causes his SEG, or rather, the Inverted Gravity Vehicle (IGV), to rise sharply into the air it begs the question, are the particles being pushed up the filament-tube jet or are they being pulled. It has always been assumed by astrophysicists that the particles that form into an astrophysical jet *push* their way into it—yet these astrophysicists cannot explain why such a jet can remain perfectly collimated for millions of miles through space unless it has negative losses. But a look at the density chart of a black hole system tells us that the jet is formed in the area of least density—so that is why mass 'falls' into it (also see the density chart for a protostar and its out-flow jets in figure 24 [Lery 2002]). Then again, if the astro filament-tube's particles are *not* being sucked out, and the Searl filament-tube's ones are, then what are the significant differences in the two systems that determine whether those particles should be pulled or whether they should be pushed?

The reader should note I am not here giving away the secret of Searl's magnetic rotors, of how their fields were configured so that they persistently rotated (mainly because I don't know how he did it...). That information is best obtained direct from John Searl himself, particularly via his current (circa. 2008) websites at: http://swallowcommand.com and http://www.searlsolution.com

The Russian website of Roschin & Godin can be found at [http://alexfrolov.narod.ru/russearl.html]

But then, when a concept is put down on paper and someone has to draw its image and put in a basic structure of magnetic field lines, then its configuration becomes more understandable. So, it then became very clear to me that when the outside rotors turn the magnetic flux lines it shares with the stationary inner ring have to be broken as the rotor rotates further around the inner ring, so for me here was a case of the artist suddenly understanding how genius works.

Note 29—Meier favors the [Conway & Murphy] geometrical structure of helical jets.

Note 30—Further information on this can be found in the [Blandford-Payne] paper (sect: 2.6) including the mathematics, but most of the work on collimation has been done by [Spruit 1999] and [Spruit 1996].

Note 31—See for this the astrophysical study by [Junor]. The mechanism that achieves collimation in almost all magnetic models proposed so far is the hoop stress of the wound-up magnetic field, which is calculated to work very well in axisymmetry. It is highly likely, however, that this nearly azimuthal field is very unstable to nonaxisymmetric modes which destroy the collimating hoop stresses. This comment though becomes rather redundant when considering the parameters of negative radial momentum (as described toward the end of this chapter).

Some other researchers have found in simulation studies that when the accretion disk mechanisms are completely stable the main reason for this is the presence of a Coriolis force—and that in some cases epicyclic oscillations produced within these mechanisms contributes to the critical stability. [Hawley-Balbus-Winters].

Note 32—Reference is from the Numerical Maryland Centrifugal Experiment [http://aries.ucsd.edu/FPA/ARC02/fpn02-30.shtml]

Note 33—[Honda-Honda] cites several cases where research into high power (petawatt) fusion laser beams support this view. And on this see the paper by [Salamin-Keitel-Faisal] which points out that by combining electron acceleration with a laser field there can occur, at a critical value, an acceleration of the electrons to several hundred GeV.

Note 34—However, because of the fact that light travels more slowly through a moving medium, and particularly through a rotating medium, one suggestion might be that in the above example the light speed will be slower than that of (c) which is normally 300 km/s. So the Mexico tube's darkening as cited above may come from the light rays becoming slowed-down and phase-shifted, and creating an interference pattern within the rotating media.

Indeed, it has been know since the early 1800's from Fresnel's discoveries regarding this drag on light when passing through a moving medium, that the faster the medium moves the more this changing refractive index effect will be amplified and the more it will drag the rays of light with it. Leonhardt in his papers on the relativistic effects on light explains that the vortex filament-tube is the perfect example of a moving medium through which light can be slowed to a great extent, and eventually even trapped [Leonhardt-Piwnicki 1999]. See the bibliography for Leonhardt's other papers.

Note 35—In one computer simulation of helical jets it has been found that there is almost zero azimuthal velocity at the center of the jet, and that the core of the jet behaves like a rigid body [Antonijoan]. This assumption also follows on from calculations that suggest that the magnetic field lines while they are dragged round by the rotational frame of a black hole move to the outside of its event horizon to the degree that its angular momentum increases. This presumably means that the charged particles rotating outside a hole create a dense wall around its perimeter, and that the mass of a hole essentially comes from this high density wall or shell [Gavrilov-Gitman]. If so then it means that the bore diameter can be varied proportionally to the particle azimuthal velocity.

Note 36 -That the gyroradius is reduced on the high-field side of the Larmor orbit, but it is increased on the low-field side, then the net result will be that the orbit does not close and will form into a spiral, or helical field.

Note 37—One such theory on black hole topology is proposed by Lery et al, [Lery 2002] (this paper also explains more of the quadrupole topology in these diagrams and has some interesting velocity and density charts relative to jet formation). The [Bicak-Ledvinka] diagrams give a good indication of the fact that magnetic field lines will approach the event horizon of a black hole but will not pass through into the black hole itself (meaning that the magnetic field lines will bunch together when nearing the event horizon and form the densest part of that magnetic field just outside the event horizon, and this is why the ergosphere forms an area of highly excited charged particles).

Note 38—Another paper on electron acceleration suggests that a similar threshold can occur and when this critical value is approached electron acceleration can reach a value as high as several hundred GeV if the uniform electric field strength approaches the critical value of $E_s{}^c =$ $(m\omega c)/(2\pi Ne)$, where m and e are the mass and charge of the electron, c is the speed of light in vacuum, N is the number of field cycles in the pulse and ω is the Doppler-shifted frequency of the laser field as seen by the electron upon initial injection. [Salamin-Keitel-Faisal].

Note 39—One of the main opponents to this has been Matt Visser who believes that the interstellar magnetic field is too stiff for any elasticity effects. But even Visser has calculated that the space-time matrix can be changed. In fact, [Visser 1990] proved that the existence of an omnipresent Lorentzian metric in spacetime is not a sufficient condition to prevent topology changes. Visser also showed that it is possible for geometrical effects to actually mimic the effects of topology change [Visser 1990]. Also on this see the [Madejski] presentation document.

Note 40—See bibliography for the full set of papers of Reva K. [Williams] theory developed from Roger Penrose's hypothesis on the extraction of energy from a rotating black hole. And also see the [De Felice-Carlotto] and [De Felice-Zanotti] studies on the pre-collimation of astrophysical jets.

Note 41—See the [Bicak] [Bicak-Ledvinka] [Punsly] [Livio] [Khanna] studies which have all raised the problems of the early Blandford-Znajek and Blandford-Payne models of the black holes and jets.

Note 42—Umstadter CURL effect is mentioned in [Mourou-Umstadter] paper (p84). And for back-scattering pulses see [Shvets 2004] and [Shvets 1998].

In fact considering that the gravito-magnetic (GM) force is supposed to be analogous to electromagnetic phenomena, what if the effects of a *non-uniform* GM field were dialed into the equation? As happens at the base of the helical-tube. In electromagnetism the *non-uniform (converging) magnetic field* is what produces the attractive force of magnetism, and so the equivalent gravitomagnetic force would also be an ATTRACTIVE force... Indeed, this *non-uniform factor* has been noted in a black hole's accretion disk and is alluded to in the paper by [Piran-Shaham] which suggests that the gravitational (GM) potential becomes greater towards the horizon.

Note 43—This geometry-induced form of collimation or setting of the helical pitch angle will be dependent also on the type and energy of the scattered particles, for example [Williams 2001] cites 1 to 30 degrees for Penrose-Compton scattered particles and 0.5 to 25 degrees for negative-positive pairs. The [De Felice-Carlotto] paper gives the basic formula for the law of geometry-induced collimation.

The gravitational force component from the Lense-Thirring effect is referred to by Williams as the "gravitomagnetic force" (GM)—a term which I am loath to use because it is so readily associated with being such a miniscule force, which in the case of black holes and the jets produced around them is clearly a gross misrepresentation. But as Williams explains this black hole force appears to be quite different from the gravitoelectromagnetism (GEM) delineated in General Relativity (and whose component fields are taken to be analogous to electric and magnetic fields) and which are only understood by Einstein to generate extremely weak forces. As Williams points out a black hole's field is mostly made up of gravitational binding energy (released from the abundant particle collision processes) which is then rotated into the black hole [Williams 2003], [Williams 2004b] and consequently radiated away through ejected particles. While it can still be the gravitational analog of an electromagnetic field, with the additional mass-density imbued into the particles and their angular momentum, this Williams' configuration of GM expresses a much more powerful influence over spacetime which it distorts and which acts on the space momentum vector of a particle (ie with respect to the negative radial momentum which collimates it into a jet). For best extrapolations of GEM though see [Mashhoon], and [Chiao 2002], and [Kiefer-Weber] .

Notes for Chapter 3

Note 1—These UFOs are similar to the Andreasson UFOs in that they also have groupings of spheres located in the lower part of their hulls, and these are the four-sphere Adamski UFO (USA), the four-sphere Piatan UFO (Brazil), the Russian three-sphere UFO (Eisk, Russia), the Jeff Savage *Onion Drive System* (USA), and the '611' craft at Dalnegorsk (Russia).

Note 2—This was in the very clever report upon an alleged new discovery about quartz by two scientists named as Kowsky & Frost which was reviewed in "Science and Invention" (for Sept 1927) unfortunately the original report it cheekily reviews was written like an April-fools-day type of ruse! This hoax, which was perpetrated to show how gullible some researchers were, was later exposed in the October 1927 issue of the same magazine. This quartz enlargement story still had some mischievousness left in it though, because it was later found being referred to, as being factual, in a book about Viktor Schauberger published in 1996 called *Living Energies* (on p283) by Callum Coats. And I have even seen this Kowsky-Frost story about quartz expanding displayed as if it were factual on Keelynet and on other websites in 2006! On a more constructive note the above mentioned ruse obviously spiraled out of the buzz of claims made in the 1920s about quartz doing strange and wonderful things, mostly to do with what we know now as acoustic phonon wind and ferroelectricity. When properly researched articles subsequently appeared they were quite informative though. See A. Crossley's article in "Proceedings of the Institute of Radio Engineers" for April 1928, and J.W. Harding & F.W.G. White's excellent article in "Philosophical Magazine" vol 8 (1929) pp169-78. But as we progress through the twenty-first century all manner of opto-electrical research is being carried out in earnest on quartz for its semiconductor and dielectric qualities, especially in China and Japan.

Note 3—I feel the word 'vacuum' is a nonsensical misrepresentation of what comprises space if it is compared to the vacuum generated in a laboratory environment. It is interesting that space was pronounced *empty* long before the first rockets were launched and long before radio telescopes showed scientists the dynamics of the cosmos, which as we know, caused Einstein to have a major re-think about his General Theory of Relativity model of a static universe where he had to formulate his cosmological constant (the Lambda force) in order to provide the much needed balance which prevented the universe from collapsing under its own gravitational pull. However, scientists now know that the universe is expanding because of its negative energy density, and is not static as Einstein originally claimed, so therefore, the only reason now for space being ascribed a vacuous condition is to do primarily with the way negative spatial geometry provides our section of the galaxy with expansive gravity, and because of that space *appears* to be vacuous when in reality it is not. So, space does not have the same vacuum that is obtained inside a bottle in a laboratory, all it has is expansion of its space-time geometry.

Moreover, scientists also now know that space has electric fields, magnetic fields, gravitational fields, virtual particle energy, cosmic microwave background (CMB) energy [Lineweaver], Higgs matter fields, Planck energy, scalar wave energy [Bearden-Rosenthal], zero-point fluctuation fields, electron radiation, gamma and X-ray radiation.

Indeed, it might just be that space is periodically filled with TOO MUCH energy, as might well be the case in our solar system's vicinity where, inside our region of space which astrophysicists refer to as the Local Bubble, there are magnetic pressure differentials, plasma density differentials, and gas cloudlets, all accompanied by a background high pressure flow of gas which has a temperature of around 10^6 K (see the paper by P.C. Frisch & J.D. Slavin *The Sun's Journey Through the Local*

Interstellar Medium: The PaleoLISM and Paleoheliosphere Astrophysics and Space Sciences Transactions (2005) preprint: arXiv:astro-ph/0606743). Moreover, the very reason WHY the extraterrestrials have come here to earth seems to be because they think there will soon be too much energy coming into our part of space – and too suddenly!

So I go along with what [Unruh 1974] says of the 'vacuum' of space in his note 6 of his *Second Quantization in the Kerr Metric* Phys. Rev. D10 (1974) 3194-3205, and I would quantify what Unruh says by suggesting that space or rather space-time is an unbridled sea of electromagnetic energies and of virtual particles constantly popping in and out of existence. And as it can be seen from the huge jets produced by black holes, that huge sea of virtual particle energy is only a condition of that energy, and merely requires an electromagnetic-gravitational mechanism to stress it to process those virtual particles into real electron-positron particles which will do work.

Note 4—See Sol-Gel research work from the Department of Inorganic Chemistry at Stockholm University at (http://www.fos.su.se/~gw/) also from Chemat Technology Inc (http://www.chemat.com/html/solgel.html) and from the very informative paper by [Yakovlev].

Note 5—See the [Morris-Thorne] paper on anisotropic stresses. Indeed, a factor strongly featured in this present study will be how dielectric materials can be used to great effect by taking advantage of the 2000-fold-plus mass differential between the negative charges and positive charges when oscillated above critical frequencies.

Note 6—The beat frequency transducer can be found in Mead and Nachamkin's patent (US 5590031—Dec 31 1996).

Note 7—The number of sphere-sets in each assembly is given in the drawings made by Betty Luca as being three or four, and three or four would work equally as well. I would think five would be too many because then the fields they would control would be too small (and the lower spheres would be too small for the job they have to do). Two sphere-sets wouldn't work at all. The complete series of books giving the account of Betty (Andreasson) Luca's experiences with Greys and extraterrestrial elders and of course UFO craft are: R.E. Fowler (1979) *The Andreasson Affair*; R.E. Fowler (1982) *The Andreasson Affair—Phase Two*; R.E. Fowler (1990) *The Watchers*; R.E. Fowler (1995) *The Watchers II*; R.E. Fowler (2000) *The Andreasson Legacy*; Betty Luca (1999) *Extraterrestrial Communications—ASBT* (Pt I & II).

Note 8—For Puthoff's thinking on the ZPF see for instance [Puthoff 1987], [Puthoff 1989a], and [Puthoff 1989b], wherein he cites the hypothesis of Sakharov which postulates that the ZPF has a non-zero renormalizing factor which can be exploited. Also see [Haisch-Rueda-Puthoff 1994]. I tend to go with the Haisch-Rueda-Puthoff understanding of the ZPF as being a fluctuation analogous to the Zitterbewegung (quivering motion) oscillations of an electron [Haisch-Rueda-Puthoff 1994], but with the modification that space-time needs to be given a curvature locally in order to bridle those fluctuations. The way I see it is that both the ZPF (and for that matter the Higgs field) are *mechanisms* through which elementary particle energy, call it Planckian masses if you like, transforms into more tangible energy or mass—but only providing the local geometry is conducive to it. See [Haisch-Rueda 2000a] and [Puthoff-Davis-Maccone] and also in the later chapters of this study where I mention the theories of Maxwell's vortices, and the rotons which in Planck's theory permeate throughout the fabric of space.

Note 9—The following comments from Betty Luca were made about the spheres and some other questions I'd raised; these are subsequent to her comments published in the Fowler books, and are from one of a series of letters received by the author from Betty over the years (my reference is Betty 56.doc dated: 29 Oct 2004).

"Re: Your question on drawing 41. The bars of light over the top sphere were just that, bars of extremely bright white light, brighter then a welders arc light striking against metal. They looked pretty solid and defined, and they did not connect to the top sphere at all, but were inches away from it as if it was standing in mid air on its own. The bands, or bars of light definitely came from the smaller, multi lighted, cut, crystal sphere. To me it looked almost cartoonish!"

It's odd you should ask about them because it seemed to me at the time that the investigators were intrigued with them! I had the feeling they knew what was causing it, and what it was doing! The small top sphere had brighter spots of extreme light in different areas around the globe. The investigators seemed extremely interested in the bars of light. It just appeared to me that some of them were aware of their appearance. The investigators were the group of scientific people present, when I was under hypnosis and being interrogated about the crafts appearance and what it did."

Yes, the bars did hover all around the top sphere. My drawing was sort of two dimensional, so it may have looked like they were in a straight line, but I know they were in every direction. It might have been confusing to put an abundance of the bars in the drawing."

The spheres were glasslike. I just thought it was like a cut crystal type of glass because of the globes outer design, and the way the light and brighter points of light appeared from it, like quartz or lead crystal."

Concerning the stems, yes, I've always thought the outside is just an aluminum casing that covers the shape of coils inside. The outer rounds seemed too tightly knit together to have individual movement. Seemed as if it were one piece. Also there was what looked like an aluminum cap that covered the top smaller sphere but seemed a little different then the stems casing at first sight, it looked thinner and duller than the bottom stem casing. And I think it was somehow removed. However I never saw how they removed it! Not seeing it come off has always bothered me! Do you think the cap could have stayed on and the bright light and bars were shining right through it, like it became transparent? I saw the pictures of the stem and cap in the thin blue book they [the ETs] allowed me to keep for ten days. There were some numbers and writings that I couldn't understand that were with the picture. Also they only showed the stem and top, no clamps, no wheels, and the body of it was leaning to the right about one o'clock! Yes, the lower large sphere rotates at times."
Betty Luca (personal correspondence 2004/6)

Note 10—In another report of this intensity of brightness observed in a UFO's power drive this phenomenon has been compared to magnesium burning, which I have also seen and, yes, that's ultra bright too, but not as damaging to the eyes as the arc welding. For the magnesium reference see the Iron Bridge Ontario, Canada case in J. Vallee (1966) *The UFO Enigma* pp221; and [Yousif].

Note 11—The basic starting kit for the study of crystal energy is [Hopfield 1958]. But this whole subjected is covered quite extensively in chapter 4 and 5 of this book.

Note 12—For an interesting summary of some further *well known* mechanisms of ionization, such as electron beam plasmas producing 10^{11} electrons per cm^{-3} at 4 to 10 Pa pressures, see the following study by [Bogaerts]. And for an interesting summary of some of the *lesser known* mechanisms of ionization see [Busch]. The ultimate examinations of avalanche mechanisms occurring in out in space of electron-positron pairs can be found in the papers by [Blandford-Znajek] and [Ruderman-Sutherland].

Note 13—Electron cascades or avalanches can occur around dielectric materials and fluids especially when a high frequency, high voltage, oscillating field is applied to them. This effect can also occur as a differential polarization, when a polarity differential is created between an oscillating dielectric and the environment surrounding it, such that the neutral and charged air molecules become affected by an electric field and are accelerated at extremely high speed toward the dielectric. When the dielectric has a relatively small volume it can set up a non-uniform electric field so that air-borne molecules converging toward the center of that oscillating potential will collide with other molecules also rushing inward knocking electrons off those other molecules. These freed electrons then smash into other molecules to free more electrons, and so on and so on, resulting in a cascade or avalanche of electrons. And just as Frank-Kamenetskii has illustrated in his study, in this way electron multiplication proceeds in a geometrical progression, rather like a chain reaction (D.A. Frank-Kamenetskii *Plasma—Fourth State of Matter* (trans. J. Norwood 1972) p10). This effect is featured in astrophysics papers mostly under the term 'dissipation'.

Note 14—See for instance [Citrin] and further studies on excitons.

Note 15—Electron avalanches occurring around semiconducting materials was shown by P. Flanagan, and was detailed in his patent (US4,743,275 of May 10 1988) of *The Electron Field Generator* which used resin blocks embedded with paramagnetic inclusions, which, when oscillated generated in the surrounding air abundant electrons through avalanching.

Note 16—Just as Mahmoud Yousif calculates for his craft. And because of the high intensity of magnetic forces in the vicinity of the spheres in the aaUFO so also will the distance between electrons be reduced to the Fermi range, meaning that the enforced interaction of their spinning magnetic fields (SMF) will create an electron-electron interaction and spinning force which will lead to the electron-fusion mentioned in this paper (as given in section 5:2 of *The Universal Energies* [Yousif]).

M.E. Yousif of Nairobi University has an avid interest in explaining how UFO-type propulsion drives can be developed and has written several studies on this subject. Yousif's papers can also be viewed at http://exmfpropulsions.com (circa. 2008).

Note 17—With respect to such laminar flow, insulating fluids can generate large amounts of electrostatic charges simply by moving them through insulated curved containers (as with the case of petroleum storage and distribution—where a lot of research is actually done on *eliminating* these large and potentially hazardous electric charges, created when they are pumped, by dispersing them through grounding conductors). Further on from the example of layering is when suspensoids are introduced into an insulating fluid, because in the presence of the electric and magnetic fields around the toroid the suspensoids in its fluid will move at differing speeds relative to whether they are dielectrics, paraelectrics or ferroelectrics with differing magnetic susceptibilities (and permeabilities), and the fluid in the toroid will behave much like a solid dielectric does with a variety of positive and negative charge domains, and those factors will subsequently effect the fluid's electrostatic charging. Of course, how the liquid is composed and how that composition reacts to electric and magnetic fields greatly effects its electrical permittivity and the way electromagnetic waves are diffracted and slowed when they pass through it (this factor of permittivity is further alluded to in chapter 9). See, for instance, Patrick Flanagan's research into the surface tension of negatively-charged colloidal water, extracting from a small volume of water when spun in a vortex an electric charge of 10,000 volts, in P. Tompkins & C. Bird *Secrets of the Soil* (1991) p109. But also see R.D. Kleeman & W. Fredrickson *Experiments on the Sign of Electric Charge Assumed by a Metal Immersed in a Liquid* Phys. Rev. 22 (1923) p134-6; and the Reynolds fluid research in R.L. Street et al (1996) *Elementary Fluid Mechanics* p231-5; and in R.P. Feynman et al *The Feynman Lectures on Physics* 2 (1970) sect. 41-6 to 41-10.

My original theory (1999) about what flowed inside the toroid shell of this aaUFO was that it was water into which were mixed suspensoids and the whole fluid would then be polarized. But because it was reasoned that water might prove too heavy I did succumb to the hypothesis that the working fluid might be a gas flowing inside the toroid (ie a gas such as deuterium that could be derived from water). I have now re-assessed my original (1999) assumption and in section 6.8 of this study there follows a presentation of the clues which show what the fluid inside the toroid should be. That many of the principles of electrical charge migration and transfer will still apply whether the fluid will be found to be a gas or a liquid, means that the basic assumption of a Lorentz force being created inside the toroid, and that a magnetic field will be generated outside that toroid will still hold through either hypothesis.

Note 18—Electrohydrodynamic (EHD) actuators can be used to generate movements in suitably prepared fluids by setting up current flows in them, by using radio frequency impulses (see for instance [Roy] reference in bibliography). But then by using specific polymer fluids and gels in the inner ring such a magnetic fluid could provide moving islands of magnetic potential, and these could be bunched together or spread apart depending upon whether the magnetic field they control needs to be intensified or dissipated (see chapter 5 and 6).

Note 19a—For the stem coils grabbing the toroid fields it is best to see an animated version.

Note 19b—For this breaking of magnetic field lines it is best to see an animated version.

Note 20—For electron production through reconnection of magnetic field lines see note 13 of chapter 2. There will also be apparent viscous shear forces acting on this gyrating plasma much in the same way that such forces act upon an accretion disk (see [Piran-Shaham]. At this micro-scale though what will be more prominent will be the quasi-dielectrophoresic and quasi-electrophoresic forces from the non-uniform (inhomogeneous) fields, which will generally pull neutral and positive particles (respectively) in waves towards the central axis.

Note 21—The basic explanation of the mechanism of magnetic flux reconnection can be found in [Ruderman-Sutherland]. Although the principles of reconnection theory can also be found in [Priest-Forbes]; in D. Biskamp's *Magnetic Reconnection in Plasmas* (2000); in [Furusawa-Sakai] and in [Di Matteo]. And a review of reconnection studies can be found in [Bulanov-Sakai]. In the Piran paper [Piran-Shaham] (especially in pp6-7) the disk's plasma fluid moves in magnetic 'blobs' due to the magnetic viscosity of the fluid (whereby the strongest viscosity will occur at the points where field line reconnection occurs). These blobs will be held together by their magnetic field (see also where this phenomenon is known as pair-production for an inhomogeneous electromagnetic field [Kim-Page].

Note 22—The constructive and destructive wave patterns in 3 dimensions might look something like a spherical cage around the spheres. But with the Bessel fields it could well appear star-like.

Note 23—I am presuming that the image of these light bars set up for Betty Luca by the Greys was a special configuration to show off these intense photon emissions, so that she would be greatly impressed by them and faithfully remember what these top spheres are able to do. Because, on closer examination it would not necessarily work like that with all the gyrating fields up and running around them. When those were in operation the emissions coming out of the top spheres would be in dynamic motion and not standing still in the air, but would look like a blur of bright light being dragged by the gyrating field (the photon orbit ring) circling around the center of the craft, where these emissions would be drawn into the central vortex (where they will essentially become the fuel for the next stage of this UFO's engine).

I also believe, in light of some recent discoveries, that such bars of light would represent only a small fraction of the energy emission intensity that could be generated and radiated out of this aaUFO's central vortex once it does become fully energized. However, I am always willing to admit that there might be some additional factors which I haven't thought about, and that these non-local physicists with their advanced minds haven't allowed us to see all there is to see about them. Perhaps there will be more things to discover about them at a later time.

Note 24—This particular outline of the birth, feeding and up-keep of this gyrating vortex so that it eventually grows up to be a strong healthy quasi-black hole is probably not all there is to this power system. Certainly, and late into this study, there has been a significant breakthrough into how this gyrating field will need to be conceived before it is maintained and powered up at the heart of the aaUFO. This latest discovery is nowhere mentioned in astrophysical studies and so it affords some rather unique information about quasi-black hole dynamics and I would like to keep it as a companion to the above outline, and rather than replace the above with it here I shall introduce the later conception-outline merely as an alternate hypothesis in chapter 7. Having said that, I am well aware of the Unruh / Leonhardt / Visser discussion about what constitutes a bona fide black hole in miniature, and so what will be found through chapter 7, and through chapter 8 and 9, promises to put a whole new perspective into that miniaturization discussion.

Note 25—Well, if all particles have a wave function—and a group velocity—if the particle-wave component which would become less and less helical and more and more azimuthal, and eventually even spin toward the state of a closed loop; would the universal laws of physics allow it to actually remain as a closed loop or would its group velocity REVERSE instead..?

Note 26—Indeed, their *azimuthal velocity* can be faster than the speed of light (c) while at the same time their *linear velocity* might only be approaching (c). It might also be interesting to conjecture what happens to light photons when they get caught and dragged faster than (c) along with the charged particles in such a filament-tube, for if one assumes that the charged particle's mass can get 'lost' into the tube what happens next? Would that mass be converted into mass-less angular momentum for instance, much in the same way that electromagnetic energy transforms into angular momentum through the agent of a black hole..? Or does more gravitational force become generated and consequently imbued into that filament-tube..?

Note 27—The general consensus in astrophysical studies is that when a black hole is miniaturized it becomes even more intensified. I think it remarkable that more attention wasn't paid to Penrose's paper on extracting energy from a black hole [Penrose 1969] from the day it was first published nearly forty years ago, for this 'news' has still not quite filtered down into the physics text books of schools! Hawking's realization that a black hole can radiate power came after his initial criticism of a paper by J.D. Bekenstein in 1972 (Nuovo Cim. Lett. 4, 737) claiming that black holes have entropy. For, Hawking argued, with entropy they must have temperature, if they have temperature they must radiate, and every fool knows black holes can't radiate! But in 1975 Hawking published his revelation that Black holes do indeed radiate energy. S.W. Hawking *Black Hole Explosions?* Nature 248 (1 Mar 1974) 30-1; S.W. Hawking *Particle Creation by Black Holes* Commun. Math. Phys. 43 (1975) 199. Ironically, Hawking is now deemed the expert in black hole miniaturizations, and he suggests that the black hole mechanism extends all the way down through to particle physics, as also does Reva K. Williams in her [Williams 2002b] paper. On Planck-scale black holes see [Barrau-Feron-Grain] and the sources therein.

Note 28—An accretion disk is used by an astrophysical black hole to break down matter into smaller particles and to direct those mass particles toward that black hole; and the accretion disk will provide the angular momentum for particles to be ejected into a jet (as in pulsar and quasar jets). However, surrounding a black hole's event horizon immediately inside the accretion disk will be an ergosphere, which the black hole uses to break down matter into even smaller particles. The ergosphere will provide greater shearing momentum on the particles and will be ripping atoms apart producing high energy photons, gravitational binding energy and sub-atomic particles. While the gyrating storage field in the aaUFO can be likened to an accretion disk, I will prefer to call it an ergosphere. And in all there are four ergospheres associated with the aaUFO (inside the toroid, inside the planar waveguide as mentioned above, and outside the craft around the lower hull): The most important of the craft's ergospheres though will be that contained inside the center of the aaUFO immediately surrounding the rotating sphere-set assembly.

Note 29—See section 2.7 and its note 19 about the J x B torquing force. Energy generation from a black hole was of course first noticed by Penrose and Zeldovich, but see [Blandford-Begelman-Rees]. Specifically though this idea is through the [Punsley-Coroniti 1990a] rendition of the [Blandford-Znajek] theory on the extraction of energy from a black hole. On the 'bulging' of the photon orbit around the four top spheres (as masses orbiting an event horizon) see for example [Hughes].

Note 30—The classical paper on electron-positron pair production is [Ruderman-Sutherland]. For pseudo-Schwinger pair-production (similar to orbital viscosity) see [Jacobson]. Also see [Corley-Jacobson]. And the rest of the ionizing mechanisms will be detailed throughout the following chapters.

Note 31—I made a point of asking Betty through our correspondence many specific questions relating to the motions of these top spheres and the lower spheres. And Betty further noted that the top spheres were revolving more quickly than the lower spheres, although it was not known whether this revolving was *caused* to occur or whether it was an induced effect from the black hole's angular momentum together with all the acoustic vibrations going on (acoustic vibration alone will be enough to rotate these spheres the way they are located into the base disc).

Note 32—For the tidal and cannibal theory of galaxies see [Ostriker-Tremaine] and also [Hausman-Ostriker]. Another mention of the cannibal theory of galaxies can be found in [Blandford-Begelman-Rees].

Note 33—Of all the hundreds of papers and books written on black holes that I have studied I have never come across this clarification. Maybe its so obvious that no-one else has bothered to delineate it, and I'm just stating the obvious. Either way its a valuable point to make here.

My tentative estimation of gravitational force is that it equates to a differential in radii of angular momentum... and is analogous to wave trapping by a spiraling flow (similar to how acoustic waves are trapped by a spiraling flow of greater velocity than those acoustic waves—see chapter 10 note 17). So all the Weber and Gertsenshtein methods which attempt to generate spin-2 gravitational waves from combinations of stationary orthogonal magnetic fields, and spin-1 electromagnetic waves, are forever going to struggle to get any sort of result... until those endeavors consider bringing a rotational force gradient into their equations.

Note 34—For the *possibility* that what is occurring here is parametric amplification see [Andersson-Glampedakis]. The children-on-swings analogy explaining parametric autoresonance can be found on page 16310-1 of [Assaf-Meerson].

Notes for Chapter 4

Note 1—See for instance the [Bondarev-Hyodo] study on dielectrics.

Note 2—For reference see [McKnight], and [Magnetostriction and Magnetostrictive Materials], and [Calkins-Flatau]. Terfenol-D is made up of terbium, iron, and dysprosium (Ter-Fe-D). The Grangemouth UFO report mentions that as the UFO passed overhead is was basically silent, except for a slight audible hum, and this can be seen at http://www.ufoevidence.org/cases/case1047.htm),

Note 3—Ferroelectric electron emission occurs at the microscopic interfaces within certain dielectric materials, and above a certain frequency the oscillating electric field bunches up on itself at those interfaces and causes in the dielectric material a repulsion of the bunched-up electrons. The phenomenon occurs where the positive and negative polarizing charges position themselves opposite each-other along the dielectric's interface surfaces, resulting in each polarized layer of charges being screened (compensated and neutralized) by charges of the opposite polarity when an alternating electric field is applied to them; so the overall electric charge is neutral; and what happens with ferroelectrics below their critical frequency is that when an alternating electric field is *slowly applied* to that material, to change those polarities the screening charges (of neutralizing electrons) move back and forth between interfaces to that of the opposite polarity interface, following the alternating polarities of the electric field, which ensures that the dielectric's overall charge remains neutral. However, when the applied electric field is switched *suddenly* from one polarity to the opposite polarity this sluggish screening arrangement fails to transfer fast enough along with the electric field, resulting in a fresh influx of screening electrons piling-up onto the remnant screening electrons (which haven't yet moved out of the way), and so instead of neutralizing the interface these screening charges create a surplus of electron charges—and it is these doubled-up charges that are repulsed away from the material as electron emissions (from the like-for-like Coulomb repulsion force). See H. Riege *Electron Emission from Ferroelectrics—A Preview* Nucl. Instr. and Meth. in Physics Res. A340 (1994) 80-89.

It has been found in ferroelectric spontaneous emission tests of PZT doped with lanthanum and with neodymium that the resulting emission current density was found to be *almost twice as high* for lanthanum (PLZT) as it was for neodymium (PNZT) [Zhang]. This same paper also makes the valuable point that, the greater number of domains that spontaneously switch (ie the more un-screened charges involved in the process) the higher will be the densities of electrons that can be emitted.

The effect is somewhat different to dielectric absorption where the dielectric's positive charges (holes) move much slower than the negative electrons (charge carriers), and when an alternating field is applied to that dielectric while the more mobile electrons are dragged back and forth by the oscillating field the positive charges remain in their domains, so the dielectric is then left with a positive charge overall, and sets up an electric field with the surrounding air to continually pull more electrons toward it to compensate, which, through their energetic motions produce an electron cascade effect in the air around them, as in [Harrop] and [Flanagan].

Note 4a—Well, the Lamb-Retherford shift is the emission and re-absorption of virtual photons and the conversion of virtual electron-positron pairs into real electron-positron pairs through the perturbation of the background field of space (or the aether), which some scientists have suggested is a process which extracts useable energy out of the zero point fluctuating (ZPF) field, in a fashion similar to how breaking and reconnecting magnetic flux lines produce electron-positron pairs [Wang]. [Sakharov] refined this hypothesis by suggesting that it is through the renormalization process that the actual gain occurs, because that process doesn't quite bring the perturbation all the way back to zero—which suggests to me that the ZPF needs a rectifying process in order to harvest from it (also see Sakharov's theory in [Misner-Thorne-Wheeler] sect. 17.2 pp427). Also see W.E. Lamb & R.C. Retherford *Fine Structure of the Hydrogen Atom by a Microwave Method* Phys. Rev. 72 (1947) 241-243. Energy production in space all comes down to the stressing of the space-time fabric, and as I mention in chapter 2 and chapter 3 its not that difficult to understand how it works. What is difficult to understand is why Science is making such a screwed-up face about passing this simple technology through its system.

Note 4b—For the Searl Effect see John Searl's website at http://www.sisrc.com/c1.htm and also the webpages at: (http://searleffect.com/free/articles/artextra.html). Looking at this process from a slightly different angle, the mechanism that triggers the ferroelectric process of electron emission from the neodymium in Searl's rotors can ALSO be related to the phenomenon called 'magnetic bubbles' (which was later coined *magnetic quanta* by Searl's research team—see image of magnetic domain patterns on a nanometer scale in the studies by [O'Dell] and in [Carey-Isaac]). This *magnetic quanta* phenomenon discovered by Searl provides an important explanation of the reconnection process (as alluded to in chapters 2 and 3 of this study)—whereupon as the magnetic flux lines are broken and reconnected they induce a back-emf process that extracts electrons out of the system (and out of the surrounding air)—because it shows how a magnetic field (flux line) can dissipate into these magnetic quanta and then reform back together when the conditions for that field's re-establishment present themselves again. And of course, as in the case of Searl's SEGs and IGVs the generated electrons were put to good use!

An interesting insight was given to John Searl once when he was setting up one of his rotors to be filmed for a TV program, and when the cameras were switched on and the moment came for that rotor to do its party-piece it fizzled out like a lame duck. The astonished Searl was dumbfounded and said it had never happened before, but afterwards he realized that here was the very control mechanism which would improve his later models! For in the past Searl rotors once they started to rotate would suddenly take to the air, hover, and then shoot off skyward and never be seen again, so they were an expensive invention! Searl figured out that the reason why that camera-shy rotor failed was because of the oscillating signal being radiated from the TV camera, which was trained so closely upon his rotor. That this signal had somehow interfered with the oscillation of the rotor's electromagnetic field and nullified it. And if I could throw my three-pennyworth in here, I would say that the camera's oscillating signal would have interfered with the shock waves being produced in the reconnection process of that rotor. That signal would have blanketed the virtual particle field around that rotor—and would have prevented that virtual particle field being stressed. Looking at it another way, in Searl's rotors the magnetic flux will be snapping and reconnecting at their own rate of oscillation, and those flux lines will therefore have their own break-remake time-periods in which to complete their cycles of reconnection, so therefore when the TV camera's oscillations took precedence over the break-remake time-periods the rotor's reconnection cycles weren't allowed to be completed. They would only have had enough time to been broken and nothing more! Hence the reconnection process collapsed altogether because the magnetic fields weren't being regenerated. But, as I say, out of this *faux pas* Searl could later devise a governing mechanism that ensured that his rotors didn't shoot off into space, and they could return back down to the ground.

Note 5—One of the earliest alloys used by Searl (circa. 1946) was a compound of aluminium, silicon, sulphur, titanium, neodymium, and iron. From these titanium and neodymium are ferroelectric elements; and the iron is magnetostrictive.

Nylon 66 was the element Prof. Searl chose to use and he used it because it had a high electron content and because it was composed of atoms joined by double covalent bonds. Its interesting that Searl chose nylon 66, because it has subsequently been found that odd-numbered nylons seem to show more promise in this field, and that nylon 6,6 was the most widely used of all nylons. Odd-numbered nylons were recognized as ferroelectric polymers in the 1960's, with nylon 7 and nylon 11 exhibiting the most stable ferroelectric properties (and they soon became used as ultrasonic transducers [Neagu]). Polymer studies are very extensive and researchers have obtained internal electric fields in nylon 11 thin-films in the order of MV/cm from the ferroelectric polarization reversal effect [Tsutsumi-Mizutani-Sakai].

Note 6—For ion migration materials see the interesting paper of Padma Kumar, P. & Yashonath, S. *Ionic Conduction in the Solid State* J. Chem. Sci. 118 (2006) 135-154. Another candidate for powerful emissions are the newly discovered pyroelectric crystals such as lithium tantalate ($LiTaO_3$) and lithium niobate ($LiNbO_3$) that emits electrons with mild temperature changes, while gently heated (or cooled) within a deuterated gas can respond with an electric field strength of 1.35×10^7 V/cm—that can eject electrons, positive ions, and excitons from their surfaces, which can be accelerated up to 170 keV. See papers by [Naranjo], and [Geuther-Danon].

Note 7—The results of a study by [Benedek-Boscolo] suggest that as well as by Coulomb repulsion forces, electrons can also be ejected from a crystal through their excitations to higher energy levels through Auger collisions (electron cascades), although as this paper points out that for the Auger process to trigger the electron density must be at least as high as 10^{20} electrons cm^{-3}.

Note 8—For the fundamental papers on ferroelectric emissions see [Riege 1998], [Rosenman], [Benedek 1997], [Gundel 1989], [Ivers], and [Gundel 1991] in the bibliography pages of this book. Although, if I were asked which paper gave the most informative description of the ferroelectric emission process I would have to suggest the paper of H. Riege *Electron Emission from Ferroelectrics—A Preview* Nucl. Instr. and Meth. in Physics Res. A340 (1994) 80-89.

Research into Ferroelectric materials was quite extensive during the 1990s and work was carried out on many different ceramic compounds; such as lead-lanthanum-zirconium-titanate (PLZT), barium titanate ($BaTiO_3$), triglycine sulfate (TGS), lead-zirconate-titanate (PZT), lead-magnesium-niobate (PMN), lead-erbium-zirconate-titanate (PEZT), and lead-neodymium-zirconate-titanate (PNZT). Notably strontium-doped lead-zirconate titanate (PSZT) and barium strontium titanate (BST) yield very high piezoelectric and ferroelectric properties.

Some ferroelectric studies have led to the development of high-emission ferroelectric cathodes and to ferroelectric electron guns, see for instance [Rosenman] (pp6148).

It is even speculated that at the Currie temperature point of $PbTiO_3$ an external field of around $100/\varepsilon_r$ GV/m can be expected to be emitted by spontaneous polarization. See [Gundel 1989], [Gundel 1991], and [Riege 1998].

Note 9—The repetition rate will depend upon what sort of ceramic is used and on what emission triggering mechanism is being used, but the top end frequency will probably rise higher and higher as research continues on these various materials. One study into field excited electron emissions (FEEE) for TGS ceramics, for instance, has come up with a repetition rate as high as 2 MHz (See H. Riege *Electron Emission from Ferroelectrics—A Preview* Nucl. Instr. and Meth. in Physics Res. A340 (1994) 80-89).

Very high current outputs (of 100 A/cm^2) were found to come from PZT and PLZT ferroelectrics by researchers who noted that theoretical current densities of up to 10^5 A/cm^2 could be achieved. See H. Riege et al *A New Beam Source Based on Electron Emission from Ferroelectric Ceramics* Bull. Am. Phys. Soc 34 (1989) 193; and also see [Zhang]. Another study which achieved outputs of 70 A/cm^2 from lead-zirconate-titanate (LTZ-2) proposed that with suitable modifications a current density of 1 kA/cm^2 could be generated in short 100 ns pulses [Ivers]. This compares to the plasmonic condition where in most metals the free electron density is 10^{22} cm^{-3}, but at the extremely high frequencies of the UV, X-ray, and gamma-ray bandwidth. Metamaterials will do the same thing but have a plasma frequency in the much lower GHz bandwidth [Wu 2003].

Note 10—The chemical bonding in glasses is very similar to that in crystals, the most important difference being that glasses have a significantly lower charge carrier mobility. Color centers are also generated in glasses from excited particles relaxing to their ground state—this is also a part of the exciton process of recombination. Also see [Vogel] pp14; and possibly L.B. Glebov et al *Luminescence of Lead in Silicate Glasses* Sov. J. Glass Phys. Chem. vol 16 (1990) p133.

Note 11—In *Andreasson Affair* p35 fig. 8 Betty indicated that the large lower spheres were what created the pulsating emissions of red, blue, green and white light that she'd earlier seen as coming from below the craft. In *ASBT* bk II drawing 41 Betty gives a better description of both the top and lower spheres and indicates that the lower ones were facetted on their outer surfaces and had bright centers inside them, some of which glowed more brightly than others. Also in *ASBT* bk II in drawings 42 and 43 Betty noted that she saw energy lines or long wires flowing inside the crystals.

The complete series of books giving the account of Betty (Andreasson) Luca's encounters with Greys and extraterrestrial elders and of course UFO craft are: R.E. Fowler *The Andreasson Affair* (1979); R.E. Fowler *The Andreasson Affair—Phase Two* (1982); R.E. Fowler *The Watchers* (1990); R.E. Fowler *The Watchers II* (1995); R.E. Fowler *The Andreasson Legacy* (2000); Betty Luca *Extraterrestrial Communications—ASBT* (Pt I & II—1999).

Note 12—See for instance the study by [Trukhin].

Note 13—A similar effect was noted in porous silicon, mostly as a result of impurity atoms of C, O, H, and F mixed into the silicon. See for instance the studies by [Qin-Jia] and [Lin].

Note 14—For blue and green in quartz see the studies by [Song 2000a] and [Luff]; for brown in quartz see *Oxford Dict. Chemistry* (2000) p156; for laser irradiated quartz see [van Ginhoven]; and for erbium see Thierry d'Almeida *Cathodoluminescence of Er3+-Doped Crystals* High Temperatures High Pressures vol 30 no 3 (1998) p351-356. Crystallography research into color centers seems to have been quite extensively carried out in Russia's scientific academies.

Note 15—See [Kapustin] p20. Crystal compounds and metallic alloys experience eutectic crystallization (reduction of grain size) when radiated with acoustic frequencies (op. cit. p19). It would be interesting to conjecture that this may be how additional strength might be manufactured into a UFO craft's structural materials.

Note 16—From *Oxford Dict. Chemistry* (2000) p267.

Note 17—See for instance the study by [Snoke-Negoita], and [Talbot].

Note 18—This data is based on the reports of Jeff Savage on the motive power unit he observed inside a UFO craft.

Note 19—Betty Luca's description of spheres (in her *ASBT* booklets and in her correspondence with me) suggests that while it is positively known that the lower spheres are facetted, the top spheres apparently shone so brightly that it was hard to tell if they were facetted or not. However, from the apparent effectiveness of the Ginzburg-Grigoryan external beam effect (as will be seen in section 5.8), and from the fact that the photon orbit ring of the black vortex will induce such powerful rotational forces upon any particles around these top spheres, it will be assumed throughout this study that the top spheres were facetted. And R/D will have to ascertain the appropriate size of the facet flats which would surface these top spheres (the facet dimensions will induce a specific frequency upon energy fields which rotate around them, close to their surfaces, and this factor will be used to get those energy fields to oscillate at the most appropriate frequency, therefore the facet size will most probably need to be deduced empirically).

Note 20—The depictions are given in *Andreasson Affair* (fig.14) on p43: in *Andreasson Affair—Phase Two* (fig.13) p70: and in *ASBT* bk II (fig.46) p49.

With reference to the holographic-like images, that these images seemed to have been engineered between the two hatches it would suggest that the ETs were establishing some sort of acoustic-plasma matrix between them, so that these various 3-D images could then be 'etched' into that plasma, with one hatch transmitting the acoustic-plasma graphic and the other buffering it.

Note 21—Excitons can form in bulk and can migrate out to the surface of the crystal, see for instance [Song 2000a]. By using a technique known as phase-coherence phonons can generate in piezoelectric crystals acoustic waves to 10^{10} Hz. X. Hu & F. Nori *Phonon Squeezed States: Quantum Noise Reduction in Solids* Physica B 263-4 (1999) 16-29.

Phonon wind was observed (and its effect photographed) in experiments with quartz crystal as early as the 1920's. This interesting little paper may help resolve the curved-surface versus facetted-surface issue. J.W. Harding & F.W.G. White *On The Modes of Vibration of a Quartz Crystal* Phil. Mag. vol 8 no 49 (Aug 1929) p169.

See also [Shilton] and [Ilisavskii]. This technology can also used to take charges which lase onto the surface of a material and to transport them onward, through the surface acoustic waves, so that electric charges can be transported from one side of a dielectric material to the other, as in the components used in light beam propulsion featured in *Laser Propulsion* (p65/6) by L.D. Maurer & M.E. Miller (2004) (published by Unitel Aerospace).

Note 22—Obviously fabrication of working components from these crystal and ceramic materials is a manufacturing art in itself—see for instance the Sol-Gel research work from the Department of

Inorganic Chemistry at Stockholm University at http://www.fos.su.se/~gw/—(Sol-Gel doping.htm). And see http://www.chemat.com/html/solgel.html—(Sol-Gel Technology.htm).

Note 23—This section is getting complicated enough and so I will cover this important Ginzburg-Grigoryan aspect in more detail (and with diagrams) in chapter 5.

Note 24—The various effects of exciton scattering can be seen in [Koch], while early theoretical studies of exciton formation can be found in [Hopfield 1958] and [Hopfield-Thomas 1963]. Also see an interesting study on exciton energies in the paper of T. Ishihara *Photonic Crystal Slab with Exciton Resonance: A Novel Research Field of Nanotechnology* (Exciton Engineering Laboratory) Riken Review 37 (2001) 38-42. And a paper on Planck scale energies in space of N. Arkani-Hamed et al *Phenomenology, Astrophysics and Cosmology of Theories with Sub-Millimeter Dimensions and TeV Scale Quantum Gravity* (1998) preprint: arXiv:hep-ph/9807344.

Note 25—The Penrose-Williams pair-production processes of ionizing collisions will be explained in chapter 7 of this study.

This breaking down of energy can be likened to the *oligodynamic* dissociation process as alluded to by Viktor Schauberger which occurred through the shearing forces of the rotating fields in his turbines. See the English translations of his works by C. Coats, especially *The Energy Evolution* pp70-3; pp100-3; p190.

Note 26—It would be extremely ungrateful of me to suggest that the ETs only allowed scant few details of how these spheres work, particularly as the ET's remit seems to be to encourage whoever is *ethically and selflessly* interested in these mechanisms to communicate with them directly (somewhat telepathically). If the reader thinks that is a rather exotic thing to aspire to then I would suggest they have lost their will to experience their higher worth. But then, personally, I like the challenge of learning new ways to communicate! Having said that there have been occasions where I have needed extra input from Betty Luca, who has previously gone through all her experiences with the extraterrestrial beings.

For instance, I shall give here Betty Luca's recent thoughts about what she was shown many years ago regarding the top spheres: *"Concerning the stems, yes, I've always thought the outside is just an aluminum casing that covers the shape of coils inside. The outer rounds seemed too tightly knit together to have individual movement. Seemed as if it were one piece. Also there was what looked like an aluminum cap that covered the top smaller sphere but seemed a little different than the stems casing at first sight, It looked thinner and duller than the bottom stem casing. And I think it was somehow removed. However I never saw how they removed it! Not seeing it come off has always bothered me! Do you think the cap could have stayed on and the bright light and bars were shining right through it, like it became transparent? I saw the pictures of the stem and cap in the thin blue book they [the Greys] allowed me to keep for ten days. There were some numbers and writings that I couldn't understand that were with the picture. Also they only showed the stem and top, no clamps, no wheels, and the body of it was leaning to the right about one o'clock! Yes, the lower large sphere rotates at times."*
Betty Luca (personal correspondence Oct 2004).

Notes for Chapter 5

Note 1—In the paper by [Bekenstein-Schiffer] (pp8) a theory is proposed to explain how gravitational radiation issuing from a black hole will effect the atoms in a polarization cloud, in such a way as to cause the more massy nuclei to sag away from the center of the electron orbit. Such a displacement will necessitate the production of an electrical displacement force within the atom so as to prevent it from being ripped apart. Also see [Saa-Schiffer] paper regarding the polarization cloud and the generation of an inter-atomic electric field to neutralize the gravitational pull.

Note 2—This action would increase the electron's continuous zitterbewegung jittering action of absorption of energy from the virtual energy field (ZPF), and therefore increase its emission of photon energy. This would have implications in the Lamb-Retherford shift, meaning that this electron resonance would then be stressing that virtual energy field—above its relaxation frequency—to produce electron-positron pairs.

Note 3—One of the most fundamental studies of crystal structure and ferroelectric crystals can be found in [Kittel 1996] (pp393-398). The displacement of an atom away from the ground state is also featured in [Kittel 1996] (p411), and will become an interesting energizing factor when such atoms circulate around the lower spheres (to produce an electric field between 'in' facing and 'out' facing sides of lower spheres). Also see the [Ang-Yu] paper for the polarization effect on ferroelectric materials. In the paper by [Bottger] it is suggested that the best ferroelectric materials are the perovskite ones where the ions are situated near crossover points, which would improve the switching (as in Titanium, Zirconium, Niobium, Tantalum, and Hafnium; and where the cation is polarizable lead and bismuth).

The polarization force is proportional to the mass-difference between the electron and the positive nuclei—so, the more massive the materials embedded into these spheres the greater the electric field produced [Saa-Schiffer].

At least one experimenter has discovered that quartz crystal can have a strain imbued into it, and that a 'stress field' in its lattice can remain for some 30 minutes after the source of the strain is removed [Gallimore]

Note 4—The best examples of this bunching up of the magnetic field lines can be found in [Punsly 2001] and [Blandford-Znajek], [Punsly-Coroniti 1990b].

Note 5—See C. Kittel (7th ed: 1996) *Introduction to Solid State Physics* (pp393-398).

Note 6—See the very detailed paper by [Yakovlev] which gives extensive details about the doping of lanthanoids into lead-based sol-gels; and the paper by [Scott] which further suggests that ferroelectric cells will collectively orientate in domains with adjacent cells in a common polarization alignment.

Note 7—Crystalline materials are featured quite prominently within many of the *Andreasson Affair* books, and it would seem that the ETs have a wide variety of uses for them. It would make sense as there is such a prominent anisotropic (one directional) feature to them which is polarizable. The upper and lower spheres featured in the aaUFO play the part of energy providers and energy accumulators, but other crystals have been observed performing very different tasks, and obviously these are configured quite differently for their quite different purposes.

For instance, one of the Greys whom appeared in Betty Luca's first recollection of her abductions held a crystal sphere (which would appear to be about 7 inches in diameter), which rolled around the Grey's hand. So that crystal must have had its own gravitational field imbued into it. But to some extent that crystal was 'controllable' by the Grey's mind, possibly through a form of psychokinesis, for want of a better word (*Andreasson Affair* p124/5). But later into Betty's recollections it transpired that their main purpose (as explained in *Watchers II* pp134) was to record information, that they were used as intelligence recorders, possibly through a technology involving psychometrics.

And some spheres were used in what appears to have been an electrical charge-draining process, when they were attached to high-voltage stakes that the Greys placed into the ground all around the perimeter of one of their craft during a recharging process, the result of which arrangement was that those stakes successfully directed the air-born high voltage electrostatic charges generated by that UFO safely into the ground (*Watchers* p66-67 figs 13 & 14).

Other spheres were used by the Greys to shine visible light and colored light in different part of their ship, meaning that those spheres developed some sort of phosphorescent luminescence in order to shine their bright light, just as we would use light bulbs (particularly those spheres that were used to illuminate the Grey's operating theatres, which hovered in the air above the patients (*Watchers* p161 figs 35 & 36). These spheres may also have transmitted energy to the patients on the operating tables.

Similar to the previous example was the pair of crystal globes (about 20 inches diameter) set hovering in the air, one beyond Betty's head and the other beyond her feet (while she was reclined flat on her back—whilst hovering in midair!). And judging by what else she was going through these spheres appear to have been pumping energy into her at the time (*Andreasson Affair—Phase Two* pp100-101 figs 22 & 23).

A somewhat different configuration of sphere, and which admittedly may prove to involve an entirely dissimilar technology, was witnessed by Betty aboard a mothership and on a journey from it, where a group of human-like Elders sat Betty body down on a seat and then brought Betty's soul out of her body, and then showed Betty while she was out-of-body a 'plasma-like' sphere which emitted a lavender-purple glow from what looked like a moving field inside it (*Watchers II* pp132-138).

Note 8—For perovskite atomic structure polarization reversals see studies by [Scott] and [Yakovlev].

Note 9—It is interesting that the aaUFO's black vortex mechanism that radiates gravitational forces through which the glass-crystal spheres turn into electrical energy, phonons and photons, is almost the reverse of what Baker (referred to in chapter 3) has set out to achieve with his quartz spheres trying to down-convert the terahertz gravitational phonons of Gertsenshtein waves. Some physicists believe that Gertsenshtein (gravitational) waves are emitted by a variable electric current in the presence of a uniform and static magnetic field (and that conversion process is similar, in principle, to the recent introduction of 'grasers' that behave somewhat like FEL lasers but which generate a *helical* electromagnetic field, and in theory, a gravitational radiation beam when that EM wave is passed through a static magnetic field, see [Bessonov]). Also of interest on this same subject is the inter-conversion between gravitons and electromagnetic photons that occurs naturally out in the cosmos, and the possible conversion of EM waves to gravitons through a beat frequency process (resulting from the constructive interference produced from interlacing an EM wave with a stationary EM field), see [Fargion]. But conversely gravitational radiation will induce photons in the micro-cavities of a dielectric, especially in a non-homogenous sphere. Moreover, that the photons will be created by differences in the refraction index throughout the glass or crystal it would agree with how Betty Luca described seeing them in the UFO, for she describes them as having glowing centers inside them and what looked like long lines deep inside the lower spheres (Betty Luca *ASBT-II* diags 41-3).

Note 10—This is analogous to the *Membrane Paradigm* for black holes and quasars as explained in [Price-Thorne], which was derived from the [Blandford-Znajek] treatment of an idea proposed in the [Ruderman-Sutherland] study of black hole dynamics.

Note 11—Towards the end of this little known paper by the Russian physicist [Veselago 1971] (p11), but also in [Veselago 1967b] (p680), there is an interesting observation that when waves are propagated through a material with a negative refractive index then instead of those radiated waves having light *pressure*, they will have a light *attraction*. That conjecture relates of course to electromagnetic waves; for gravitational waves it would be the other way round, for the inference would be that instead of the gravitational wave carrying a positive *attractive force*, it would carry a negative *repulsive force*. This backs up Unruh's hypothesis that a black hole can propagate gravitational waves having the properties of negative energy density [Unruh 1974].

Backward waves were first suggested by H. Lamb in 1904 (see H. Lamb *On Group-velocity* Proc. London Math. Soc. 1 (1904) 473-479).

Note 12—This paper explores the potentially interesting phenomenon of negative group velocity. Mostly it suggests a tunneling effect that can occur just inside a metamaterial's surface where a resonance can occur to trap the incoming waves. See [Pendry 1996], and [Pendry 2004] [p11-12].
Note 13—For experiments showing existence of GHz frequency anomalous behavior of crystal metamaterials leading to plasmonic transmission of energy see the paper by [Wu].

Incidental evidence of this technology might also be found in some energy fields which Betty Luca described as appearing to look like windows of the aaUFO, from when she was inside the craft being taken up to the control center by one of the Greys; when Betty described some small window-shaped oval areas in the internal walls as being overlaid by energy ripples which made their oval light sources blend into the wall's surface with no sharply-defined edges (*Andreasson Affair* p39 fig.12). This could be a NRI-type of effect where the material's permittivity-permeability can be configured to different physical permutations by a variable oscillating field, to make them either opaque or transparent. The one major conclusion I came to about metamaterial research carried out during the late 1990s was that it still had a long way to go, and when reading through [Pendry 1998], and through [Sievenpiper-Sickmiller-Yablonovitch] (also see note 20 below), one gets the feeling that this field has so much more potential.

An extension to the premise that negative refraction materials, by reconstituting their permittivity and permeability, can be made to change from opaque or transparent at light frequencies, might have been involved in the case of a change to a material's refractive index which occurred inside the UFO that Bob Lazar worked on at Area 51. For when it was discovered how that craft's auxiliary equipment could be switched on several of the walls changed from being dull-skinned and opaque to brightly lit and covered with alien glyphic writing (see *Alien Contact* by Timothy Good (1993)). The author has been informed recently (too recent to gather any further information about it, unfortunately) that there is a special glass known as 'electro-chrome glass' which has been produced that will electrically switch from opaque to transparent. Just to clarify something about NRI materials, NRI metamaterials achieve RF (radio frequency) transparency not optical transparency; and although NRI research never quite achieved *optical transparency* this particular goal will be reached one day, NRI materials, and the associated development of metallodielectric photonic crystals, do theoretically show much promise in the area of optical transmission. See Shvets' University of Texas Physics Dept. webpage on metamaterials at http://www.ph.utexas.edu/~shvetsgr/lens.html

Note 14—Conductivity is attenuated at low RF frequencies but in the microwave range the plasmonic state switches on. See R. Marques et al *Role of Bianisotropy in Negative Permeability and Left-handed Metamaterials* Phys. Rev. B65 (2002) 144440.

Note 15—See [Smith 1999] on the dependency of the final plasmonic frequency on inductance. Also see [Pendry 1996].

Note 16—Pendry notes that electrons in a magnetic field through their extra *momentum contribution* (in classical mechanics) take on more mass than their positive counterparts, and after calculating those changes in the effective mass of both the electron and the proton as being equivalent (for aluminum atoms) to m_{eff} = 2.4808 x 10^{-26} kg, Pendry then calculated that this figure equates to the quantities 2.7233 x 10^4 m_e (of electrons) and only 14.83 m_p (of protons); so in other words their effective mass rations are in the proportion of 2.7233 x 10^4: 14.83. Hence, the electron's mass becomes nearly four orders of magnitude *more massive* than the proton's mass! For the mathematics of this see [Pendry 1998] (p4). Also note that some researchers have seen this method of mass enhancement as a link to the Higgs boson mechanism for the way mass is brought into being.

Note 17—Normally, ultra thin wire cannot be touched with fingers as the wire (when under 10 microns) is difficult to work with because it's coils readily stick together from oxidation, and so that's why I'd suggest that its installation inside quartz sleeving would make it much more workable. Some background information about ultra thin wires can be found at Toshiba's materials webpage http://www.toshiba.co.jp/tmat/eng/material/e_sonota_2.htm (It is known that synthetic silica glass can be manufactured by spraying distilled silicontetrachloride into a detonating gas fire—see W. Anwand et al *A Study of Positron Properties in Quartz Crystals and Synthetic Silica Glass* Acta Physica Polonica A99 (2000) 321-328).

Note 18—Apparently monatomic gold has high-spin states, and with closed electron shells can be superconductive at room temperature.

Note 19—This was a study carried out on a 3-D carbon-diamond-type lattice [Sievenpiper-Sickmiller-Yablonovitch]. A 3-D mesh waveguide is obviously something 'different' but it would suggest that a homogeneous frequency would be obtained within it at a much lower frequency than its lattice spacing would otherwise represent.

Note 20—When an RF oscillating energy is directed through a rectangular horn that energy will be shaped by the dimensions of that horn, this is how microwaves are shaped and propagated. But waves of even smaller wavelength (even as small as infra-red wavelengths) are transmitted over flat-edged wave-guides *that merely have grooves cut into their surface*, mostly cut in a saw-tooth shape. So the principle is sound, especially because of the fact that those circulating particles will be imbued with negative radial momentum inside this ergosphere—so that they will hug closely around the surface of the sphere.

Note 21—The Ginzburg External Beam effect from V.L. Ginzburg in Compt. Rend. Acad. Sci. USSR 56 (1947) 583. A brief note on the effect can be found in a 1958 lecture by I. E. Tamm, see [Ginzburg]. Also see [Bekenstein-Schiffer].

Grigoryan's theory is that synchrotron and Cherenkov radiation is given off when a particle rotates around a small sphere when some of that particle's electromagnetic field penetrates into that sphere [Grigoryan 2005a]. And also see [Grigoryan 2005b].

There will also be a correlation here to a dielectric absorption effect where the microwave oscillations at the facets will cause a net polarization around the surface of the lower spheres [Goodge]. Hence they will attract to them streams of electrons (from outside the craft). Interestingly, those particles will be bremsstrahlung braked at the edges of those facets and so there would appear to be an emission particle collision mechanism to consider also.

Note 22—The Western websites for the Russian Dalnegorsk (hill 611) UFO are: http://www.stealthskater.com/Documents/Dalnegorsk_01.doc (in English) http://home.pacbell.net/joerit/docs2/crash/soviet.htm (in English). The Russian websites are: http://vlad.kp.ru/2006/02/22/doc103572/ (in Russian) http://www.gorizonto.ru/mag43.html (in Russian, and I'd recommend the Google foreign-language translator which takes only a few minutes to translate from Russian into English.

Note 23—When I checked this fact with Dvuzhilny he told me that these 'iron' balls were in actual fact metal balls comprised of different combinations of rare-earth metals mixed with (but not *alloyed* to) nickel, magnesium, and only a small amount of iron; and some even contained small pockets of fluorine and argon gases! Where the Western websites say these iron balls were so hard they could not be cut and analysed, this is complete nonsense! So it would seem to me that the Western websites which carry details of the Dalnegorsk UFO landing are wholly misrepresentative in their translations. That some of these balls are comprised of elements from opposite ends of the periodic table (ie the most electronegative and the most electropositive) – and that they are made up mostly of magnetostrictive and ferroelectric compounds says volumes about how they would work inside a thin-wire network.

Note 24—Dvuzhilny was in fact one of the scientific advisors for the CCCP (Russian) military on UFO detection up until 1991, which is why he managed to find out so much about this UFO's artifacts – and discover an anomalous 'triangle' around the Dalnegorsk area where both UFOs and civilian passenger aircraft cannot be detected on conventional radar sweeps. That UFOs seem to use this triangle to enter into and leave from our atmosphere it is not surprising that Dvuzhilny also discovered (and this he included in his initial report), that during his interviews with some of the older residents of Dalnegorsk he was informed that this was not the first time UFOs had been seen in that local area. And, it would appear that this particular area has had so much UFO activity, both before 1986 and since, that the locals have established exactly when the area has peaks of activity each year (UFO flights peak here during the months of January, February, and November), hence, these witnesses were not completely surprised by seeing a UFO fly—and land—so close to their town!

Note 25—The reader will be reminded here that in the USAF report of the UFO scanned by the crew of a RB-47H reconnaissance aircraft as it flew over Texas in the 1950s, it also mentions that the UFO that they chased and tracked had an RF signature of 3000 megacycles, which of course, in present day parlance is 3 GHz [McDonald].

The microwave frequency used most commonly in microwave ovens is 2.45 GHz. And at 3 GHz this is the frequency also known for its ionization factor upon air-borne gases.

Note 26—This hypothesis might at first seem to fit in with the Roswell UFO, for if after it had been hit by lightning, whatever damage was sustained to the Roswell UFO's hull could have stopped any one of its spheres from revolving, and this would have caused one or more of them to overheat and to ignite. But actually, this doesn't quite tally with the evidence which came forward from the people of Roswell, for we know from eyewitness accounts that just after the Roswell craft was hit by lightning it careered to the ground, skimmed along the ground, and then rose back into the air again. When this skimmed area of ground was searched later an oval the size of a football pitch was found where the ground had been blackened from intense heat-radiation and where some of its sand had crystallized into solid glassy streaks: And this would be consistent with that ground being flashed by a very large expanse of heated plasma, probably much larger than the hull of that stricken UFO, meaning that the whole of the Roswell UFO's exterior energy field of gases and charged particles would have been enveloped with a high-temperature plasma that was being trapped inside its electromagnetic envelope, and so it would have been the UFO's heated exterior energy field which would have singed the ground and crystallized the sand at Roswell, not just one of its spheres, and nor any one section of its hull.

In fact, in September 1947, when Master Sgt. Lewis B. Rickett joined up with Dr. Lincoln La Paz (from Univ. New Mexico, Albuquerque) to scan the whole area of ground where the Roswell UFO came down, they found more debris from that UFO that was still laying on the ground, and it was they who discovered the heat-crystallized sand where the UFO initially, and briefly, skimmed the ground [Schmitt-Carey].

Note 27—Yes, a piece of news that circulated around the world, but which wasn't allowed to circulate through the news media of America, was that when Chernobyl underwent its disaster on April 26, 1986, several witnesses reported seeing a UFO presiding over that nuclear power station taking into its surrounding energy field some of the dangerous emissions of atomic radiation being ejected from its stricken power-plant. Witnesses and engineers dealing with the Chernobyl disaster watched one of the UFO craft fix two beams of crimson energy on the affected power unit, and saw that the UFO become surrounded by an 8-meter diameter ball of swirling plasma, which it later took away with it. A similar event took place on September 16, 1989, when the same nuclear power station again developed a problem, and for several hours yet another UFO was witnessed hovering over that power station, absorbing into its surrounding electromagnetic field an amber-like energy (sourced at Pravda website for November 17, 2006, from the article *"UFO Prevents Blast at Chernobyl Nuclear Plant,"* and subsequently reported at the Rense.com website, and Whitley Strieber's Unknown Country website). See the Russian website at: http://ufoinv.pochta.ru/ua/chernobl.htm

Note 28—Straightaway here are clues to what the aliens are using these little iron-lanthanoid balls for, strontium (Sr) and barium (Ba) are at the bottom end of the electronegative scale, and both these gases of fluorine (F) and argon (Ar) are at the highest end of the electronegative scale! And as I mention in my chapter 4's note 8, strontium-doped lead-zirconate titanate (PSZT) compounds and barium strontium titanate (BST) compounds have been found to yield the most pronounced piezoelectric and ferroelectric spontaneous reversal reactions.

Strangely, you can find a different list of elements for the Dalnegorsk UFO on almost every Western website that has reported this UFO! These American websites give numerous chemical element lists for the Dalnegorsk UFO which mostly include silicon, aluminum, iron, zinc, titanium, magnesium, silver, and also minor traces of copper, lanthanum, praseodymium, calcium, sodium, vanadium, cerium, chrome, cobalt, nickel and molybdenum. Now, even though these form a good list and some of these elements are magnetostrictive (iron, nickel, cobalt); some are ferroelectric (zinc, titanium, lanthanum, cerium, praseodymium); and some are group 14 carbon elements (silicon, titanium, lead); and while the iron, nickel and silicon have very high magnetic permeabilities... this list of elements bears only a passing resemblance to the actual elements discovered by the analysts in the Russian universities mentioned in the reports that Dvuzhilny got back from those laboratories!

Note 29—The inclusion of these iron-ferroelectric pellets echoes the ceramic high-permittivity pellets inserted into wire-mesh photonic crystal materials as was proposed in the [Sievenpiper-Sickmiller-Yablonovitch] study (which I have mentioned in note 19 above). Iron, and silicon will have a high relative permeability (about 5 times greater than ferrites) so their inclusion would greatly benefit the alien's metamaterial. Indeed, the whole point about metamaterials (as artificial metals) is that they develop high-intensity magnetic fields, which are then used to amplify the electrical charges (which is quite the opposite to the way we on earth generate our electromagnetic energies).

Note 30—The full list of elements found in the UFO's debris, according to Dvuzhilny, included even more metals (such as zirconium, neodymium, germanium and tungsten). Interestingly both Zr and Nd are also favored elements used in ferroelectric emission researches—as alluded to in this study's ch.4 note 8.

Strangely, the academic reports received by Valeri Dvuzhilny on the mesh's wire indicated that this gold-silver-nickel alloy in spectroscopy tests, when heated beyond 1500 C degrees in vacuum, shifted from having gold-silver-nickel frequencies to the spectral frequencies of α-titanium-rhenium-molybdenum (but while this is certainly interesting, I would suggest this high temperature change-point would be outside the normal operational temperatures of the inside of that UFO). Similarly, the fact that in conductivity tests this alloy mesh showed an electrical resistance at room temperature, but a high-conductivity when warmed with a small amount of heat (indicating the alloy had a low Curie threshold temperature as well as a high temperature one)—is not the most significant factor relative to the plasmonic mechanism as mentioned in the main text.

Note 31—As a result of tests conducted by scientists from the Ukraine University of Geochemistry and Physics of Minerals it has been estimated that the UFO's site was influenced by a source of power which at times generated 100 million volts at currents of 100,000 Amperes (although, this magnitude of power probably wouldn't have come from these spheres alone. As I will explain throughout this study for the Andreasson UFO, the lower spheres of these UFOs supply electrical power which is then routed into that UFO's black vortex, and it will be the black vortex which will raise the UFO's power output to such grandiose levels—this power rating equates to a collossal 10,000 Gigawatts and compares to a nuclear power station's 1 Gigawatt power output! So it could just be that the Dalnegorsk UFO also had a shearing energy field which broke down its atomic and subatomic mass-particles into raw energy, just like the Andreasson UFO and the Lazar 'sfour' UFO have been found to have).

Should the reader be surprised to learn that research into ferroelectrics enjoyed an Indian summer very shortly after this 1986 UFO sighting? For significant advances were made in this field afterwards leading to numerous papers on this subject being published in 1988. This is to say, that before these 1988 publications artificial dielectric researches were in the doldrums, they were a known field of research but that field wasn't going anywhere because those dielectrics weren't producing anywhere near high-enough yields. But then after the late 1980s and early 1990s impressive results suddenly started coming out of laboratories from physicists, and the whole science of metamaterials rose into a higher level of existence—or is this merely a coincidence?

Note 32—Carbon is one of the best molecules for joining with atoms of other compounds because of its interconnecting bonds that organize (catenate) those other compounds, for instance into a diamond-like lattice (check carbon at Wikipedia website for carbon glass). Interestingly, carbon materials mostly have a high thermal conductivity but above that threshold temperature carbon will burn very rapidly and very fiercely. This is because once carbon has started to burn it rapidly oxidizes, so the whole structure starts to burn from the inside-out. As a result of being burned in such a way a carbon structure becomes completely perforated, full of holes, and then its structure simply breaks apart [Davis, M 2001].

Note 33—Highly pressurized carbon dioxide glass, also known as a-carbonia, doesn't yet exist in a stable state on this planet, although it may exist deep within a planet not too far away in some high-pressure stratas. But work is currently proceeding to find ways of producing this special carbon glass, which is hoped will have a hardness similar to diamond. With its greater stiffness it will presumably mean that on a cellular level it would be more ideally suited to initiate emissions from any ferroelectric elements embedded into them, because its stiffer structure would provide a more robust polarity switching of any electric field permeating through such a dielectric. However, there is more about the ET's carbon-glass than meets the eye (as will be shown from details that will be made available at a later date).

Note 34—I shall be saying more about this Dalnegorsk UFO in chapter 11 of this book and in a later publication (simply because there is so much scientific data to sift through and correlate). It was also interesting that some of the wood found at the Dalnegorsk UFO site showed signs of high temperature 'fusing' (which some reports worded in such a way as to suggest that the wood showed signs of being melted or welded). This might at first sound absurd, but it would be relative to the process of silicon production, of when wood (carbon) is used in the manufacturing process to transform silica into silicon. The carbon that had melted onto the wire mesh was actually diagnosed as being very similar to a carbon derived from mosses, lichens and wood! Considering all the trace elements found at the Dalnegorsk site, while these do point to a carbon-glass sphere disintegrating at that site, they may also suggest that a repair was performed by the extraterrestrials to that stricken craft, and that they manufactured a quantity of lead and silicon (and the latter product might explain the electric arc flashes seen by witnesses while the UFO was grounded).

Note 35—Some of the other samples of UFO alloys in Valeri Dvuzhilny's possession, which have an almost total (99%) composition of lanthanoid (rare earth) elements (with only traces of Fe, Co, Si, Ti, Cu, W and Mn), were discovered to be extremely prone to ignition when tested. So, while the extraterrestrials are using such ceramics for electron emissions those materials would be highly susceptible to friction and abrasion (and in our normal oxygenated atmosphere they would readily burn).

Notes for Chapter 6

Note 1—An informative paper on the dynamics of a Taylor-vortex flow is that of [Marcus]. The reader should note that while the target might be vortical flows what he or she will find is studies of cylindrical flows. Vortexes don't like mathematicians and vice versa.

For the counter-rotating vortical fields, technically known as the hypocycloidal field, see [Binder 2002a] and [Binder 2002b]. I did find Binder's studies more relevant to the aaUFO's vortical flows than Berry's own papers.

Note 2—The paper by [Stone-Balbus] is especially useful to this study because it gives a good overview of the kinetic energy of a rotating system of fluid typically found inside an ergosphere around a black hole. It also addresses the issue of eddy currents (from the viewpoint of them being epicyclic oscillations). Also that such a rotating field of energy will have both inward and outward convection, so that there will occur an oscillatory interchange between the two.

Note 3—Interestingly, where an electrified body is encased by a wall with a concave curve the mid-point of that curve will always possess a zero voltage. Inside the toroid there will be a complex series of factors (explained through later sections) which will confirm a Lorentzian polarization of this fluid, and which will also point to the mid-section of the fluid being zero-charged, and therefore it can be regarded for the purpose of this study as a current sink.

Note 4—I have bunched all the repetition rates and all the different frequencies involved around this engine into section 7.6 and its notes. These microwave beams won't make that much sense until the reader sees the whole process of what they are doing, so this is why I will put more details about them in the next three chapters.

Note 5—These MHz frequencies may be useful though, John Kanzius, of Pennsylvania, in 2007 found that by oscillating salt-water at 13.56 MHz the sodium and oxygen atoms would separate from the hydrogen in the water and to produce a heating effect in that water.

Such MHz frequencies would ensure that those electromagnetic waves travel on the outside of the thick wires between the wire's outer surfaces and the glass or quartz material surrounding them, so possibly they might take advantage of some sort of metal-dielectric interface phenomenon, which would be conducive to the generation of a highly energized exciton field around the wires which would then emit photon luminescence of various colors through the wheels. However, there was no mention of these wheels ever glowing with any light or any colored emissions, rather, they were described as being particularly dull-looking crystal-like glass.

Note 6—Research into hollow one-millimeter metal wires for plasmon propagation has taken off very recently (since 2004-2005). So far the bandwidth covered ranges from 50 GHz to 150 GHz [Deibel]. These studies deal with single wire configurations, but multiple wires joined together would increase the surface area to lower the attenuation factor even more.

Note 7—In her booklets Betty describes them as being diamond shaped ends *ASBT-II* p27-8, but as a result of our correspondence Betty further described these crystals as being 'box' shaped, or in her words, *"Yes, it could be wires, tubes, or rods. There were some kind of irregular box type things at the outer ends. It was difficult to see. They seemed to be glass-like."*

Note 8—Depending upon what material is used in these box-shaped ends such energy transmissions by surface acoustic waves (SAWs) could be initiated up to 20 GHz (See report on SAWs by Seidel, Hesjedal, Perepelittchenko and Koch).

Note 9—The idea here would be for the acoustic oscillations to generate acoustic shock waves that would transport phonons (as phonon wind) through the crystal box-cylinder, whereby on the outer surface of that crystal cylinder where the phonons and electron-hole excitons would form as a plasma they would then be coupled into the acoustic oscillations (as carrier pulses) and be transported onward from the other side of that material [Shilton].

Note 10—See [Hopfield 1958], for example the Wannier coupling of weak-binding excitons (p1561). See also the study by [Keldysh] on transportation of excitons.

Note 11—I was rather surprised when following through the paper trail on gravitational wave detection to find that no gravitational waves have yet been found. The hundreds of different theories on this so-far elusive goal are grand and as varied as they are very profuse. Zeldovich follows on from Gertsenshtein and gives the explanation of electromagnetic waves being converted to gravitational ones because of the additional stress tensor which occurs when the EM wave interacts with the static magnetic field, even though he fails to mention the final configuration for how those gravitational waves, after being generated, might actually DO WORK. This photon-to-graviton conversion theory was later taken up and further developed in an excellent paper by [Fargion] though.

Note 12—Well, in gravitational theory it had been postulated that while gravitational radiation is a spin-2 field and electromagnetic waves are spin-1 [Kaminsky] [Chiao 2003] [Kiefer-Weber] that while a direct conversion between them cannot happen, there is theoretically a way in which the conversion from one to the other can be made *through a medium* such as a quantum Hall fluid. And it is thought that such a superfluid together with an applied DC magnetic field (which is strong enough to add sufficient angular momentum to the spin-1 waves) *might* trigger the conversion to gravitational waves. With respect to the aaUFO model the central vortex could been seen as a quantum Hall superfluid, so many of the ingredients exist inside this UFO.

Indeed, the main ingredients for this process have already been noted in the main text but it has for a long time been conjectured (since Weber in the 1960s) that *quartz crystal* (in spherical form) might also facilitate this conversion. This essentially is a postulated process whereby quartz crystal is thought to be one of the mediums able to couple electromagnetic waves to a static magnetic field in order that they become circularly polarized, because the quartz lattice will also produce a birefringence effect when it has a static magnetic field applied to it, so by this theory the two *should* marry up and produce cylindrical gravitational waves. Unfortunately there has been little success in actually producing, or even detecting these gravitational waves. So perhaps it should be asked, is there a missing ingredient to the existing theories on gravitational forces? Certainly, from Raymond Chiao's experiments it is difficult to see why there can be no detection of gravitational waves, unless his formula omits one or two components, such as angular momentum and a method by which the gravitational force once captured or converted can remain so (such as by an anisotropic stress). He might also do better by working from a much stronger provider of gravitational waves such as comes through a rotating field, such, for example, as exists within this aaUFO model of gravitational force generation. Also see [Chiao 2002] to see how this theory has been formulated.

Moreover, contrary to the above spin-2 field speculative theory there is a lot more going on inside this miniature ergosphere of the aaUFO with regard to available magnetic fields; for the magnetic fields in this part of the craft are not static ones per se. They are sequentially pulsed (although by these pulses generating back-electromotive forces they will end up alternating). But there is angular momentum in abundance and of course actual gravitational forces too. So there is a much greater potential for these rotating fields to absorb gravitational forces, and there is also more potential to convert electromagnetic waves to gravitational ones, and possibly the other way around.

Note 13—See [Bessonov] for a rundown on Graser beam technology.

Note 14—See [Chu] and [Ronald]. Another factor which can assist the intensifying of the magnetic field aligned parallel to the axis of these microwave tubes will be to make the hollow tubes out of a high permeability material in such a way as to concentrate the magnetic flux lines close around the tubes.

Note 15—I think it should be said that trying to determine the precise format and frequency for these waves is outside the remit of this present study, because in later chapters it will be mentioned that there may be multiple tasks for these beams to perform, which will have to be taken into consideration to ascertain the final working frequency. That microwaves are propagated through hollow waveguides means that the frequency is roughly proportional to the bore size, but because there are different modes of wave propagation it means that the frequency is variably proportioned; for instance, that these microwaves will be flowing through these tubes as helical waves brings with them a quite different set of parameters. I have seen studies where a dielectric-lined tube of 10 mm bore diameter will transmit a helical field at 11 GHz (which usually has a free-space wavelength 27 mm). But then the same bore sized tube used in a gyrotron in an over-moded configuration will work as a waveguide for 140 GHz microwaves—so it depends upon how the tubes are configured.

In electron cyclotron heating (ECH) of water gyrotrons are used in a range of frequencies between 100 GHz to 170 GHz [Chu], although I have seen studies which use ECH at 2.45 GHz [Yamada]. Water heating by other microwave methods (lower hybrid) are used in the 4 GHz to 7 GHz range. So possibly the hollow tubes are to transfer electron energy into what must be a specific microwave frequency bandwidth where *one of its tasks* will be to heat the water.

Note 16—See [Granatstein] for the graser configuration. But also see [Bourgeois] from the point of view of the way this model uses a cavity to shape these circular waves (as this might be useful in the aaUFO where the electromagnetic radiation passes from the arm axles into the circular cavity of each telemeter wheel).

Note 17—Gyrotron cavities are usually much larger than conventional microwave tubes for that frequency (because they have to fit inside them the epicyclic rotations which generate the final microwave beam within their center). For this reason such configurations can handle greater levels of power than ordinary microwave tubes.

Otherwise gyrotrons operating in the TE mode (not whispering-gallery mode) can develop an intense beam that when it passes through a powerful magnetic field (9 Tesla) oriented parallel to the beam axis, which would be the case in the aaUFO model, this gyrotron mode makes the electrons gyrate in a small orbit along a helical trajectory. And interestingly, such gyrotrons use a fused-silica window for the beam to pass through [Vieregg] (p22).

Note 18—It would have been extremely helpful to know that these terminating crystals were colored, for if they were bluish they would have been sapphire, or if pinkish they would have been doped with chromium ions (as in ruby maser crystals), but Betty did not notice them having any color about them, just that they looked like clear crystal.

On the glass-crystal material used in the wheel, it may well be crystal used for its birefringence characteristics (wave rotation in crystal is inversely proportional to the square of the wavelength, and proportional to the thickness of the crystal) [Kaminsky]. Electromagnetic waves through quartz rotate 21° 44' (per mm) (W.A. Wooster (1949) *Crystal Physics* pp156).

Note 19—See R. Penrose (2004) *The Road to Reality* pp500-501. Note also the two-slit experiment, where an electron can arrive at this famous experiment as a particle, pass through the slit as a wave, and then arrive at the screen as a particle again (op. cit. pp504-9). How can this be? Well if you consider that in quantum electrodynamics electrical charge is defined as a fixed rate of emission of photons from a charged particle [Ostoma-Trushyk]: And that an electron at rest is comprised of 918 photon pairs [York]: And so, if this string of 918 photon pairs (in each electron) is considered analogous to the beads of a necklace then it can be envisaged that—the necklace can track around itself and coil up into a ball (to be an electron particle)—or, this necklace can spread out and be an undulating wave.

And consider Phillipp Lenard's discovery of electrons (in cathode rays) that those electron charges would pass straight through solid matter [Lenard], suggesting that the atomic space occupied by even a massive metal such as platinum is as empty as space is of stars beyond earth (F. Close (2004) *Particle Physics* p28).

Note 20—Half way through this study I did wonder whether the toroid fluid could be either a gas or a liquid. I did toy with the idea that deuterium might be used. Deuterium is hydrogen water without oxygen and can be produced by electrolysis from normal water.

Why this idea was so appealing was because there were many reports from people who have seen UFOs hover over lakes or over the sea drawing huge quantities of water into them and then dumping back a similar quantity of water. And from this my initial thought was that the UFO's were taking in water, processing it for deuterium, and then dumping the residue that they didn't need.

Deuterium would become an electric fluid because when pressurized this gas breaks down into atoms and converts to a conductive fluid—but a fully conductive fluid would not be useful to this craft—it would not be what was wanted

But as the reader will find out toward the end of this chapter (and through chapters 8 and 9), the evidence points to the fluid would necessitate being much heavier, so a liquid like water (not seawater as it is too lossy) would be perfect.

Note 21—The Greys landed by the side of a lake during one of Betty Luca's abduction experiences, they then ran a bunch of different hoses between the UFO and the lake and pumped back and forth water from the lake. The hoses while doing this became charged up electrically, and in an 'apparent mistake' a very high voltage discharge occurred which blew the whole of the UFO's electrodynamic circuits out and disabled it (the flash-over also temporarily blinded Betty who was standing outside the craft (*Watchers* pp53-85). Another ET contactee Herbert Shirmer was told by the ETs whom abducted him that UFO ships take water from our reservoirs, rivers and lakes in order to draw power from it *(Beyond Earth: Man's Contact With UFO's* by Ralph and Judy Blum (1974) p117).

Note 22—Water is rather unique in that chemically it possesses a structure that can be both an acid or a base, in that it can either loose a proton or gain one. By a phenomenon known as autoprotolysis the water molecules can separate into positive hydronium (hydroxonium) and negative hydroxyl ions. The hydroxonium component of water moves through the hydrogen bonded structure of water as a series of excess proton transfer diffusions that tunnel from one water molecule to the next. These excess protons can have a high mobility (high diffusion rate) in water. Apparently the diffusing structure within water begins as a ($H_9O_4^{plus}$) cluster but then converts to dihydronium ($H_5O_2^{plus}$) and then by proton transfer settles at (H_3O^{plus}).

In seawater there is slightly more conductivity than in fresh water from the fact that its hydrogen bonds only weakly hold the water molecules together. But if it is a 'lossy' fluid then it won't hold electric charge so well and it will be more susceptible to heating – which I am presuming is not what is required in these UFOs. Dr. Rustom Roy (of Penn State University) a noted authority on water structure, when he was testing John Kanzius' theory in 2007, found that the negative hydrogen could be separated away from the positive oxygen and sodium molecules of salt water because the 13.56 MHz frequency oscillation used by Kanzius was a harmonic frequency of the sodium ion, and at this frequency the sodium and oxygen atoms literally shook themselves away from the hydrogen. But I don't think the ETs would be doing this (precisely for the reason mentioned in note 29 below). So, I would still suggest that the actual process and frequency used in the telemeter beams of this UFO would be that which would separate the positive hydroxonium ions (H_3O^+) from the negative hydroxyl ions (OH^-) ions.

Note 23—See a study by Russian physicist [Maksimov] using a laser pulse (of 6 nanoseconds duration) to heat the seawater.

Note 24—And so one of the things this will do is create a temperature-difference in the water at each of the four locations where the beams pierce into that fluid and this will induce movement by convection (buoyancy effect) into the toroid's fluid. And so there would possibly be provided a way to increase the flow speed of the toroid's fluid proportional to the pulse rate of the beams fired into the water. At microwave frequencies the hydrogen bonds of water are broken which result in the generation of heat. The optimum temperature for the toroid's water should be 25-30 degrees Celsius (77-86 F) (see section 9.4).

A similar flow will be induced into the fluid by the permittivity gradient process which occurs from a low temperature fluid to a high temperature one, where the conducting gradient produces a Coulomb force or dielectric force. [Scharrer].

Note 25—Well, it would be a 9.5 Hz repetition rate if all four were fired simultaneously, and this synchronization would be highly beneficial (as will be made clear in chapter 10). If not and the telemeter rods aligned in a staggered sequence then it would mean that the fluid would receive 38 pulses per second. These pulse duration figures may be compared with those of microwave beams used in gyrotrons which are used to heat water (through electron cyclotron resonance heating), with those 200 kw gyrotrons firing 1 millisecond bursts (of 140 GHz microwaves) into water to heat it [Granatstein]. In the same paper mention is made of gyroklystrons (a rotating beam version of the old klystron) which are used in the much lower 4.5 GHz range (but with a pulse duration of only 100 nanoseconds). In note 23 above it mentions that a laser pulse duration of just 6 nanoseconds will heat seawater.

Note 26—In electrochemistry, in the ion-exchange treatment of water, resins are normally added into the water in order to contribute additional cations or anions as required. Additional cation contributors such as manganese (atomic mass 54), potassium (39), magnesium (24), sodium (22), or lithium (6), for instance, might be used in water to enhance it electro-positively.

But also, to reiterate the point made earlier, that the positive oxygen molecules in water will have an almost 1000 times higher magnetic susceptibility than the negative hydrogen molecules. So because the magnetic field passing around and through this water will be so prominent its influence will work on keeping the positive and negative molecules of the water apart (through the Lorentz force). The negatively charged molecules (of hydroxyl) will also be attracted to the INSIDE of the top face of the toroid because of the flow of positively charged particles rushing over the top face of the toroid moving toward the vortex.

Note 27—The lab report stated that the amount of zinc mixed in with the magnesium was 3%, and it is quite normal for zinc to be added to magnesium (anywhere between 3%-8%) as a hardener. Magnesium alloys are usually die-cast into their finished shape, and can be very strong and are even used as engine blocks for formula 1 racing cars, so this alloy will have ample strength for its task in a UFO. But although the high-mass material in this sample was named as bismuth, in the report it was noted that it had an isotope composition more like lead, and so I have assumed that this would put it into the ballpark of recent researches into lead-magnesium-niobate (PMN) which has been found to be highly electrostrictive (also see notes 28 and 40 below). See Linda Moulton Howe *Glimpses of Other Realities* Vol. II p11-22; or the article *Fragments of Evidence* by Linda Moulton Howe NEXUS magazine Aug/Sep 1997 pp64; and also Linda's original report at http://earthfiles.com/

Note 28—This rating of dielectric constant (for 38 degrees C) is extremely high, considering that the best ferroelectrics only range between 6000—14,000 K. Such a compound interleaved with electronegative bismuth would provide the ultimate capacitor because there would be very little leakage of charges. Research on lead-magnesium-niobate (PMN) has flourished since the late 1990s, for instance see Zhao, J. et al *Electromechanical Properties of Relaxor Ferroelectric Lead Magnesium Niobate-lead Titanate Ceramics* Jpn. J. Appl. Phys. 34 (1995) 5658-5663; and Li, J.Y. and Rao N. *Dramatically Enhanced Effective Electrostriction in Ferroelectric Polymeric Composites* Appl. Phys. Lett. 81 (2002) 1860-1862.

There has been a lot of interesting research on bismuth since the 1880s, and even Kapitza and Shoenberg (of Cambridge, England) circa.1935 found that bismuth and magnesium both change their dimensions and resistance in powerful magnetic fields (by magnetostriction).

Just the same, it might also be worthwhile looking at the slightly different compound of manganese-bismuth which apparently has a better bunching effect on flux lines than cobalt, on this see Bobeck, A.H. & Scovil, H.E.D. *Magnetic Bubbles* "Scientific American" 224 (Jun 1971) 78-90.

Note 29—Because hydrogen can become a by-product of this polarization process within the UFO's water (as shown with salt-water by John Kanzius (see notes 5 and 22 above), then indeed, this could have been the case in the Roswell UFO, whereby, if an excessive amount of this gas was accumulated within its toroid shell then this might explain why the lightning striking upon its hull, at precisely the wrong place, caused such a devastating effect upon that UFO and caused its lower hull to explode. It may even explain the explosion in the air around the aaUFO during Betty Luca's lake incident (mentioned in note 21 above)!

Indeed, this may be an important factor which determines when the UFO's fluid must be replenished—when it has produced and accumulated too much hydrogen. Such a by-product of the polarization process would obviously be governed by a safety limit on exactly how much hydrogen can be separated from the water and allowed to accumulate under what conditions, beyond which safety limit the UFO would be required to land somewhere on earth, and to release the excess hydrogen from its toroid (as well as to pump out the old depleted water), and then get its toroid fluid replenished with fresh water. Obviously, if too much hydrogen gas is allowed to pressurize inside its toroid the UFO then becomes a very combustible flying object, and this (if the aaUFO has similarities in design with the Roswell UFO) may be *precisely* what happened to that greatly denied aircraft on that stormy night as it was doing its reconnaissance flight over Roswell's nuclear arsenal. In fact, Mac Brazel, the rancher who was out in that storm on the night of July 2nd 1947 did report that he heard an explosion in the distance just after a flash of lightning (see [Schmitt-Carey] report, and also the excellent *The Roswell Crash* video (50th Anniversary Special X-Factor edition) on these recollections).

Otherwise, it may have been that the Roswell UFO was dissociating its water into hydroxide (OH-) ions and hydrogen (H+) ions, as through a simple electrolysis process. If so, then a negatively charged lightning bolt striking that UFO's hull, 'grounding' against the positively-charged (H+) volume inside the toroid, would have instantly converted those hydrogen ions into hydrogen gas, while at the same time would have caused a rupture between the positive and negative polarized fluids, and so then the whole contents of the toroid would have suddenly become highly volatile, and any further lightning strikes would have caused those hydrogen gases to explode. Such an inherently dangerous process of polarization is worth mentioning though, not least as a deterrent. This is why I have suggested in this study that the water would be separated into hydroxyl (OH-) and hydroxonium (H3O+), on the grounds of it being somewhat safer basis to work from.

Either way, I would conjecture that the debris that was found at the first Roswell site, on the day following the craft's initial explosion that was heard by Mac Brazel, was probably only the debris from the lower hull of that UFO, from its outer casing having been blown-out by that explosion. Moments later the excess hydrogen would have poured out of its toroid shell and would have then spilled into the craft's surrounding magnetosphere, and would certainly have been aflame—and so the UFO would have probably looked like a huge ball of fire falling to the ground. If the initial blast went upward then the crew would have perished instantly and the top of the Roswell craft would have been ripped open, or completely severed from the lower section of the craft while it was still in the air. However, we know from the witness reconstructions that the UFO, or the upper deck section of it at least, rose back into the air and stayed in the air for a further three miles or so. Obviously, the upper hull section would have been where the craft's crew would have been located at the time of the initial explosion. This would also explain why there were two crash sites at Roswell, and that the second site of debris found on July 8, 1947 (where the four aliens were also found), was where the civilian witnesses described seeing only the top half of that UFO (of the control center and upper decks) partially wedged into the ground (as has already been alluded to in chapter one's note 2).

This upper-deck-and-lower-deck separation conjecture may be supported by a recent disclosure report made by an ex-aerospace employee who claims he was detailed to do some 'special work' under strict security procedures, which involved the reception of, sorting of, and categorizing of a stream of sealed (and some not-so-sealed) packages being routed through his 'special vault' from NASA, NORAD, MJ-12, MAJIC, and from what seems to have been Corso's original investigations back in the 1980's—with some packages bearing the signature of President Truman on them. Some of the packages that he had to deal with and categorize (of reports, military photographs of UFOs, and every conceivable AV format for video and film) were marked with the words 'Roswell' or 'Los Alamos,' and other packages had on them the words 'upper' and 'lower', as if to denote the upper and lower decks of that UFO craft. Either way, for the Roswell UFO the word *lower* would have denoted that UFO's electro-dynamic structure, and the *upper* designation would have related to the control section and medical rooms of that UFO.

Note 30—Just to clarify, to increase the polarization is to hyperpolarize, while to decrease it is to depolarize.

They say that to err is only human... But was this short-circuit incident (mentioned in note 21 above) an un-solicited 'accident,' don't forget whatever these Greys do when they are down here interacting with us is overseen by the Elders, the Elders effectively see through the eyes of the Greys (and of course whatever the Elders do and think is known to the entities even more highly evolved in the God pyramid that oversee them). So I would regard this incident as a lesson rather than a mistake (which in Betty's case was a painful lesson because the flash-over left permanent scars on her retinas). Otherwise, why would that UFO's whole crew of Greys (and at least two other crews of two other UFOs) go to all that trouble to show Betty and the other abductees what can go wrong in this replenishment process? If the process needed doing then they could easily have off-loaded the abductees and then come back to this, or any other lake, to perform this process under the cover of darkness without anyone knowing... This is why I say in chapter one, that this was the royal tour that Betty underwent and much of it was especially pre-arranged for her to record, and what Betty has been allowed to recall of it delineates quite a comprehensive instruction manual for UFO electro-dynamics. Certainly, and I hope the reader will agree, this way of back-engineering UFOs through abductee recollections is so much more enterprising than how the guys at Area 51, allegedly, would have been doing it!

Note 31—The complete series of books giving the account of Betty (Andreasson) Luca's encounters with Greys and extraterrestrial elders and of course UFO craft are: R.E. Fowler *The Andreasson Affair* (1979); R.E. Fowler *The Andreasson Affair—Phase Two* (1982); R.E. Fowler *The Watchers* (1990); R.E. Fowler *The Watchers II* (1995); R.E. Fowler *The Andreasson Legacy* (2000); Betty Luca has published a series of booklets called *Extraterrestrial Communications—A Step Beyond Tomorrow (ASBT)* (Bk I & II—1999).

Note 32—There are ferromagnetic-fluids of grained cobalt and magnetite; water-based magnetic fluids; and magnetic polymer gels; see Berkovsky, B.M. (1978) *Thermomechanics of Magnetic; Fluids* Berkovsky, B.M. et al (1993) *Magnetic Fluids*; Rosenweig, R.E. (1985) *Ferrohydrodynamics*.

Note 33—These ferro-gels can then be manipulated by an exterior (non-uniform) magnetic field to change their shape by up to 40% of their original dimensions [Zrinyi].

Note 34—Betty described this set-up as follows; *"Two arms extending out from the central part of the stem supported and held a rotating crystal wheel, that pressed and revolved against a large, clear, soft plastic like donut filled with gray fluid. This inner tube donut encircled all three [or four] machines."*

(From *ASBT-II* p27-28). And in further correspondence (in 2002) when I asked Betty more about the fluid in this inner ring-tube she said, *"Yes, there appeared to be a gray colored moving-liquid-like-gel, (at least it looked like a gel) in the tube that the wheels rotated upon."*

Note 35—See the flow actuation study by [Roy], and the electro-osmosis study by [Kim, M.J].

Note 36—I must admit I initially thought the movement of this inner ring-tube's fluid and its generation of a strong (orthogonal) magnetic field might be used to rotate the telemeter wheels, and therefore drive the sphere-set assembly which in turn would spin up the central power-drive vortex and get the whole UFO craft energized, but now I can clearly see that its the other way round. Because I did assume that there was some sort of magnetic Lorentz force link between the inner ring-tube's fluid with the telemeter wheels, that would act on the wheel's arrays of conducting tubes forcing them to turn the wheels, as in a typical electric motor mechanism, but it is now so obvious that the vortex will generate such a powerful rotating force all around it that the sphere-sets will simply be dragged around by that force (and faster and faster). So its obvious the ring's magnetic field is needed to BRAKE these wheels which rotate above it. This is of course perfect for it facilitates the use of just one single transmission line to control the craft's acceleration.

Note 37—It might also be reasonable to suppose that the helical force induced on the microwave beams passing through the telemeter tubes can also be controlled by the inner ring tube (by it controlling the orthogonal magnetic field which surrounds the tubes). Meaning, the microwave pulses could be given a variable intensification.

Note 38—Q-factor in an electrical circuit is the circuit's ability to sustain a resonant frequency with the least amount of energy input. So, the less power needed to sustain that oscillation the higher the Q. Soft plastic can be fabricated out of a polymer such as polybutadiene, there are a whole range of solid polymers which extend from brittle ones (polystyrene) to softer ones. See Interrante, L.V. *Silicon-based Ceramics from Polymer Precursors* Pure Appl. Chem. 74 (2002) 2111-2117.

Note 39—Throughout my studies I have never quite understood why Col. Corso waxed so lyrical about the Roswell craft's outer hull needing to be made of ultra-low resistance silver-copper metal—why was it deemed necessary by his analysts to have a supa-dupa silver-copper metal compound for that UFO's hull? Such a compound may be advantageous for the upper shell but it would be disadvantageous for the lower hull. Was this something that the American analysts back in the 1970s *presumed* was the best and most exotic material of their time, so they consequently assumed it would be the most appropriate metal for a whole UFO craft to be made of? I say this because *superconducting* metals either exclude magnetic flux lines or only allow them to penetrate at skin-depth—but the aaUFO specifically routes its magnetic field lines THROUGH its hull (so does Lazar's sfourUFO, the *Onion Drive* craft, and the Travis Walton UFO) and those magnetic field lines are used to carry gravitational force. A high-conductivity metal might be a suitable requirement for the aaUFO's base disc where it is required to concentrate the magnetic field lines around the hole through which the central vortex flows, but then the rotating flow of the vortex will do that anyway!

I have an aversion to anything termed *superconducting* or *cryogenic* in these craft because superconducting means temperature control, and I don't see a UFO craft working within a narrow range of temperatures when such craft fly in our atmosphere, in our sea, and in our local space. And so I have thus far avoided the need to include any such materials in this UFO's design (apart from gold wire of the metamaterials delineated in chapter 5 possibly being superconductive, and the craft's black vortex being a superfluid).

Note 40—Further to the note about Terfenol-D (in chapter 4: note 2) a material which exhibits 'giant magnetostriction' will, at a critical level of magnetization, undergo a 'burst effect' and suddenly change its magnetic orientation. This 'jump' in the material's surface magnetization will then react against the applied magnetic field and induce a kinetic movement into that material (see for instance Clark, A.E. et al *Magnetostriction "jumps" in twinned $Tb_{0.3}Dy_{0.7}Fe_{1.9}$* J. Appl. Phys. 63 (1988) 3910-3912). This is why I say that such materials would be ideal embedded inside the glass spheres to trigger a piezoelectric-type of strain in those spheres. The reader will notice how similar this kinetic effect is with the electrostrictive effect mentioned above.

I did read somewhere of a physicist, claiming to be from Edwards Air Force Base, who announced that this Roswell magnesium-bismuth compound will generate an electrogravitic effect—and an antigravity force, to boot... which of course is absolute hogwash!

Note 41—Ionization of the toroid's fluid will also change its refraction index. An optimum increase in permittivity will come from a low GHz microwave polarization (ie 6 GHz for water at 20 degrees Celsius). Also, through the upper range of microwave frequencies the molecules in the toroid's fluid may not rotate fast enough to remain in phase with the applied microwave field, meaning that the fluid's permittivity in those localized areas will drop and steady down to a lower factor of permittivity than by using a lower microwave range. Therefore, getting the most beneficial ionization frequency from the telemeter beams is crucial.

Notes for Chapter 7

Note 1—As space-time curvature is defined by the mass of a rotating black hole, this ergosphere will generate a curvature of space-time. For Lense-Thirring inertial-frame-dragging see [Ostoma-Trushyk].

Note 2—The azimuthal speed of a particle in a rotating flow (its angular velocity) can be faster than the speed of light (c) because, as in the aaUFO's vortical flow, this flow is helical. The speed of light is a linear value for the propagation velocity of electromagnetic waves and light waves as they would travel along a straight-line path [Abramowicz 1997]. Therefore there is a variable flavor to what many people assume is a non-variable.

Note 3—What also precludes a gravitational force to exist directly above the mouth of a black vortex is the fact that magnetic field lines will be excluded from the core of a quasi-black hole, so there will be no agent through which to transfer any gravitational or rotational forces, other than at the outer wall of that vortex where the magnetic field lines bunch up just outside the event horizon. This factor is also reflected in Einstein's observation that mass curves space-time.

The old Walt Disney movie *Black Hole* comes to mind here where the crew are frantically trying to stop their spaceship 'falling' into that big black hole. But it would never happen like that. If a spaceship happened upon a black hole out in space that ship would face nothing other than a shearing force speeding faster than the speed of sound, which would rip that spaceship (and its crew) into a zillion pieces before it got any nearer to that black hole. That spaceship's (and that crew's) subatomic particles would then go on to feed extra energy into that black hole, giving it more rotational speed, which it would accumulate until eventually that rotational speed would become greater than the speed of light, hence even light would not then escape from it.

Note 4—For a black hole system the consensus is that the bound photons at the photon orbit ring will be blue-shifted due to the rotational forces and curvature effects of the black hole, with some achieving energies as high as GeV [Williams 2003]. That a particle is bound means that it is gravitationally bound to that black hole (or as in the case of the aaUFO to its black vortex).

While the atoms and molecules which are broken apart at the vortex's photon orbit ring would initially be divided into subatomic particles, some of which would be ejected away from the photon ring by Penrose-Williams collisions, the subatomic components which pass irretrievably into the UFO's black vortex would presumably be further broken down as they pass over the event horizon and become directed toward the vortex's gravitational core. For following the example of the astrophysical black hole the whole point of the UFO's black vortex would be to further dismember those subatomic particles, so that the atomic binding energy (which holds atomic nuclei together) would be released and quickly captured by that black vortex core. See also [Di Matteo].

Note 5—There exists an interesting Schauberger hypothesis which formulates the breaking down of atomic energies into smaller components, through a process that Schauberger termed *oligodynamic* dissociation (C. Coats *The Energy Evolution* (2000) pp72-73; 103; 192).

Note 6—See for instance the groundbreaking paper by [Penrose 1969]. This hypothesis was made all the more understandable by the superb drawings Penrose rendered to illustrate his concepts... physicists please note what a benefit good drawings are when proposing complicated new ideas.

Note 7—Where Williams points out that the reversed particle ALSO takes with it an additional gravitomagnetic force from the black hole's pull on the original excited photon, which Williams attributes to the Lense-Thirring effect [Williams 2003] (pp16), I would see that particle taking with it a gravitational force and an amount of Lorentzian force. I would not use these terms *gravitomagnetic* and *gravitoelectromagnetic* mainly because they can, by some physicists, be regarded as trivial forces (just as the gravitational forces around earth are also seen as trivial forces). But gravitational forces in close proximity to a black hole or quasi-black hole (ie a black vortex) are extremely powerful forces.

The same energy amplification theme is carried further in the paper by [De Felice-Carlotto] where it is suggested that when these blue-shifted *gravitoinertial* particles do escape out of the quasi-black hole system, because of the inertial space-time frame-dragging imbued into their orbits (which can also be encompassed into the term *negative radial momentum*) they would escape from that system along a vortical trajectory. It should be noted that protons as well as electrons may be ejected with negative energy trajectories out of the system [Leiter-Kafatos].

The Williams-Penrose mechanism can be applied to all-mass-size black holes (even miniature ones), and Williams cites as a good example of this transfer of escaping negative radial momentum into the helical filament-tube of the M87 jet [Williams 2003].

Note 8—This is how the Penrose-Pair-Production mechanism produces electron-positron pairs out of the ZPF virtual-particle field of space.

Note 9—For an interesting comparison of energy levels produced by a rotating field Williams has calculated that photon energies can be boosted up to GeV levels, and electron energies up to 4 GeV [Williams 2003]. But also see [Punsly-Coroniti 1989] [Leiter-Kafatos] and [Williams 2002a].

This is slightly different from the negative energy hypothesis of [Unruh 1974] which will be the subject of further investigation through chapter 9.

This sudden reversing of a particle's orbital trajectory so that it would be imbued with extra energy featured strongly in the Repulsine turbines engineered by Schauberger in the 1940's. Its a shame Schauberger wasn't versed in twentieth-century astrophysics as he would have been able to explain his gravitational manipulation technology much more clearly and with a perfectly suited nomenclature.

Note 10—On the efficiency of black holes and the extraction of energy from them see [Carilli-Barthel]; [Penrose 1969]; [Li 2002]; and [Li 2000]. On the power of a hole's electric field acceleration force see [Bicak-Ledvinka] (In this intriguing paper there is also offered the viewpoint that gravitational perturbations (waves) are coupled to electromagnetic perturbations (waves).

Note 11—See [Hawking 1974] and [Hawking 1975] papers. Particle emissions from black holes is attended to in numerous studies, an interesting one of those studies is by [Traschen].

Note 12—See [Unruh 1995] on the sonic analogue of black holes. Also see [Berti 2005] and [Basak-Majumdar]. And see the comprehensive study of [Schutzhold-Unruh]. And see collaborative work in [Jacobson]. For recent interest in creating a small scale black hole in the laboratory see [Dimopoulis-

Landsberg]. For recent successes in creating plasma jets in the laboratory see [Lebedev 2002a], [Lebedev 2002b], [Lebedev 2005].

Of course, the present study into quasi-black hole dynamics might accelerate the creation of the highly efficient mini black hole—although the physicists will have to acknowledge the extraterrestrials and recognize their brushstrokes through this present study.

Note 13—Leonhardt's delineation of a miniature black hole, and how to calculate the critical radius of a vortex below which light waves cannot escape, can be found in [Leonhardt-Piwnicki 2000a]. But also see [Magueijo] on singularity theory.

Visser has studied this problem very rigorously and is keeping a tight reign on just what can and can't be regarded as a black hole [Visser 2000].

Note 14—What the ETs call the 'cyclonetic trowel' (*Watchers* p85) is most likely the central vortex because by using the word 'trowel' they are referring to something in the singular, and so it must be something to do with the central vortex, whereby a trowel is a mechanism which scoops and spreads. Etymologically, trowel comes from the word trua (from Latin) which means *stirring spoon*, while trulla (in Latin) means *scoop*, which broadly speaking is essentially what the frame-dragging action is of a vortex for it scoops-up all particles of matter and drags them along with its rotational force (the extraterrestrials that communicated with Betty Luca had a penchant for using earth's ancient languages in many of the terms they used, using in particular Latin and Greek words, as I found out when I endeavored to translate some of their warnings about the energy field building around our sun).

So the final resting place for *cyclonetic trowel* has to be with the craft's central vortex (within its frame-dragging ergosphere) particularly as the full expression used by the Greys and which they impressed into Betty Luca's mind, was that they were *"purging and lining the cyclonetic trowel"* (*Watchers* p76), and the only rotating mechanism that needed to be lined in the aaUFO was its central vortex.

Note 15—In conjunction with the control of the inner ring-tube both the *tick-over* and the *power-down* modes of operation may be initiated through the adjustable permeability covering caps around the top spheres (mentioned in section 4.12 above). Because, as was conjectured in the earlier chapter, these covering caps could be made opaque or transparent to the magnetic fields running through the top spheres. One study on plasmonic metamaterials which may be interesting here is [Alu-Engheta]. But as I say elsewhere, the whole field of NRI metamaterials studies from Veselago through to Pendry would be suited to the sort of variable permeability gradient needed for these caps.

Note 16—I have made the following calculations based on the central sphere-set assembly rotating at one revolution every two seconds (0.5 Hz), and based on my arbitrary guesstimates of the aaUFO's proportional dimensions. Needless to say these are only estimated values and can be improved where necessary; and while each wheel has four microwave tube conductors, at 90° intervals, I have given the firing angle (the duration of arc where the tubes line up with the axle cavity and conduct) of each microwave tube as being $\delta = 10°$ of arc of those telemeter wheels.

therefore at 0.5 Hz rotation of all four sphere-sets equates to 9.51 firings per wheel per second equates to a firing repetition time of 105.14 milliseconds equates to a firing pulse length of 11.68 milliseconds ($\delta = 10°$ of arc)

This gives the approximate pulsing rate of 9.5 Hz for the central sphere set assembly, and assumes that all four wheels are synchronized to fire their pulses simultaneously. If the wheels can be made to fire in a way that is precisely staggered one after the other then the four wheels will give a pulsing rate of 38 Hz (4 x 9.5 Hz). For the craft's acceleration the sphere-set assembly would need to turn faster, for instance, if the sphere-set assembly is accelerated to one revolution per second then the firing repetition time would be doubled to 19 Hz (again, this assumes all four wheels are set up to fire at exactly the same time, otherwise they'd fire at 76 Hz); but the trade-off would be that the firing pulse duration would be halved. And obviously, if the sphere-sets could be wound up to 2 revolutions per second then that would amount to a synchronized firing of 38 Hz.

Note 17—Well, one of the mechanisms will be to generate heat [Yamada]. This can be seen within the phenomenon of electron cyclotron resonance (ECR), where charged particles are generated from the interaction between the electric field oscillating at a microwave frequency, and a superimposed magnetic field. This mechanism will act on the particles gyrating around the planar waveguide when passing through the magnetic fields superimposed around the rim, and also which curve around the outside of the rim. The above cited paper gives as an example a magnetic field of 8.75×10^{-2} tesla and a microwave frequency of 2.45 GHz (with a wavelength of 12.25 cm) which would initiate the angular cyclotron frequency resonance. Also see [Bogaerts]. And there might also be an intriguing link between microwave frequency electromagnetic waves and gravitational radiation that should be further investigated, as in [Chiao 2003].

Note 18—This is mentioned in *Andreasson Affair* pp122-123. Which Betty remembered all the more because it wasn't noticed when she was brought into that UFO earlier. In a letter answering some of my questions about this Betty explained, *"The quarter bubble room where I entered the craft in 1967, had an edge that dipped downward like a gully around the inner rim of the floor. I think the stairs folded up into the door somehow because they were gone and the gully continued unobstructed all around the outer rim. After my return from the experience when they were taking me home, we passed the changing room area through another door. We were in the quarter bubble room again. But this time the floor was different. The floor went upward. It was raised, and we went over the raised area which also had a gully around the side of the floor that continued around the inner edge to the exit door. I did not see this in the beginning of the trip when I first entered that room. It was as if the floor could move, shift and possibly even rotate. There may have been a slight hum, but I'm not positive. We were stopped. I don't remember a rushing energy beneath, and was not aware of any danger there. However, beside other feelings I was going through, I felt very agitated and impatient while standing there."* (Betty Luca, March 2004)

Note 19—Again, polymers exist in a huge variety which are activated by either electric fields or magnetic fields. Magnetically activated polymers are mentioned in the [Zrinyi] study.

Note 20—The field circulating around the outside of a shallow cylinder shaped with a conical top will be a frustum field (in other words a conical-shaped toroid). On the upper part of that frustum the curvature will have a smaller radius than at the base of it on the outside, so that there will exist a magnetic potential-difference and there will result a magnetic drift on charged particles to move toward the smaller radius of curvature. So charged particles will always curve inward toward the axis of a frustum, and indeed, will always cause that frustum-shaped magnetic field to draw inward towards its central axis, so that the charged particles will naturally curl into a cone shape. By conservation of energies as those charged particles curl through smaller and smaller radius of curvature they will accelerate with greater angular momentum—to facilitate the *spinning ice-skater effect*.

Note 21—A very similar example of this current ring plasma field is given in the 1958 lecture by [Tamm].

I would think sensors will relay the configuration of any electrical field and local conditions around the UFO craft to the controlling pilot (by computerized visual imaging), so that the configuration of it could by precisely controlled by the craft's pilot.

Note 22—Well, this is not quite true, Betty did mention standing in the UFO's waiting chamber and just starting to feel weightless, before everything normalizes and she is allowed to move into the upper decks of the UFO, although, strangely, she then described when passing though one particular area of the UFO that for a moment she felt much heavier (*Andreasson Affair* pp38).

Notes for Chapter 8

Note 1—Throughout the *Andreasson Affair* Betty Luca made several observations about the nature of the rotating flow in the lower section of the UFO she saw and travelled in. For the attributes of the sonic analogue of a black hole see [Visser 1998a], [Visser 1998b], [Liberati-Sonego-Visser], [Basak 2003], [Prix], [Unruh 1995], [Federici], and [Leonhardt-Piwnicki 2000a]. Where the perturbation of the fluid flow is similar to the curving of spacetime, and where the flow is inviscid, barotropic and irrotational (such as tornadoes), see [Jacobson].

Around the event horizon the group velocity of the sound waves can drop lower and lower toward a roton minimum, and then reverse [Jacobson], [Liberati-Sonego-Visser]. Rotons can be viewed as small vortex rings associated with negative pressure energy [Winterberg 2002a] [Winterberg 2002b]). As will receive more attention in chapter 11.

This mimics the fact that phonons cannot escape once they enter the event horizon, see [Visser 1998a].

Note 2—This sort of stealth is not in any way associated with a technology called 'noise canceling' whereby an electronic amplifier is configured to reproduce an exact copy of the sound coming from a particular area such as an engine. When given a 180-degree phase-shift and directed back toward the sound's source anyone listening—in that same localized area—will not hear that engine and only hear silence. Well, that's the theory anyway, but in practice it is very difficult to *completely* cancel out all of the sound by this crude method. Saab did do trials with such a sound limiting system on their commercial aircraft for a while, but they found it doesn't produce complete silence. An interesting and poignant example of noise canceling is associated with certain reported cases of cattle mutilation in America. Britain's Nick Cook, in a TV documentary *Billion $ Secret* on Channel 5 (UK) (in 1999 and 2000) reporting on Black Budget enterprises in the US using such technologies, interviewed John Harr, a cattle rancher in Colorado who described 'perceiving' a black helicopter hovering around his ranch house one night that gave him the impression something sinister was going on (and sure enough, the next morning he found some of his cattle had been killed and mutilated by its crew). Because he could feel the down-pressure of its revolving rotor-blades as that unlighted craft hovered only a few meters above him in the darkness, he knew it was a helicopter (rather than a craft engineered by beings with advanced minds). But then such unmarked helicopters, that push the sounds of the rotors upward rather than downward were developed for the US FDA and have been is service for many years. Disguised helicopters as used by the military have also been mentioned in *Glimpses of Other Realities Vol II* [Howe].

Note 3—The localized velocity of sound around a black hole is a variable dependant on the equation of state (ω), the ratio of pressure to energy density [Schutzhold-Unruh], so the localized sound velocity would also be lower than that of normal atmospheric air.

Note 4—The potential energy of a pressure or density displacement will be converted into kinetic energy to ensure that buoyancy returns the envelope to an equilibrium with its outside surroundings. So the UFO would develop a displacement in those outside surroundings.

Also, if it is understood how the earth's atmospheric gravity waves are structured in the atmosphere, it should also be possible to ride them, or even to propagate them to create a differential, as happens within earth's weather mechanisms (considering for instance, that gravity waves in the atmosphere will induce momentum into a cold front). Perhaps by creating these localized changes in the atmosphere, by using the same sort of pressure-differentials or density-differentials that govern wind movements and movements of clouds the UFO then actually becomes a distributor of localized atmospheric electric charges, thereby explaining why thunderstorms and lightning strikes are known to occur in the vicinity of a UFO.

Note 5—In an interesting aside, Ewa Wisnierska, a paraglider (in February 2007) while paragliding at Manilla, New South Wales, Australia in the hope of picking up good thermals, suddenly connected into the updraft of a huge thunderstorm that took hold of her chute and dragged it through a rotating column of air to an incredible 32,000 ft—at an average speed of 65 feet per second! And she survived the exciting ordeal (*Daily Mail* Feb 17, 2007).

Note 6—When I contacted Phil Trevis, who took the Grangemouth UFO photograph, to ask him about this UFO he told me that after the craft moved slowly beyond where he and his friend was standing (and was a safe distance from them) only then did it accelerate faster and faster and then finally disappear into the night sky at incredible speed. The photograph of it (http://www.ufoevidence.org/cases/case1047.htm) taken when the UFO was only 200 to 300 feet in the air clearly shows a heat trail from a patch of energy glowing on the underside of its lower hull. In the original photograph the stars in the sky beyond the UFO are quite visible. Heat trails are a very visible phenomenon in the video footage of the UFOs over Mexico in the 1990s, and too in the UFOs featured in the Patrick Uskert DVDs of California UFOs from 2005.

Note 7—Radio presenter Xan Phillips in those days had a lively program at GU2 (Surrey University) radio station in 2004 and broadcast this story while interviewing local people with interesting accounts to tell.

Note 8—The most recent website to carry the Bob Lazar package is at http://www.zamandayolculuk.com/cetinbal/UFOTEKNOLOJII.HTM

Note 9—For years it has been assumed that this aside was too fanciful to be understood and so like many other things not readily understood it wasn't given much importance, but I now see it as wholly relevant in showing scientifically how these crystal or glass spheres can be imbued with gravitational force, through what seems to involve quantum processes.
Note 10—The Lorentz spin force or angular momentum induced into any charged particles will be accelerated by this magnetic field. Also, when the magnetic field of a rapidly rotational field is anchored then a large amount of magnetic stress will be generated, and this magnetic stress will go directly into the charged particles spinning out of the craft [Lebedev 2005].

Note 11—Zeldovich and Novikov noted that a body of matter, which has passed over the event horizon of a black hole, will go through a contraction which compresses it down to a maximum density, after which it will expand. This will be because that body of mass while electrically charged will have the same electric fields inside and outside of it as it contracts down in size, and this electric differential will help explain why mass cannot contract down to a zero dimension. Because, on reaching a maximum density its electric fields will reverse and that body then takes on a negative density and a repulsive force. Then, its inertia of expansion will determine that when it does expand out again it will expand as a negative mass. Although, the effects of this are more imaginable when viewed through the mass's spatial curvature being contracted down to its minimum (minimum roton curvature), and then re-expanding through more and more negative spatial curvature [Zeldovich-Novikov] (pp144-148).

Note 12—This 'dink-dink-dink' sound has always reminded me of a recording made by Steven Greer (as issued out to his members at CSETI) coming from a UFO's propulsion system as the UFO flew over him. His recording sounds like the propulsion pulses of the UFO flying above him had a repetition rate of about 8Hz, so in other words the energy was building up and bursting out of the UFO around 8 times per second.

Notes for Chapter 9

Note 1—This current ring would act like a flywheel in an internal-combustion engine. There is no real equivalent in micro-electronic circuitry, because a damping capacitor can actually fully discharge and then reverse-charge; and an inductor in a circuit can make that circuit oscillate backward and forward, so it bears little or no comparison to these (unless they were also flanked by diodes, but even then that circuit's voltage will only flow toward a potential-difference in order to balance it out, once it is balanced that flow stops): It can't be compared to a thyristor-rectifier (even if it were imagined to be coupled to a huge array of capacitors) because thyristors work only in pulses. Considering how the UFO's vortex, and its gyrating storage field (which effectively means both ergosphere A and ergosphere B) all rotate in the same direction together with the current ring, then all this Lense-Thirring momentum (kinetic-inertia), which in some places will be turning at relativistic speeds, will cohere together and possess a tremendous amount of 'grunt,' and this flywheel effect will act as a substantial stabiliser to keep the electro-dynamic energy moving in the same direction and to remain fairly stable. Interestingly, when discharges *do occur* then shearing will happen around different diameters of that craft's overall flywheel and these shearings of rotation will then produce more electron-positron pairs—so it will have a self-sustaining flavor to it to boot...

Note 1a—From the CRED's descriptions, shown in the DIA documents published in *Exempt from Disclosure* [Collins-Doty-Cooper], this micro-sized black box (2cm x 2cm) of spinning bosons (of photons, or quark-composites, or gluons) probably works like a matrix of micro-Planck vortexes (like the Thomson-Maxwell-Planck model of vortices in the aether, or similar to quantum dots, or to geon-solitons which generate their own gravitational fields, or like Hawking's miniature black holes even). What you get is a 3-D matrix of these string-like vortexes all encased inside a sphere permeated by a strong magnetic field and anchored at the bottom on a crystal base (somewhat like how a quantum dot matrix can be made up of hundreds of micro-vortices which extend out of that crystal semiconductor's surface) [Quantum Dot] [Winterberg 2002b]. I would think the ETs are allowing the H4 nuclei to spread in amongst these vortices—so that they become a macro-atom-condensate. And similar to how the spheres are polarized in the aaUFO the gravitational pull from these vortices is used to pull the positive protons slightly away from the negative electron cloud. So straightaway you get the potential for a small amount of displacement current to be generated within each H4 atomic configuration. What the ETs would have designed into the CRED would have been a way that this matrix assembly would assume a neutral power output whenever it was unloaded (ie the H4 proton-electron configurations would be balanced).

The matrix, according to the Los Alamos input in the DIA documentation is made up of 6366 spinning bosons. The sum total of how many of those spinning bosons become affected, and to what degree, will depend upon how much load there is (how much of the 6366 matrix becomes deformed). So, a small load will affect a small amount of deformation and a large load will affect a large deformation. It basically follows along the line of the Abrikosov vortex lattice where, *"Deformations of the ionic lattice caused by the Abrikosov vortices have been studied from several aspects. For example, if a vortex moves, the motion of such deformation demands a motion of ions which contributes to the inertial mass of the vortex. In the static case, forces evoked by vortices create a tension which modifies the total volume of the sample and which was observed as magnetostriction."* [Lipavsky].

Its very clever what the ETs are doing and it can be seen as analogous to dissociating the H4 atoms, and then reconstituting them again after introducing an additional 'lattice structure' of artificial strong nuclear force to re-organize the H4 nuclei. Or in other words, because there will be a gravitational force centered on each and every one of these 6366 vortices this matrix will act much like a lattice of nuclear strong force upon the protons of the hydrogen. But the electrons being 1836 times less massive will be much less affected by the gravity of these vortices. So, the hydrogen protons will be clumped together in the matrix in the center, but between the matrix and the wall of the sphere will be where the electrons will orbit maintaining their electrical balance with the protons, but also they will be moving into and through the matrix around those protons (zigzagging in and around the matrix).

I would venture that they are using the output load on this device to deform the matrix, the documentation actually says that the lattice-matrix (which I would see as the trailing ends of the vortices) twists CCW when it is loaded, and this twisting-deformation creates a sort of compression-repulsion of the protons (like a van der Waals effect) within that matrix. The resulting repulsion-potential then puts a de-neutralizing effect on the heavy hydrogen (H4) atom configurations and throws them out of electrical balance with the electron cloud. Much like mis-alignment of the proton-electron balance, or ionization, of a single atom, and when thousands or millions of these H4 'quasi-atoms' unbalance and each contribute their displacement currents in unison, then the resulting accumulation of electric charge could be substantial.

The initial load would draw on the outer electron-swarm (surrounding the matrix) and that deficit would be what would alter the proton configuration and deform the shape of the matrix's vortices against the magnetic flux lines, thereby accelerating the bosons to generate more gravitational pull on the H4 protons. The more the vortices polarize the protons the more displacement current will be generated—to give the required output potential and current. This would also be why the weight of the device changes (because the deformations would strengthen the matrix's collective gravitational force, and they would change the overall gravitational vector of the matrix).

Furthermore, this would also be why the Lorentz forces and the Lenz electromotive forces are not as expected because really its not an electromagnetic effect taking place here – the end result is an inter-atomic, or quantum, effect. For the same reasons, its not an effect which will generate heat, so no significant temperature changes would take place. Again, it sort of validates what happens inside the aaUFO's spheres (where the ETs are gravitationally polarizing the ferroelectric ceramics to trigger spontaneous reversals in them and drawing off electrical current).

As for manufacturing such a CRED, well, these documents divulge an enormous amount of information, they really do—indicating that whoever penned them must have known basically how the CRED actually does work! Anyway, the process would be similar to manufacturing a quantum dot crystal semiconductor, the other elements spoken of in these documents, some of which are magnetostrictive compounds and others have high dielectric constants, are fairly self-explanatory for an electric generator.

Just think, in the early days of electricity when Lord Kelvin (William Thomson) was starting to discover more and more about vortical energies, if Nicola Tesla hadn't come along to propose his alternating current theory, and the vortex atom had been fully developed instead, and by English scientists, then we might all now be using rotating field electricity generators—with no need whatsoever for barrels of oil, fossil fuel, and nuclear power generators..!

Note 2—See [Schutzhold-Unruh].

Note 3—The constriction of a flow channel is called a *vena contracta* which, more precisely is an abrupt contraction which is used to decelerate the flow of a fluid (*Elementary Fluid Mechanics* Ed: R.L. Street et al, (1996) pp362-7).

Is there a similar mechanism happening here where groups of charged particles are bunched up, as happens in ECR beams. In this gyrating field when it strikes the white wall suddenly, surely the momentum of the massive charged particles will bunch them into the leading edge of that gyrating field, to amplify the electric field.

Note 4—This amplification process might more accurately be called the Williams-Starobinskii amplification process. Because, while Penrose (in 1969) postulated the amplification of particles evolving out of black hole rotational collisions, it was the Russian physicist Zeldovich who first formulated the amplification of electromagnetic waves through rotating fields in 1971 and whom suggested that a reflecting mirror would enhance that amplification (in Y.B. Zeldovich, JETP Lett. 14 (1971) 180), and also [Zeldovich 1972]. Although in the book by [Zeldovich-Novikov] (sect. 1.12, pp40) it was noted that mass and angular momentum could be carried away by gravitational waves. But is was Russian physicist Starobinskii (in 1973) who computed that electromagnetic and gravitational waves AND angular momentum could be extracted from a rotating black hole [Starobinskii].

Note 5—In astrophysical studies of superradiance it is assumed that the reflected out-going wave's amplitude will always be larger than the in-going wave (and so there would occur 'step-like' increases to the continual amplification process [Schutzhold-Unruh]). But in the aaUFO the opposite might be true, in that the reflected in-going gyrating wave-trains would be amplified to a greater amplitude, from them having to shrink down to a smaller radius of curvature (across the dense flux lines of the toroid's field, which action alone would feed more momentum into that vortex) before those wave-trains strike the vortex's photon ring to impart their greater rotational force on it. So here, the inductive-coupling between the various magnetic field intensities would suggest a high working efficiency.

Note 6—In the case of an acoustic (Kerr) black hole, or rotating black hole, the waves bouncing off the ergosphere would be reflected back with an amplitude that exceed that wave's original incident amplitude, leading to a *superresonance* effect, see [Basak-Majumdar].

A black hole with an ergosphere has been found to reflect from its rotating field a scalar wave which amplifies a certain range of frequencies. This amplification is relative to the angular momentum of the vortex. But Unruh says evaporation is not restricted to high frequency waves. If though those waves are too long to fully pass over the point of no-return they will be gyrated away from the black vortex. Also see [Schutzhold-Unruh] and [Unruh 1995].

Note 7—The term *superradiance* was first penned by Dicke (R.H. Dicke, Phys. Rev. 93 (1954) 99), but was first applied to reflective scattering off a rotating black hole by Misner in 1972 (C.W. Misner, Phys. Rev. Lett. 28 (1972) 994). But it was through the Press-Teukolsky use of an almost fully reflecting 'mirror-like' sphere surrounding the black hole that the concept of a field being amplified, by it bouncing back and forth between it and the black hole, that this effect gained interest. Although, this very basic model assumed that only a small amount of amplification could come from this process; but many factors were not then taken into account, mainly because it was not then envisioned how much energy could come from the Penrose mechanism of particle ejection and production, see [Press-Teukolsky].

Note 8—See [Schutzhold-Unruh] for the best model of a black hole's outflow capabilities, and [Berti 2005]. Although, the study by [van Putten 1999] offers a clear description of the superradiance process where electromagnetic waves interact between a black hole and a surrounding torus field.

Note 9— Of course all sorts of materials can form a reflecting surface around a black hole, it could be a plasma caught in a magnetic bottle-neck, it could be a fluid such as water rotating around a hollow cylinder. Electromagnetic waves, such as light waves, passing through such a medium (on their way to being reflected by that medium) will have their wavelength divided by the refractive index of that medium). Interestingly, by the factor that the medium can be ionized so that medium will modify its refraction index, see [Antonsen-Bian].

As has already been noticed in this study, while the electrodynamic systems of this UFO craft can be pumped up to a great degree then the process of being able to extract as much rotational energy as one likes from the craft's quasi-black hole drive would suggest a very appealing type of engine to use, see [Cardoso].

Note 10—In the astrophysical magnetosphere surrounding a black hole where an ergospheric cavity is formed, and where the scalar (spin-0) electromagnetic (spin-1) and gravitational (spin-2) waves are caught in such a cavity these will rebound back and forth inside that cavity, causing a continuous production of electron-positron pairs out of the background field of virtual (Planckian) particles.

Note 11—This is an interesting study in the storing of a huge amount of potential energy, see [Karas].

Note 12—see note 9 above.

Note 13—The wave's amplitude would grow to 10^7 times its initial amplitude in those 13 seconds, while its energy content would grow 10^{14} times [Cardoso] (p6). For reflection coefficients proportional to the black hole's rotational velocity see [Berti 2004], Berti also suggests that the system would benefit from a variation in the reflecting medium's dispersion relation (of its refractive index). For chart showing reflected amplification of wave packets see [Schutzhold-Unruh].

Note 14—The most informative papers on the formulation of a reflecting wall, with calculations for its possible dimensions see [Bekenstein-Schiffer], [Berti 2004], and [Cardoso].

Note 15—On the ionization of water see above sections 6.6 through 6.10; also see dielectric ionization in [Kouropoulos] and at the excellent [ISBU water] website. Pure water almost has no conductivity, it is only through the solution of salts in water that water can become a conductor of electricity, whereby the salts will dissociate into charged particles (as noted by Swedish chemist Svante Arrhenius) and change the water's dielectric constant (or permittivity).

I did wonder if seawater might be suitable for the toroid's fluid but seawater is considered 'extremely lossy' [Somaraju-Trumpf], and by lossy it denotes a dielectric that dissipates or leaks away electric charge more than is considered satisfactory. In seawater this would produce a phase difference in the water relative to the applied field, so that the field would lag behind as it traversed through the water, and at microwave frequencies this will result in a conversion into heat.

Sound waves (longitudinal waves) will speed up when they pass through water but plane waves slow down. It is also known that radio waves are slowed when they travel through an ionized medium.

Note 16—At around 0 degrees Celsius water's dielectric constant will peak at 3 GHz microwave frequency, while at 20 degrees C it will peak at 6 GHz, and at 40 degrees C it will peak at around 12 GHz. And generally the most useful temperatures are between 0 degrees and 40 degrees C, the lower the temperature the greater the dielectric constant [ISBU water]. But the ultimate study of this complicated subject, and one which is used as the dielectric reference standard for the permittivity of water (covering the frequency range 0—1000 GHz and temperature range of 0º—100ºC), is the one by W. J. Ellison, K. Lamkaouchi & J.-M. Moreau, *Water: A Dielectric Reference* Journal of Molecular Liquids 68 (1996) 171-279.

Note 17—With this sort of superradiance amplification it can be quite understandable that most UFO power drive's are essentially cylindrical. Because if a UFO is stripped down to its very bare essentials what it predominantly utilizes is a series of cylinders to both amplify its energies and to shape its energy wave-packets into propulsive fields.

Note 18—See Wikipedia webpage on 'Speed of Light,' and see the [Hau] paper on the reduction of the group velocity of light down to just 17 m/sec.

To illustrate the dispersive effect in another way Leonhardt suggests a method used in *Electromagnetically Induced Transparency* (also used by [Hau] to bring down the group speed of light) whereby if the dielectric medium were radiated by an orthogonal beam which could result in the transition of the atomic energy states (to form a susceptibility change) then there would result an interference effect between the light oscillations and those atomic oscillations, thus substantially reducing the group velocity of the light waves. In the paper of [Leonhardt-Piwnicki 2000a] it has been conjectured that a group speed of light may even come down to 1 cm/sec at the core of a vortex.

In fact what our scientists know of as the speed of light (c) is only relative to the absolute vacuum—which has a specific amount of permittivity and permeability. Einstein would have created much more of a stir than he did through his fascination of the speed of light had he conjectured what speed light *could* be capable of if it were passed through a super-vacuum of variable permittivity and permeability.

Note 19—In [Zeldovich 1972] (p1085) the term (μ) here is referred to as the *multipole moment* of an electromagnetic wave which would be the angular momentum of that wave; while in [Bekenstein-Schiffer] it is referred to as the *azimuthal quantum number* (the angular momentum with respect to the axis of rotation).

Note 20—The fact that the fluid is moving introduces a Doppler effect, which alters the wavelength of the light rays. See [Leonhardt-Piwnicki 2000a]. This is borne out by Einstein, saying that gravitational force is mediated by changes in the curvature of space-time, suggesting that space-time can act like a dispersive medium.

Note 21—See [Leonhardt-Piwnicki 2001] and [Leonhardt-Piwnicki 1999]; in this latter paper it is suggested that a rotating cylinder of fluid will act upon light rays as would a homogeneous magnetic field upon a charged particle. Taking this concept further, it would suggest that the flow a vortex would act as an inhomogeneous magnetic field upon a charged particle, in other words, light would drift toward the strongest convergence of that fluid—at the vortex's core. The wave-front dislocation due to the Aharonov-Bohm effect can be seen very clearly in the photographs of light travelling through a rotating fluid, which show that the light waves curl with the flow of the fluid, see [Berry] paper (p161) for instance.

Note 22—See [Leonhardt-Piwnicki 2000a]. The theme of a reversal of waves around the black vortex will be continued in chapter 10.

Note 23—Although [Punsly-Coroniti 1990b] (pp587-588) has made a tentative attempt to suggest where one could be.

Note 24—A similar process occurs in solar flares around the sun [Plyusnin] and [Di Matteo]. Also see [Jones-Ellison], and [Priest-Forbes].

An extensive study has been carried out by [Priest-Forbes] which lists around a dozen mechanisms of magnetic field reconnection (pp461). The reconnection principle is otherwise known as *runaway electron production*, or the *Dreicer phenomenon*. Reconnection *per se* is when current sheets with oppositely directed magnetic fields merge together, and the magnetic energy consumed in this process is then converted into heat, out-flowing kinetic energy and charged particle acceleration, and is a typical mechanism in the creation of solar flares. The phenomenon has received much attention by physicists reproducing it in the laboratory [Cothran].

In the case of the aaUFO when the flux lines of the toroid quadrants are pulled around the central sphere-set assembly when that assembly rotates, and those flux lines are loaded with gyrating electrons in a high state of excitation, rather than continuing to coil around the top spheres indefinitely those flux lines will individually break apart one after the other. In breaking those separate flux lines will collapse back on themselves and will then reverse their flow direction momentarily (ie as back-emf momentum), so that when a subsequent flux line breaks there is provided an opportunity for two oppositely directed flux lines to marry up. In the above mentioned reconnection process which generates heat, kinetic energy and particle acceleration the already excited electrons repulse-scattered from the breaking of the flux line are then further accelerated and energized by the generated magnetic forces and electric fields. This leads to the generation of high-energy runaway electrons [Plyusnin]. In a plasma field when a further current sheet is generated by the reconnection this is known as the collisionless effect (or Hall MHD effect) [Ma 2005]. But when that effect occurs in a confined volume, such as would be the case in the narrow planar waveguide the high-energy accelerations will cause electron-electron collisions and as a result create secondary high energy electron avalanches [Plyusnin]. A similar process occurs in Active Galactic Nuclei (AGN) and black hole accretion disks where the MHD shock waves from magnetic reconnections can heat plasma to over 10^9 K, which is hot enough to trigger hard-photon (X-ray) emissions from the plasma [Di Matteo]. The strength of the MHD shock wave can be directional depending on the predominant configuration of the magnetic fields [Gontikakis]. Either way this is essentially an additional pre-amplification of the energy gyrating in the planar waveguide which will join into the waves of energy being bounced between the white wall and the black vortex in the superradiance mechanism! Because this is the middle section 'exploding' in shock waves both inward toward the central vortex and outward toward the white wall!

Note 25—Interestingly, there will be a differential factor in the longitudinal and transverse components of these shock waves, as the longitudinal seem to be quite unrestricted, while the transverse will be restricted by the walls of the planar waveguide.

Note 26—As in [Andersson-Glampedakis], but the superradiant waves are only carriers, they could be carrying gamma-rays or X-rays and still not be readily visible, so in order to actually see the superradiance there would need to be a huge amount of visible-light photons carried out of the black hole. Such an effect brings to mind the launch site scene in the movie *Contact* where at the center of the magnetic ring-trains a ball of light is generated and pulsates with huge amounts of photon energy.

Note 27—As pointed out by [Ferrari] in his study of an astrophysical black hole; that the rotations of that black hole will bounce vertically up and down in response to these emissions (particularly if there is very little electrodynamic damping). This electrodynamic damping will be an important factor to consider regarding the superradiance being expelled from the aaUFO's ergosphere.

Note 28—Because in the astrophysical model what causes the creation of a black hole system will be a 'failed jet' mechanism; of when a star's explosive energy field fails to actually explode and so the helical structure of its outflow jet suddenly shrinks back down into the star, and like a set of stretched rubber coils that collapse back in on themselves, that collapse begins the process of implosion which continues to form into a black hole [Wheeler-Meier-Wilson]. So, again, if there might be a one-way valve damping mechanism somewhere in this UFO's black vortex circuit then the black vortex's potential will always remain fairly stable—and this would come from the Lense-Thirring rotation within the ergosphere A, from the current ring and the rim field circulating around the rim, both of which are inductively coupled to the central vortex.

Notes for Chapter 10

Note 1—See astrophysical jet production in section 2.10 of this study, and see [Lebedev 2005] whom in his development of miniaturized jets suggests that the toroid field develops out of the quasi-black hole's poloidal field, and eventually elongates into what Lebedev calls *magnetic towers.*

Note 2—For her rotating field's *supplementary* gravitomagnetic force Williams establishes a similarity to the Lorentz electromagnetic force with the understanding that $F_{GM} \sim m\,(v \times H)$, where (m) is the mass particle, (v) is its velocity, and (H) is the gravitomagnetic vector field [Williams 2002a] (p11/17), and [Mirza-Saleem] (p3). On this subject also see [Khanna]. For non-magnetically induced rotating fields see [Williams 2002b], and [Williams 2003]; and also see [Bardeen]. For the geometry-induced collimation of a helical field see [De Felice-Carlotto]. Another way of describing this pre-collimation curling force is by reference to the helical flow's pitch-angle.

Note 3—Firstly, the gravitational force in a rotating field will depend upon the frame dragging velocity [Williams 2002b], and this factor can be extended back to [Zeldovich 1972] (p1086) wherein it is suggested that a rotating field will amplify and generate both electromagnetic and gravitational waves as it accelerates, meaning that if those gravitational force carrying particles are guided into a smaller orbital radius, into a smaller radius of curvature, then there will result an increase in the frame dragging velocity and therefore an increase in that gravitational force. But what should be considered also is that the volume of energy which in this case will be gravitational force will be compressed, hence, the effectiveness of that force will automatically increase.

Note 4—But I think Zeldovich [Zeldovich-Novikov] is wrong here, yes the higher order and quadrupolar waves can disperse angular momentum away from a black hole, but not mass, simply by the reasons stated in section 5.1 (note 1), regarding the gravitational polarization effect. I think Zeldovich here was presenting a bridging theory to appease those who still favored the Newtonian belief that gravitational force was proportional ONLY to the mass, and the relativistic theories of Einstein. This is substantiated by Zeldovich himself by noting that left-hand and right-hand polarized photons experience different red-shifts in a gravitational field (op. cit. p38). Also see [Bicak-Janis] (p900) on the dragging of magnetic field lines.

Note 5—See for instance the paper by [Rabinowitz] on the entropy of a black hole. Another factor which can be gleaned from this study is that the filament tube will form a modified version of a black hole, obviously not in every way, but as an enclosed rotating field with mass components it will generate its own pressure gradient and a density gradient. Also see [Mirza-Saleem].

In the Jeff Savage model of the *Onion Drive UFO*, what its helical field does to govern its "collective" (of electromagnetic force, gravitational forces, and angular momentum) is expand or reduce the radius of rotation of its helical field (as well as increase or reduce that collective's angular velocity) so as to amplify the gravitational force it develops.

Note 6—This is based on a study by [Sushkov-Khriplovich] on gravitational energy emitted by a rotating particle in an electromagnetic field (which in turn is based on the assumption by [Gertsenshtein] that a resonant interference can be formed between light waves and gravitational waves. See also [Portilla-Lapiedra].

Paul Davies has noted something similar by inferring that when a body accelerates through a quantum vacuum it faces a shower of thermal radiation similar to the Hawking radiation that comes from black holes [Davies 2001b].

Note 7—There is a very interesting paper on this phenomenon [Ferrari] of when mass transits through an acute turning point. This is mostly what is seen around the rim of a UFO, its not the energy field the craft is using, it is the photon emissions coming from that energy field. These colored emissions can be seen in the UFOs that were seen hovering over Mexico back in the 1990's and were filmed by dozens of local people.

Note 8—The Onion-drive, so named from the shape of the energy field it produces (of a bulbous energy envelope around the craft from which it extends a long narrow filament-tube—so it looks like an onion stem), has an even more compact engine unit than this aaUFO. Instead of having a cylindrical ergosphere to build up its rotational energies it uses a two-part cone structure, one inside the other. It produces its charged particles by curvature ionization mainly, and then uses a cone-shaped accelerator to generate a gravitational force. This Onion-drive design caused much nodding of heads when it was introduced to physicists at Stanford in 2003...and since then its design has been racing around the propulsion labs. But I think the astrophysicists would have a field-day with its design, particularly William Unruh at the University of British Columbia, Vancouver; and Ulf Leonhardt at the University of St. Andrews, Scotland whom are both searching for a way to miniaturize a black hole. The common theme between the Onion-drive and the aaUFO is that they both retain their energies within an enclosed structure and develop the *induced* energies from outside of those enclosures; and the Onion especially doesn't release any energy whatsoever through any of its electrodynamic mechanisms, it induces *energies from outside of it* to form its propulsion drive.

Note 9—The curves that would express this dome field trajectory best would be Fibonacci (phi) curves, so the wave-packets would be bounded by two Fibonacci curves, the leading edge and trailing edge of the gyrating wave-packet.

Note 10 – For the math on the inertial frame-dragging limit of a rotating field (for when it would become solid in the air) see [Williams 2003] (p11), and [Williams 2002a] (p3/4). For the Lense-Thirring frame-dragging enhancement of the gravity effect around rotating fields see [Ostoma-Trushyk]; and also see [Greyber 2005b] on the establishment of a filament-tube through the wrapping of layers of air (causing the particle density of the sheath's wall to increase—which, of course, denotes a Meissner effect). The high speed twisting of the magnetic field lines (as well as their breaking and reconnection) in filament-tubes will produce electron-positron pairs (as alluded to in earlier chapters).

There is an interesting hypothesis regarding the scalar wave field of longitudinal potential in space, spoken of by Lt. Col. Thomas Bearden, and is much like the field of potential hypothesised by Planck and Wheeler [Bearden-Rosenthal]. There would have to be some sort of natural hyperspace resonance throughout space for these hypothesis to be useful. Then, it would be possible to cause a localized excitation of these energies (ie produced out of the virtual fields of energies of space), which would then engineer itself into a channel of viscoelastic fluid (with a little help from Sakharov's elastic spacetime curvature). If the virtual fields of energy in space can be made to produce real electron-positrons by using magnetic stresses (ie which happens around black hole systems), and matter can be engineered through the Higgs procedure, seemingly from some omnipresent virtual mass matrix, then why should there not exist a way to engineer a cylindrical filament-tube through space—by using energy harvested from space itself? See the [Audretsch-Skarzhinsky] study for pair-production in space through twisting flux tubes (but then, the reconnection mechanism will produce copious charged particles too).

Note 11—From the many times that Betty Luca was taken away by this particular design of UFO craft, from when she observed them from the outside before and after her journeys in them, there was at least one occasion when she recalled that this canopy section was extended upward several feet above the dome shell. And I would speculate that once the craft had landed, either on earth or inside a mothership, this would be done to let the old air out and to re-balance the craft's internal density-pressure levels.

Note 12—It would be interesting to discover further what happens to the different particles that make up the wave-packets and whether any sort of plasma absorption interchange occurs within the helical filament-tube to accelerate them, for where the velocity of the particles might initially be greater than the helical waves the lighter particles (ie the electrons) may be separated and be accelerated ahead of the other positive particles, riding on the crests of the waves of the helical field (like a wake-field) [Tamm] (p478). Electromagnetically, such a differential acceleration will set up a substantial electric field between those leading electrons and the trailing positive charges [Mourou-Umstadter 2002]. It would also be interesting to see which particles win the battle to be the lead field, because when they separate the greater gravitational forces carried by the more massive positive ions would suggest THEY would be accelerated faster than the electrons in a helical field.

Note 13—This is derived from [Morris-Thorne]. As I have said previously I am not an advocate of traversable wormholes or wormhole travelers, mainly because I'm not an advocate of having people fragmented into subatomic particles through such a set-up of extreme forces.

Note 14—If the universe were a sealed container which kept on expanding then, yes, it would engineer a vacuum like that inside a bottle, but there is no evidence that this is the case. Put another way, if you sealed off our solar system from the rest of the universe's curvature effect, by surrounding it in its own little bubble, there would then be no vacuum in the space surrounding us inside our solar system.

Note 15—This diagram is based on a number of inputs; from the suggestions that our universe is formed as a 4-dimensional membrane which is expanding in a balloon shape [von Weber], or that it may be formed as a horn-shaped hyperbole [Ostriker-Steinhardt], or that it is formed in a spherical shape [Abbott] and [Bucher-Spergel]. And it must be said, this model runs parallel with the concerns among cosmological physicists who now think that the *big bang theory* is giving more problems than answers (for instance, if the universe started with a bang that bang's energy would have calmed down and died away billions of years ago, so how come the universe is expanding so gently now, and how come that expansion after all this time is presently accelerating...). So, no, the sums just don't add up for the big bang theory to still hold water [Magueijo-Albrecht]. So this is why I say that the universe must be generating its own power, on a constant basis, and the accelerating expansion that we are experiencing is the result of the geodesics of space-time curvature—not some explosive force that physicists claim has never stopped since the beginning of time.

Note 16—Again, see [Magueijo-Albrecht]. For the interesting effects of variations in the dielectric constant of the cosmos the reader should look at table 1 and 2 in [Puthoff-Little-Ibison 2002b];

whose hypothesis is that the *cosmic dielectric constant* is relative to mass density, electromagnetic energy density, and the cosmic polarization energy density.

According to conformal cosmology a negatively spatially curved universe acts like a diverging refractive medium, where (k) can be negative [Mannheim]. But if negative spatial curvature can be seen as a diverging refractive medium, and in such a medium if k becomes less than 1 (k < 1) then the velocity of light will be greater than c and material objects will expand; and conversely, if k becomes greater than 1 (k > 1) light waves slow down and material objects shrink: this is based on the Puthoff model which has for its reference frame at infinity K = 1 [Puthoff 2002a] and [Puthoff-Little-Ibison 2002b]. The polarizable vacuum (PV) approach of Puthoff-Little-Ibison relates to Einstein's equation for the curvature of space-time established by the mass-energy stress tensor, and the Yilmaz approach to gravity. The PV (polarizable vacuum) theory was first established by H.A. Wilson in his paper *An Electromagnetic Theory of Gravitation* Phys. Rev. 17 (1921) 54-59; and later developed by R.H. Dicke in *Gravitation Without a Principle of Equivalence* Rev. Mod. Phys. 29 (1957) 363-376.

Note 17—Based on Wheeler's observation in [Misner-Thorne-Wheeler] (p5).

Here I will draw the readers attention to the (Potter) characterization of how a gravitational field works (which I started to delineate in chapter 2—note 20) (and in chapter 3 note 33), and that is through the differential rotation of mass-energy. My tentative estimation of gravitational force is that it equates to a differential in radii of angular momentum... and is analogous to wave trapping by a spiraling flow (similar to how acoustic waves are trapped by a spiraling flow of greater velocity than those acoustic waves), because in a converging flow if looked at as a series of different angular momentums one will be trapped by the next and so on. Similar to Einstein-Rosen cylinders, but with a rotational force gradient, so that where there is a difference in rotational force there will result a gravitational force. Depending on which way the force gradient runs determines whether the gravitational force is attractive (positive) or expansive (negative). That being the case, of course, where the difference is only minor, for instance, deep within the earth's core, the resulting gravitational force will also only be a minor force. However, as this force would be a relative force, relative to the differential of rotations then what Einstein formulated about gravitational force in his GTR is a nonsense, and furthermore, it also makes a nonsense of Newton's premise the gravitational force depends wholly, and only, upon mass density.

Note 18—Well, the best description of this occurring in a black hole is in [Jacobson], but the idea is based on [Schutzhold-Unruh]. And as Jacobson so rightly points out, this evaporation of expansive gravitational force coming away from a black hole is the negative energy version of Hawking black hole radiation. But also see [Unruh 1974], and from the point of view of the lambda cosmological constant being a repulsive force of gravity see [Greyber 2005a], [Ostriker-Steinhardt], [Davis-Puthoff], and [Mannheim].

The reader should be careful how they view negative energy (and this is why I drew this cosmological diagram). Negative energy pressure can be described as tension: And when viewed as an expanding ring of energy in an upright funnel shaped container this tension can be felt as an expanding force from above, but a vacuous force from below. A brief summary of the thoughts of Hawking, Einstein, Weber, and Guth on the negative density-pressure-gravitational forces can be found in the [Greyber 2005a] paper.

Note 19—The Lazar UFO's system of gravitational manipulation is far more advanced than the technology which Bob Lazar attributed to it, as will be explained in the comprehensive rework of the original Lazar hypothesis in chapter 12 of this study and its supplement.

Note 20—For Chirped-Pulse Amplification (CPA) see [Mourou-Umstadter 1992] and [Mourou-Umstadter 2002]. For backscattering pulses see [Shvets 1998] and [Shvets 2004].

Note 21—For a quantum insight into negative energies see [Ford-Roman 1999].

Note 22—For negative energy density and the attempts to circumvent the quantum observation see [Ford-Roman 2000] (p53).

Note 23—A similar mechanism happens when a jet of air produces turbulence and is heated by that turbulence, so that it then pulls more of the surrounding air toward it [Omma].

Note 24—Judging from UFO reports, and when there are any sounds from the audio recordings of UFOs (as in the recordings made by Dr. Steven [Greer] at CSETI), I would assume that the wave-packets can be ejected from such craft at repetition rates of between 7 Hz to 60 Hz. The 'dink-dink-dink' sounds that Betty Luca heard her UFO make as it left from her backyard in 1967 were around 8 Hz, the same as in the Greer recordings. This correlates with Bob Lazar's UFO, from him learning that the UFO he worked on at Area 51 operated through gigaherz waves being pulsed at 7.46 Hz. This low frequency pulsing of higher frequency waves is obviously saying a lot about the amplification processes involved (in that because they are being pulsed, as *groups of waves*, those waves can be slowed down and REVERSED, and then speeded up again, wherever and whenever required). The reader will also recall how feasible it has been for the Andreasson UFO to operate with microwave frequency waves pulsed in wave-packets in the 9 Hz to 38 Hz range.

Note 25—Low temperature electromagnetically induced transparency is alluded to in the articles published by [Hau] and [Leonhardt-Piwnicki 2000b]. The material prepared by the orthogonal beam will lower the group velocity of the electromagnetic waves passing through it. But metamaterial science also addresses this transition state, such as in the interesting study of transparency in plasmonic materials [Alu-Engheta]. The reader will recall that in metamaterials and photonic crystals when the permittivities and the permeabilities are BOTH switched to negative that material becomes transparent to radiated waves, and as in the [Pendry 2000] study this transition from (RF) opaque to (RF) transparent occurred at 5 GHz for the particular material he used. More on induced transparency in the next chapter.

Note 26—The duty cycle, of the time-periods of those 38-per-second pulses may be adjustable from how the orthogonal magnetic fields are bunched, or spread out, at the four locations around the telemeter wheels, as I mentioned in section 6.11, because then the degree of radial arc employed in the firing process could be varied to make the actual pulse duration longer.

Note 27—See the paper on parametric resonance of standing gravitational waves and autoresonance (similar in some ways to a superradiance effect), for when the repetition rate of those wave amplifications can be chirped down into wave-packets [Assaf-Meerson], and [Evans]. There is a useful child-on-a-swing analogy explaining parametric autoresonance that can be found on page 16310-1 of the [Assaf-Meerson] paper.

Note 28—This field of research is most concentrated in laser technology but the principle is the same because Chirped-Pulse Amplification (CPA) works on electromagnetic waves placed into wave-packets. In the [Mourou-Umstadter 1992] paper where this mechanism is used in a laser beam that laser beam's stored energy can be substantially intensified by first stretching its energy pulses and then amplifying them and then re-compressing them into a highly concentrated repetition of pulses (ie this is the chirp technique). The CPA laser dissects a laser pulse according to its frequency components and re-orders them into a time-stretched and lower-peak-intensity pulse of those components [Davis-Puthoff]. Same in the [Mourou] paper where this form of amplification can be taken to unprecedented power levels.

A more simplified system is to use long pulse propagation and compress the long pulse with a backscattering pulse [Shvets 2004]. In an inertial confinement system a laser pulse can be backscattered and lead to superradiant amplification, which is like a back-intensification [Shvets 1998].

Note 29—There may be another way to do this being suggested in the drawings published in the *ASBT* booklets [Luca] (*ASBT-I* pp64-65), in that the ETs are using some sort of energy field similar to that which emanates from a black hole. For optical density and positive-to-negative gravity, see [Mannheim]. Also see [Greyber].

Note 30 – This facility of UFO craft to manufacture their own density environment around them will be dealt with more thoroughly in chapter 11, but a clue as to the way in which a UFO engineers its own channel of media can be found in the old World War II sightings of UFOs. For instance, there was a sighting made by the crew of an RAF Halifax on 26/27 May 1943, of a cylindrical UFO several times larger than their aircraft which appeared ahead of them. This UFO was sharply defined and clearly visible as it was hanging in the air stationary, but as the Halifax approached it in one instant it shot-off at a speed estimated in the thousands of mph. The most intriguing factor about this UFO's movement was that as it accelerated its shape *blurred* and *foreshortened* before it finally disappeared into the distance – and this portrays the signature of a manufactured energy environment within which that UFO sped through, at an altered mass-density [Good 2006] (p20).

Note 31—Physicist Paul Davies has said, *"that rapidly changing gravitational fields can create particles from the vacuum"* [Davies 2001b]. But also, spontaneous pair creation will occur by electromagnetically stressing the virtual field of energy below the UFO, in much the same way as the Searl rotors stress the air surrounding them.

For the sea of virtual particles spontaneously popping in and out of existence and the Dirac or Planckian energies of the vacuum see [DeWitt] (p111), [Penrose 2004] (p624-632), and [Ford 2000]. Also see the phonon-roton conjecture of Maxwell's and Planck's aether in [Winterberg 2002a and 2002b].

If the Planck field is viewed as a sea of rotons it will be easier to see how all this works. The zig-zag rotons which are microscopic vortices are perfectly balanced and are a ZPF field that permeates spacetime. Its just like the virtual electromagnetic field, through which magnetic flux lines 'appear' out of seemingly nowhere to interact with electric charge (as mentioned in section 4.4 above), when for instance a generator transfers electric current along a wire. In order to get the vacuum to do work the roton energy just has to be unstabilized, and that can be done by stressing the electromagnetic field [Penrose 2004] (p624-644).

Interestingly, in the [Gavrilov-Gitman] paper it is suggested that particle production and the transfer of kinetic momentum through those particles is proportional to a hole's gravitational force. What is also striking in this paper is the general recognition that the electron-positron energy production (of the accretion disk) is drawn out of the electromagnetic sea of space. This is a differently worded rendition of the ZPF force (such as occurs in the Lamb shift) which appears to have much more acceptance than ZPF in astrophysics circles.

Note 32—Presumably, this would be one way in which the UFO could manufacture around it a surrounding energy field of variable permittivity. Having an outside energy field where that energy can have its permittivity and permeability adjusted would be useful for the craft's gravitational potentials, as well as for a variable refractive index (to determine how much light is reflected off that UFO).

Notes for Chapter 11

Note 1—Well actually Betty did see something more, she says the second and third levels were smooth and slanting which, bearing in mind that there would have been sound-trapping in those lower levels, it may have been that she was looking at vibrant fields of energy rotating toward the black vortex, her additional description of the lower levels which she published in *ASBT-II* pp27-28 goes like this;

"The whole craft I saw in 1967 was about thirty feet in diameter, and once inside and past the waiting chamber it appeared much larger than first expected. When I was escorted to an upper room I could see partially open levels below. The lowest level appeared dark like a black hole. Second and third levels revealed portions of round smooth slanted floors. The fourth level where the waiting chamber was located is where I entered the stairway to the fifth floor. The exterior of this craft did not seem as large as the interior, however the Watchers mentioned one of their words and its meaning to me. They said they could "opulate and deopulate," which they revealed meant, they could make more or less of, increase or decrease size, appear and disappear!"

Note 2—Metamaterials will change to radio frequency (RF) transparency, whereby while below a certain threshold frequency they will not allow oscillations to pass through them, but when a radio frequency signal above that threshold is applied to them they will allow other RF signals to pass through them: this is basically one step onward from transistor technology (see chapter 10's note 25 for details). I did hear that Japanese physicists have gotten to an advanced stage in this area though. But the extraterrestrials seem to have this optical transparency technology down to a fine art. As in Betty's description of how the Greys turned the WHOLE of the lower section (ie the toroid shell) transparent, which at first seemed to me to be some sort of holographic trick, but perhaps not. The Greys did this when their UFO was parked on the ground in 1967 to show her the workings of the central sphere-sets and the glowing bars around the top spheres, Betty was standing outside the craft with some of the UFO's crew and suddenly she could see past the outer surface and into the center. This may beg the question do the extraterrestrials predominantly use metal in their constructions or some exotic semiconductor or photonic crystal arrangement... A corollary is that Bob Lazar described seeing the interior walls of the UFO he helped to back-engineer suddenly become transparent, when one of the other physicists powered up the craft's electronic circuit and an electric field was applied to those wall materials.

Our scientists on earth have started to work in this optical transparency area now though. The new technology of terahertz wave oscillations (in the infra-red bandwidth) is fast becoming a new rival (to X-rays, for instance) to render physical objects invisible. Is this what the Greys were doing to Betty's UFO in 1967, irradiating its toroid shell with terahertz wavelength oscillations to make it transparent?

Note 3—Corso said that the flight suits or skins used by the aliens who piloted the Roswell UFO were made of a flexible substance whose molecular structuring was elongated, so that it offered electrical resistance in one direction but lengthwise it offered no resistance to electric current [Corso] (p96).
Note 4—Changes to the fine structure constant (alpha) have been noted in distant quasars [Webb] as has also been conjectured in section 10.7 of this study. This has led Joao Magueijo to formulate his theory that light's velocity was greater in earlier times (see my figure 80).

Note 5—There is also, if the reader cares to look at the various shades of the Polarizable Vacuum (PV) representation of Einstein's theories on relativity, a different way of looking at the dimensional constant of matter. In the PV formulation the (c) in ($e=mc^2$) is viewed as (c/k), whereby the speed of light is proportional to the spatial dielectric constant (ie the refractive index of the medium through which it passes), and indeed the same goes for the values of energy (e) and mass (m) in that they too are subject to a variable proportioning from changes to that environment's electric permittivity and magnetic permeability (refractive index).

The PV (polarizable vacuum) theory was first established by H.A. Wilson in his paper *An Electromagnetic Theory of Gravitation* Phys. Rev. 17 (1921) 54-59; and later developed by R.H. Dicke in *Gravitation Without a Principle of Equivalence* Rev. Mod. Phys. 29 (1957) 363-376. For recent updates to this theory see the [Depp] [Desiato] treatments of it which have provided solutions to some of that theory's problems.

The basic idea can be seen in Puthoff's mass-distribution tables for the space-time metric effects of PV-GR, of how this works when variations are made to the permittivity and permeability (to change the spatial dielectric constant k) of a medium you change all the $e=mc^2$ components. Mass decreases and the ambient frequency blue-UV-shifts up from visible to UV when the ambient permittivity-permeability drops below unity (from this world's customary k<1 value). For instance see [Puthoff 2002a] (table on p21) and [Puthoff-Little-Ibison 2002b] (tables 1 & 2) for the space-time metric effects attributable to the PV approach to General Theory of Relativity. Also see the [Rabinowitz] paper on black hole mass-density effects.

Note 6—If the explosion which occurred inside the Roswell UFO on that night of July 2nd, 1947 blew a gaping hole in the upper deck, or if as I favor, that the explosion caused the upper deck to become severed from the lower deck, then the crew's environmental space on the upper decks would have been breached very suddenly. Therefore it is highly likely that those extraterrestrials died in mid-air because of the rapid changes to the mass-density of their bodies, rather than from the crash landing which occurred soon after the explosion. And the effect on their bodies would have been analogous to the effects that sudden decompression would cause in deep-sea divers resulting in the bends. Those bodies would have been of a much higher mass-density inside the UFO craft, and then all of a sudden their surrounding environment would have become exposed to earth's normal (more rarified) density, so those beings must have suffered an explosive expansion of every cell in their body. If one did survive then he wouldn't have survived for long. A post-note to the above comment: after reading through Corso's report again in *The Day After Roswell* (pp93-94) I read that Corso says that when the US military guards secured the area around the second section of debris (on July 8th, 1947) when the upper part of the craft and the alien bodies were found, those guards reported that TWO of the aliens were still alive, but that they both had extreme difficulty in breathing... There you have it, I rest my case.

Note 7—Obviously, it wasn't known then how this mass-density scaling effect occurs, so Fred Youngren's technical diagram featured on p225 of *Andreasson Affair* used 1:1 'logical' scaling throughout, and so the UFO's shell in his drawings had to be shaped almost as a spherical shell to get everything to fit in and still have rooms which were at least 7 feet high. Even then his UFO equates to being almost forty-feet in diameter—which is larger than the thirty-foot diameter dimension remembered by Betty. And of course, Betty's drawings of the outside of the UFOs show them as having acute curves, not quite saucer-shaped, but certainly not near-spherical as in the Youngren drawings; and so the shape Betty remembers (and which is featured throughout this study) has even less volume inside it than the Youngren version. Note that according to this original Youngren diagram the lower glass spheres would each have had to have been at least five-feet in diameter.

Note 8—For more accounts of these *topological diminution* observations, of where witnesses have observed small exteriors to the UFO craft yet huge interiors once inside those craft see the Vallee-NIDS (National Institute for Discovery Science at www.nidsci.org) reports. Also see [Davis, E.W. 2001] (p45).

Note 9—The skirt could be created by neutralizing the upper current-ring field and energizing a lower rim current-ring field around the perimeter of the toroid shell, that way energy from the gyrating storage field would exit the rim and go downward instead of upward, and not in wave-packets but probably as an oscillating electromagnetic-gravitational energy. In this way the skirt field would be made up of charged particles gyrating around the thin-walled shell of the skirt (much like in a magnetosphere). Notice also that the central part of this field comes from the gap between the toroid and the base disc, not from the vent hole in the center of that disc (as if its coming from the anti-vortices, as mentioned in section 11.4 of the main text).

Note 10—The closest I've come to this word is the Latin opulens > opes which denotes abundance, wealth (to increase). A greater abundance would presumably be a reference to a gradient scale for mass-density or an effective-mass spectrum, so I would surmise that by opulate the extraterrestrials are speaking of increasing their effective mass; and by deopulating they are talking about decreasing their effective mass. Why should the ETs use Latin the reader may ask, well for one thing a word will always impinge upon a researcher when that researcher has to do some work to find out what that word means. But I believe I have in the past proved very successfully that the extraterrestrials do have a penchant for using earth's more ancient languages, like Latin, Greek, Egyptian and Gaelic-Celtic to communicate with us from when I translated the ET's warnings about our sun, and about the fields of high-energy particles drifting into our solar system now, via the Local Interstellar Material—LISM, and heating up the earth's atmosphere from the outside, and, most importantly, what they have seen coming our way from deep space – in the form of a supernova shock wave – with their offer of getting us out of here). The term *Watchers* is an interesting one, Betty Luca does not refer to the ETs as extraterrestrials, she refers to them as *Watchers* because that was the term they used for themselves. The use of this term would correspond with what other abductees have been told by the ETs about their involvement with human-kind, and particularly that they have been around our earth-realm for thousands possibly millions of years. The *Watchers* that Betty was involved with, and I'm sure a lot of other abductees were involved with them too, seem to be the good guys (see also the [Howe] study and the [Jacobs] studies on this), and they seem to have come forward now as our guardians, and possibly, they were or are the original genetic engineers that were instructed to modify the human body into the form which we are clothed in at present...and this indeed would explain why and how they know infinitely more about our physiologies than we do! That these *Watchers* have now returned suggests its own particular story... This particular field of study will definitely NOT have a "don't go there" tag attached to it, which some people have been saying about it, its just that it is very profound and does not fit into this present book.

Note 11—This opulating-deopulating of UFOs may go back a long way, even before Roswell, to the *Foo Fighters* seen by many aircraft pilots during the second world war. These were small balls of intense light about the size of an aircraft's engine and were extremely agile units which flew around and with the B-17s and Lancasters. Were these the early examples of enhanced mass-density vehicles (EMDVs)?

Note 12—The security patrolmen who witnessed these scenes near RAF Woodbridge (according to a subsequent report by USAF intelligence officers) did not mention that the aliens they saw were much smaller than expected, but they must have been if the craft they came out of and were endeavoring to repair was only 2 meters in diameter! How would you explain that to the newspapers?

Note 13—I am emphasizing this not to be pedantic but essentially so as to clarify some confusion about how these effects should be perceived. Because it is claimed in Puthoff's papers, in table 1 of [Puthoff 2002b] for example, that for a k-greater-than-one mass density (where the effective mass increases so that the objects shrink) that the time component will be affected in such a way that "clocks run slower," but whose clocks, internal or external? What I'm saying is there might be a better way of indicating from what perspective timed events are being slowed down, and this is where graphic examples are more useful – more so than a whole bunch of numbered equations.

Note 14—The environmental control mechanism discoverable inside the sealed shell of a UFO could be transferred into our world with relative ease. It would offer, for instance, a high-tech revamp of our hospitals so as to improve post-operative recovery times. Surgeons could operate through an entirely new breed of rapid-physical-healing facilities. But the positive ramifications from this advanced technology are truly endless – and the possibilities are essentially limitless for *anything that takes time to grow* – food crop growing times could be drastically improved and reduced from months down to days! As I have said earlier into this study, this UFO technology is not just about engineering a flying craft – its an open door to a completely different world...

Note 15—This may also be the same sort of mechanism that was used on Travis Walton when he was lifted off the ground and pulled up into a UFO. It is certainly known in fluid dynamics that there are up-flowing vortices and down-flowing ones, known as positive vortices and negative vortices respectively [Lin 2002] and see Schecter, D.A. & Dublin, D.H.E. *Vortex Motion Driven by a Background Vorticity Gradient* Phys. Rev. Lett. 83 (1999) 2191-2194. Otherwise such a manufactured and contained environment—with its own mass-density gradient and its own gravitational forces—could be used by the crew to affect repairs on their craft with a certain amount of impunity from outside interferences, as seems to have been the case with the Rendlesham Forest UFO in 1980, whose alien engineers were observed floating beneath that UFO inside a beam.

Note 16—Again, this is something I noticed in the manipulation of gravitational force in the anti-vortex hypothesis and in fluid dynamics, and which has been alluded to in both the papers of [Lin 2002] and [Visser 1998a].

Note 17—This is from Planck's law which covers everything up from electromagnetic photons. But see [Penrose 2004] (pp500-501) and [Krogh] [Spears] [Magueijo-Albrecht] [Magueijo 2003]. And also, through an effect called the *Einstein coefficient* an increase in the Planck constant will equate to a lower frequency for when spontaneous photon-emission occurs (and so this would be very advantageous regarding how the UFO's black vortex engine breaks down its incoming energies and is powered).

Note 18—Interestingly, alien species are known to enter into our dimension through the barriers of our physical walls as if they are using the mass of those walls in some frequency transition process. Can we discover how this task is performed—and develop a whole new science for ourselves from it?

And would this new-found skill then correlate with the next dimension that some of us are being informed about? Might this be the next stage in our spiritual evolution, the stage being spoken about by some of the higher evolved beings that have come here for our event, as delivered and discussed in the Dana [Redfield] *The ET-Human Link* book and in Julie Soskin's *Transformation* book. Can we lead ourselves through to this higher form of existence by developing the extraterrestrial's wisdom about bridging the dimensions? The most obvious apparency has been that the ETs have been showing us their scientific technology so we may build advanced flying craft like theirs... but. If there is a side issue to the whole extraterrestrial communication, which has mostly been delivered through the experiences of abductees, that issue seems to be something to do with the more spiritual beings of this Earth realm working out for themselves exactly what doors of opportunity will be opening for them very soon. Personally, I think their underlying communication is all about FREQUENCIES, and in particular the bunching of frequencies. Another dimension but no time no looking back, just a new existence a new stage of evolution...

Notes for Supplement Chapter 12

Note 1—See some of the recent researches into the exotic phenomena of giant halos in high-mass elements (like zirconum-122 and calcium-60) [Meng 2006]. The most recent (2008) Bob Lazar website is at http://www.zamandayolculuk.com/cetinbal/UFOTEKNOLOJII.htm (although, I don't think its actually Bob that runs it). Another excellent appraisal of Bob Lazar can be found on Ken Wright's website at http://www.gravitywarpdrive.com/Anti-Matter_Reactor.htm

Note 2—The negative or expansive gravitational force in the Lazar UFO is based on a flow-through force on the opposite end of a quasi-black hole, and a model of this is briefly explained in sections 8.5 and 10.7. This being a premise which follows on from the Zeldovich-Novikov hypothesis for an imploding energy field (as detailed in note 11 for ch. 8 above). However, the reader would do well to study the basic principles of group velocity reversal in order to get a better idea of an expansive gravitational force. One of the best descriptions of this occurring on the border of a black hole is in [Jacobson], and this is based upon the hypothesis of [Schutzhold-Unruh]. And as Jacobson so rightly points out, this evaporation of expansive gravitational force coming away from a black hole is the negative energy version of Hawking black hole radiation. But also see [Unruh 1974], and from the point of view of the lambda cosmological constant being a repulsive force of gravity see [Greyber 2005a], [Ostriker-Steinhardt], [Davis-Puthoff], and [Mannheim].

The alternative way of getting negative energy, or negative gravitational force is to copy how the black hole generates negative energy out in space. Essentially this would be done by firing subatomic particles into the path of the gravitationally-bound particles rotating in the quasi-black hole. This would follow the Penrose-Williams hypothesis for extracting energy from a black hole system, to produce and direct reverse-flowing waves of gravitational force (of expansive gravitational force) out of that system (which in this case would the gravity amplifier heads) and to channel them into rotating beams of force.

The reader should be careful how they view negative energy density (and this is why I drew the universe diagram—see figure 80). Negative energy density can be described as tension: And when viewed as an expanding ring of energy in an upright funnel shaped container this tension can be felt as an expanding force from above, but a vacuous force from below.

Note 3—As for the gluon serving as a quark modifier see [Dunne].

That there may be found no difference between strong nuclear force and gravitational force see [Goradia 2000a]. Indeed, perhaps what Isaac Newton *should* have said was... that relative to mass objects *in their rest state* gravitational force tends to show its influence most through those mass objects in proportion to their massiveness.

Note 4—See [Carilli-Barthel] on a black hole's 32% efficiency. Also, following along the [Penrose] process is the assumption that if the adjoining magnetic field is strong enough then rotational energy can be extracted from the hole at its transition region (just outside its event horizon), and obviously, if this region is also where charged particle production is most prominent then the two will go hand in hand to produce rotating charged particles. The study by [Li 2000] claims that up to 29% of a black hole's total energy can be extracted.

Note 5—This *oligodynamic* dissociation is a very descriptive term for the breaking down process of matter. Oligo-dynamic was actually the term used by Viktor Schauberger, from the Greek *oligo* (meaning little, small) to denote disintegration, dissociation or decomposition of matter into lesser constituent parts. The antiproton-proton annihilation decay process to pions, mesons, pair-productions and gamma-ray production is described in [Borowski].

Note 6—It has long been known that a black hole will only bring in mass of a density that is appropriate for the forces of its accretion disk [Penrose 1969]. The perfect example of this is the black hole W50 which, although it has 'captured' the SS 433 star it can only 'consume' it by breaking it down bit by bit (through its rotational shearing forces). This is logical when one considers that the black hole needs to be protected from being too gluttonous by the spiraling action of its accretion disk. If this suggests a gravitational polarization effect, and that the overall force of gravity generated by a black hole takes on a spiral path and not linear one, then so be it.

Note 7—The first successful annihilations of antiprotons were carried out in 1955/6 where the result was a total energy output estimated at 1400 MeV [Goldhaber]. NASA's estimate for its antiproton-proton reactions is 1876 MeV [Borowski] (although in NASA's case it would still be designing that antiproton-proton reaction power around an explosive force to be used in a space-craft which utilizes the same-old-same-old rocket technology of out-the-back prop-ulsive thrust). In the industry standard *Atomic Data and Nuclear Data Tables* it indicates that the basic binding energy figure for the energy released from a 115 proton / 184 neutron element would be just over 2126 MeV [Moller-Nix-Kratz].

On the other hand though, in the Penrose pair-production process around a photon orbit ring (as developed in the Andreasson UFOs with quasi-black holes), the relativistic electron-positron pairs created and ejected as a result of hard photon collisions at those rotating rings, could have energies of up to 4 GeV (4000 MeV) [Williams 2004b]. And because with the Penrose-process *ejected* particle energy is proportional to *consumed* energy then it means that mass-energy particles with up to 4 GeV of energy will be constantly fed into the quasi-black holes of those UFOs. I know it might seem a little facitious to suggest here that the Lazar UFO, while it has THREE quasi-black holes, might be more power-ful than the Andreasson UFO which only sports ONE quasi-black hole, from these ballpark figures there may one day, hopefully, be developed by scientists a proper range of energy generation calculations for these UFO drive units (when those scientists recognise how energy-efficient these black vortexes are).

Note 8—See the studies of ununpentium 115 and other superheavy elements in [Meng 2000] [Meng 2006] [Geng 2003] [Geng 2004] [Geng 2006].

Note 9—See [Meng 2006] and the very informative [Cwiok 2005] study on superheavies. Also see the excellent presentation by [Heenen-Nazarewicz].

Note 10—See the [Saa-Schiffer] paper on gravitational polarization; and the [Kittel] book on dielectrics (in that book's section 13 on "Dielectrics and Ferroelectrics", and its fig.25).

Note 11—Because so many of the unstable superheavy elements have outer shells of weakly bound nuclei this region of elements is also called the drip line, where nucleons quite literally drip away from those atoms. See drip-line charts in [Meng 2006] and [Heenen-Nazarewicz]. Obviously, I am not privy to whether the skin around the extraterrestrial's ununpentium was of protons or neutrons. A neutron skin comes up as more prevalent in research laboratories so perhaps the intention is to push neutrons out from the wedge's point and rely on neutron collisions to trigger some sort of spontaneous decay where the protons drip off, or rely on the wedge's free-neutrons to decay, which they will do after a minute or so, down to a proton (plus an electron and a neutrino), and then hit those protons in order to produce the antiprotons. In fact, because the Lorentz force of the magnetic field lines inside the reactor will separate the protons from the neutrons, and will spin any free protons around the outside of the wedge, these protons would end up spinning directly into the path of the bullet-protons coming up from the cyclotron. Hmm, next time I see Bob Lazar I'll have to ask him about this...

Note 12—For the astute reader who has spotted that this halo effect is normally only ascribed to lighter nuclei, here is the more exotic part to this theory, as promised in the main text... How can it be that a superheavy element can be 'assembled' to have stability but at the same time have a nucleon halo?

Well, going back to the extraterrestrials existing in an environment with a greater density than ours, there is mounting evidence to suggest that the ETs can engineer a change in the dimensional constant, also known as the fine structure constant (α), which proportions the effectiveness of the electromagnetic spectrum (ie the speed of light), and because this constant also decides the effective density of all mass objects then it also proportions the size of those objects, and too the size of atoms.

Most of the discussion about this newly discovered differential between extraterrestrial controlled environments and our present environment can be found in my chapter 11, but something very pertinent that I should repeat here is that when Betty Andreasson was inside one of the UFOs that abducted her, she saw a group of Greys watching over a cylinder in the middle of the UFO and when she asked what they were doing, she was told they were watching molecules. This would indicate that the environment those ETs (and Betty) were then existing in for the duration of that journey inside that UFO, was a manufactured environment where the velocity of electromagnetic waves was considerably SLOWER than the electromagnetic wave velocity we know of on earth and always expect to be at (c). And as I have already alluded to in chapter 11 above, this ability of altering the dimension constant, inside UFOs, would explain why the *interiors* of UFOs have been observed to be smaller than the UFO's apparent *exterior* size when viewed from the outside...

The implication of this mass-density differential most relevant here, and which could well be exploited for our use in this world's mass-density, is that there will exist a different inter-atomic spacing with different degrees of Coulomb and nucleon-nucleon strong-force repulsion forces inside every atom. And this ununpentium 115's atomic structure could be seen as a perfect example of the exploitability of this effect. For it may be possible to manufacture a quantity of such an element as the superheavy 115, in that higher-mass density (where atomic materials will have a 'slower' frequency and will exhibit a lesser degree of excitation) and give it a greater degree of stability than it can be manufactured with here. On bringing that material into this 'faster' mass-density environment that material would still be affected by an increase in its inter-atomic forces—but if that element was to be brought in as a stable element it may remain stable. And this would be a great improvement, because at the moment the physicists of this world can only synthesize a form of ununpentium 115 which only has a half-life of 87ms (mainly because it can only be produced with 173 neutrons). This quantity of neutrons is 11 shy of what is required, from the fact that physicists have calculated that the 'magic numbers' for this element is 115 protons and 184 neutrons. So, possibly, this is where altered density manufacturing might be very useful, for it would lessen the inter-atomic forces of repulsion during the manufacturing process to allow superheavy elements to be synthesized with stability.

Note 13—Negative radial momentum is a geometry-based mechanism which gives ejected particles a curling trajectory, and obviously this gives parallel structure (collimation) to those particles so that they can organize themselves into a beam. This geometry-induced form of collimation or setting of the helical pitch angle will be dependent also on the type and energy of the scattered particles, for example Williams (2001) cites 1 to 30 degrees for Penrose-Compton scattered particles and 0.5 to 25 degrees for negative-positive pairs. The De Felice-Carlotto paper gives the basic formula for the law of geometry-induced collimation. Another rendition of this curling effect is that which is known as the Umstadter CURL effect, which is mentioned in [Mourou-Umstadter] paper (on p84). I get the idea also that this up-travelling force will need to be braked before it exits from the end of this central tube, and so this braking will add to the azimuthal velocity of this rotating field.

Note 14—Referring here in particular to the UFOs filmed by Mexican Department of Defense aircraft in March 2004 during their surveillance flight over the Yucatan Peninsula. And, of course, this is the area taken up by R.M.L. Baker Jr (with his US patents 6,417,597 granted Jul 9 2002; and 6,784,591 granted Aug 31 2004) where he is developing converters of high frequency gravitational waves in the 10^{15} Hz region (infra-red to visible to UV) by using quartz phonon converters.

Note 15—Black holes, and even miniaturized ones, are known to produce oscillating gravitational forces [Saa-Schiffer] [Cui-Zhang-Chen] [Press-Teukolsky] [Teukolsky-Press] [Bardeen] [Kiefer-Weber] (and to radiate them through the agent of the magnetic flux fields), and so from a combination of their electric fields and these gravitational force pulsings, the more massive nucleons are being accelerated relative to their accompanying electrons, and are consequently creating micro-currents within those heavy-mass materials.

This follows a relatively recent finding that strong electric fields or short pulses move electrons from their normal locations inside the atoms of a semiconductor. Then, because these movements are sudden their accelerations generate electric currents within the semiconductor, and these accelerations emit electromagnetic radiation in the infra-red region—at terahertz wavelengths [Souza-Egues] [Hasselbeck] [Leitenstorfer].

In the case of the base disc structuring in the Andreasson and Roswell UFOs, where the base disc has undulations of bismuth sandwiched through that magnesium-zinc disc, what those undulations of the bismuth are doing is this; when the gravitational polarization force radiates outward from the core through the base disc it exerts its influence up one length of the undulation, which generates a localized electric current, and in its effort to produce a bulk effect and spread through the adjoining atoms of that material that perturbation then goes *down* through the next undulation of that bismuth material, and this moves the electric current radially. This process repeats continually all through the rest of that material, so that the electric field expands radially through the circular base disc to its outer perimeter. This up-down influence creates an oscillating frequency—but here is the twist—those oscillations won't be moving at the normal speed of electromagnetic waves they will be traveling much, much slower. So while it expands to the perimeter the electric field as it spreads radially has to jump up and down, and this produces the exact same perturbation as with the displacement currents of electrons excited and accelerated within the semiconductor in the example above. So when large amounts of these movements repeatedly radiate from the base disc's central hole they collectively produce emissions of terahertz oscillations (at infra-red wavelengths but at speeds which will depend on what material is used) and carry with them the UFO's gravitational forces radiating from its black vortex.

Note 16—It should be noted that Huntsville's Ning Li does say, in her 1992 paper, that no physicist can substantiate whether or not gravitational waves do actually travel at light speed; and that the phase velocity of gravitational force is, as yet, completely unknown to any physicist [Li-Torr]. Of course, if gravitational perturbations do not travel at the same velocity as electromagnetic waves, then this would explain why the physicists who have postulated that spin-1 waves (electromagnetic) can be converted into spin-2 waves (gravitational waves), as did Raymond Chiao, for instance, have been getting zero results from their experiments to create and to detect gravitational waves (the reader may like to see my thoughts on 'gravitational wave spin' in my chapter 6 notes 11, and 12). And surely, that the experiments of LIGO and VIRGO in the USA and in Italy, respectively, to detect gravitational waves, which are also based upon this fundamental assumption that gravitational waves SHOULD travel with light speed, it may help to explain why these two projects have also not yet been able to detect any of those gravitational waves. Not that I would have the audacity to tell them they're wrong in any way, of course.

Note 17—The Roswell UFO's base disc, or remnants of it that were retained by the local towns-people, showed that its undulations are 1-4 microns thick with a period of repetition set into them of approximately 6-10 microns—well within the terahertz range.

Interestingly, if this acoustic phonon effect can be generated in the outer layers of the craft, throughout the upper and lower hulls, it could produce a double-layer effect or double skin all around that craft. With an acoustic double-layer which would effectively be OUTSIDE the material layer, then, when any incident wave strikes that craft then TWO waves will be refracted-reflected off and away from that surface (and beyond it to an observer). But, should those two refracted-reflected waves interfere between eachother, and become destructively matched, then that craft will become almost invisible [Oh 2004].

Note 18—The basic SAW acoustic wavelength (λ) is calculable by $\lambda_o = v/f_o$ where (v) is the SAW velocity, and (f_o) is the fundamental frequency [Ebbecke] [Oh 2004]

Bibliography of References
(Where the reference is to a book I have tried to indicate this by placing the published date directly after the authors name).

Abbott, L. *The Mystery of the Cosmological Constant* Sci. Am. 258 (May 1988) 82-88.

Abramowicz, M.A. *Black Holes and the Centrifugal Force Paradox* Sci. Am. (Mar 1993) 26-31.

Abramowicz, M.A. et al *Gravitational Waves from Ultracompact Stars: The Optical Geometry View of Trapped Modes* Class. Quantum Grav. 14 (1997) L189-L194.

Adams, A. *Helical Particle Waves of J.L. Gaasenbeek* (1986). Webpage: http://www.geocities.com/aliadams/helicalwave.htm

Air Pressure webpage: http://www.research.umbc.edu/~tokay/chapter5.html

Alfvén, H. & Falthammer, C. (1963) *Cosmical Electrodynamics*.

Alu, A. & Engheta, N. *Achieving Transparency with Plasmonic Coatings* Phys. Rev E72 (2005) 016623. preprint: http://arxiv.org/pdf/cond-mat/0502336

Andersson, N. & Glampedakis, K. *A Superradiance Resonance Cavity Outside Rapidly Rotating Black Holes* Phys. Rev. Lett. 84 (2000) 4537-4540. preprint: http://arxiv.org/pdf/gr-qc/9909050

Andreasson Affair books are: R.E. Fowler (1979) *The Andreasson Affair*; R.E. Fowler (1982) *The Andreasson Affair—Phase Two*; R.E. Fowler (1990) *The Watchers*; R.E. Fowler (1995) *The Watchers II*; R.E. Fowler (2000) *The Andreasson Legacy*; Betty Luca (1999) *Extraterrestrial Communications—ASBT* (Pt I & II).

Ang, C. & Yu, Z. *DC Electric-Field Dependence of the Dielectric Constant in Polar Dielectrics: Multipolarization Mechanism Model* Phys. Rev. B69 (2004) 174109.

Annino, G. et al *Study on Planar Whispering Gallery Dielectric Resonators. II—A Multiple-Band Device* (Accepted for Int. J. Infrared and Millimeter Waves 2002). preprint: http://arxiv.org/pdf/physics/0203010

Antonijoan, J. et al *Non-Linear Spirals in the Taylor-Couette Problem* Phys. Fluids 10 (1998) 829-838.

Antonsen Jr, T.M. & Bian, Z. *Ionization Induced Scattering of Short Intense Laser Pulses* Phys. Rev. Lett. 82 (1999) 3617-3620.

Assaf, M. & Meerson, B. *Parametric Autoresonance in Faraday Waves* Phys. Rev. E72 (2005) 016310. preprint: http://arxiv.org/pdf/physics/0506026 Website: http://www.phys.huji.ac.il/~assaf/index_files/faraday_autoresonance.pdf

Audretsch, J. & Skarzhinsky, V.D. *Quantum Processes Beyond the Aharonov-Bohm Effect* Found. Phys. 28 (1998) 777-788. preprint: arXiv:hep-th/9709095

Balbus, S.A. & Hawley, J.F. *A Powerful Local Shear Instability in Weakly Magnetized Disks—Linear Analysis* Astrophys. J. 376 (1991) 214-222.

Bardeen, J.M. *Rotating Black Holes: Locally Nonrotating Frames, Energy Extraction, and Scalar Synchrotron Radiation* Astrophys. J. 178 (1972) 347-369.

Barrau, A. & Feron, C. & Grain, J. *Astrophysical Production of Microscopic Black Holes in a Low—Planck-scale World* Astrophys. J. 630 (2005) 1015-1019. preprint: http://arxiv.org/pdf/astro-ph/0505436

Basak, S. & Majumdar, P. *Reflection Coefficient for Superresonant Scattering* Class. Quant. Grav. 20 (2003) 2929-2936. preprint: http://arxiv.org/pdf/gr-qc/0303012

Basak, S. & Majumdar, P. *'Superresonance' from a Rotating Acoustic Black Hole* preprint: http://arxiv.org/pdf/gr-qc/0203059

Bearden, T.E. & Rosenthal, W. *On a Testable Unification af Electromagnetics, General Relativity, and Quantum Mechanics* Association of Distinguished American Scientists (1991). Webpages: http://www.enterprisemission.com/hyper2.html and http://www.geocities.com/Area51/Shadowlands/9654/bearden/testunify.html

Bekenstein, J.D. & Schiffer, M. *The Many Faces of Superradiance* Phys. Rev. D58 (1998) 064014. preprint: http://arxiv.org/pdf/gr-qc/9803033

Benedek, G. & Boscolo, I. *A Model for Photoemission from Prepoled Ferroelectric Ceramics* Appl. Phys. Lett. 72 (1998) 522-524.

Benedek, G. et al. *Electron Emission from Ferroelectric/Antiferroelectric Cathodes Excited by Short High-Voltage Pulses.* J. Appl. Phys. 81 (1997) 1396-1403.

Berkovsky B.M. (1978) *Thermomechanics of Magnetic Fluids*

Bernstein, H.J & Phillips, A.V. *Fiber Bundles and Quantum Theory* Sci. Am. 245 (July 1981) 95-109.

Berry, M.V. et al *Wavefront Dislocations in the Aharonov-Bohm Effect and its Water Wave Analogue* Eur. J. Phys. 1 (1980) 154-162. Website: http://www.phy.bris.ac.uk/people/berry_mv/the_papers/Berry096.pdf
Berti, E. *Black Holes in a Bathtub* J. Phys. Conf. Ser. 8 (2005) 101-105.

Berti, E. et al, *Quasinormal Modes and Classical Wave Propagation in Analogue Black Holes* Phys. Rev. D70 (2004) 124006. preprint: http://arxiv.org/pdf/gr-qc/0408099

Bessonov, E.G. *Grasers Based on Particle Accelerators and on Lasers* (1998). preprint: arXiv:physics/9802037

Bicak, J. & Janis, V. *Magnetic Fluxes Across Black Holes* Mon. Not. R. Astr. Soc. 212 (1985) 899-915.

Bicak, J. & Ledvinka, T. *Electromagnetic Fields Around Black Holes and Meissner Effect* Il Nuovo Cimento 115 (2000) 739-750. preprint: http://arxiv.org/pdf/gr-qc/0012006

Binder, B. *Berry's Phase and Fine Structure* (2002a) PITT-PHIL-SCI00000682. Website: http://www.quanics.com/alfa137MN6.pdf (updated version) or http://philsci-archive.pitt.edu/documents/disk0/00/00/06/82/index.html or http://eprints.anu.edu.au/archive/00000481/00/alfa137MN5pc1.pdf

Binder, B. *Geometric Phase Locked in Fine Structure* (2002b) PITT-PHIL-SCI00000782. Webpage: http://philsci-archive.pitt.edu/documents/disk0/00/00/07/82/index.html

Birula, B. et al *The s-wave Implosion and Explosion of Free Particles* (2001). preprint: http://arxiv.org/pdf/quant-ph/0110116

Bibliography

Blandford, R.D. & Begelman, M.C. & Rees, M.J. *Cosmic Jets* Sci. Am. (May 1982) 84-94.

Blandford, R.D. & Payne, D.G. *Hydromagnetic Flows from Accretion Discs and the Production of Radio Jets* Mon. Not. R. Astr. Soc. 199 (1982) 883-903.

Blandford, R.D. & Znajek, R.L. *Electromagnetic Extraction of Energy from Kerr Black Holes* Mon. Not. R. Astr. Soc. 179 (1977) 433-456.

Blum, R. & Blum, J. (1974) *Beyond Earth: Man's Contact With UFO's*

Bogaerts, A. et al. *Gas Discharge Plasmas and their Applications* Spectrochemica Acta Part B57 (2002) 609-658 (sect. 2.2). Website: http://www.phys.tue.nl/EPG/epghome/papers/2002/science.pdf

Bondarev, I.V. & Hyodo, T. *Sensitivity of Positronium Momentum Distribution to Phase Transitions in Crystalline Dielectrics* Acta Physica Polonica A107 (2005) 673-684.

Borowski, S.K. *Comparison of Fusion/Antiproton Propulsion Systems for Interplanetary Travel* 23rd Joint Propulsion Conf. San Diego (Jun 29-Jul 2, 1987). Website: http://gltrs.grc.nasa.gov/reports/1996/TM-107030.pdf

Boswell, R.W. *Very Efficient Plasma Generation by Whistler Waves Near the Lower Hybrid Frequency* Plasma Phys. and Contr. Fusion 26 (1984) 1147-1162.

Bottger, U. *Dielectric Properties of Polar Oxides* (in *Polar Oxides: Properties, Characterization, and Imaging* (2005) eds: Waser, R. & Bottger, U. & Tiedke, S. 11-38). Website: http://media.wiley.com/product_data/excerpt/21/35274053/3527405321.pdf

Bourgeois, P.-Y. et al *Maser Oscillation in a Whispering-Gallery-Mode Microwave Resonator* (2005). preprint: arXiv:quant-ph/0506076

Bucher, M.A. & Spergel, D.N. *Inflation in a Low-Density Universe* Sci. Am. (Jan 1999) 63-69.

Bulanov, S.V. & Sakai, J-I. *Magnetic Field Line Reconnection in Weakly Ionized Plasmas* Astrophys. J. Supp. 117 (1998) 599-625.

Busch, K.L. *Desorption Ionization Mass Spectrometry* Jnl. of Mass Spectr. 30 (1995) 233.

Calkins, F.T. & Flatau, A.B. *Terfenol-D Sensor Design and Optimization* Journal of Noise Control Engineering (submitted 4/97). Website: http://www.aerosmart.umd.edu/TechPubs/calkins_1.pdf

Cardoso, V. et al, *The Black Hole Superradiant Instabilities* Phys. Rev. D70 (2004) 044039. preprint: http://arxiv.org/pdf/hep-th/0404096

Carey, R. & Isaac, E.D. (1996) *Magnetic Domains*

Carilli, C.L. & Barthel, P.D. *Cygnus A* Astron. Astrophys. Rev. 7 (1996) 1-54.

Carmon, T. et al *Rotating Propeller Solitons* Phys. Rev. Lett. 87 (2001) 143901.

Carter, B. *Charge and Particle Conservation in Black Hole Decay* Phys. Rev. Lett. 33 (1974) 558-561.

Carter, T.A. *Measurement of Lower-Hybrid Drift Turbulence in a Laboratory Reconnection Experiment* Dept. Phys. & Astron, UCLA (April 15, 2003) (reconnect diagram is from website http://www.physics.ucla.edu/~tcarter/lhdi_talk.pdf and his website URL is http://www.physics.ucla.edu/~tcarter/)

Chernobyl UFO involvement. See the Russian website at: http://ufoinv.pochta.ru/ua/chernobl.htm

Chiao, R.Y. *Superconductors as Quantum Transducers and Antennas for Gravitational and Electromagnetic Radiation* (2002). preprint: arXiv:gr-qc/0204012

Chiao, R.Y. et al *Search for Quantum Transducers Between Electromagnetic and Gravitational Radiation: A Measurement of an Upper Limit on the Transducer Conversion Efficiency of Yttrium Barium Copper Oxide* (2003). preprint: arXiv:gr-qc/0304026

Chu, K.R. *The Electron Cyclotron Maser—Relativity in Action* AAPPS Bulletin 15 No.1 (Feb 2005) (also: Rev. Mod. Phys. 76 (2004) 489-540).

Citrin, D.S. *Excitonic Radiative Dynamics in Multiple Quantum Wells* Phys. Stat. Sol. B188 (1995) 43.

Clark, P.A. & Clark, A. *Radiative Damping of Trapped Gravity Waves in the Solar Atmosphere* Solar Phys. 30 (1973) 319-325.

Coats, C. (1996) *Living Energies*

Coherence of Electromagnetic Waves (Wikipedia) webpage: http://en.wikipedia.org/wiki/Coherence_(physics)

Collins, R.M. & Doty, R.C. & Cooper, T.S. (2008) *Exempt from Disclosure (2nd Ed)*

Conway, J.E. & Murphy, D.W. *Helical Jets and the Misalignment Distribution for Core-Dominated Radio Sources* Astrophys. J. 411 (1993) 89-102.

Cook, N. (2001) *The Hunt for Zero Point*

Corley, S. & Jacobson, T. *Black Hole Lasers* Phys. Rev. D59 (1999) 124011. preprint: arXiv:hep-th/9806203

Corso, Coll. P.J. (1997) *The Day After Roswell*

Cothran, C.D. et al *Three-dimensional Structure of Magnetic Reconnection in a Laboratory Plasma* Geophys. Res. Lett. 30 (2003) 17.1-17,4.

Cui, W. & Zhang, S.N. & Chen, W. *Evidence for Frame-dragging Around Spinning Black Holes in X-ray Binaries* Astrophys. J. 492 (1998) L53-L57.

Cwiok, S. et al *Shape Coexistence and Triaxiality in the Superheavy Nuclei* Nature 433 (2005) 705-709.

Cwiok, S. et al *Structure of Odd-N Superheavy Elements* Phys. Rev. Lett. 83 (1999) 1108-1111.

Cyclones and anticyclones. Webpage: http://www.uwsp.edu/geo/faculty/ritter/geog101/textbook/circulation/cyclones_and_anticyclones.html

Dalnegorsk "611" UFO webpages: http://www.stealthskater.com/Documents/Dalnegorsk_01.doc
The Russian websites are: http://vlad.kp.ru/2006/02/22/doc103572/ (in Russian)
http://www.gorizonto.ru/mag43.html (in Russian, and I'd recommend the Google foreign-language translator as it takes only a few minutes to translate the Russian).

Dar, A. & De Rujula, A. *The Threat to Life from Eta Carinae and Gamma-Ray Bursts* Astrophysics and Gamma Ray Physics in Space (eds. A. Morselli and P. Picozza), Frascati Physics Series XXIV (2002) 513-523. preprint: http://arxiv.org/pdf/astro-ph/0110162

Davies, P. (2001a) *How To Build a Time Machine*

Davies, P. *Paradox Lost* New Scientist 157 (Mar 21 1998) pp26. Webpage: http://www.fortunecity.com/emachines/e11/86/paradox.html

Davies, P.C.W. *Quantum Vacuum Noise in Physics and Cosmology* Chaos 11 (2001b) 539-547.

Davis, E.W. *UFO Phenomenology* and *UFO-Extraterrestrial Intelligence Hypothesis (ETH)* Aerospace Physics & Astrophysics Div. National Institute for Discovery Science (NIDS) (2001). Website: http://www.nidsci.org/pdf/davis_mufon2001slides.pdf

Davis, E.W. & Puthoff, H.E. *Experimental Concepts for Generating Negative Energy in the Laboratory* Space Technology and Applications International Forum STAIF (2006). Website: http://www.earthtech.org/publications/davis_STAIF_conference_1.pdf

Davis, M. *Small Worlds* (on carbon-silicon) Southern Illinois University Carbondale (2001) webpage: http://www.siu.edu/~perspect/01_sp/carbon.html

Dawson, J.M. *Plasma Particle Accelerators* Sci. Am. (Mar 1989) 34.

Deardorff, J. & Haisch, B. & Maccabee, B. & Puthoff, H.E. *Inflation-Theory Implications for Extraterrestrial Visitation* Journal of the British Interplanetary Society (JBIS) 58 (2005) 43-50.

De Felice, F. & Carlotto, L. *Jet Dynamics in Black Hole Fields: A Collimation Mechanism* Astrophys. J. 481 (1997) 116-126. Website: http://cdsaas.u-strasbg.fr:2001/ApJ/journal/issues/ApJ/v481n1/33804/33804.pdf

De Felice, F. & Zanotti, O. *Jet Dynamics in Black Hole Physics: Acceleration During Subparsec Collimation* (1999). preprint: arXiv:astro-ph/9912413 Website: http://arxiv.org/abs/astro-ph/9912413

De Rujula, A. *Gamma-ray Bursts and the Sociology of Science* to be published in the Proceedings of the Neutrino Telescopes Workshop (Venice 2003). preprint: http://arxiv.org/pdf/hep-ph/0306140

Deibel, J.A. et al *Frequency-dependent Radiation Patterns Emitted by THz Plasmons on Finite Length Cylindrical Metal Wires* Optics Express 14 #19 (2006) 8772-8778.

Depp, J.G. *Electron Charge Renormalization and the PV-RN Model* (May 2, 2006). Webpage: http://www.geocities.com/psistar@sbcglobal.net/Renormalization.pdf

Desiato, T.J. & Storti, R.C. *Electro-Gravi-Magnetics (EGM) Practical Modeling Methods of the Polarizable Vacuum -1* Journal of Advanced Propulsion Methods (May 14, 2003). Webpage: http://65.108.189.168/Docs/EGM_I_v1.pdf

DeWitt, B. *Quantum Gravity* Sci. Am. 249 (Dec 1983) 104-115.

Dielectrics webpage: http://hyperphysics.phy-astr.gsu.edu/HBASE/electric/dielec.html

Di Matteo, T. *Magnetic Reconnection: Flares and Coronal Heating in Active Galactic Nuclei* Mon. Not. R. Astr. Soc. 299 (1998) L15-L20. preprint: http://arxiv.org/pdf/astro-ph/9805347

Dimopoulos, S. & Landsberg, G. *Black Holes at the [Large Hadron Collider] LHC* Phys, Rev. Lett. 87 (2001) 161602. preprint: arXiv:hep-ph/0106295 Webpage: http://arxiv.org/abs/hep-ph/0106295

Drewsen, M. *Laser Hazards* (2002)

Dunne, P. A. *Reappraisal of the Mechanism of Pion Exchange and its Implications for the Teaching of Particle Physics* Physics Education 37 (May 2002) 211-222.

Ebbecke, J. et al *Quantized Charge Pumping Through a Quantum Dot by Surface Acoustic Waves* (2003). preprint: arXiv:cond-mat/0312304

ElectricAccel3 note: *Acceleration With Alternating Electric And Magnetic Fields* Webpage: http://agni.phys.iit.edu/~vpa/electromagnetic.html

ElectricAccel4 note: *Electric Fields-The Source of Particle Acceleration in Cosmic Plasma* Webpage: http://public.lanl.gov/alp/plasma/elec_fields.html

Electron mass.htm webpage: http://qd.typepad.com/27/2005/10/is_a_faster_ele.html

Ellison, W. J. & Lamkaouchi, K. & Moreau, J.-M. *Water: A Dielectric Reference* Journal of Molecular Liquids 68 (1996) 171-279.

Etreed Induction webpage (on UFOs): http://www.softcom.net/users/etreed/index/induct-8.htm

Evans, J. et al *Matter Waves in a Gravitational Field: An Index of Refraction for Massive Particles in General Relativity* Am. J. Phys. 69 (2001) 1103-1110. preprint: arXiv:gr-qc/0107063

Fargion, D. *Prompt and Delayed Radio Bangs at Kilohertz by SN1987A: A Test for Graviton-Photon Conversion* (1995). preprint: arXiv:astro-ph/9604047

Federici, F. *Superradiance from BEC Vortices: a Numerical Study* (2005). preprint: arXiv:gr-qc/0503089

Ferrari, V. *Different Approaches to the Study of the Gravitational Radiation Emitted by Astrophysical Sources* Annalen der Phys. 1 (2000) 3-17. preprint: http://arxiv.org/pdf/gr-qc/9912074

Feynman, R.P. et al (1970) *The Feynman Lectures on Physics*

Finlay, C.C. *Alfvén Waves* (May 31 2005) Inst. Geophys. Tectonics Univ. Leeds

Flanagan, P. US Patent # 4,743,275 (1988) *Electron Field Generator*. Also US Patent # 4,391,773 (1983).

Ford, L.H. & Roman, T.A. *Negative Energy, Wormholes and Warp Drive* Sci. Am. (Jan 2000) 46-53.

Ford, L.H. & Roman, T.A. *The Quantum Interest Conjecture* Phys. Rev. D60 (1999) 104018. preprint: http://arxiv.org/pdf/gr-qc/9901074

Foteinopoulou, S. *Refraction in Media with a Negative Refraction Index* Phys. Rev. Lett. 90 (2003) 107402.

Fowler, R.E. (1979) *The Andreasson Affair*; (1982) *The Andreasson Affair—Phase Two*; (1990) *The Watchers*; (1995) *The Watchers II*; (2000) *The Andreasson Legacy*.

Frank-Kamenetskii, D.A. (trans. J. Norwood 1972) *Plasma—Fourth State of Matter*.

Freund, H.P. & Parker, R.K *Free-electron Lasers* Sci. Am. (Apr 1989) 56-61.

Frisch, P.C. *The Journey of the Sun* (Enrico Fermi Institute Report 97-23) (1997) 1-20. preprint: http://arxiv.org/abs/astro-ph/9705231

Bibliography

Frisch, P.C. & Slavin, J.D. *The Sun's Journey Through the Local Interstellar Medium: The PaleoLISM and Paleoheliosphere* Astrophysics and Space Sciences Transactions (2005). preprint: arXiv:astro-ph/0606743

Froning Jr, H.D. *Requirement for Rapid Transport to the Further Stars* J.B.I.S. 36 (1983) 227-230.

Furusawa, K. & Sakai, J-I. *Simulation of the Collision of Magnetic Flux Tubes in the Quiet Solar Photosphere* Astrophys. J. 540 (2000) 1156-1171.

Gaasenbeek, J.L. *Helical Particle Waves* (1990) Webpage: http://www2.rideau.net/gaasbeek/spap1.html

Gallimore, J.G. *Hydrostatic Space-Time and Gravitational Monitors Monitoring Space/Time Stresses* (March 8, 1992) published on Keelynet

Gato-Rivera, B. *Brane Worlds, the Subanthropic Principle and the Undetectability Conjecture* (2003). preprint: http://arxiv.org/pdf/physics/0308078

Gavrilov, S.P. & Gitman, D.M. *Vacuum Instability in External Fields* Phys. Rev. D53 (1996/2005) 7162. preprint: http://arxiv.org/pdf/hep-th/9603152

Geng, L.S. et al *Alpha-decay Chains of $^{288}_{173}115$ and $^{287}_{172}115$ in the Relativistic Mean Field Theory* Phys. Rev C68 (2003) 061303. preprint: http://arxiv.org/pdf/nucl-th/0310032

Geng, L.S. et al *Proton and Neutron Skins of Light Nuclei Within the Relativistic Mean Field Theory* Nucl. Phys. A730 (2004) 80-94. preprint: http://arxiv.org/pdf/nucl-th/0309009

Geng, L.S. *The Stability and the Shape of the Heaviest Nuclei* J. Phys. G32 (2006) 573-582. preprint: http://arxiv.org/pdf/nucl-th/0602030

Gertsenshtein, M.E. *Wave Resonance of Light and Gravitational Waves* Sov. Phys. JETP 14 (1962) 84-85.

Geuther, J.A. & Danon, Y. *Electron and Positive Ion Acceleration with Pyroelectric Crystals*. J. Appl. Phys. 97 (2005) 074109.

Ginzburg, V.L. Compt. Rend. Acad. Sci. USSR 56 (1947) 583. A brief note on the effect can be found in a 1958 lecture by I. E. Tamm. Website: http://nobelprize.org/physics/laureates/1958/tamm-lecture.pdf

Goldhaber, G. *The Observation of Antiproton Annihilation* Antiproton Symposium (Oct.28 2005) Website: http://inpa.lbl.gov/pbar/talks/F4_Goldhaber.pdf

Golubitsky, M. et al *Hopf Bifurcation from Rotating Waves and Patterns in Physical Space* J. Nonlinear Science 10 (2000) 69-101.

Good, T. (1993) *Alien Contact*

Good, T. (2006) *Need to Know*

Goodge, A.C. *Lens Focused Microwave Reflectometry Concepts for Ceramic Coating Characterization* M.S. Thesis (University of Virginia Engineering Dept) (1997), Chapter2 on Microwave Fundamentals [and Permittivity]. Website: http://www.ipm.virginia.edu/process/PVD/Pubs/thesis4/chapter2.pdf

Goradia, S.G. *Classical Equation of Gravity to Quantum Limit* (2000a). preprint: http://arxiv.org/pdf/math-ph/0009025

Goradia, S.G. *Consistent Equation of Classical Gravitation to Quantum Limit and Beyond* (2000b). preprint: http://arxiv.org/pdf/physics/0011066

Gorder, P.F. *Vortex Drive* New Scientist (23 Oct 2004) 31-34.

Granatstein, V.L. et al *A Quarter Century of Gyrotron Research and Development* IEEE Trans. Plasma Sci. 25 (1997) 1322-1335.

Granatstein, V.L. *Status of High Power Gyrotron Technology* Lab Plasma & Fusion Energy Studies (1987) 1696-1700.

Green, J.L. et al *Water and Solutions at Negative Pressure: Raman Spectroscopic Study to -80 Megapascals* Science 249 (1990) 649-652.

Greer, S. CSETI disclosure project and Zero-Point-Energy research. Webpage: http://www.cseti.com

Greyber, H. D. *On the Electrodynamics of Cosmic Repulsion* (Aug 21 2005a). preprint: http://arxiv.org/pdf/astro-ph/0509222

Greyber, H. D. *Powering and Structuring the Universe Starting at Combination Time—On the Electrodynamics of the Big Bang Universe* [presented at the 22nd Texas Symposium on Relativistic Astrophysics at Stanford University, Dec 13-17, 2004] (Sep 8 2005b). preprint: http://arxiv.org/pdf/astro-ph/0509223

Grigoryan, M.L. (2005a) *Intense Radiation from a Relativistic Electron Rotating About a Dielectric Ball* [presented at the RSNE NANO-2005 Conference Moscow 14-19 Nov, 2005]. preprint: http://arxiv.org/pdf/hep-th/0512080

Grigoryan, M.L. et al. (2005b) *High Power Cherenkov Radiation From a Relativistic Particle Rotating Around a Dielectric Ball* [presented at the Workshop RC2005, Frascati, Italy 25-28 July, 2005]. preprint: http://arxiv.org/pdf/hep-th/0512106

Gundel, H. et al. *Fast Polarization Changes in Ferroelectrics and Their Application in Accelerators* Nucl. Instr. Meth. in Physics Res. A280 (1989) 1-6.

Gundel, H. et al. *Time-Dependent Electron Emission from Ferroelectrics by External Pulsed Electric Fields* J. Appl. Phys. 69 (1991) 975-981.

Gyrotron webpage: http://www.cpii.com/mpp/products/primer.shtml

Haisch, B. & Rueda, A. *On the Relation Between a Zero-Point-Field-Indiced Inertial Effect and the Einstein-de Broglie Formula* Phys. Lett. A268 (2000a) 224-227. preprint: arXiv:gr-qc/9906084

Haisch, B. & Rueda, A. *Toward an Interstellar Mission: Zeroing in on the Zero-Point-Field Inertia Resonance* STAIF-2000 (2000b). preprint: arXiv:physics/9909043

Haisch, B. & Rueda, A. & Puthoff, H.E. *Inertia as a Zero-Point Field Lorentz Force* Phys. Rev. A49 (1994) 678-694.

Harding, J.W. & White, F.W.G. *On The Modes of Vibration of a Quartz Crystal* Phil. Mag. 8 (1929) 169-178.

Harrop, P.J. (1972) *Dielectrics*

Hasselbeck, M.P. et al *Emission of Terahertz Radiation from Coupled Plasmon-phonon Modes in InAs* Phys. Rev. B65 (2002) 233203.

Hau, L.V. et al *Light Speed Reduction to 17 Metres per Second in an Ultracold Atomic Gas* Nature 397 (1999) 594-598.

Hausman, M.A. & Ostriker, J.P. *Galactic Cannibalism. III. The Morphological Evolution of Galaxies and Clusters* Astrophys. J. 224 (1978) 320-336.

Hawking, S.W. *Black Hole Explosions?* Nature 248 (1974) 30-31.

Hawking, S.W. *Particle Creation by Black Holes* Commun. Math. Phys. 43 (1975) 199.

Hawley, J.F. *Three-Dimensional Simulations of Black Hole Tori* Astrophys. J. 381 (1991) 496-507.

Hawley, J.F. & Balbus, S.A. & Winters, W.F. *Local Hydrodynamic Stability of Accretion Disks* (1998). preprint: http://arxiv.org/pdf/astro-ph/9811057

Heenen, P.H. & Nazarewicz, W. *Quest for Superheavy Nuclei* Europhysics News 33 (2002). Webpage: http://www.europhysicsnews.com/full/13/article2/article2.html

Honda, M. & Honda, Y.S. *Self-collimation and magnetic field generation of astrophysical jets* preprint: http://arxiv.org/pdf/astro-ph/0204048

Hopfield, J.J. *Theory of the Contribution of Excitons to the Complex Dielectric Constant of Crystals* Phys. Rev. 112 (1958) 1555.

Hopfield, J.J. & Thomas, D.G. *Theoretical and Experimental Effects of Spatial Dispersion on the Optical Properties of Crystals* Phys. Rev. 132 (1963) 563.

Howe, L.M. (1998) *Glimpses of Other Realities Vol II—High Strangeness* Website: http://www.earthfiles.com/shop.cfm

Hu, X. & Nori, F. *Phonon Squeezed States: Quantum Noise Reduction in Solids* Physica B263-4 (1999) 16-29.

Hughes, S.A. *Gravitational Waves from Extreme Mass Ratio Inspirals: Challenges in Mapping the Spacetime of Massive, Compact Objects* Class. Quantum Grav. 18 (2001) 4067-4073. Website: http://web.mit.edu/sahughes/www/Papers/lisaIII.pdf

Ilisavskii et al Phys. Rev. Lett. 87 (2001) 146602. Also see The American Institute of Physics Bulletin of Physics News No 557 (Sep 20, 2001). Webpage: http://newton.ex.ac.uk/aip/physnews.557.html

Inertial Alfven—*Inertial Alfvén waves, mode conversion, resonance cones* Cal Tech. Webpage: http://ve4xm.caltech.edu/Bellan_plasma_page/inertial.htm

Interrante, L.V. et al *Silicon-based Ceramics from Polymer Precursors* Pure Appl. Chem. 74 (2002) 2111-2117.

Ivers, J.D. et al *Electron-beam Diodes Using Ferroelectric Cathodes* J. Appl. Phys. 73 (1993) 2667-2671.

Jacobs, Dr. D.M. (1998) *The Threat*

Jacobs, Dr. D.M. The International Center for Abduction Research (ICAR) webpage: http://www.ufoabduction.com/

Jacobson, T. *Trans-Planckian Redshifts and the Substance of the Space-time River* Prog. Theor. Phys. Suppl. 136 (1999) 1-17. preprint: http://arxiv.org/pdf/hep-th/0001085

Jensen, A.S. & Riisager, K. & Fedorov, D.V. *Structure and Reactions of Quantum Halos* Rev. Mod. Phys. 76 (2004) 215-261.

Ji, H. & Carter, T. & Hau, S. & Yamada, M. *Study of Local Reconnection Physics in a Laboratory Plasma* Earth Planets Space 53 (2001) 539-545.

Jones, F.C. & Ellison, D.C. *The Plasma Physics of Shock Acceleration* Space Sci. Rev. 58 (1991) 259-346.
Joseph, L.E. (2007) *Apocalypse 2012*

Junor, W. et al *Formation of the Radio Jet in M87 at 100 Schwarzschild Radii From the Central Black Hole* Nature 401 (1999) 891.

Kaminsky, W. *Experimental and Phenomenological Aspects of Circular Birefringence and Related Properties in Transparent Crystals* Rep. Prog. Phys. 63 (2000) 1575-1640

Kapustin, A.P. (trans from Russian 1962) *The Effects of Ultrasound on the Kinetics of Crystallization*

Karas, V. *Asymptotically Uniform Magnetic Field Near a Kerr Black Hole* Phys. Rev. D40 (1989) 2121-2123.

Keefe, D. *Collective-effect Accelerators* Sci. Am. 226 (Apr 1972) 23-33.

Keldysh, L.V. *Correlations in the Coherent Transient Electron-Hole System* Physica Status Solidi B188 (1995) 11-27.

Khanna, R. *Generation and Evolution of Magnetic Fields in the Gravitomagnetic Field of the Kerr Black Hole* Proceedings of The Third William Fairbank Meeting on The Lense-Thirring Effect (June 1998) Ed. R. Ruffini et al. preprint: arXiv:astro-ph/9903091

Kiefer, C. & Weber, C. *On the Interaction of Mesoscopic Quantum Systems with Gravity* (2004). preprint: arXiv:gr-qc/0408010

Kim, M.J. et al *Electro-osmosis-driven Micro-channel Flows: A Comparative Study of Microscopic Particle Image Velocimetry Measurements and Numerical Simulations* Experiments in Fluids 33 (2002) 170-180.

Kim, S.P. & Page, D.N. *Remarks on Schwinger Pair Production by Charged Black Holes* Proceedings of the 8th Italian-Korean Symposium on Relativistic Astrophysics, Pescara, Italy (2003). preprint: arXiv:gr-qc/0401057

Kino, S.G. & Shaw, J. *Acoustic Surface Waves* Sci. Am. 227 (Oct 1972) 51-68.

Kirk, J.G. & Lyubarsky, Y. *Reconnection in Pulsar Winds* Publ. Astron. Soc. Aust. 18 (2001) 415-420.

Kirk, J.G. *Particle Acceleration in Relativistic Current Sheets* Phys. Rev. Lett. 92 (2004) 181101. preprint: http://arxiv.org/pdf/astro-ph/0403516

Kirtley, J.R. & Tsuei, C.C. *Probing High-temperature Superconductivity* Sci. Am. (Aug 1996) 50-55.

Kittel, C. (5th ed: 1976)(7th ed: 1996)(8th ed: 2005) *Introduction to Solid State Physics*

Kleeman, R.D. & Fredrickson, W. *Experiments on the Sign of Electric Charge Assumed by a Metal Immersed in a Liquid* Phys. Rev. 22 (1923) 134-136.

Klimov, V.V. *Spontaneous Emission of an Excited Atom Placed Near a "Left-Handed" Sphere* Optics Comm. 211 (2002) 183-196.

Koch, M. et al *Exciton Scattering with Bare Electrons* Physica Status Solidi B188 (1995) 485.

Kouropoulos, C.P. *Classically Bound Electrons EVs, Exotic Chemistry & 'Cold Electricity'* (2003). Webpage: http://web.archive.org/web/2003 1003232232/http://www.mypage.bluewin.ch/Bizarre/EVs.htm

Krogh, K. *Gravitation Without Curved Space-time* (2005) preprint: arXiv:astro-ph/9910325 Website: http://arxiv.org/abs/astro-ph/9910325.pdf

Kudoh, T. & Matsumoto, R. & Shibata, K. *Magnetically Driven Jets from Accretion Disks III – 2.5 Dimensional Nonsteady Simulations for Thick Disk Case* Astrophys. J. 508 (1998) 186-199.

Kuramitsu, Y. & Krasnoselskikh, V. *Gyroresonant Surfing Acceleration* Phys. Rev. Lett. 94 (2005) 031102.

Lakhtakia, A. & Mackay, T.G. *Towards Gravitationally Assisted Negative Refraction of Light by Vacuum* J. Phys. A37 (2004) L505-L510.

Lazar, R. *The Sportsmodel UFO Back-engineered at Area 51*. Currently (August 2008) being hosted at website: http://www.zamandayolculuk.com/cetinbal/UFOTEKNOLOJII.HTM

Lebedev, S.V. et al *Experiments with Radiatively Cooled Supersonic Plasma Jets Generated in Conical Wire Array Z-pinches* BEAMS-DZP, Albuquerque, NM (June 2002a).

Lebedev, S.V. et al *Laboratory Astrophysics and Collimated Stellar Outflows: The Production of Radiatively Cooled Hypersonic Plasma Jets* Astrophys. J. 564 (2002b) 113-119.

Lebedev, S.V. et al *Magnetic Tower Outflows from a Radial Wire Array Z-pinch* Mon. Not. R. Astr. Soc. 361 (2005) 97-108.

Lec-24-25.pdf website: http://www.physics.rutgers.edu/ugrad/342/notes/lec-24-25.pdf

Leitenstorfer, A. et al *Femtosecond Charge Transport in Polar Semiconductors* Phys. Rev. Lett. 82 (1999) 5140-5143.

Leiter, D. & Kafatos, M. *Penrose Pair Production in Massive, Extreme Kerr Black Holes* Astrophys. J. 226 (1978) 32-36.

Lenard, P.E.A. *On Cathode Rays* (Nobel Lecture May 28, 1906). Webpage: http://nobelprize.org/nobel_prizes/physics/laureates/1905/lenard-lecture.pdf

Leonhardt, U. & Piwnicki, P. *Light in Moving Media* (2001). Webpage: http://www.st-andrews.ac.uk/~ulf/media.html

Leonhardt, U. & Piwnicki, P. *Optics of Nonuniformly Moving Media* Phys. Rev. A60 (1999) 4301. preprint: http://arxiv.org/pdf/physics/9906038

Leonhardt, U. & Piwnicki, P. *Relativistic Effects of Light in Moving Media with Extremely Low Group Velocity* Phys. Rev. Lett. 84 (2000a) 822-825. preprint: arXiv:cond-mat/9906332

Leonhardt, U. & Piwnicki, P. *Slow Light in Moving Media* (2000b). preprint: http://arxiv.org/pdf/physics/0009093

Leonhardt, U. *Space-time Geometry of Quantum Dielectrics* (2000c). preprint: http://arxiv.org/pdf/physics/0001064

Lery, T. et al *A Global Jet / Circulation Model for Young Stars* (2002). preprint: http://arxiv.org/pdf/astro-ph/0203090

Lery, T. et al *Magnetised Protostellar Bipolar Outflows* Astron. Astrophys. 350 (1999) 254-274. preprint: http://arxiv.org/pdf/astro-ph/9902362

Li, H. & Lovelace, R.V.E. et al. *Magnetic Helix Formation Driven by Keplerian Disk Rotation in an External Plasma Pressure: The Initial Expansion Stage* Astrophys. J. 561 (2001) 915-923.

Li, Li-Xin *Extracting Energy From a Black Hole Through the Transition Region* Astrophys. J. 540 (2000) L17-L20.

Li, Li-Xin *Toy Model for the Magnetic Connection Between a Black Hole and a Disk* Phys. Rev. D65 (2002) 084047. Website: http://cfa-www.harvard.edu/~lli/professional/html/papers/2002/circle.pdf

Li, N. & Torr, D.G. *Gravitational Effects on the Magnetic Attenuation of Superconductors* Phys. Rev. B46 (1992) 5489-5495.

Liberati, S. & Sonego, S. & Visser, M. *Unexpectedly Large Surface Gravities for Acoustic Horizons?* Clas. Quant. Grav. 17 (2000) 2903. preprint: arXiv:gr-qc/0003105

Light, M. et al. *Axial Propagation of Helicon Waves* Phys. Plasmas 2 (1995) 4094.

Lin, H. et al *Vortex Dynamics and Angular Momentum Transport in Accretion Disks* Center for Turbulence Research Annual Research Briefs (2002) 289-299. Website: http://ctr.stanford.edu/ResBriefs02/lin.pdf

Lin, J. *Ultraviolet Light Emission from Oxidized Porous Silicon* Sol. State Comm. 97 (1996) 221-224.

Lineweaver, C.H. *Gold in the Doppler Hills: Cosmological Parameters in the Microwave Background* preprint: arXiv:astro-ph/9702042

Lipavsky, P. et al *Surface Deformation Caused by the Abrikosov Vortex Lattice* (2008). preprint: http://arxiv.org/pdf/0802.0831.pdf

Livio, M. et al *Extracting Energy from Black Holes: The Relative Importance of the Blandford-Znajek Mechanism* [to be published in Astrophysical Journal]. preprint: http://arxiv.org/pdf/astro-ph/9809093

Lovelace, R.V.E. *Dynamo Model of Double Radio Sources* Nature 262 (1976) 649-652.

LSBU Water (London South Bank University website on water structure and behavior). Webpage: http://www.lsbu.ac.uk/water/microwave.html

Luca, B.A. (1999) *ASBT Extraterrestrial Communications* (Pt I & II)

Luff, B.J. et al *Cathodoluminescence of Synthetic Quartz* J. Phys. Condens. Matter 2 (1990) 8089-8097.

Lyubarsky, Y. & Kirk, J.G. *Reconnection in a Striped Wind* Astrophys. J. 547 (2001) 437-448. preprint: http://arxiv.org/pdf/astro-ph/0009270

Ma, Z.W. *MHD Simulation of Magnetic Reconnection and Magnetic Substorms* Proceedings of ISSS-7 (26-31 Mar, 2005). Website: http://www.rish.kyoto-u.ac.jp/isss7/CDROM/CONTENTS/DATA_PDF/T-ZWMA.PDF

Mackay, T.G. & Lakhtakia, A. *Negative Refraction in Outer Space?* (2004) preprint: http://arxiv.org/pdf/physics/0405103

Madejski, G. *Review of the Paper by Semenov, Dyadechkin, and Punsly on 'Simulations of Jets Driven by Black Hole Rotation'* (Aug 13 2004) (PowerPoint presentation). Webpage: http://www.slac.stanford.edu/exp/glast/ground/GlastScience/year2004/GregMadejski/GregMadejski_aug04.ppt

Magnetic drifts note: *Magnetic Drifts* webpage: http://farside.ph.utexas.edu/teaching/plasma/lectures/node17.html

Magnetostriction and Magnetostrictive Materials UCLA-Active Materials Lab. Webpage: http://aml.seas.ucla.edu/research/areas/magnetostrictive/mag-composites/Magnetostriction%20and%20Magnetostrictive%20Materials.htm

Magueijo, J. & Albrecht, A. *A Time Varying Speed of Light as a Solution to Cosmological Puzzles* Phys. Rev. D59 (1999) 043516. preprint: http://arxiv.org/pdf/astro-ph/9811018

Magueijo, J. *New Varying Speed of Light Theories* (2003). preprint: http://arxiv.org/pdf/astro-ph/0305457 Also see Magueijo archive at: http://eprintweb.org/S/authors/astro-ph/ma/Magueijo

Makhnovskiy, D.P. & Panina, L.V. *Field Dependent Permittivity of Composite Materials Containing Ferromagnetic Wires* J. Appl. Phys. 93 (2003) 4120-4129.

Maksimov, A.O. et al *Structure of Acoustic Shock Waves Emitted by Optical Breakdown in Water with Laser Pulse* XVI Session of the Russian Acoustical Society (Nov14-18 2005) 129-32. Website: http://www.akin.ru/Docs/Rao/Ses16/F34.PDF

Maleev, I.D. & Swartzlander.Jr, G.A. *Composite optical vortices* J. Opt. Soc. Am. B20 (2003) 1169.

Mallett, R.L. *Weak Gravitational Field of the Electromagnetic Radiation in a Ring Laser* Phys. Lett. A269 (2000) 214-217.

Mannheim, P.D. *Implications of Cosmic Repulsion for Gravitational Theory* Phys. Rev. D58 (1998) 103511. preprint: http://arxiv.org/pdf/astro-ph/9804335

Mansuripur, M. *Angular Momentum of Circularly Polarized Light in Dielectric Media* Opt. Express. 13 (2005) 5315-5324.

Mansuripur, M. *Radiation Pressure and the Linear Momentum of the Electromagnetic Field* Opt. Express. 12 (2004) 5375-5401.

Marcus, P.S. *Simulation of Taylor-Couette Flow. Pt.1—Numerical Methods and Comparison with Experiment & Pt.2—Numerical Results for Wavy-Vortex Flow with One Travelling Wave* J. Fluid Mech. 146 (1984) 45-64.

Markos, P. & Soukoulis, C.M. *Structures with Negative Index of Refraction* Physica Status Solidi A197 (2003) 595-604.

Mashhoon, B. *Gravitoelectromagnetism: A Brief Review* (2003). preprint: http://arxiv.org/pdf/gr-qc/0311030

Maurer, L.D. & Miller, M.E. (2004) *Laser Propulsion* (privately published by Unitel Aerospace—Portland, Oregon)
McDonald, James *Air Force Observations of an Unidentified Object in the South-Central U.S., July 17, 1957* Astronautics & Aeronautics (AIAA) July 1971 66-70.

McKnight, G.P. *Magnetostrictive Materials Background* UCLA-Active Materials Lab. Webpage: http://aml.seas.ucla.edu/research/areas/magnetostrictive/overview.htm

Meier, D.L. & Koide, S. & Uchida, Y. *Magnetohydrodynamic Production of Relativistic Jets* Science 291 (2001) 84-92.

Meier, D.L. & Nakamura, M. (JPL Cal. Inst. Tech) *Poynting Flux Dominated Jets in Decreasing Density Atmospheres (The Non-Relativistic Current-Driven Kink Instability and the Formation of "Wiggled" Structures* (2004). preprint: http://arxiv.org/pdf/astro-ph/0406405

Meier, D.L. & Nakamura, M. (JPL Cal. Inst. Tech) *The Production and Propagation of Variable Cosmic Jets* Variable Radio Universe Workshop (Jul 10-11 2003).

Meier, D.L. (JPL Cal. Inst. Tech) *Magnetically-Dominated Jet and Accretion Flows* Ultra-Relativistic Workshop Banff (Jul 12 2005).

Meintjes, P.J. & Venter, L.A. *The Diamagnetic Blob Propeller in AE Aquarii and Non-Thermal Radio to Mid-Infrared Emission* Mon. Not. R. Astr. Soc. 360 (2005) 573.

Meng, J. et al *Relativistic Continuum Hartree Bogoliubov Theory for Ground State Properties of Exotic Nuclei* Prog. Part. Nucl. Phys. 57 (2006) 470-563. preprint: http://arxiv.org/pdf/nucl-th/0508020

Meng, J. et al *The Structure of Superheavy Elements Newly Discovered in the Reaction of ^{86}Kr with ^{208}Pb* Phys. Rev. C61(2000) 064319. preprint: http://arxiv.org/pdf/nucl-th/9908040

Mirza, B.M. *Charged Particle Dynamics in the Field of a Slowly Rotating Compact Star* (2004). preprint: http://arxiv.org/pdf/astro-ph/0408579

Mirza, B.M. & Saleem, H. *Plasma Frequency Shift Due to a Slowly Rotating Compact Star* (May 10 2005). preprint: http://arxiv.org/pdf/astro-ph/0501003

Misner, C.W. & Thorne, K.S. & Wheeler, J.A. (1971) *Gravitation*

Moller, P. & Nix, J.R. & Kratz, K.L. *Nuclear Properties for Astrophysical and Radioactive-ion-beam Applications* in the Atomic Data and Nuclear Data Tables 66 (1997) 131-343.

Morris, M.S. & Thorne, K.P. *Wormholes in Spacetime and Their Use for Interstellar Travel: A Tool for Teaching General Relativity* Am. J. Phys. 56 (1988) 395-412.

Morris, M.S. & Thorne, K.S. & Yurtsever, U. *Wormholes, Time Machines, and the Weak Energy Condition* Phys. Rev. Lett. 61 (1988) 1446-1449.

Mourou, G. *The Ultrahigh-peak-power Laser: Present and Future* Appl. Phys. B65 (1997) 205-211.

Mourou, G. & Umstadter, D. *Development and Applications of Compact High-intensity Lasers* Phys. Fluids B4 (1992) 2315-2324.

Mourou, G. & Umstadter, D. *Extreme Light* Sci. Am. (May 2002) p81-86.

Murad, P.A. *An Electromagnetic Rocket Stellar Drive...Myth or Reality? Fluid Dynamic Interactions and an Engine Concept* (1995) AIAA95-2894 presented at the July 10-12, 1995/San Diego, CA. 31st AIAA/ASME/SAE/ASEE Joint Propulsion Conference and Exhibit.

Murad, P.A. *Torsion Physics* (Nov 28 2005). Website: http://www.trinitas.ru/rus/doc/0231/008a/02310063.pdf

Naranjo, B. et al. *Observation of Nuclear Fusion Driven by a Pyroelectric Crystal* Nature 434 (2005) 1115-1117.

Nature of Mass (Calphysics Institute). Webpage: http://www.calphysics.org/mass.html

Neagu, R.M. et al *Dielectric Studies of Dipolar Relaxation Processes in Nylon 11* J. Phys. D33 (2000) 1921-1931.

Nimtz, G. *Superluminal Signal Velocity* Ann. Phys (Leipzig) 7 (1998) 618. preprint: http://arxiv.org/pdf/physics/9812053

Numerical Maryland Centrifugal Experiment. Webpage: http://aries.ucsd.edu/FPA/ARC02/fpn02-30.shtml

O'Dell, T.H. (1974) *Magnetic Bubbles*

Oh, E. et al *THz Radiation and Acoustic Phonon Pulse Wave from GaN-based Light Emitting Diode Structures Generated by Ultra Short Pulse Lasers* Proc. of SPIE 5352 (2004) 180-187.

Omma, H. et al *Heating Cooling Flows with Jets* Mon. Not. R. Astr. Soc. 348 (2004) 1105. preprint: arXiv:astro-ph/0307471

Ostoma, T. & Trushyk, M. *A Simple Physical Interpretation of the Lense-Thirring Effect Based on EMQG Theory* (Mar 16 1999). preprint: arXiv.org:physics/9903025 Website: http://arxiv.org/ftp/physics/papers/9903/9903025.pdf

Ostriker, J.P. & Steinhardt, P.J. *The Quintessential Universe* Sci. Am. (Jan 2001) 46-53. Website: http://astro.berkeley.edu/~arons/Ay84/ostriker_steinhardt_sciam.pdf

Ostriker, J.P. & Tremaine, S.D. *Another Evolutionary Correction to the Luminosity of Giant Galaxies* Astrophys. J. 202 (1975) L113-L117.

Ouellette, J. *Free-Electron Lasers: A Radical Alternative* The Industrial Physicist (Mar 1997) 18-21.

Paschmann, G, N. Sckopke et al., *Observations of Gyrating Ions in the Foot of the Nearly Perpendicular Bow Shock* Geophys. Res. Lett. 9 (1982) 881-884.

Pendry, J.B. et al, *Extremely Low Frequency Plasmons in Metallic Mesostructures* Phys. Rev. Lett. 76 (1996) 4773-4776. Website: http://ceta-p5.mit.edu/metamaterials/papers/external/pre2k/Pendry.HSY.pdf

Pendry, J.B. et al, *Low Frequency Plasmons in Thin-wire Structures* J. Phys. Condens. Matter 10 (1998) 4785-4809. Website: http://ceta-p5.mit.edu/metamaterials/papers/external/pre2k/pendry_jphys_cond_matt_1998.pdf

Pendry, J.B. *Negative Refraction* Contemporary Physics 45 (Jan-Feb 2004) 191-202

Pendry, J.B. *Negative Refraction Makes Light Run Backwards in Time* (07 April 2000)

Penrose, R. (2004) *The Road to Reality* (Jonathan Cape publishers)

Penrose, R. *Gravitational Collapse: the Role of General Relativity* Rivista Del—Nuovo Cimento 1 Numero Speciale (1969) 252-276.

Piran, T. & Shaham, J. *Production of Gamma-Ray Bursts Near Rapidly Rotating Accreting Black Holes* Astrophys. J. 214 (1977) 268-299.

Plyusnin, V.V. *Study of Runaway Electron Generation Process During Major Disruptions in JET* (International Atomic Energy Agency) (2004) EX/P2-27. Website: http://www-pub.iaea.org/MTCD/Meetings/PDFplus/fusion-20-preprints/EX_P2-27.pdf

Podkletnov, E.E. *Weak Gravitation Shielding Properties of Composite Bulk $Y Ba_2Cu_3O_{7-x}$ Superconductor Below 70 K Under E.M. Field* (1997). preprint: http://arxiv.org/pdf/cond-mat/9701074

Polarization Drift note: *Comment on Gravitational Damping of Alfvén Waves* Astrophys. J. 488 (1997) 895-897. Webpage: http://www.journals.uchicago.edu/ApJ/journal/issues/ApJ/v488n2/35875/sc1.html

Portilla, M. & Lapiedra, R. *Generation of High Frequency Gravitational Waves* Phys. Rev. D63 (2001) 044014.

Povinelli, M.L. & Pendry, J.B. et al *Toward Photonic-crystal Metamaterials: Creating Magnetic Emitters in Photonic Crystals* Appl. Phys. Lett. 82 (2003) 1069-1071.

Press, W.H. & Teukolsky, S.A. *Perturbations of a Rotating Black Hole. II. Dynamical Stability of the Kerr Metric* Astrophys. J. 185 (1973) 649-673.

Pretorius, F. *Quantum Interest for Scalar Fields in Minkowski Spacetime* Phys. Rev. D61 (2000) 064005. preprint: arXiv:gr-qc/9903055

Price, R.H. & Thorne, K.S. *The Membrane Paradigm for Black Holes* Sci. Am. (Apr 1988) 45-55.

Priest, E.R. & Forbes, T. (2000) *Magnetic Reconnection*—(MHD Theory and Applications)

Prix, R. *Covariant Vortex in Superconducting—Superfluid—Normal Fluid Mixtures with Stiff Equation of State* (2005). preprint: http://arxiv.org/pdf/gr-qc/0004076

Punsly, B. & Coroniti, F.V. *Electrodynamics of the Event Horizon* Phys. Rev. D40 (1989) 3834-3857.

Punsly, B. & Coroniti, F.V. *Ergosphere-Driven Winds* Astrophys. J. 354 (1990b) 583-615.

Punsly, B. & Coroniti, F.V. *Relativistic Winds from Pulsar and Black Hole Magnetospheres* Astrophys. J. 350 (1990a) 518-535.

Punsly, B. (2001) *Black Hole Gravitohydromagnetics* (Springer-Verlag publisher)

Punsly, B. *Force Free Waves and Black Hole Magnetospheric Causality* (2002). preprint: http://arxiv.org/pdf/astro-ph/0210103

Puthoff, H.E. *Gravity as a Zero-Point-Fluctuation Force* Phys. Rev. A39 (1989a) 2333-2342.

Puthoff, H.E. *Ground State of Hydrogen as a Zero-Point-Fluctuation-Determined State* Phys. Rev. D35 (1987) 3266-3269.

Puthoff, H.E. *Polarizable-Vacuum (PV) Representation of General Relativity* Found. Phys. 32 (2002a) 927-943. preprint: http://arxiv.org/pdf/gr-qc/9909037

Puthoff, H.E. *Source of Vacuum Electromagnetic Zero-point-energy* Phys. Rev. A40 (1989b) 4857-4862.

Puthoff, H.E. & Davis, E.W. & Maccone, C. *Levi-Civita Effect in the Polarizable Vacuum (PV) Representation of General Relativity* Gen. Relativ. Gravit. 37 (2005) 483-489. preprint: http://arxiv.org/pdf/physics/0403064

Puthoff, H.E. & Little, S.R. & Ibison, M. *Engineering the Zero-Point Field and Polarizable Vacuum for Interstellar Flight* Journal of the British Interplanetary Society (JBIS) 55 (2002b) 137-144. preprint: http://arxiv.org/pdf/astro-ph/0107316 Website: http://www.earthtech.org/publications/JBIS_55_137-144.pdf

Bibliography

Quantum Dots article by Henri Saarikoski, Laboratory of Physics, Helsinki University of Technology. Webpage: http://www.fyslab.hut.fi/epm/Qdots/vortex.html

Qin, G.G & Jia, Y.Q. *Mechanism of the Visible Luminescence in Porous Silicon* Sol. State Comm. 86 (1993) 559.

Rabinowitz, M. *Black Hole Radiation and Volume Statistical Entropy* Int. J. Theor. Phys. 45 (2006) 851-858. preprint: arxiv.org/abs/physics/0506029

Redfield, D. (2001) *The ET-Human Link*

Rendlesham 1980 UFO Incident (eye-witness accounts of Larry Warren, John Burroughs and Lt Col Charles Halt compiled by Dave Cosnette). Website: http://www.ufos-aliens.co.uk/cosmicrend.html

Riege, H. et al. *Electron Emission from Ferroelectrics—A Review* Nucl. Instr. and Meth. in Physics Res. A340 (1994) 80-89.

Riege, H. et al. *Features and Technology of Ferroelectric Electron Emission* J. Appl. Phys. 84 (1998) 1602-1617.

Ronald, K. et al, *Observations of Dynamic Behaviour in an Electron Cyclotron Maser Oscillator* J. Phys. D34 (2001) L17-22.

Rosenweig, R.E. (1985) *Ferrohydrodynamics*

Rosenman, G. et al. *Electron Emission from Ferroelectrics*. J. Appl. Phys. 88 (2000) 6109-6161.

Rosu, H. *Classical and Quantum Inertia: A Matter of Principles* Gravitation and Cosmology 5 (1999) 81-91. preprint: arXiv:gr-qc/9412012

Roswell UFO: *The Roswell Crash—50th Anniversary Special Edition* (video) Bud Finley Productions (1997). Also see the Schmitt and Carey Roswell Report.

Roy, S. *Flow Actuation Using Radio Frequency in Partially Ionized Collisional Plasmas* Appl. Phys. Lett. 86 (2005) 101502. Website: http://cpdl.kettering.edu/apl-dbd.pdf

Rozas, D. et al. *Propagating dynamics of optical vortices* J. Opt. Soc. Am. B14 (1997) 3054.

Ruderman, M.A. & Sutherland, P.G. *Theory of Pulsars: Polar Gaps, Sparks, and Coherent Microwave Radiation* Astrophys. J. 196 (1975) 51-72.

Rueda, A. & Haisch, B. *Inertia as Reaction of the Vacuum to Accelerated Motion* Phys. Lett. A240 (1998) 115-126. preprint: arXiv:physics/9802031

Saa, A. & Schiffer, M. *The Gravitational Vavilov-Cherenkov Effect* Mod. Phys. Lett. A13 (1998) 1557. preprint: http://arxiv.org/pdf/gr-qc/9805080

Sakharov, A. D. *Vacuum Quantum Fluctuations in Curved Space and the Theory of Gravitation* Soviet Physics DOKLADY 12 (May 1968) 1040-1041.

Salamin, Y.I. & Keitel, C.H. & Faisal, F.H.M. *Electron Acceleration in Combined Laser and Uniform Electric Fields* J. Phys. A: Math. Gen. 34 (2001) 2819-2837.

Scharrer, M. et al *Manipulation of Microbeads Using DC/AC Electrical Fields*

Schauberger, V. *The Secret of the Eggform* Implosion Magazine 112 (Aug 1995) 56-60.

Schmitt, D.R. & Carey, T.J. *The Roswell Report* (the most comprehensive investigation into the US Army's, and the local civilian, involvement in both the contemporary events and the subsequent cover-up by the US government) SCIFI.com UFOlogy Resource Center. Webpage: http://www.scifi.com/ufo/roswell/articles/014.html

Schutzhold, R. & Unruh, W.G. *Gravity Wave Analogue of Black Holes* Phys. Rev. D66 (2002) 044019. preprint: arXiv:gr-qc/0205099

Scott, J.F. *Temperature Dependent Permittivity and Phase Transitions in Ba Sr TiO Ferroelectric Thin Film Oxides* Website: http://cantab.jkut.com/Barium%20Strontium%20Titanate.pdf

Searl, J.R.R. Searl Energy Generator (SEG). Websites: http://swallowcommand.com and http://www.searlsolution.com and http://www.sisrc.com/c1.htm

Sengupta, S. *Charged Particle Trajectories in a Toroidal Magnetic and Rotation-Induced Electric Field Around a Black Hole* (1997). preprint: arXiv:gr-qc/9707014

Setiawan, S. *A Short Note on Climbing Gravity Improbable: Superradiant Ring Fellowship of Launching the High Lorentz-Factor Outflows/Jets* (2005). preprint: http://arxiv.org/pdf/astro-ph/0506557

Shilton, J.M. et al. *On The Acoustoelectric Current in a One-Dimensional Channel* J. Phys: Condens. Matter 8 (1996) L337-L343.

Shkunov, V.V. & Zeldovich, B. *Optical Phase Conjugation* Sci. Am. 253 (Dec 1985) 40-45.

Shock Wave webpage: http://neumann.dph.aber.ac.uk/research/what/shockinfo.html

Shvets, G. *Design of Metamaterials with Exotic Electromagnetic Properties* (University of Texas Physics Dept.) webpage at: http://www.ph.utexas.edu/~shvetsgr/lens.html

Shvets, G. *Nonlinear Pulse Compression in Plasma: Beyond Chirped Pulse Amplification* Lake Tahoe School of Plasma Physics (Mar 25 2004).

Shvets, G. *Superradiant Amplification of an Ultrashort Laser Pulse in a Plasma by a Counterpropagating Pump* Phys. Rev. Lett. 81 (1998) 4879-4882. Website: http://w3.pppl.gov/~gena/sra.pdf

Sievenpiper, D.F. & Sickmiller, M.E. & Yablonovitch, E. *3D Wire Mesh Photonic Crystals* Phys. Rev. Lett. 76 (1996) 2480-2483. Website: http://www.ee.ucla.edu/~photon/pubs/ey1996prl7614.pdf

Slatyer, T.R. & Savage, C.M. *Superradiant Scattering from a Hydrodynamic Vortex* (2005). preprint: http://arxiv.org/pdf/cond-mat/0501182

Slusher, R.E. & Yurke, B. *Squeezed Light* Sci. Am. 258 (1988) 32-38.

Smith, D.F. *Particle Acceleration by Strong Plasma Turbulence. II. Acceleration of Nonrelativistic Electrons in Solar Flares* Astrophys. J. 217 (1977) 644-656.

Smith, D.R. et al, *Loop-wire Medium for Investigating Plasmons at Microwave Frequencies* Appl. Phys. Lett. 75 (1999) 1425-1427.

Snoke, D.W & Negoita, V. *Pushing the Auger limit: Kinetics of Excitons in Traps of Cu_2O* Phys. Rev. B61 (2000) 2904.

Bibliography

Sol-Gel work from the research labs of the Department of Inorganic Chemistry at Stockholm University, Sweden. Webpage: http://www.fos.su.se/~gw/

Sol-Gel another webpage on sol-gel fabrication (in US): http://www.chemat.com/html/solgel.html

Somaraju, R. & Trumpf, J. *Electromagnetic Wave Propagation and the Permittivity of Seawater* (2005). Website: http://users.rsise.anu.edu.au/~trumpf/somaraju_trumpf_permittivity.pdf

Sonett, C.P. & Morfill, G.E. & Jokipii, J.R. *Interstellar Shock Waves and Be 10 from Ice Cores* Nature 330 (1987) 458-460.

Song, J. et al *Self-Trapped Excitons at the Quartz (0001) Surface* Faraday Discuss. 117 (2000a) 303-311. Website: http://www-theory.chem.washington.edu/~hannes/paperSiO2-STE-Surf/STE-surfFaraday01.pdf

Song, J. et al *Self-trapped Excitons in Quartz* Nucl. Instrum. Methods Phys. Res. B166-167 (2000b) 451-458. Website: http://www.hi.is/~hj/paperSTEnib00.pdf

Soskin, J. (1995) *Transformation*

Souza, F.M. & Egues, J.C. *Bare LO-Phonon Peak in THz-Emission Signals: a Dielectric-Function Analysis* Brazilian J. Phys. 32 (2002). preprint: arXiv:cond-mat/0210049

Sparks, J. (2006) *The Keepers: An Alien Message for the Human Race*

Spears, M.F. *Permittivity Creates Dark Matter and an Older Universe with Accelerating Expansion* (July 8, 1998). Website: http://members.aol.com/tigermfs/mfsgrav2.pdf

Spruit, H.C. et al *Collimation of Magnetically Driven Jets from Accretion Discs* Mon. Not. R. Astr. Soc. 288 (1997) 333-342.

Spruit, H.C. *Jets from compact objects* Highly Energetic Physical Processes and Mechanisms for Emission from Astrophysical Plasmas—IAU Symposium 195 (1999). preprint: http://arxiv.org/pdf/astro-ph/0003043

Spruit, H.C. *Magnetohydrodynamic Jets and Winds from Accretion Disks* Physical Processes in Binary Stars—NATO ASI series (1996). preprint: http://arxiv.org/pdf/astro-ph/9602022

Starobinskii, A.A. *Amplification of Waves During Reflection from a Rotating "Black Hole"* Soviet Physics JETP 37 (1973) 28-32.

Stone, J.M. & Balbus, S.A. *Angular Momentum Transport in Accretion Disks via Convection* Astrophys. J. 464 (1996) 364-372.

Street, R.L. et al (1996) *Elementary Fluid Mechanics*

Strieber, W. (2001) *The Key* (obtainable from http://www.unknowncountry.com)

Supernova (*What will happen to Earth when Eta Carinae goes SuperNova*) Webpage: http://www.valdostamuseum.org/hamsmith/13Mar41.html

Surface Acoustic Waves (Selected Brief Reports). Website: http://www.pdi-berlin.de/jabe02/saw_02.pdf

Sushkov, O.P. & Khriplovich, B. *The Gravitational Radiation Emitted by an Ultrarelativistic Charged Particle in an External Electromagnetic Field* Soviet Physics JETP 39 (1974) 1-3.

Talbot, J.B. *Phosphors and Other Emissive Materials* (2006). Webpage: http://ftp.wtec.loyola.edu/loyola/word/fsudisp/06_ch5.doc

Tamm, I.E. *General Characteristics of Radiations Emitted by Systems Moving with Super-light Velocities with some Applications to Plasma Physics* (Nobel Lect. 1958) 470-482. Website: http://nobelprize.org/physics/laureates/1958/tamm-lecture.pdf

Teukolsky, S.A. & Press, W.H. *Perturbations of a Rotating Black Hole. III—Interaction of the Hole with Gravitational and Electromagnetic Radiation* Astrophys. J. 193 (1974) 443-461.

Thierry d'Almeida *Cathodoluminescence of Er3+-Doped Crystals* High Temperatures High Pressures 30 (1998) 351-356.

Thomson, P. *Charge Sheath Vortex Basics For Tornado*. Webpage at: http://www.peter-thomson.co.uk/tornado/fusion/Charge_sheath_vortex_basics_for_tornado.html

Tompkins, P. & Bird, C. (1991) *Secrets of the Soil*

Traschen, J. *An Introduction to Black Hole Evaporation* (2000). preprint: arXiv:gr-qc/0010055 Website: http://arxiv.org/abs/gr-qc/0010055.pdf

Trukhin, A.N. et al *Elementary Electronic Excitations in Pure Sodium Silicate Glasses* Phys. Stat. Sol. B99 (1980) 155.

Tsinganos, K. et al. *Collimation of Astrophysical MHD Outflows* Astrophys. & Space Sci. 287 (2003) 103-108.

Tsutsumi, N. & Mizutani, T. & Sakai, W. *Internal Electric Field and Second-Order Optical Nonlinearity of Ferroelectric Nylons* Macromolecules 30 (1997) 1637-1642.

Uchida, Y. et al *Distribution of Faraday Rotation Measure in Jets from Active Galactic Nuclei I. Prediction from our Sweeping Magnetic Twist Model* Astrophys. J. 600 (2004) 88-95. preprint: http://arxiv.org/pdf/astro-ph/0309605

Uchida, Y. & Shibata, K. *Magnetodynamical Acceleration of CO and Optical Bipolar Flows from the Region of Star Formation* Publ. Astron. Soc. Japan 37 (1985) 515-535.

UFOphysics website, yes its still available! http://web.archive.org/web/*/http://www.ufophysics.com

Unruh, W.G. *Second Quantization in the Kerr Metric* Phys. Rev. D10 (1974) 3194-3205.

Unruh, W.G. *Dumb Holes [Sonic Analogue of Black Holes] and the Effects of High Frequencies on Black Hole Evaporation* Phys. Rev. D51 (1995) 2827-2838. preprint: http://arxiv.org/pdf/gr-qc/9409008

Uskert, P. Producer of DVDs of California UFOs (2005).

van Ginhoven, R.M. et al *An Ab Initio Study of Self-Trapped Excitons in Alpha-Quartz* J. Chem. Phys. 118 (2003) 6582-6593. Website: http://www.hi.is/~hj/paperSiO2clust03.pdf

van Putten, M.H.P.M. & Levinson, A. *Detecting Energy Emissions from a Rotating Black Hole* Science 295 (8 Mar 2002) 874.

van Putten, M.H.P.M. *Superradiance in a Torus Magnetosphere Around a Black Hole* Science 284 (1999) 115.

Veselago, V.G. & Rudashevskii, E.G. *On the Amplification of Electromagnetic Waves in Electrically Conducting Ferromagnetic Materials* Sov. Phys. Solid State 8 (1967) 2290-2293.

Veselago, V.G. *Electrodynamics of Substances with Simultaneously Negative Electrical and Magnetic Permeabilities* (1971) 5-13. Website: http://data.ufn.ru//news/eng/2000/veselago.pdf

Veselago, V.G. *Properties of Materials Having Simultaneously Negative Values of the Dielectric (ε) and the Magnetic (μ) Susceptibilities* Sov. Phys. Solid State 8 (1967a) 2854-2856.

Veselago, V.G. *The Electrodynamic Properties of a Mixture of Electric and Magnetic Charges* Sov. Phys. JETP 25 (1967b) 680-681.

Veselago, V.G. *The Electrodynamics of Substances with Simultaneously Negative Values of ε and μ* Sov. Phys. USPEKHI 10 (1968) 509-514. Website: http://ceta-p5.mit.edu/metamaterials/papers/external/pre2k/Veselago.pdf

Vieregg, J.R. *A 250 GHz Gyrotron Oscillator for use in Dynamic Nuclear Polarization* Plasma Science and Fusion Center MIT(2000).

Vigier, J-P. & Amoroso, R.L. *Can One Unify Gravity and Electromagnetic Fields?* (2002) Gravitation and Cosmology: From the Hubble Radius to the Planck Scale. Website: http://www.mindspring.com/~noetic.advanced.studies/Amoroso17.pdf

Visser, M. *Wormholes, Baby Universes, and Causality* Phys. Rev. D41 (1990) 1116-1124. Webpage: http://prola.aps.org/abstract/PRD/v41/p1116_1

Visser, M. *Acoustic Black Holes* (1998b). preprint: http://arxiv.org/pdf/gr-qc/9901047

Visser, M. *Acoustic Black Holes: Horizons, Ergospheres, and Hawking Radiation* Clas. Quant. Grav. 15 (1998a) 1767. preprint: http://arxiv.org/pdf/gr-qc/9712010

Visser, M. *Comment on "Relativistic Effects of Light in Moving Media with Extremely Low Group Velocity"* Phys. Rev. Lett. 85 (2000) 5252. preprint: http://arxiv.org/pdf/gr-qc/0002011

Vogel, W. (1965) *Structure and Crystallization of Glasses*

von Weber, S.F. *Cosmic Membrane Theory of Gravitation* (Aug 2006) Webpage: http://webuser.fh-furtwangen.de/~webers/membthe1.htm [25/Sep07/vonWeber.htm] and NASA's webpage article on Multiple Universes at: http://rst.gsfc.nasa.gov/Sect20/A10.html

Wang, X-H. *Giant Lamb Shift in Photonic Crystals* Phys. Rev. Lett. 93 (2004) 073901.

Webb, J.K. et al, *A Search for Time Variation of the Fine Structure Constant* Phys. Rev. Lett. 82 (1999) 884-887. preprint: http://arxiv.org/pdf/astro-ph/9803165

Weber, J. *Detection and Generation of Gravitational Waves* Phys. Rev. 117 (1960) 306-313.

Weber, J. *Gravitational Radiation From the Pulsars* Phys. Rev. Lett. 21 (1968) 395-396.

Wheeler, J.C. & Meier, D.L. & Wilson, J.R. *Asymmetric Supernovae from Magnetocentrifugal Jets* Astrophys. J. 568 (2002) 807-819. preprint: http://arxiv.org/pdf/astro-ph/0112020

Widom, A. et al, *Gravitational Wave Dispersion in Condensed Matter Systems* J. Phys. A: Math. Gen. 14 (1981) L213-L215.

Williams, R.K. *Collimated Energy-Momentum Extraction from Rotating Black Holes in Quasars and Microquasars Using the Penrose Mechanism* (2001). preprint: http://arxiv.org/pdf/astro-ph/0111161 Website: http://xxx.arxiv.cornell.edu/PS_cache/astro-ph/pdf/0111/0111161.pdf

Williams, R.K. *Collimated Escaping Vortical Polar $e^- e^+$ Jets Intrinsically Produced by Rotating Black Holes and Penrose Processes* Astrophys. J. 611 (2004b) 952-963. preprint: http://arxiv.org/pdf/astro-ph/0404135 Website: http://xxx.arxiv.cornell.edu/pdf/astro-ph/0404135

Williams, R.K. *Extracting X-rays, Gamma-rays, and Relativistic $e^- e^+$ Pairs from Supermassive Kerr Black Holes Using the Penrose Mechanism* Phys. Rev. D 51 (1995) 5387-5427. Webpage: http://prola.aps.org/abstract/PRD/v51/i10/p5387_1

Williams, R.K. *New Energy Source Controlled by Gravity Alone?* (2002b). preprint: http://arxiv.org/pdf/astro-ph/0210139 Website: http://xxx.lanl.gov/pdf/astro-ph/0210139

Williams, R.K. *Production of the High Energy-Momentum Spectra of Quasars 3C 279 and 3C 273 Using the Penrose Mechanism* (2003). preprint: http://arxiv.org/pdf/astro-ph/0306135 Website: http://xxx.lanl.gov/pdf/astro-ph/0306135

Williams, R.K. *The Gravitomagnetic Field and Penrose Processes* (2002a) preprint: http://arxiv.org/pdf/astro-ph/0203421 Website: http://xxx.arxiv.cornell.edu/PS_cache/astro-ph/pdf/0203/0203421.pdf

Winterberg, F. *Planck Mass Rotons as Cold Dark Matter and Quintessence* Z. Naturforsch. 57a (2002a) 202-204. Website: www.znaturforsch.com/aa/v57a/s57a0202.pdf

Winterberg, F. *Maxwell's Aether, The Planck Aether Hypothesis and Sommerfeld's Finestructure Constant* University of Nevada, Reno (2002b).

Wu, D. et al, *Terahertz Plasmonic High Pass Filter* Appl. Phys. Lett. 83 (2003) 201-203. Website: http://ceta-p5.mit.edu/metamaterials/papers/external/2003/wu_apl_2003.pdf

Wu, N. *Gravitational Shielding Effects in Gauge Theory of Gravity* Commun. Theor. Phys. 41 (2004) 567-572. preprint: arXiv:hep-th/0307225

Yakovlev, S. *Lanthanide Ions Doping Effects on Structural, Electrophysical and Functional Properties of Sol-gel Fabricated PbTiO3 Thin Films (dissertation)* (2004). Website: http://e-diss.uni-kiel.de/diss_1199/d1199.pdf

Yamada, M. et al *Experimental Studies of Collisionless Reconection Processes in Plasmas* Phys. Plasmas 4 (1997) 1937. Website: psfcwww2.psfc.mit.edu/vtf/jegedal/reconnection/main2.pdf

York, W. *YGEM (York-Geier Electron Model)*. Website: http://webpages.charter.net/pubmaster/FinalTheory.pdf

Yousif, M.E. of Nairobi University—*The Universal Energies* (sect. 5:2). Websites: ExMF Propulsions (http://exmfpropulsions.com) and at www.americanantigravity.com (http://www.americanantigravity.com/articles/493/1/External-Magnetic-Field-Propulsion-&-Flying-Objects) and Mahmoud's previously published paper at the web-based Journal of Theoretics: http://www.journaloftheoretics.com/Links/Papers/Yousif.pdf

Zank, G.P. & Frisch, P.C. *Consequences of a Change in the Galactic Environment of the Sun* Astrophys. J. 518 (1999) 965-973. preprint: arXiv:astro-ph/9901279

Zeldovich, Y.B. & Novikov, I.D. (1971) *Relativistic Astrophysics—Stars and Relativity* (trans. from Russian by E. Arlock).

Zeldovich, Y.B. *Amplification of Cylindrical Electromagnetic Waves Reflected from a Rotating Body* Sov. Phys. JETP 35 (1972) 1085-1087.

Zeldovich, Y.B. *Electromagnetic and Gravitational Waves in a Stationary Magnetic Field* Sov. Phys. JETP 38 (1974) 652-654.

Zhang, S. et al. *X-ray Studies of Ferroelectric Cathodes* J. Am. Ceramic Soc. 83 (2000) 1317-1319.

Zheng Yihua *Reflection and Shock Drift Acceleration* (1967). Webpage: http://www-ssg.sr.unh.edu/Physics954/Notes/Yihua.doc

Zheng, Q. et al *Liquids at Large Negative Pressures: Water at the Homogenous Nucleation Limit* Science 254 (1991) 829-832.

Zhongzhou Ren et al *Ground State Properties of Odd-Z Superheavy Nuclei* Phys. Rev. C67 (2003) 064302.

Zrinyi, M. et al *Direct Observation of Discrete and Reversible Shape Transition in Magnetic Field Sensitive Polymer Gels* Webpage: http://www.kfki.hu/~cheminfo/hun/olvaso/zrinyi/polymgel.html

ANCIENT ALIENS ON MARS
By Mike Bara
Bara brings us this lavishly illustrated volume on alien structures on Mars. Was there once a vast, technologically advanced civilization on Mars, and did it leave evidence of its existence behind for humans to find eons later? Did these advanced extraterrestrial visitors vanish in a solar system wide cataclysm of their own making, only to make their way to Earth and start anew? Was Mars once as lush and green as the Earth, and teeming with life? Chapters include: War of the Worlds; The Mars Tidal Model; The Death of Mars; Cydonia and the Face on Mars; The Monuments of Mars; The Search for Life on Mars; The True Colors of Mars and The Pathfinder Sphinx; more. Color section.
252 Pages. 6x9 Paperback. Illustrated. $19.95. Code: AMAR

ANCIENT ALIENS ON THE MOON
By Mike Bara
What did NASA find in their explorations of the solar system that they may have kept from the general public? How ancient really are these ruins on the Moon? Using official NASA and Russian photos of the Moon, Bara looks at vast cityscapes and domes in the Sinus Medii region as well as glass domes in the Crisium region. Bara also takes a detailed look at the mission of Apollo 17 and the case that this was a salvage mission, primarily concerned with investigating an opening into a massive hexagonal ruin near the landing site. Chapters include: The History of Lunar Anomalies; The Early 20th Century; Sinus Medii; To the Moon Alice!; Mare Crisium; Yes, Virginia, We Really Went to the Moon; Apollo 17; more. Tons of photos of the Moon examined for possible structures and other anomalies. 8-Page Color Section.
248 Pages. 6x9 Paperback. Illustrated. $19.95. Code: AAOM

QUEST FOR ZERO-POINT ENERGY
Engineering Principles for "Free Energy"
by Moray B. King
King expands, with diagrams, on how free energy and anti-gravity are possible. The theories of zero point energy maintain there are tremendous fluctuations of electrical field energy embedded within the fabric of space. King explains the following topics: Tapping the Zero-Point Energy as an Energy Source; Fundamentals of a Zero-Point Energy Technology; Vacuum Energy Vortices; The Super Tube; Charge Clusters: The Basis of Zero-Point Energy Inventions; Vortex Filaments, Torsion Fields and the Zero-Point Energy; Transforming the Planet with a Zero-Point Energy Experiment; Dual Vortex Forms: The Key to a Large Zero-Point Energy Coherence. Packed with diagrams, patents and photos. With power shortages now a daily reality in many parts of the world, this book offers a fresh approach very rarely mentioned in the mainstream media.
224 PAGES. 6x9 PAPERBACK. ILLUSTRATED. $14.95. CODE: QZPE

TAPPING THE ZERO POINT ENERGY
Free Energy & Anti-Gravity in Today's Physics
by Moray B. King
King explains how free energy and anti-gravity are possible. The theories of the zero point energy maintain there are tremendous fluctuations of electrical field energy imbedded within the fabric of space. This book tells how, in the 1930s, inventor T. Henry Moray could produce a fifty kilowatt "free energy" machine; how an electrified plasma vortex creates anti-gravity; how the Pons/Fleischmann "cold fusion" experiment could produce tremendous heat without fusion; and how certain experiments might produce a gravitational anomaly.
180 PAGES. 5x8 PAPERBACK. ILLUSTRATED. $12.95. CODE: TAP

THE FREE-ENERGY DEVICE HANDBOOK
A Compilation of Patents and Reports
by David Hatcher Childress
A large-format compilation of various patents, papers, descriptions and diagrams concerning free-energy devices and systems. *The Free-Energy Device Handbook* is a visual tool for experimenters and researchers into magnetic motors and other "over-unity" devices. With chapters on the Adams Motor, the Hans Coler Generator, cold fusion, superconductors, "N" machines, space-energy generators, Nikola Tesla, T. Townsend Brown, and the latest in free-energy devices. Packed with photos, technical diagrams, patents and fascinating information, this book belongs on every science shelf. With energy and profit being a major political reason for fighting various wars, free-energy devices, if ever allowed to be mass distributed to consumers, could change the world! Get your copy now before the Department of Energy bans this book!
292 PAGES. 8x10 PAPERBACK. ILLUSTRATED. BIBLIOGRAPHY. $16.95. CODE: FEH

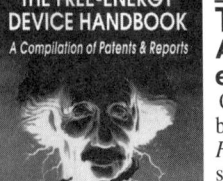

THE TIME TRAVEL HANDBOOK
A Manual of Practical Teleportation & Time Travel
edited by David Hatcher Childress
Childress takes us into the weird world of time travel and teleportation. Not just a whacked-out look at science fiction, this book is an authoritative chronicling of real-life time travel experiments, teleportation devices and more. *The Time Travel Handbook* takes the reader beyond the government experiments and deep into the uncharted territory of early time travellers such as Nikola Tesla and Guglielmo Marconi and their alleged time travel experiments, as well as the Wilson Brothers of EMI and their connection to the Philadelphia Experiment—the U.S. Navy's forays into invisibility, time travel, and teleportation. Childress looks into the claims of time travelling individuals, and investigates the unusual claim that the pyramids on Mars were built in the future and sent back in time. A highly visual, large format book, with patents, photos and schematics. Be the first on your block to build your own time travel device!
316 PAGES. 7x10 PAPERBACK. ILLUSTRATED. $16.95. CODE: TTH

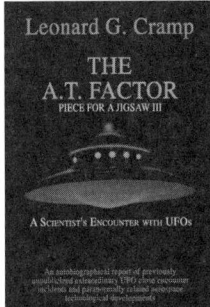

THE A.T. FACTOR
A Scientists Encounter with UFOs: Piece For A Jigsaw Part 3
by Leonard Cramp
British aerospace engineer Cramp began much of the scientific anti-gravity and UFO propulsion analysis back in 1955 with his landmark book *Space, Gravity & the Flying Saucer* (out-of-print and rare). His next books (available from Adventures Unlimited) *UFOs & Anti-Gravity: Piece for a Jig-Saw* and *The Cosmic Matrix: Piece for a Jig-Saw Part 2* began Cramp's in depth look into gravity control, free-energy, and the interlocking web of energy that pervades the universe. In this final book, Cramp brings to a close his detailed and controversial study of UFOs and Anti-Gravity.
324 PAGES. 6x9 PAPERBACK. ILLUSTRATED. BIBLIOGRAPHY. INDEX. $16.95. CODE: **ATF**

COSMIC MATRIX
Piece for a Jig-Saw, Part Two
by Leonard G. Cramp
Leonard G. Cramp, a British aerospace engineer, wrote his first book *Space Gravity and the Flying Saucer* in 1954. Cosmic Matrix is the long-awaited sequel to his 1966 book *UFOs & Anti-Gravity: Piece for a Jig-Saw*. Cramp has had a long history of examining UFO phenomena and has concluded that UFOs use the highest possible aeronautic science to move in the way they do. Cramp examines anti-gravity effects and theorizes that this super-science used by the craft—described in detail in the book—can lift mankind into a new level of technology, transportation and understanding of the universe. The book takes a close look at gravity control, time travel, and the interlocking web of energy between all planets in our solar system with Leonard's unique technical diagrams. A fantastic voyage into the present and future!
364 PAGES. 6x9 PAPERBACK. ILLUSTRATED. BIBLIOGRAPHY. $16.00. CODE: **CMX**

UFOS AND ANTI-GRAVITY
Piece For A Jig-Saw
by Leonard G. Cramp
Leonard G. Cramp's 1966 classic book on flying saucer propulsion and suppressed technology is a highly technical look at the UFO phenomena by a trained scientist. Cramp first introduces the idea of 'anti-gravity' and introduces us to the various theories of gravitation. He then examines the technology necessary to build a flying saucer and examines in great detail the technical aspects of such a craft. Cramp's book is a wealth of material and diagrams on flying saucers, anti-gravity, suppressed technology, G-fields and UFOs. Chapters include Crossroads of Aerodymanics, Aerodynamic Saucers, Limitations of Rocketry, Gravitation and the Ether, Gravitational Spaceships, G-Field Lift Effects, The Bi-Field Theory, VTOL and Hovercraft, Analysis of UFO photos, more.
388 PAGES. 6x9 PAPERBACK. ILLUSTRATED. $16.95. CODE: **UAG**

THE TESLA PAPERS
Nikola Tesla on Free Energy & Wireless Transmission of Power
by Nikola Tesla, edited by David Hatcher Childress
David Hatcher Childress takes us into the incredible world of Nikola Tesla and his amazing inventions. Tesla's rare article "The Problem of Increasing Human Energy with Special Reference to the Harnessing of the Sun's Energy" is included. This lengthy article was originally published in the June 1900 issue of *The Century Illustrated Monthly Magazine* and it was the outline for Tesla's master blueprint for the world. Tesla's fantastic vision of the future, including wireless power, anti-gravity, free energy and highly advanced solar power. Also included are some of the papers, patents and material collected on Tesla at the Colorado Springs Tesla Symposiums, including papers on: •The Secret History of Wireless Transmission •Tesla and the Magnifying Transmitter •Design and Construction of a Half-Wave Tesla Coil •Electrostatics: A Key to Free Energy •Progress in Zero-Point Energy Research •Electromagnetic Energy from Antennas to Atoms •Tesla's Particle Beam Technology •Fundamental Excitatory Modes of the Earth-Ionosphere Cavity
325 PAGES. 8x10 PAPERBACK. ILLUSTRATED. $16.95. CODE: **TTP**

THE FANTASTIC INVENTIONS OF NIKOLA TESLA
by Nikola Tesla with additional material by David Hatcher Childress
This book is a readable compendium of patents, diagrams, photos and explanations of the many incredible inventions of the originator of the modern era of electrification. In Tesla's own words are such topics as wireless transmission of power, death rays, and radio-controlled airships. In addition, rare material on German bases in Antarctica and South America, and a secret city built at a remote jungle site in South America by one of Tesla's students, Guglielmo Marconi. Marconi's secret group claims to have built flying saucers in the 1940s and to have gone to Mars in the early 1950s! Incredible photos of these Tesla craft are included. The Ancient Atlantean system of broadcasting energy through a grid system of obelisks and pyramids is discussed, and a fascinating concept comes out of one chapter: that Egyptian engineers had to wear protective metal head-shields while in these power plants, hence the Egyptian Pharoah's head covering as well as the Face on Mars! •His plan to transmit free electricity into the atmosphere. •How electrical devices would work using only small antennas. •Why unlimited power could be utilized anywhere on earth. •How radio and radar technology can be used as death-ray weapons in Star Wars.
342 PAGES. 6x9 PAPERBACK. ILLUSTRATED. $16.95. CODE: **FINT**

THE ENERGY GRID
Harmonic 695, The Pulse of the Universe
by Captain Bruce Cathie.
This is the breakthrough book that explores the incredible potential of the Energy Grid and the Earth's Unified Field all around us. Cathie's first book, *Harmonic 33*, was published in 1968 when he was a commercial pilot in New Zealand. Since then, Captain Bruce Cathie has been the premier investigator into the amazing potential of the infinite energy that surrounds our planet every microsecond. Cathie investigates the Harmonics of Light and how the Energy Grid is created. In this amazing book are chapters on UFO Propulsion, Nikola Tesla, Unified Equations, the Mysterious Aerials, Pythagoras & the Grid, Nuclear Detonation and the Grid, Maps of the Ancients, an Australian Stonehenge examined, more.
255 PAGES. 6x9 TRADEPAPER. ILLUSTRATED. $15.95. CODE: TEG

THE BRIDGE TO INFINITY
Harmonic 371244
by Captain Bruce Cathie
Cathie has popularized the concept that the earth is crisscrossed by an electromagnetic grid system that can be used for anti-gravity, free energy, levitation and more. The book includes a new analysis of the harmonic nature of reality, acoustic levitation, pyramid power, harmonic receiver towers and UFO propulsion. It concludes that today's scientists have at their command a fantastic store of knowledge with which to advance the welfare of the human race.
204 PAGES. 6x9 TRADEPAPER. ILLUSTRATED. $14.95. CODE: BTF

THE HARMONIC CONQUEST OF SPACE
by Captain Bruce Cathie
Chapters include: Mathematics of the World Grid; the Harmonics of Hiroshima and Nagasaki; Harmonic Transmission and Receiving; the Link Between Human Brain Waves; the Cavity Resonance between the Earth; the Ionosphere and Gravity; Edgar Cayce—the Harmonics of the Subconscious; Stonehenge; the Harmonics of the Moon; the Pyramids of Mars; Nikola Tesla's Electric Car; the Robert Adams Pulsed Electric Motor Generator; Harmonic Clues to the Unified Field; and more. Also included are tables showing the harmonic relations between the earth's magnetic field, the speed of light, and anti-gravity/gravity acceleration at different points on the earth's surface. New chapters in this edition on the giant stone spheres of Costa Rica, Atomic Tests and Volcanic Activity, and a chapter on Ayers Rock analysed with Stone Mountain, Georgia.
248 PAGES. 6x9. PAPERBACK. ILLUSTRATED. BIBLIOGRAPHY. $16.95. CODE: HCS

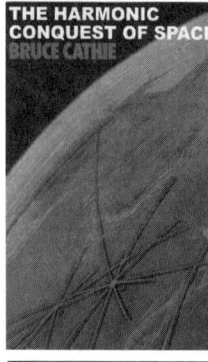

LEY LINE & EARTH ENERGIES
An Extraordinary Journey into the Earth's Natural Energy System
by David Cowan & Chris Arnold
The mysterious standing stones, burial grounds and stone circles that lace Europe, the British Isles and other areas have intrigued scientists, writers, artists and travellers through the centuries. How do ley lines work? How did our ancestors use Earth energy to map their sacred sites and burial grounds? How do ghosts and poltergeists interact with Earth energy? How can Earth spirals and black spots affect our health? This exploration shows how natural forces affect our behavior, how they can be used to enhance our health and well being. A fascinating and visual book about subtle Earth energies and how they affect us and the world around them.
368 PAGES. 6x9 PAPERBACK. ILLUSTRATED. BIBLIOGRAPHY. INDEX. $18.95. CODE: LLEE

DARK MOON
Apollo and the Whistleblowers
by Mary Bennett and David Percy
•Was Neil Armstrong really the first man on the Moon?
•Did you know a second craft was going to the Moon at the same time as Apollo 11?
•Do you know that potentially lethal radiation is prevalent throughout deep space?
•Do you know there are serious discrepancies in the account of the Apollo 13 'accident'?
•Did you know that 'live' color TV from the Moon was not actually live at all?
•Did you know that the Lunar Surface Camera had no viewfinder?
•Do you know that lighting was used in the Apollo photographs—yet no lighting equipment was taken to the Moon?
All these questions, and more, are discussed in great detail by British researchers Bennett and Percy in *Dark Moon,* the definitive book (nearly 600 pages) on the possible faking of the Apollo Moon missions. Bennett and Percy delve into every possible aspect of this beguiling theory, one that rocks the very foundation of our beliefs concerning NASA and the space program. Tons of NASA photos analyzed for possible deceptions.
568 PAGES. 6x9 PAPERBACK. ILLUSTRATED. BIBLIOGRAPHY. INDEX. $32.00. CODE: DMO

THE ANTI-GRAVITY HANDBOOK
edited by David Hatcher Childress, with Nikola Tesla, T.B. Paulicki, Bruce Cathie, Albert Einstein and others

The new expanded compilation of material on Anti-Gravity, Free Energy, Flying Saucer Propulsion, UFOs, Suppressed Technology, NASA Cover-ups and more. Highly illustrated with patents, technical illustrations and photos. This revised and expanded edition has more material, including photos of Area 51, Nevada, the government's secret testing facility. This classic on weird science is back in a 90s format!
- **How to build a flying saucer.**
- **Arthur C. Clarke on Anti-Gravity.**
- **Crystals and their role in levitation.**
- **Secret government research and development.**

230 PAGES. 7x10 PAPERBACK. ILLUSTRATED. $16.95. CODE: AGH

ANTI–GRAVITY & THE WORLD GRID

Is the earth surrounded by an intricate electromagnetic grid network offering free energy? This compilation of material on ley lines and world power points contains chapters on the geography, mathematics, and light harmonics of the earth grid. Learn the purpose of ley lines and ancient megalithic structures located on the grid. Discover how the grid made the Philadelphia Experiment possible. Explore the Coral Castle and many other mysteries, including acoustic levitation, Tesla Shields and scalar wave weaponry. Browse through the section on anti-gravity patents, and research resources.

274 PAGES. 7x10 PAPERBACK. ILLUSTRATED. $14.95. CODE: AGW

ANTI–GRAVITY & THE UNIFIED FIELD
edited by David Hatcher Childress

Is Einstein's Unified Field Theory the answer to all of our energy problems? Explored in this compilation of material is how gravity, electricity and magnetism manifest from a unified field around us. Why artificial gravity is possible; secrets of UFO propulsion; free energy; Nikola Tesla and anti-gravity airships of the 20s and 30s; flying saucers as superconducting whirls of plasma; anti-mass generators; vortex propulsion; suppressed technology; government cover-ups; gravitational pulse drive; spacecraft & more.

240 PAGES. 7x10 PAPERBACK. ILLUSTRATED. $14.95. CODE: AGU

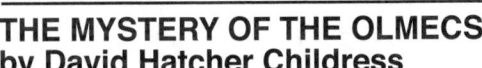

THE MYSTERY OF THE OLMECS
by David Hatcher Childress

Lost Cities author Childress takes us deep into Mexico and Central America in search of the mysterious Olmecs, North America's early, advanced civilization. The Olmecs, now sometimes called Proto-Mayans, were not acknowledged to have existed as a civilization until an international archeological meeting in Mexico City in 1942. At this time, the megalithic statues, large structures, ceramics and other artifacts were acknowledged to come from this hitherto unknown culture that pre-dated all other cultures of Central America. But who were the Olmecs? Where did they come from? What happened to them? How sophisticated was their culture? How far back in time did it go? Why are many Olmec statues and figurines seemingly of foreign peoples such as Africans, Europeans and Chinese? Is there a link with Atlantis? In this heavily illustrated book, join Childress in search of the lost cites of the Olmecs!

432 Pages. 6x9 Paperback. Illustrated. Bibliography. $20.00. Code: MOLM

PATH OF THE POLE
Cataclysmic Pole Shift Geology
by Charles H. Hapgood

Maps of the Ancient Sea Kings author Hapgood's classic book *Path of the Pole* is back in print! Hapgood researched Antarctica, ancient maps and the geological record to conclude that the Earth's crust has slipped on the inner core many times in the past, changing the position of the pole. *Path of the Pole* discusses the various "pole shifts" in Earth's past, giving evidence for each one, and moves on to possible future pole shifts. Packed with illustrations, this is the sourcebook for many other books on cataclysms and pole shifts.

356 PAGES. 6x9 PAPERBACK. ILLUSTRATED. $16.95. CODE: POP

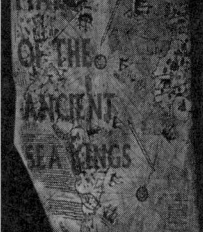

MAPS OF THE ANCIENT SEA KINGS
Evidence of Advanced Civilization in the Ice Age
by Charles H. Hapgood

Charles Hapgood's classic 1966 book on ancient maps produces concrete evidence of an advanced world-wide civilization existing many thousands of years before ancient Egypt. He has found the evidence in the Piri Reis Map that shows Antarctica, the Hadji Ahmed map, the Oronteus Finaeus and other amazing maps. Hapgood concluded that these maps were made from more ancient maps from the various ancient archives around the world, now lost. Not only were these unknown people more advanced in mapmaking than any people prior to the 18th century, it appears they mapped all the continents. The Americas were mapped thousands of years before Columbus. Antarctica was mapped when its coasts were free of ice!

316 PAGES. 7x10 PAPERBACK. ILLUSTRATED. BIBLIOGRAPHY & INDEX. $19.95. CODE: MASK

THE LAND OF OSIRIS
An Introduction to Khemitology
by Stephen S. Mehler
Was there an advanced prehistoric civilization in ancient Egypt who built the great pyramids and carved the Great Sphinx? Did the pyramids serve as energy devices and not as tombs for kings? Mehler has uncovered an indigenous oral tradition that still exists in Egypt, and has been fortunate to have studied with a living master of this tradition, Abd'El Hakim Awyan. Mehler has also been given permission to present these teachings to the Western world, teachings that unfold a whole new understanding of ancient Egypt. Chapters include: Egyptology and Its Paradigms; Asgat Nefer—The Harmony of Water; Khemit and the Myth of Atlantis; The Extraterrestrial Question; more.
272 PAGES. 6x9 PAPERBACK. ILLUSTRATED. COLOR SECTION. BIBLIOGRAPHY. $18.00 CODE: LOOS

REICH OF THE BLACK SUN
Nazi Secret Weapons and the Cold War Allied Legend
by Joseph P. Farrell
Why were the Allies worried about an atom bomb attack by the Germans in 1944? Why did the Soviets threaten to use poison gas against the Germans? Why did Hitler in 1945 insist that holding Prague could win the war for the Third Reich? Why did US General George Patton's Third Army race for the Skoda works at Pilsen in Czechoslovakia instead of Berlin? Why did the US Army not test the uranium atom bomb it dropped on Hiroshima? Why did the Luftwaffe fly a non-stop round trip mission to within twenty miles of New York City in 1944? *Reich of the Black Sun* takes the reader on a scientific-historical journey in order to answer these questions. Arguing that Nazi Germany actually won the race for the atom bomb in late 1944, *Reich of the Black Sun* then goes on to explore the even more secretive research the Nazis were conducting into the occult, alternative physics and new energy sources. The book concludes with a fresh look at the "Nazi Legend" of the UFO mystery by examining the Roswell Majestic-12 documents and the Kecksburg crash in the light of parallels with some of the super-secret black projects being run by the SS. *Reich of the Black Sun* is must-reading for the researcher interested in alternative history, science, or UFOs!
352 PAGES. 6x9 PAPERBACK. ILLUSTRATED. BIBLIOGRAPHY. $16.95. CODE: ROBS

THE GIZA DEATH STAR
The Paleophysics of the Great Pyramid & the Military Complex at Giza
by Joseph P. Farrell
Physicist Joseph Farrell's amazing book on the secrets of Great Pyramid of Giza. *The Giza Death Star* starts where British engineer Christopher Dunn leaves off in his 1998 book, *The Giza Power Plant*. Was the Giza complex part of a military installation over 10,000 years ago? Chapters include: An Archaeology of Mass Destruction, Thoth and Theories; The Machine Hypothesis; Pythagoras, Plato, Planck, and the Pyramid; The Weapon Hypothesis; Encoded Harmonics of the Planck Units in the Great Pyramid; High Freqquency Direct Current "Impulse" Technology; The Grand Gallery and its Crystals: Gravito-acoustic Resonators; The Other Two Large Pyramids; the "Causeways," and the "Temples"; A Phase Conjugate Howitzer; Evidence of the Use of Weapons of Mass Destruction in Ancient Times; more.
290 PAGES. 6x9 PAPERBACK. ILLUSTRATED. $16.95. CODE: GDS

THE GIZA DEATH STAR DEPLOYED
The Physics & Engineering of the Great Pyramid
by Joseph P. Farrell
Farrell expands on his thesis that the Great Pyramid was a chemical maser, designed as a weapon and eventually deployed—with disastrous results to the solar system. Includes: Exploding Planets: The Movie, the Mirror, and the Model; Dating the Catastrophe and the Compound; A Brief History of the Exoteric and Esoteric Investigations of the Great Pyramid; No Machines, Please!; The Stargate Conspiracy; The Scalar Weapons; Message or Machine?; A Tesla Analysis of the Putative Physics and Engineering of the Giza Death Star; Cohering the Zero Point, Vacuum Energy, Flux: Synopsis of Scalar Physics and Paleophysics; Configuring the Scalar Pulse Wave; Inferred Applications in the Great Pyramid; Quantum Numerology, Feedback Loops and Tetrahedral Physics; and more.
290 PAGES. 6x9 PAPERBACK. ILLUSTRATED. BIBLIOGRAPHY. $16.95. CODE: GDSD

THE GIZA DEATH STAR DESTROYED
The Ancient War For Future Science
by Joseph P. Farrell
Recapping his earlier books, Farrell moves on to events of the final days of the Giza Death Star and its awesome power. These final events, eventually leading up to the destruction of this giant machine, are dissected one by one, leading us to the eventual abandonment of the Giza Military Complex—an event that hurled civilization back into the Stone Age. Chapters include: The Mars-Earth Connection; The Lost "Root Races" and the Moral Reasons for the Flood; The Destruction of Krypton: The Electrodynamic Solar System, Exploding Planets and Ancient Wars; Turning the Stream of the Flood: the Origin of Secret Societies and Esoteric Traditions; The Quest to Recover Ancient Mega-Technology; Non-Equilibrium Paleophysics; Monatomic Paleophysics; Frequencies, Vortices and Mass Particles: The Topology of the Aether; "Acoustic" Intensity of Fields; The Pyramid of Crystals; tons more.
292 pages. 6x9 paperback. Illustrated. $16.95. Code: GDES

ANCIENT TECHNOLOGY IN PERU & BOLIVIA
By David Hatcher Childress
Childress speculates on the existence of a sunken city in Lake Titicaca and reveals new evidence that the Sumerians may have arrived in South America 4,000 years ago. He demonstrates that the use of "keystone cuts" with metal clamps poured into them to secure megalithic construction was an advanced technology used all over the world, from the Andes to Egypt, Greece and Southeast Asia. He maintains that only power tools could have made the intricate articulation and drill holes found in extremely hard granite and basalt blocks in Bolivia and Peru, and that the megalith builders had to have had advanced methods for moving and stacking gigantic blocks of stone, some weighing over 100 tons.
340 Pages. 6x9 Paperback. Illustrated.. $19.95 Code: ATP

THE ENIGMA OF CRANIAL DEFORMATION
Elongated Skulls of the Ancients
By David Hatcher Childress and Brien Foerster
In a book filled with over a hundred astonishing photos and a color photo section, Childress and Foerster take us to Peru, Bolivia, Egypt, Malta, China, Mexico and other places in search of strange elongated skulls and other cranial deformation. The puzzle of why diverse ancient people—even on remote Pacific Islands—would use head-binding to create elongated heads is mystifying. Where did they even get this idea? Did some people naturally look this way—with long narrow heads? Were they some alien race? Were they an elite race that roamed the entire planet? Why do anthropologists rarely talk about cranial deformation and know so little about it?
250 Pages. 6x9 Paperback. Illustrated. $19.95. Code: ECD

ROSWELL AND THE REICH
By Joseph P. Farrell
Farrell here delves ever deeper into the activities of this nefarious group. In his previous works, Farrell has clearly demonstrated that the Nazis were clandestinely developing new and amazing technologies toward the end of WWII, and that the key scientists involved in these experiments were exported to the Allied countries at the end of the conflict, mainly the United States, in a move called Operation Paperclip. Now, Farrell has meticulously reviewed the best-known Roswell research from UFO-ET advocates and skeptics alike, as well as some little-known source material, and comes to a radically different scenario of what happened in Roswell, New Mexico in July 1947, and why the US military has continued to cover it up to this day. Farrell presents a fascinating case that what crashed may have been representative of an independent postwar Nazi power—an extraterritorial Reich monitoring its old enemy, America, and the continuing development of the very technologies confiscated from Germany at the end of the War.
540 pages. 6x9 Paperback. Illustrated. $19.95. Code: RWR

TECHNOLOGY OF THE GODS
The Incredible Sciences of the Ancients
by David Hatcher Childress
Popular *Lost Cities* author David Hatcher Childress takes us into the amazing world of ancient technology, from computers in antiquity to the "flying machines of the gods." Childress looks at the technology that was allegedly used in Atlantis and the theory that the Great Pyramid of Egypt was originally a gigantic power station. He examines tales of ancient flight and the technology that it involved; how the ancients used electricity; megalithic building techniques; the use of crystal lenses and the fire from the gods; evidence of various high tech weapons in the past, including atomic weapons; ancient metallurgy and heavy machinery; the role of modern inventors such as Nikola Tesla in bringing ancient technology back into modern use; impossible artifacts; and more.
356 PAGES. 6X9 PAPERBACK. ILLUSTRATED. BIBLIOGRAPHY. $16.95. CODE: TGOD

ATLANTIS & THE POWER SYSTEM OF THE GODS
Mercury Vortex Generators & the Power System of Atlantis
by David Hatcher Childress and Bill Clendenon
Atlantis and the Power System of the Gods starts with a reprinting of the rare 1990 book *Mercury: UFO Messenger of the Gods* by Bill Clendenon. Clendenon takes on an unusual voyage into the world of ancient flying vehicles, strange personal UFO sightings, a meeting with a "Man In Black" and then to a centuries-old library in India where he got his ideas for the diagrams of mercury vortex engines. The second part of the book is Childress' fascinating analysis of Nikola Tesla's broadcast system in light of Edgar Cayce's "Terrible Crystal" and the obelisks of ancient Egypt and Ethiopia. Includes: Atlantis and its crystal power towers that broadcast energy; how these incredible power stations may still exist today; inventor Nikola Tesla's nearly identical system of power transmission; Mercury Proton Gyros and mercury vortex propulsion; more. Richly illustrated, and packed with evidence that Atlantis not only existed—it had a world-wide energy system more sophisticated than ours today.
246 PAGES. 6X9 PAPERBACK. ILLUSTRATED. $15.95. CODE: APSG

ORDER FORM

10% Discount When You Order 3 or More Items!

One Adventure Place
P.O. Box 74
Kempton, Illinois 60946
United States of America
Tel.: 815-253-6390 • Fax: 815-253-6300
Email: auphq@frontiernet.net
http://www.adventuresunlimitedpress.com

ORDERING INSTRUCTIONS

- ✓ Remit by USD$ Check, Money Order or Credit Card
- ✓ Visa, Master Card, Discover & AmEx Accepted
- ✓ Paypal Payments Can Be Made To:
  info@wexclub.com
- ✓ Prices May Change Without Notice
- ✓ 10% Discount for 3 or More Items

SHIPPING CHARGES

United States

- ✓ Postal Book Rate { $4.50 First Item / 50¢ Each Additional Item }
- ✓ POSTAL BOOK RATE Cannot Be Tracked! Not responsible for non-delivery.
- ✓ Priority Mail { $6.00 First Item / $2.00 Each Additional Item }
- ✓ UPS { $7.00 First Item / $1.50 Each Additional Item }
  NOTE: UPS Delivery Available to Mainland USA Only

Canada

- ✓ Postal Air Mail { $15.00 First Item / $3.00 Each Additional Item }
- ✓ Personal Checks or Bank Drafts MUST BE US$ and Drawn on a US Bank
- ✓ Canadian Postal Money Orders OK
- ✓ Payment MUST BE US$

All Other Countries

- ✓ Sorry, No Surface Delivery!
- ✓ Postal Air Mail { $19.00 First Item / $7.00 Each Additional Item }
- ✓ Checks and Money Orders MUST BE US$ and Drawn on a US Bank or branch.
- ✓ Paypal Payments Can Be Made in US$ To:
  info@wexclub.com

SPECIAL NOTES

- ✓ RETAILERS: Standard Discounts Available
- ✓ BACKORDERS: We Backorder all Out-of-Stock Items Unless Otherwise Requested
- ✓ PRO FORMA INVOICES: Available on Request
- ✓ DVD Return Policy: Replace defective DVDs only

ORDER ONLINE AT: www.adventuresunlimitedpress.com

10% Discount When You Order 3 or More Items!

Please check: ✓

☐ This is my first order ☐ I have ordered before

Name
Address
City
State/Province   Postal Code
Country
Phone: Day   Evening
Fax   Email

Item Code	Item Description	Qty	Total

Please check: ✓

☐ Postal-Surface
☐ Postal-Air Mail (Priority in USA)
☐ UPS (Mainland USA only)

Subtotal ▶
Less Discount-10% for 3 or more items ▶
Balance ▶
Illinois Residents 6.25% Sales Tax ▶
Previous Credit ▶
Shipping ▶
Total (check/MO in USD$ only) ▶

☐ Visa/MasterCard/Discover/American Express

Card Number:
Expiration Date: Security Code:

✓ SEND A CATALOG TO A FRIEND: